Human Tumor Cloning

Proceedings of the

Fourth Conference on Human Tumor Cloning

Tucson, Arizona
January 8-10, 1984

Grune & Stratton Rapid Manuscript Reproduction

Human Tumor Cloning

Edited by

Sydney E. Salmon, M.D.
Professor of Medicine
Section of Hematology and Oncology
Department of Internal Medicine
Director, Cancer Center
University of Arizona College of Medicine
Tucson, Arizona

Jeffrey M. Trent, Ph.D.
Associate Professor of Medicine
Section of Hematology and Oncology
Department of Internal Medicine
Director, Basic Research at the Cancer Center
University of Arizona College of Medicine
Tucson, Arizona

Grune & Stratton, Inc.
(Harcourt Brace Jovanovich, Publishers)
Orlando San Diego San Francisco
New York London Toronto Montreal
Sydney Tokyo São Paulo

Grune & Stratton, Inc.
Orlando, Florida 32887

Distributed in the United Kingdom by
Grune & Stratton, Ltd.
24/28 Oval Road, London NW 1

Library of Congress Catalog Number 84-48119
International Standard Book Number 0-8089-1671-8

Printed in the United States of America
84 85 86 87   10 9 8 7 6 5 4 3 2 1

Contents

III. STUDIES OF SPECIFIC TUMOR TYPES

IV. PHARMACOLOGY

V. CLINICAL TRIALS

Preface

In vitro colony assays for human tumors have become increasingly important to basic and clinical cancer research, and may have major future impact — almost certainly in new drug development, and hopefully in the selection of anti-cancer drugs — for individual patient chemotherapy. Marked tumor heterogeneity has been observed in such assays, leading to the belief that the study of primary tumor material may represent a more realistic model against which to test biological and chemotherapeutic agents. The University of Arizona Cancer Center has had a major research effort in human tumor cloning since 1975. An important component of this program has been ongoing close interaction between laboratory scientists and clinical oncologists.

When we held our first Human Tumor Cloning Conference in 1979, its major goal was to provide "technology transfer" by training interested workers in the basic methodology for agar culture of human tumors. The second and third conferences (in 1980 and 1981) had similar primary goals. By the time of our third conference in 1981, over 1000 laboratory workers had been trained at our Center through this conference series. We felt that this had provided a sufficient impetus for development and testing, and considered that time would be needed for biologists, pharmacologists, and clinicians to study these and other methods in their own Institutions.

Thus, at the time of the fourth conference, held again in Tucson on January 8 – 10, 1984, our goal was quite different. On this occasion, we wished to "test the water," and determine the state of scientific development in the field of human tumor cloning. In order to have a balanced representation of the sciences, the program was developed with a limited number of invited participants, plus an international call for competitive abstracts for oral and poster presentations. We were most gratified with the enthusiastic response to the call for participation in this fourth conference, and feel that the papers and posters presented provided an accurate assessment of developments in this field in leading laboratories. A number of excellent papers were presented by internationally recognized experts in laboratory and clinical cancer research, who have focused on various applications of tumor cloning. Over 200 registrants came from 18 countries to hear the invited presentations, the competitively selected papers, and the poster presentations that are published in this text. Scientific interest was high and discussion lively throughout the three-day conference. We therefore believe that this conference achieved its stated scientific goal.

As conference co-chairmen, we owe a special debt of thanks to Dr. Bridget Hill, who agreed to take on the Herculean task of providing a summary overview of the conference. Her remarks, which are the final chapter in this book, were well received by all participants and we believe can be viewed as representing a reasonable consensus.

We also wish to offer special thanks to the Conference Coordinator, Mrs. Mary Humphrey, as well as our dedicated staff, who provided invaluable assistance during the meeting.

In order to publish these proceedings quickly (when they are of greatest value), we chose to use the camera-ready format. In that regard, we are grateful to the fine staff at Grune & Stratton for enabling this volume to appear so quickly.

Sydney E. Salmon, M.D.
Jeffrey M. Trent, Ph.D.
Tucson, Arizona, January 11, 1984

Human Tumor Cloning

I. TUMOR BIOLOGY

The Cell Renewal Hierarchy in Ovarian Cancer

RONALD N. BUICK

Associate Professor
Department of Medical Biophysics
University of Toronto and Ontario Cancer Institute
Toronto, Ontario, Canada

There is a growing overall awareness that the cellular organisation of human carcinomas might be based on tissue-specific differentiation similar to their normal tissue equivalents (Pierce *et al*, 1977). In such a situation the growth of tumor tissue could be attributed largely to the "tumor stem cells", and the majority of tumor cells would express tissue specific differentiation markers and would not have infinite growth potential (Mackillop *et al*, 1983a). The importance of such a model lies in the fact that the tumor stem cells must be regarded as the only target for curative tumor therapy. Direct evidence for the applicability of a stem cell model to human carcinomas is still lacking; similar concepts are widely accepted for normal tissue renewal and for growth of hemopoietic malignancies (McCulloch, 1979). There are however, a number of sources of information which indirectly support the notion of epithelial tumor stem cells. These include the clinical experience with radiation therapy which argues for a true target size much smaller than the entire tumor (Bush and Hill, 1975; Tepper, 1981), evidence of tissue-specific differentiation of tumor cells both *in vivo* and *in vitro* (Pierce *et al*, 1977) and evidence of minority sub-populations of cells with a high degree of proliferative capability (as assessed by colony formation in semi-solid culture Hamburger and Salmon, 1977).

On the basis of these biological and clinical clues to stem cell function in carcinomas, we have proposed a stem cell model of human tumor growth which attempts to explain properties of tumors through consideration of stem cell renewal and differentiation (Mackillop *et al*, 1983a). By

HUMAN TUMOR CLONING
ISBN 0-8089-1671-8

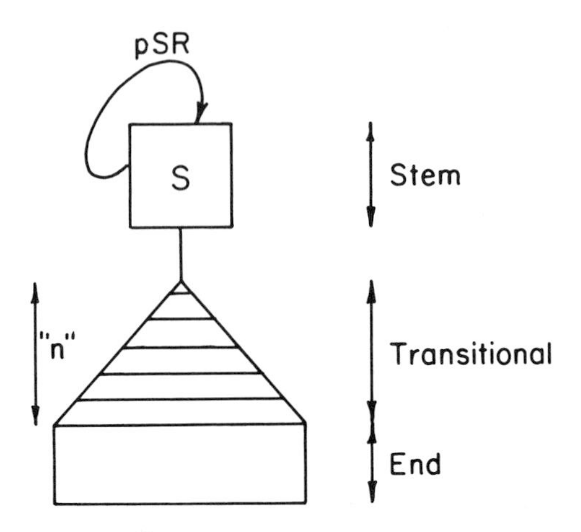

Figure 1: Schematic representation of a stem cell differentiation hierarchy. Three cell populations are predicted; stem cells (proliferative cells with renewal potential), transitional cells (proliferative, but of a determined proliferative potential) and end cells (terminally differentiated, non proliferative).

analogy to normal cell renewal systems, tumors were postulated to consist of three cell populations; stem, transitional and end cells (Figure 1) forming a hierarchy of differentiation. The proportion of cells at different stages of such a hierarchy depend on the cell renewal properties of the stem cells (pSR) and the number of cell divisions in the transitional cell compartment. This simple model has formed the basis for our experimental design in the study of tumor cell renewal and differentiation.

As a model system we have chosen to study the processes of cell renewal, proliferation and differentiation in human ovarian carcinoma. Such tumors are derived from the germinal epithelium of the ovary; a tissue which under normal conditions in pre-menopausal life, maintains cell number by stem cell renewal mechanisms. The choice of tumor model was based on practical consideration of accessibility and the nature of tissues which could be derived from the ascitic form of ovarian carcinomas. In addition, at the time of initiation of these studies in 1980, it was clear from the work of Hamburger et al (1978) that ovarian tumor cells were particularly amenable to study in colony-forming assays.

Stem Cell Renewal in Ovarian Carcinoma

In agreement with the work of Hamburger et al (1978), we have shown that a small proportion (usually less than 1%) of ovarian tumor cells have the capacity to form recognizable colonies in culture (Buick and Fry, 1980; Buick and Mackillop, 1981). Colony formation, however, is not a unique characteristic of stem cells; the process of self-renewal is the single property which distinguishes a stem cell from a cell with a limited growth potential (a clonogenic cell). In animal models, stem cell characteristics can be assessed by the ability to re-form a tumor after subcurative therapy or by the ability to generate micrometastases in lung or liver. Since such tumor repopulation studies are impossible in humans, we must rely on other less direct means of assessment of stem cell function. It has been possible for instance to demonstrate that pooled tumor colonies will form tumors when implanted in immune-deprived mice (Carney et al, 1982), indicating that such pooled cell suspensions do contain at least one cell with stem cell properties. We chose to demonstrate self-renewal by harvesting and re-culturing individual and/or pooled colonies from primary colony cultures. We were able to show that only a small proportion of methyl-cellulose-clonogenic cells had the potential for self-renewal. Marked patient-to-patient variation was demonstrated in this proportion and the variation could be tentatively related to tumor growth in vivo (Buick and Mackillop, 1981). Similar techniques have proven successful in analysing self-renewal in malignant melanoma (Thomson and Meyskens, 1982).

Cell Differentiation in Ovarian Carcinoma

The phenotypic markers used for our studies of ovarian tumor cell differentiation were chosen because, individually, they have been used in either the pathological description of human ovarian tumor tissue or have been postulated to be associated with the malignant process. Oil-red O staining is thought to detect terminally differentiated cells in the germinal epithelium of the ovary (Guraya, 1977). Expression of the antigens CEA and Ca-1 has been associated with a variety of tumor types including ovarian cancer; NB/70K and Ca125 expression has been related to tumor burden in ovarian carcinoma (Knauf and Urbach, 1981; Bast et al, 1983).

Within ovarian tumor ascites populations, there is easily demonstrable heterogeneity with respect to oil-red O staining and expression of the tumor-associated antigens Ca-1, CEA, NB/70K and Ca125 assessed by indirect immunofluorescent microscopy (Bizzari et al, 1983). The proportion of tumor cells of a given phenotype is markedly different in different patient samples. We have attempted to answer two questions relating to this heterogeneity; (a) what is the relationship of cells expressing these phenotypes to cells expressing proliferative function within the tumor? and (b) can the functional and phenotypic markers be arranged in a hierarchy describing cell differentiation in ovarian carcinoma?

The procedure which has allowed us to approach these questions is the fractionation of tumor cell pouplations on the basis of density. Ovarian tumor cell populations, like other cell systems displaying differentiation, are markedly heterogeneous with respect to density. By separating the total tumor cell population into iso-density fractions in step gradients of bovine serum albumin, it has been possible to describe the relationship of individual cell phenotype or function to cell density (Bizzari et al, 1983; Buick et al, 1983; Mackillop and Buick, 1981). It is important to add a statement regarding selectivity of samples for these experiments; in order to perform fractionation experiments we pre-selected tumor samples with 3 criteria; (a) large numbers of cells available ($\geq 10^8$) allowing multiple cryopreserved aliquots; (b) absence of non-tumor cells, and (c) collected cells were in the form of a suspension of single cells. These selection criteria were met in 12 cases by collecting approximately 250 ascitic tumors.

The density fractionation procedure has been rendered more important by the finding that cells of a particular phenotype from different patients have very similar densities (Mackillop and Buick, 1982; Bizzari et al, 1983) and that predictable changes in physical properties occur with tumor progression (Mackillop et al, 1983b).

By analysing the density distribution of cells of a particular phenotype in relation to primitive cells (detected by labelling index or clonogenicity) and terminal differentiated cells (oil-red O positive) we can make a tentative assignment of a given marker to a position in the cell differentiation hierarchy. A schematic description of

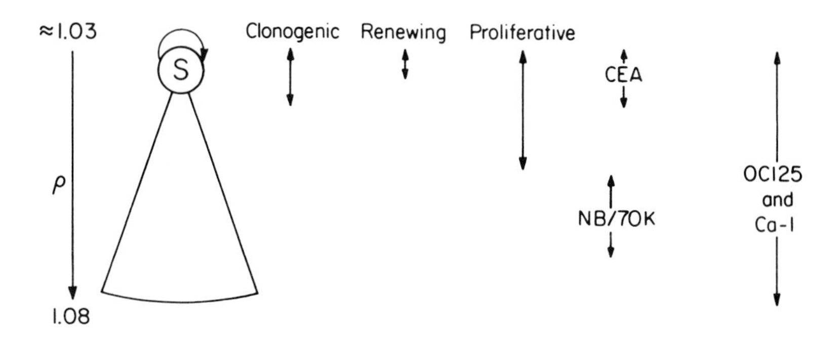

Figure 2: The relationship between cell density, function and phenotype in human tumor cell populations.

the conclusions of our accumulated studies is shown in Figure 2. The cell phenotype of expression of NB/70K is displayed on cells intermediate in terms of density while expression of CEA seems to be more closely associated with primitive cells (Bizzari et al, 1983). Clonogenic cells however, do not appear to have cell surface expression of CEA (Buick et al, 1983).

Importantly cells bearing the antigens Ca-1 or Ca125 are not restricted to one area of the density gradient (Buick et al, 1984). Rather a constant proportion of cells of all density fractions are detectable with this reagent. It may be possible therefore to classify tumor-associated antigens with respect to their involvement with tissue-specific differentiation processes. This may prove of value in selecting monoclonal antibodies as candidates for immunodiagnosis or immunotherapy.

Heterogeneity of Cell Proliferative Potential in Ovarian Carcinoma

We have recently tested the possibility that a hierarchy of proliferative potential may exist, related to the cell differentiation hierarchy proposed in the first section of this paper. The theoretical background for predicting the existence of a hierarchy derives from our mathematical

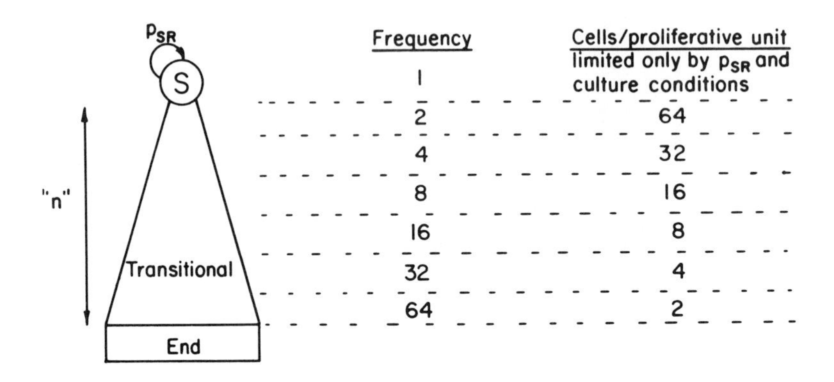

Figure 3: The relationship between differentiation state and proliferative potential of individual cells within a cell renewal hierarchy. (For derivation see Mackillop et al., 1983a).

modelling of stem cell growth (Mackillop et al, 1983a). At any point in time (i.e., when a particular tumor is biopsied) a tumor will be heterogeneous due to stem cell differentiation. It can be hypothesized that the differentiation state of an individual tumor cell is related to its intrinsic proliferative potential. This is schematically represented in Figure 3 in terms of the existing model of stem, transitional and end cells which was displayed in Figure 1.

In keeping with other investigators, we have found that when ovarian tumor cells from malignant ascites are cultures in semi-solid medium, the majority of cells are inert, a small minority form large colonies and a large number of cells give rise to clonal units of intermediate size. An assessment of frequency of clonal units of different sizes allows the derivation of a colony size distribution; an example is shown in Figure 4. The general features of the colony size distributions generated from ovarian tumor cells are summarized in Table 1. We have hypothesized that the information generated by such colony size distributions represents the cell heterogeneity imposed by stem cell differentiation, and that the cells giving rise to large colonies of non-finite growth potential may be the stem cells of the tumor.

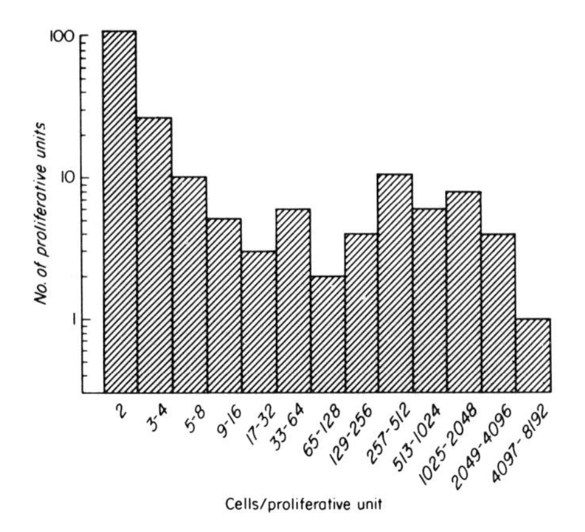

Figure 4: Clone size distribution for single cells of human ovarian carcinoma grown in agar culture. Time of incubation 21 days. Generated by assessing 200 units of $\geq$ 2 cells.

TABLE 1
Summary Characteristics of Colony Size Distributions for Ascitic Ovarian Tumor Cells in Agar

(a) All patient samples give range of colony size. Samples generating no colony growth (i.e., > 40 cell size units) do show many units of smaller size.

(b) The shape of the size distribution is common to all samples, particularly with regard to the initial exponential portion (2-32 cell size).

(c) The proportion of units in the $\geq$ 64 cell size category is very variable.

(d) The kinetics of the development of $\leq$ 64 cell size units and $\geq$ 64 cell units are distinguishable. The larger units continue to increase in size and frequency with culture time, while the smaller units remain constant in number after approximately 7 days culture.

On a practical note, the cut-off point for determination of clonogenicity which seems to maximise the likelihood of quantitating the progeny of stem cells is 64 cells for ovarian carcinoma. This number is likely to be different for tumors of different tissue types. When colony size criteria are applied to measurements of ovarian tumor cell

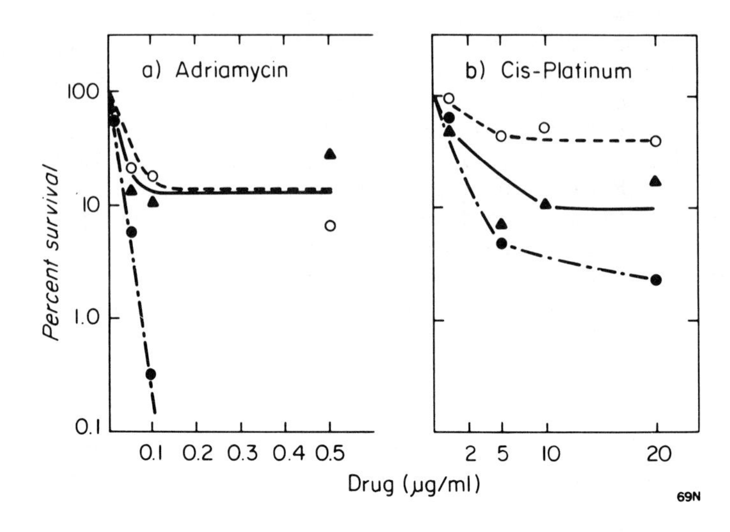

Figure 5: Fractional survival of colony-forming cells from ovarian tumor ascites (a) and (b) exposed to concentrations of adriamycin and cis-platinum respectively. Colonies assessed after 21 days culture as ≥ 20 cells (o - o); ≥ 32 cells (▲-▲) or ≥ 64 cells (● - ●).

sensitivity to chemotherapeutic agents, potentially important artefacts are exposed. Figure 5 shows examples of drug dose response curves generated by culturing ovarian tumor cells from two ascitic cell populations in the presence of adriamycin and cis-platinum respectively. Colony counts were performed after 21 days culture using 3 different criteria. Differing degrees of resistance or sensitivity are seen for the same cells dependent on the colony size cut-off used. Most sensitivity is seen when the largest cut-off (> 64 cells) point is used. This influence on the measurement of drug sensitivity is almost certainly artefactual based on limited cell division of sterilized cells analogous to that seen after radiation. It is recommended that care should be exercised in determining drug sensitivity indices in cells of limited proliferative capacity.

CONCLUSION

Ovarian carcinoma cell populations can be demonstrated to have heterogeneity on the basis of acquisiton of tissue-specific cell differentiation markers, cell renewal and cell proliferative potential. It is possible that these phenomena represent the function of a stem cell differentiation hierarchy. In this regard, it is hoped that ovarian carcinoma can be regarded as an example of a human epithelial tumor. Given the similarity of cell renewal organisation in diverse normal epithelia, it is likely that concepts of stem cell growth can be applied generally to human tumors (Mackillop *et al*, 1983a). A minimal knowledge of stem cell frequency and biological potential would seem to be required for advances in chemo- and radiation therapy of these diseases (Steel, 1977; Goldie and Coltman, 1983). Furthermore, a knowledge of stem cell phenotype derived from studies of cell differentiation could allow design of novel therapeutic and diagnostic approaches based on antibody specificity (Buick *et al*, 1983).

The molecular bases for malignant stem cell renewal and differentiation are not understood. It has become clear over the last few years that aspects of the tumor phenotype can be attributed to dominant-acting oncogenes and to tumor growth factors (TGF) (which may be oncogene products) (Land *et al*, 1983). No attempt has yet been made to integrate the molecular bases of carcinogenesis with the emerging view of tumors as stem cell differentiating systems (Buick and Pollak, 1984). Viewing oncogene expression and TGF production or response as individual cell phenotypes within a cell renewal and differentiation system, such as the one described, may lead to an understanding of this relationship.

ACKNOWLEDGEMENTS

My thanks to Ms. Rose Pullano for excellent technical assistance. The development of the concepts applied have been the result of collaboration with a number of my colleagues including, Drs. J.E. Till, A. Ciampi, I. Tannock, W.J. Mackillop and J.-P. Bizzari.

REFERENCES

Bast, R. C., Jr., Klug, J. L., St. John,, E. et al. A radioimmunoassay using a monoclonal antibody to monitor the course of epithelial ovarian cancer. New England Journal of Medicine, 1983, 309, 883-887.

Bizzari, J.-P., Mackillop, W. J. and Buick, R. N. Cellular specificity of NB/70K, a putative human ovarian tumor antigen. Cancer Research, 1983, 43, 864-867.
Buick, R. N. and Fry, S. E. A comparison of human tumor cell clonogenicity in methylcellulose and agar culture. British Journal of Cancer, 1980, 42, 933-937.
Buick, R. N. and Mackillop, W. J. Measurement of self-renewal in culture of clonogenic cells from human ovarian carcinoma. British Journal of Cancer, 1981, 44, 349-355.
Buick, R. N., Pullano, R., Bizzari, J.-P. and Mackillop, W. J. Radioimmunoimaging and Radioimmunotheray. New York, Elsevier Press, 1983.
Buick, R. N., Pullano, R. and Mackillop, W. J. Evidence for a stem cell hierarchy in human ovarian cancer. Submitted for publication, 1984.
Buick, R. N., Pullano, R., Knapp, R. C., Urbach, G. I., Chang, P.-L. and Bast, R. C., Jr. Classification of ovarian tumor-associated antigens on the basis of distribution on cell populations undergoing differentiation. Proceedings of the American Association for Cancer Research, 1984.
Buick R. N. and Pollak, M. Perspectives on human tumor clonogenic cells, stem cells and oncogenes. Cancer Research In Press, 1984.
Bush, R. S. and Hill, R. P. Biologic discussions augmenting radiation effects and model systems. The Laryngoscope, 1975, 85, 1119-1133.
Carney, D. N., Gazdar, A. F., Bunn, P. A., Jr. and Guccion, J. G. Demonstration of the stem cell nature of clonogenic tumor cells from lung cancer patients. Stem Cells, 1981, 1, 149-164.
Goldie, J. H. and Coltman, A. J. Quantitative model for multiple levels of drug resistance in clinical tumors. Cancer Treatment Reports, 1983, 67, 923-931.
Guraya, S. S. Recent advances in the morphology, histochemistry, and biochemistry of the developing mammalian ovary. International Review of Cytology, 1977, 51, 49.
Hamburger, A. W. and Salmon, S. E. Primary bioassay of human tumor stem cells. Science, 1977, 197, 461-463.
Hamburger, A. W., Salmon, S. E., Kim, M. B., Soehlan, B. J. and Alberts, D. J. Direct cloning of human ovarian carcinoma cells in agar. Cancer Research 38, 1978, 38, 3438-3444.
Knauf, S. and Urbach, G. I. Identification, purification and radioimmunoassay of NB/70K, a human ovarian tumor-associated antigen. Cancer Research, 1981, 41, 1351-1357.

Land, H., Parada, L. F. and Weinberg, R. A. Cellular oncogenes and multistep carcinogenesis. Science, 1983, 222, 772-778.
Mackillop, W. J. and Buick, R. N. Cellular heterogeneity in human ovarian carcinoma studied by density gradient fractionation. Stem Cells, 1981, 1, 355-366.
Mackillop, W. J., Ciampi, A., Till, J. E. and Buick, R. N. A stem cell model of human tumor growth: implications for tumor cell clonogenic assays. Journal of the National Cancer Institute, 1983a, 70: 9-16.
Mackillop, W. J., Trent, J. M., Stewart, S. S. and Buick, R. N. Tumor progression studied by analysis of cellular features of serial ascitic ovarian carcinoma tumors. Cancer Research, 1983b, 43, 874-878.
McCulloch, E. A. Abnormal myelopoietic clones in man. Journal of the National Cancer Institute, 1979, 63, 883-892.
Pierce, G. B., Shikes, R. and Fink, L. M. Cancer. A Problem of Developmental Biology. Englewood Cliffs, Prentice-Hall, 1977.
Steel, G. G. Growth Kinetics of Tumors: Cell Population Kinetics in Relation to the Growth and Treatment of Cancer, Oxford, Clarendon Press, 1977.
Tepper, J. Clonogenic potential of human tumors - A hypothesis. Acata Radiologica (Oncology), 1981, 20, 283-288.
Thomson, S. P. and Meyskens, F. L., Jr. Method for measurement of self-renewal capacity of clonogenic cells from biopsies of metastatic human melanoma. Cancer Research, 1982, 42, 4606-4613.

Studies of Self-Renewal of Human Melanoma Colony-Forming Cells

FRANK L. MEYSKENS, JR.

Associate Professor
Department of Internal Medicine and Cancer Center

STEPHEN P. THOMSON

Research Associate
Department of Internal Medicine and Cancer Center

Cancer Center

University of Arizona

Tucson, AZ

INTRODUCTION

We have studied in depth the biology and chemosensitivity of human melanoma colony-forming cells (HMCFU) which grow in soft agar.[1-3,9-20,22-24] We have demonstrated that the cells are malignant melanocytes as defined by special stains,[14,20] electron microscopy,[20] and karyology.[14,24] Additionally, we have shown that classical survival curves of HMCFU to ionizing radiation can be obtained if excellent single cell suspensions are used and artifactual cellular aggregates carefully

Supported in part by grants from the National Cancer Institute (CA 17094).

HUMAN TUMOR CLONING
ISBN 0-8089-1671-8

monitored.[9] We have also correlated the *in vitro* effects of HMCFU to chemotherapeutic agents to clinical response.[10,11] Both retrospective and prospective studies demonstrated that clinical progression can be estimated with 85% accuracy.

Clinical response was predicted correctly about 50% of the time, a result which was significantly better than single agent standard chemotherapy. These biological and clinical studies suggested that an in depth study of HMCFU should provide further insight into melanoma tumor cell growth.

Studies in animals suggest that colony-forming cells are related to stem cells, the key replicative unit of tumors which supports the growth and regrowth of tumors after subcurative therapy.[21] Extensive investigations of normal hematopoietic renewal systems indicate that a clonal hierarchy is present.[4] One portrayal of this concept is represented in Figure 1.

A number of methods have been used to identify stem cells. As *in situ* self-renewal studies of tumors in humans can not be performed, measurements of "stemness" for human melanoma cells must necessarily be indirect. One relatively easy way to measure high proliferative capacity is to plate individual primary colonies in microtiter wells and measure the development of permanent cell lines. This type of approach is particularly valuable in studying early phenotypic changes of cells in culture and has not been widely exploited.[8] Formation of tumors in nude mice after injection of human tumor colonies has also been used to measure "stemness".[7,15] Both these approaches are severely limited for experimental versatility inasmuch as the endpoints for measuring self-renewal are largely qualitative and use of these systems to study self-renewal in a quantitative mode is inherently restricted. We have therefore selected replating of cells from melanoma colonies as a method to quantify self-renewal. Prior workers have used two different replating procedures. Buick et al. has used cells in the upper plating layer after 7 to 14 days of culture to study human leukemic and ovarian cancer progenitor cells; both non-proliferating, host, and cells in colonies were replated.[5,6] This approach has the advantage of including host cells in the measurement of secondary colony formation. However, as we have shown elsewhere (Thomson et al., this volume), plating of large numbers of cells may well lead to an underestimate of proliferative capacity. Additionally the presumption has been made that non-colony cells in the primary cultures will not proliferate when replated. This has not been convincingly demonstrated as yet. We have therefore elected to pluck individual primary colonies and to replate cells from individual or pooled colonies to quantitate self-renewal. This approach allows one to study the formation of secondary colonies over a concentration (of cells) range. If desired, the influence of host cells on secondary colony formation can also be studied in a dose-response manner.

METHODS FOR MEASUREMENT OF SELF-RENEWAL

Previous investigators have used a hand-held Pasteur pipette to remove colonies. In general, these colonies were large and contained hundreds to thousands of cells. Human melanoma colonies (and most other human tumors) generally contain tens to several hundreds of cells and therefore we use a micropipette to remove colonies. The micromanipulator allows precise movement of the pipet in 3 axes. The calibrated pipet is connected pneumatically via Tygon tubing to a microsyringe which is hand-held during use. Details of this set-up and our approach to preparing absolute single cell suspensions, which is critical for these studies, is described elsewhere.[22]

REPLATING OF CELLS FROM POOLED COLONIES

Our initial studies used cells from patients which grew colonies greater than 100μm in diameter. Approximately 80 primary colonies were collected for each transfer. Repetitive tapping of the tube containing the colonies was sufficient to disaggregate the colonies into single-cell suspensions. Figure 2 shows the results from a representative patient. Primary colonies were formed in a linear fashion according to the number of viable tumor cells plated. Primary colonies were removed after 18 days and secondary colonies formed, which were similar in size and morphology to the primary colonies. The secondary, tertiary, and later colonies were formed in a linear

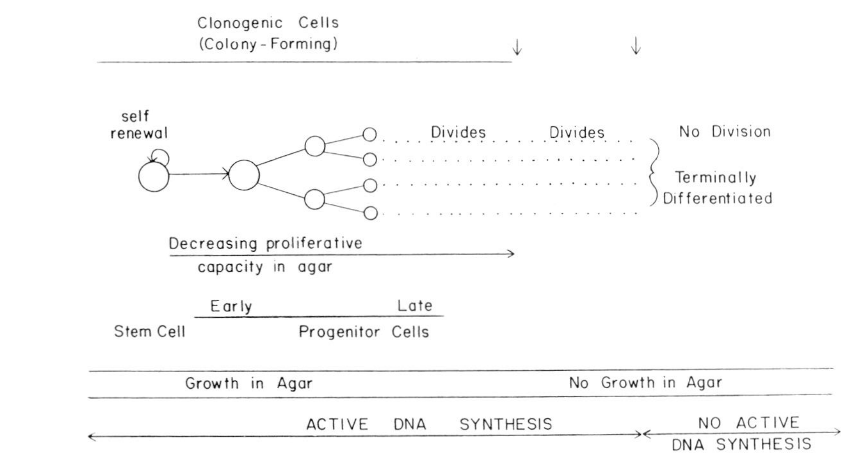

FIGURE 1. Schematic representation of clonal hierarchy. Essential features include cells which are incapable of further division (no DNA synthesis) and those which are either actively proliferating or can be induced to proliferate (DNA synthesis). A subpopulation of those cells which can divide also can form colonies in semisolid medium and are designated clonogenic or colony-forming cells. These colony-forming cells are composed of different cellular subpopulations, one of which is capable of self-renewal - the defining feature of stem cells.

fashion according to the number of cells plated and with similar frequency as primary colonies until the eighth transfer. The frequency decreased and no colonies formed after the 9th serial transfer (curve 10). We also studied in detail the replating characteristics of pooled

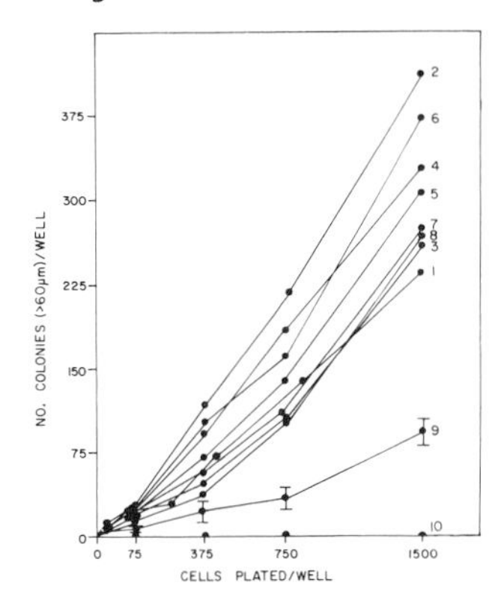

FIGURE 2. Replating of cells from primary colonies. Melanoma colony formation from cells obtained from melanoma tissue (curve 1) and serial transfer of cells from pooled colonies (curves 2 and 9). No colonies were formed after the ninth transfer (curve 10). Data are the means of 6 replicates. Our standard errors averaged 10% and were omitted for clarity except for colonies from the eight transfer (curve 9). Reproduced with permission from reference 22.

primary colonies from an additional 5 patients (and unpublished).[22] Evaluation of these results suggested the following:

1. The frequency of primary and secondary colony formation was independent. These results are similar to those obtained by Buick et al. for human myelogenous leukemia and ovarian cancer progenitor cells[5,6] and emphasize the biological importance of studying replating.
2. The cloning efficiency with serial transfers in agar varied considerably among patients. Comparable studies have not been reported in other cell renewal systems, so the meaning of these results remain to be defined.

For human leukemia, secondary colony formation has been delineated as a strong prognostic variable for survival while primary colony formation was not predictive. We have examined for the relationship of secondary melanoma colony formation to survival in our six patients and the results are shown in Figure 3.

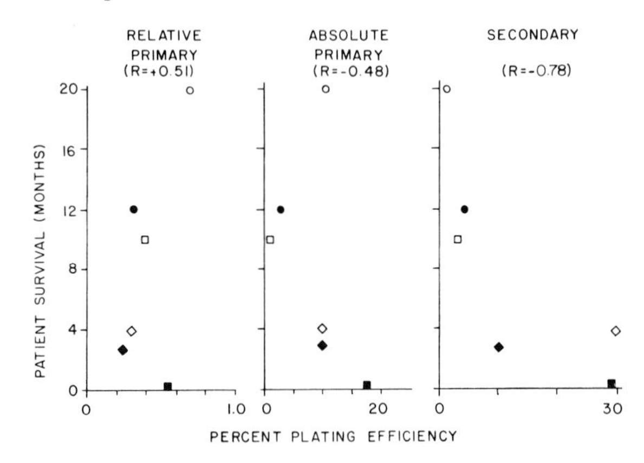

FIGURE 3. Relationship of secondary plating efficiency and patient survival. With an increased frequency of secondary colony formation patient survival decreased.

An increased frequency of secondary colony formation was associated with a poor prognosis. Confirmation of these findings in a larger series of patients will be fundamental to understanding the role of self-renewal in this disease. Additionally, a comparison of chemosensitivity of primary and secondary colony-forming cells in these different cases may allow an appreciation of the relationship between sensitivity to therapeutic agents and the development of resistance at the cellular level.

RELATIONSHIP BETWEEN COLONY SIZE (NUMBER OF CELLS) AND SECONDARY COLONY FORMATION

We have presented evidence in the previous section that cells in colonies greater than 100μm in diameter could replate in agar. A fundamental question is if cells which form primary colonies of different sizes replate at the same or different frequencies. If a classical clonal hierarchy is present, several features should be evident: (1) Small primary colonies may have limited proliferative capacity and not replate; that is, self-renew; (2) A clonal hierarchy may exist and above a certain size primary colony the replating frequency will fall or not occur. To be able to conduct such studies requires a knowledge of the number of cells within colonies of different sizes. The number of cells in hematopoietic colonies has been accurately determined by counting of cells _in situ_ using an inverted microscope as these colonies generally grow in a single plane. Accurate counting of the number of cells in tumor colonies is considerably more difficult as the cells do not grow in one or even two planes, but as a spherical-like structure. We have carefully enumerated the number of cells in individual tumor colonies of different sizes plucked with a micromanipulator and have related the maximum colony diameter in the horizontal plane of the inverted microscope and the mean diameter of the individual cells to the number of cells in the clusters or colonies. We have developed a nomogram which relates these parameters (Figure 4). Using this data we have also developed the following simple equation (discussed in detail in reference 19):

$$\text{No cells/colony} = \frac{(2.40)(\text{colony diameter})^{2.378}}{(\text{cell diameter})^{2.804}}$$

This equation illustrates several key features about the number of cells in colonies:

1. The number of cells is not linearly related to colony diameter.
2. The number is inversely related to the diameter of cells in the colonies.

As the size of cells is the same within a tumor but varies considerably from tumor to tumor, the number of cells in a colony varies markedly in colonies of the same size from patient to patient (see Figure 5 for a graphic representation).

Accurate enumeration of the number of cells in colonies has allowed us to design appropriate studies to examine the relationship between colony size and frequency of replating. These studies are detailed elsewhere and the general strategy presented here. We have initially

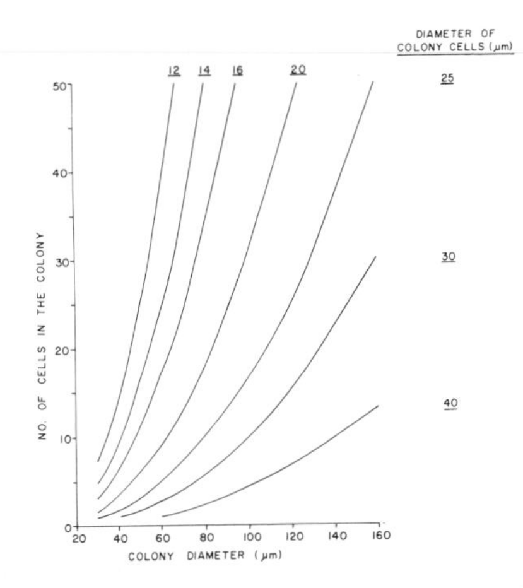

FIGURE 4. Nomogram relating cluster and colony diameter, cell diameter and the number of cells within the cluster or colony. The nomogram was constructed using data from 14 sources of tumor cells in which colonies of different diameters were plucked and the number of cells measured. Reproduced with permission from reference 19.

selected colony-forming cells from patients that showed high primary and secondary colony formation to assure that sufficient numbers of colonies would be generated to obtain representative frequency distributions.

The frequency distribution of primary colonies by diameter of primary colony was determined from two patients and a representative general pattern is shown in Figure 5. We grouped the colonies by size according to the calculated number of cells (and divisions) as shown in Table 1.

Colonies from each of four size classes were individually plucked, pooled, and dispersed into single cell suspensions; from 25 to over 1000 cells were replated in microtiter wells so that secondary colony formation would be studied in a dose-response fashion. Secondary colony formation and the frequency of colonies of different diameters was assessed in detail for primary colonies from 4 size classes in 2 patients. The results from these studies which are detailed elsewhere allows us to make the following tentative conclusions:[16]

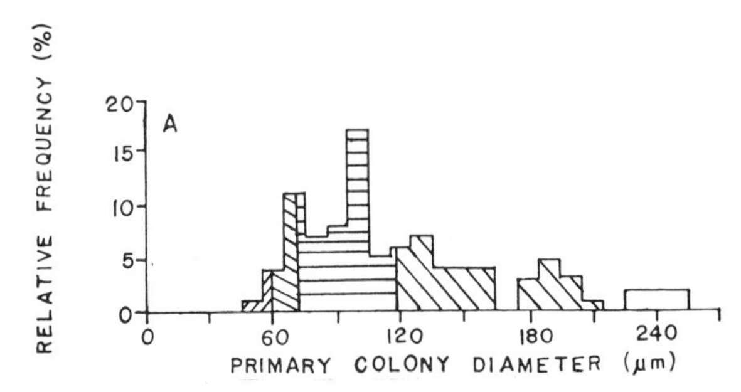

FIGURE 5. Frequency distribution of primary colonies by diameter. The colonies were grouped into 4 size classes based on the number of cells in the cluster or colonies and the number of divisions they represented.

1. Primary and secondary colony formation in agar were independent.
2. The capacity for serial replating of cells from colonies varied from patient to patient.
3. Replating of cells from primary colonies of different sizes with high primary and secondary colony formation was similar.
4. Overall, these experiments suggested that cells comprising primary colonies that underwent less than 3 to 4 divisions had diminished self-renewal capacity.

TABLE 1
Cloning Efficiencies and Median Diameter of Secondary Colonies Generated From Cells of Primary Colonies of Different Sizes

Primary Colonies			Secondary Colonies	
Diameter (Microns)	No. Cells	No. Divisions	C.E. (%)	Median Diameter (Micrometers)
50-60	13-20	4-6	x_1	y_1
60-70	20-29	5-6	x_2	y_2
70-120	29-104	5-8	x_3	y_3
120-200	104-351	8-10	x_4	y_4

CONCLUSIONS

The technology we have developed will now allow us to approach several important biological questions and to undertake a study of the experimental therapeutics of melanoma tumor stem cells. The biological questions include: What is the frequency with which melanomas give rise to colonies with self-renewal capacity? Is the presence of these feature related to the clinical biology (prognosis, survival, chemosensitivity) of melanomas? Do host cells regulate the formation of secondary colonies? Does a clonal hierarchy exist for melanoma tumors? The key experimental therapeutic question is: Do cells which form primary and secondary colonies respond differently to therapeutic agents (see Figure 6)? If self-renewing cells are as sensitive to an agent as the bulk of colony-forming cells, then the survival curve for primary and secondary colony formation will be the same. If the self-renewing cells are less sensitive, then secondary colony formation will be less affected than primary colony survival (curve A). However, if the self-renewing cells are more sensitive then survival of secondary colonies will be more decreased than primary colonies (curve B).

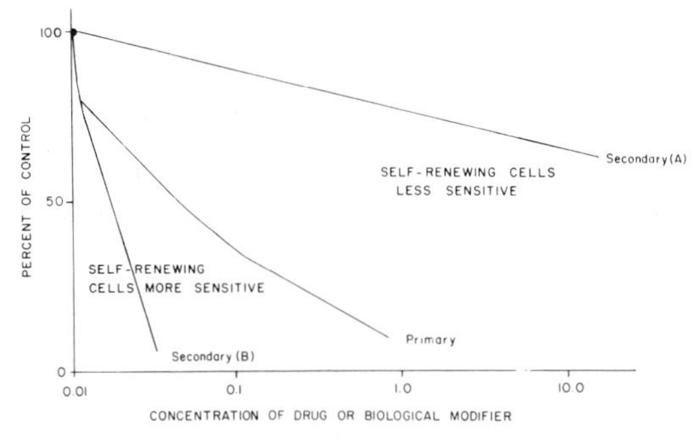

FIGURE 6. Effect of therapeutic agents on primary and secondary colony formation. The response of primary and secondary colony formation to an agent is measured.

Agents which produce curve A would be expected to have only transient and minor antitumor effect *in vivo*, since at the cellular level they are inhibiting only late progenitor, transitional, and end cells. Compounds which produce curve B may have potential to produce sustained favorable responses *in vivo*, since they are inhibiting the stem cell and early progenitor cell population. A deeper understanding of stem cells should lead to fundamental knowledge of the biology of this disease and improved therapeutics, and perhaps, extend our general understanding of drug-refractory human tumors as well.

REFERENCES

1. Bregman MD, Meyskens FL Jr.: Inhibition of human malignant melanoma colony-forming cells *in vitro* by prostaglandin A. Cancer Res 43:1642-1645, 1983.
2. Bregman MD, Meyskens FL Jr.: In vitro modulation of human and murine melanoma growth by prostanoid analogues. Prostaglandins, in press, 1983.
3. Bregman MD, Peters E, Sander D, Meyskens FL Jr.: Dexamethasone, prostaglandin A, and retinoic acid modulation of murine and human melanoma cells grown in soft agar. J Natl Cancer Inst 71:927-932, 1983.
4. Broxmeyer HE: Hematopoietic stem cells, in Trubowitz S, Davis S(ed): The Human Bone Marrow: Anatomy, Physiology, and Pathophysiology. Boca Raton, Florida, CRC Press, Inc., 1982, pp. 77-124.
5. Buick RN, MacKillop WJ: Measurement of self-renewal in culture ovarian carcinoma. Br J Cancer 44:349-355, 1981.
6. Buick RN, Maiden MD, McCulloch, EA: Self-renewal in culture of proliferative blast progenitor cells in acute myeloblastic leukemia. Blood 54:95-104, 1979.
7. Carney ON, Gazdar AR, Bunn PA, Jr., Guccion JG: Demonstration of the stem cell nature of clonogenic tumor cells from lung cancer patients. Stem Cells 1:149-164, 1982.
8. Kern DH, Campbell MA, Worth GD, Morton DL: Cloning efficiencies and drug sensitivities of freshly prepared cell suspensions, cryopreserved cells, and cell lines derived from human solid tumors. Proc. Am Assoc Cancer Res 22:274 Abs, 1981
9. Meyskens FL Jr: Radiation sensitivity of clonogenic human melanoma cells. Lancet ii:219, 1983.

10. Meyskens FL Jr, Loescher L, Moon T, Tagasugi B, Salmon SE: Relation of in vitro colony survival to clinical response in a prospective trial of single agent chemotherapy for metastatic melanoma. J Clin Oncology (Submitted).
11. Meyskens FL Jr, Loescher L, Moon T, Tagasugi B, Salmon SE: Quantitation of drug sensitivity by human metastatic melanoma colony forming units. Br J Cancer 44:787-797, 1981.
12. Meyskens FL Jr., Salmon SE: Inhibition of human melanoma colony formation by retinoids. Cancer Res 39: 4055-4057, 1979.
13. Meyskens FL Jr, Salmon SE: Modulation of clonogenic human melanoma cells by follicle-stimulating hormone, melatonin, and nerve growth factor. Br J Cancer 43:111-116, 1981.
14. Meyskens FL Jr, Soehnlen BJ, Saxe DF, Casey WJ, Salmon, SE: In vitro clonal assay for human metastatic malignant melanoma cells. Stem Cells 1:61-72, 1981.
15. Meyskens FL Jr, Thomson SP: Biology and chemosensitivity of clonogenic human melanoma tumor cells, in Costanzi J (ed): Melanoma II. New York: Raven Press, 1984, in press.
16. Meyskens FL Jr, Thomson SP: Self-renewal of melanoma colonies in agar: similar colony formation from replating primary colonies of different sizes. J Clin Invest (submitted).
17. Meyskens FL Jr, Thomson SP: Tumor growth in semisolid medium is inversely related to the number of cells plated. Br J Cancer, in press, 1983.
18. Meyskens FL Jr, Thomson SP, Moon TE: Early cellular proliferation of human metastatic malignant melanoma tumor colony forming units in soft agar. J Cell Physiol (submitted).
19. Meyskens FL Jr, Thomson SP, Moon TE: Quantitation of the number of cells within tumor colonies in semisolid medium and their growth as oblate spheroids. Cancer Res, 44:271-277, 1984.
20. Persky B, Thomson SP, Meyskens FL Jr, Hendrix MJC: Methods for evaluating the morphological and immunohistochemical properties of human tumor colonies grown in soft agar. In Vitro 18:929-936, 1982.

21. Steel GG: Growth Kinetics of Human Tumors. Oxford: Clarendon Press, 1977, 217.
22. Thomson SP, Meyskens FL Jr: Method for measurement of self-renewal capacity of clonogenic cells from biopsies of metastatic human malignant melanoma. Cancer Res 42:4606-4613, 1982.
23. Thomson SP, Wright MD, Meyskens FL Jr: Improvement of human melanoma colony formation in soft agar using boiled instead of autoclaved agar. Int J Cell Cloning 1:85-91, 1983.
24. Trent JM, Rosenfeld SB, Meyskens FL Jr: Chromosome 6q involvement in human malignant melanoma. Cancer Gen and Cytogen 9:177-180, 1983.

Studies of Estrogen Receptor Expression in MCF-7 Cells Grown in Clonogenic Assay

Fumio Kodama
Visiting Research Associate
Cancer Center
University of Arizona College of Medicine
Tucson, Arizona 85724

Geoffrey L. Greene
Ben May Laboratory for Cancer Research
University of Chicago
Chicago, Illinois 60637

Sydney E. Salmon
Professor of Medicine
Department of Internal Medicine
Director, Cancer Center
University of Arizona College of Medicine
Tucson, Arizona 85724

It is currently unknown whether hormonal agents can exert cytocidal effects on tumor stem cells. Tumor stem cells constitute the population of malignant cells capable of numerous cycles of self-renewal and are responsible for recurrence after subcurative therapy in vivo (1). It is well known that response to endocrine therapy of human breast cancer correlates with the presence of high cellular levels of estrogen receptor (ER) and that women whose primary tumors are ER negative rarely respond to endocrine therapy (2), suggesting ER-mediated effects of hormonal therapy on human breast cancer. In order to better understand the relationship of ER expression on a cellular level and to study the relation of ER to tumor stem cells, we have attempted to determine whether ER is expressed in tumor stem cells or only on more differentiated progeny in human breast cancer. As clonogenic cells in vitro are thought to be closely related to tumor stem cells in vivo, assay of clonogenic cells is generally used as an in vitro model for tumor stem cells in vivo. A major objective of our studies

HUMAN TUMOR CLONING
ISBN 0-8089-1671-8

was to determine with a clonogenic assay whether cellular expression of ER in human breast cancer cells occurred on a clonal basis (with some clones negative and others positive) or as a function of the differentiation process, wherein cells within the same clone might be either negative or positive.

We used a human breast cancer cell line, MCF-7 (3) in our model and employed an immunohistochemical stain with monoclonal antibodies for cytochemical detection of ER. The clonogenic assay system used is basically similar to the two layer semisolid agar system (in 35 mm plastic petri dishes) as described by Hamburger and Salmon (4); however, a special modification was required in order to optimize the histochemical staining for ER.

As an important element, glass fiber filters (GFF, Whatman type GF/A) were employed as a rigid but clear and porous support for clonal growth as described by Kodama et al. (5), as GFF can be readily plucked from the cultures, whole in situ clusters and colonies formed on a GFF can be easily obtained, fixed, and washed free of background agar and serum-containing medium on a Buchner's funnel. Our methods and results will be published in detail elsewhere (6).

The culture medium consisted of three layers of medium: 1.0 ml of 0.5% (w/v) Bacto agar (DIFCO), containing hard agar as a base; 0.5 ml of 0.3% (w/v) Bacto agar-containing soft agar as a middle layer, and 1.0 ml of liquid medium as a top layer. The final concentrations of the ingredients of the culture medium are 76.5% (v/v) of modified McCoy's 5A (GIBCO), 20% (v/v) fetal calf serum stripped by treatment with dextran-coated charcoal, 0.25 mM L-glutamine (GIBCO), half concentration of nonessential amino acids (FLOW) and 53 units/ml penicillin/streptomycin (GIBCO). Five thousand viable MCF-7 single cells suspended in 20 μl of medium were inoculated onto each premoistened GFF (7mm diameter circular cutouts) with a micropipette, and the GFF's were immediately transferred to the surface of the 0.5% agar baselayer for the cultures. One-half ml of molten 0.3% agar medium was gently pipetted onto the surface of the base layer (avoiding direct pipetting onto the GFF's). The layer of soft agar medium rapidly permeates and covers the GFF prior to gelatin. After gelling of soft agar medium, 1.0 ml of the liquid medium was gently poured onto the layer of soft agar medium. Triplicate cultures were then incubated in a fully-humidified 37°C incubator with a controlled low oxygen atmosphere (5% O_2, 8% CO_2, 86% N_2) and harvested on days 0, 5, 7, 9 and 11. After harvesting, each GFF containing clusters and colonies was transferred into a buffered picric acid-paraformaldehyde fixative solution, into a Buchner funnel containing prewetted filter paper, covered with an additional filter, and washed with dropwise addition of phosphate buffered saline (0.01 M, pH 7.4-PBS) until the eluate was colorless. Glass fiber filters were covered with 10% (w/v) ovalbumin in PBS and incubated for 15 minutes. After an additional wash, the GFF's were overlayed for 30 minutes with an equal mixture of two different rat monoclonal antibodies (MoAb) to ER from MCF-7 cells (clones D547 and D75) (protein concentration of 50μg/ml for each MoAb) (7). This combination of MoAb's yields better histochemical staining with the peroxidase-antiperoxidase (PAP) staining technique (8) than is obtained with either MoAb alone. Control incubations were obtained by parallel incubation of additional GFF's with the same concentration of normal rat IgG. Subsequent to an additional washing on the filter, GFF's were incubated with a 1/500 dilution of goat anti-rat IgG (Calbiochem-Behring) for 30 minutes, washed again, incubated in serum containing medium for 30 minutes at 37°C (to reduce background staining) and washed again. Glass fiber filters were then incubated for 30 minutes with a 1/80 dilution of PAP reagent (Sternberger-Meyer), washed with PBS, and stained with a mixed reagent comprised of 0.6 mg/ml 3,3'-diamino-benzidine (SIGMA) and 0.06% H_2O_2 and washed again. Counterstaining was done subsequently with 0.1% methylene blue for ten minutes followed by washing, dehydration in an ethanol-xylene series which renders the filters transparent, after which they were mounted permanently on glass slides.

Clusters and colonies formed were defined as follows: small clusters (4-20 cells), large clusters (21-39 cells), and colony (more than 40 cells in a group). Most GFF's generated usually more than 100 small clusters, large clusters and colonies.

In the clonogenic assay, MCF-7 cells exhibit progressive growth over time from single cells to clusters and eventually to colonies with early death of a number of apparently abortive clusters (figure 1).

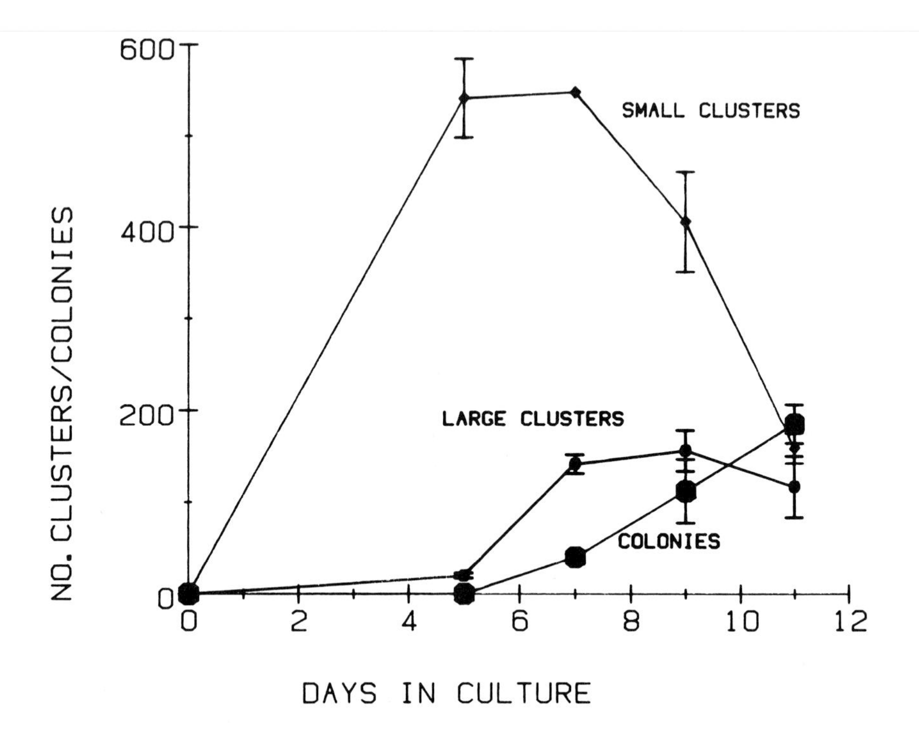

Figure 1. Growth of MCF-7 clones in the human tumor clonogenic assay (HTCA). Many small clusters extinguish by day 11 and do not proceed to the large cluster and colony stages of clonal growth. Points shown ± s.e.m.

ER was localized immunohistochemically primarily in the nuclei and very rarely in the cytoplasm of positive cells. While such localization would not have been predicted from usual "cytosol" biochemical assays (9), it is consistent with a result from a biochemical assay of ER in MCF-7 cells reported by Zava et al. (10). In order to classify ER staining of individual clones, three grades of ER positivity were defined: 0=no staining, I=>0%-50% of cells, and II=>50% to 100% of cells in a clone. More than 30 (and usually 100) each of three different sizes of clones were differentially counted per GFF as to the ER positivity grade (Table 1).

TABLE 1

Percentage Distribution of ER Positivity Grades

Clonal Growth Stage	ER Positivity Grade	Days of Culture 5	7	9	11
Small Cluster	0	10.0±2.0	5.3±0.6	14.3± 5.8	11.3±5.5
	I	24.3±2.3	29.0±3.6	37.3± 0.6	33.7±3.2
	II	65.7±4.2	65.7±4.0	48.3± 6.4	55.0±6.1
Large Cluster	0	1.0±1.7	0	0.3± 0.6	0.3±0.6
	I	5.5±4.8	10.0±2.0	22.7±10.0	23.3±3.1
	II	93.5±5.9	90.0±2.0	77.0±10.4	76.3±2.5
Colony	0	0	0	0	0
	I	0	2.4±2.4	9.0±6.0	9.3±3.1
	II	0	97.6±2.4	91.0±6.0	89.7±4.5

Two hundred cells in the single cell preparation on day 0 and 1000 cells among clones in each of the clonal size classes formed on day 5 and 11 were also differentially counted for ER positivity (Table 2).

TABLE 2

Proportion of ER Positive Cells in Single Cells and Clones

Cell Growth Stage	Days of Culture 0	5	7	9	11
Single Cell	54.8±4.8	NC	NC	NC	NC
Small Cluster	-	66.1±0.4	62.0±6.9	49.3±4.1	49.1±1.8
Large Cluster	-	±	70.1±0.8	63.9±6.3	58.2±0.6
Colony	-	±	75.9±1.4	69.0±2.6	66.3±2.5

NC=Not Counted; -=no clonal growth; ±=minimal clonal growth.

At the small cluster stage, the proportion of ER positive

cells was small and variable, and included some ER negative clusters as shown in Table 1. Once the colony stage was reached (by day 7 and thereafter), ER was expressed consistently in about 70% of cells and none of the colonies was ER negative (Table 2). The percentage of ER positive cells in clones was the highest on the 7th culture day, but as the GFF's were cultured longer, ER staining declined. The decreased proportion of ER positive cells late in culture may be due to some increase in dead cell population within clones on the later culture days in which ER is probably negative. Alternatively, there may be a preferential increase in proliferation of early precursors or stem cells in clones which have a more rapid growth rate and express little or no ER content. Recently reported cytokinetic studies support this later possibility for MCF-7 cells (11).

Our findings indicate that ER expression increases with clonal growth in clones of MCF-7 cells, suggesting that cellular expression of ER does not occur on a clonal basis (with some colonies positive while others are negative) but rather as a marker of differentiation wherein increasing proportions of cells within all clones exhibit positivity until an optimum level is reached. These findings support the concept of a proliferation-differentiation hierarchy for human breast cancer, wherein ER is not expressed in tumor stem cells but is increasingly expressed in the transit and end-cell compartments. Our findings also suggest that endocrine therapy likely acts beyond the level of the tumor stem cell and might, therefore, suppress but not eradicate neoplastic breast cancer clones. Additional studies in this model system (with MCF-7 cells and fresh tumor biopsies) using hormonal agents and antiestrogens may further elucidate the relationship between ER expression and the differentiation process in human breast cancer.

REFERENCES

1. Steel GG: Growth kinetics of tumours (ed), Oxford, Clarendon, 1977, p 217-267.

2. McGuire WL, Carbone PP, Sears ME, Eschert GC: Estrogen receptors in human breast cancer: An overview. In: McGuire WL, Carbone PP, Vollmer EP (eds), Estrogen Receptors in Human Breast Cancer, New York, Raven Press, 1974, p 1-7.

3. Soule HD, Vazquez J, Long A, et al.: A human cell line from a pleural effusion derived from a breast carcinoma. J Natl Cancer Inst 51:1409-1413, 1973.

4. Hamburger AW, Salmon SE: Primary bioassay of human tumor stem cells. Science 197:461-463, 1977.

5. Kodama F, Maruta A, Motomura S, et al.: A new system of hemopoietic colony formation for permanent slides and medium changes: Use of glass fiber filters. Blood 57:1119-1124, 1981.

6. Kodama F, Greene GL, Salmon SE: Submitted for publication, 1984.

7. Greene GL, Nolan C, Engler JP, Jensen EV: Monoclonal antibodies to human estrogen receptor. Proceedings, Natl Acad Sci, USA, 77:5115-5119, 1980.

8. Sternberger LA, Hardy PH Jr, Cuculis JJ, Meyer HG: The unlabeled antibody enzyme method of immunohistochemistry. Preparation and properties of soluble antigen-antibody complex (horseradish peroxidase-antihorseradish peroxidase) and its use in identification of spirochetes. J Histochem Cytochem 18:315-333, 1970.

9. Jensen EV, Desombre ER: Estrogen-receptor interaction. Science 182:126-134, 1973.

10. Zava DT, Chamness GC, Horwitz KB, McGuire WL: Human breast cancer: Biologically active estrogen receptor in the absence of estrogen? Science 196:663-664, 1977.

11. Jakesz R, Smith CA, Aitkin S et al.: Modification of estrogen receptor content in MCF-7 cells by cell proliferation. Proceedings, 13th International Congress of Chemotherapy. Molecular basis of hormone action and clinical aspects of endocrine therapy. Vienna, Austria, 1983, p 12-15.

Colony Size, Linearity of Formation, and Drug Survival Curves Can Depend on the Number of Cells Plated in the Clonogenic Assay

S.P. THOMSON

Research Associate
Department of Internal Medicine and Cancer Center

J.A. BUCKMEIER

Laboratory Assistant
Department of Internal Medicine and Cancer Center

N.J. SIPES

Molecular Biology Graduate Student
Department of Internal Medicine and Cancer Center

F.L. MEYSKENS, JR.

Associate Professor of Medicine
Department of Internal Medicine and Cancer Center

Cancer Center

University of Arizona

Tucson, Arizona 85724

R.A. HICKIE

Professor of Medicine
Department of Pharmacology
College of Medicine

University of Saskatchewan
Saskatoon, Sask. Canada

HUMAN TUMOR CLONING
ISBN 0-8089-1671-8

INTRODUCTION

A main interest of our laboratory is the characterization of growth in the bilayer agar clonogenic assay. We have focused on the growth of cells from human melanoma but have used several sources of cells in some studies to confirm the generality of our findings. We have done morphological studies and shown that the colonies which form from cells of melanoma biopsies actually contain melanoma cells[1], no normal cells have been found within the colonies. Cytogenetic studies have shown abnormal karyotypes; selective alteration of the long arm of chromosome 6 occurs frequently[2]. Other studies from our laboratory have shown improvement of melanoma colony formation using boiled instead of autoclaved agar[3]. We have also conducted replating experiments and showed that melanoma colony cells have extensive self-renewal and proliferative capacities[4]. Furthermore, injection of cells from melanoma colonies into nude mice gave tumors with histology similar to the original patients' tumor[5]. These data suggest that at least some clonogenic melanoma cells represent stem cells. We have also developed a technique to quantitate clonogenic growth in terms of cell numbers through the use of a general formula that relates number of cells per colony, colony diameter and colony cell diameter [7].

This paper will present and discuss our data which suggests that colony size, linearity of formation and drug survival curves depend on the number of cells plated in the bilayer agar clonogenic assay. In addition, we will present data showing that there is a marked variation of clonogenic growth between incubators and a significant loss of water from the clonogenic system during incubation.

THE LIMIT ON PROLIFERATION WITHIN CLONOGENIC SYSTEMS

Our initial observation was that the extent of individual cellular proliferation within the clonogenic system depended on the number of cells plated. We examined the relationship between the number of cells plated and colony formation. We found that both the frequency of formation and colony size were affected, with an inverse relationship between colony size and number of cells plated (Fig. 1).

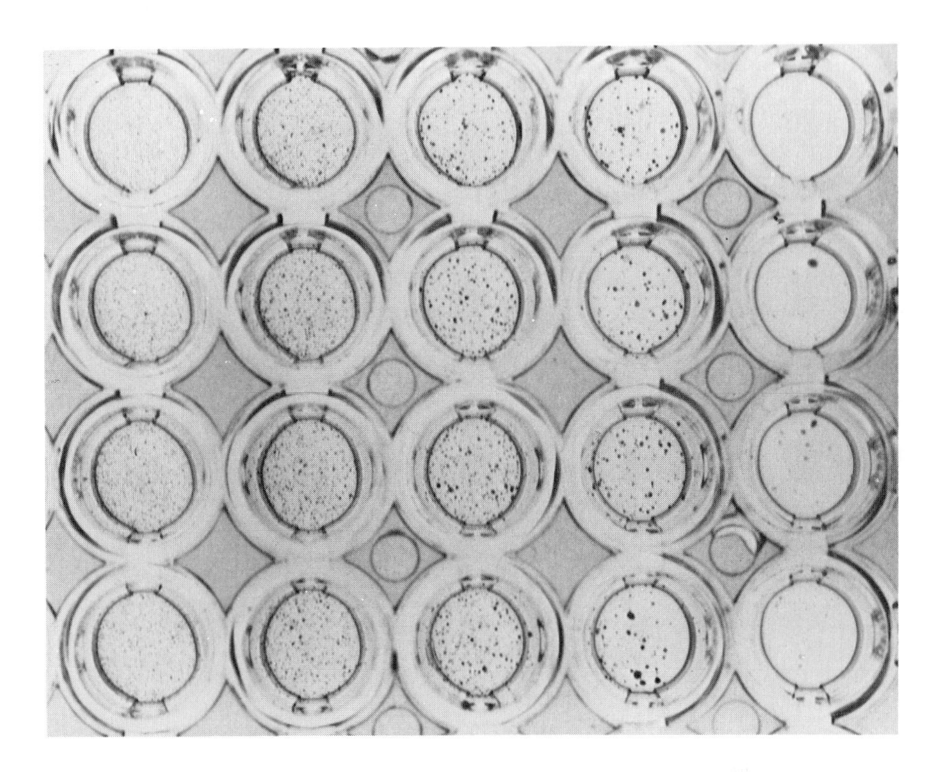

FIGURE 1. The number and size of colonies depend on the number of cells plated. The columns (top to bottom) contain 4 replicates of 5 different concentrations of cells plated in a bilayer agar microtiter system, L to R 7.5, 3.75, 0.75, 0.375 and 0.075 x 10^3 cells/well. The microtiter method is described in reference 4. Note that colony size decreased and colony number increased with the number of cells plated.

These observations support and reflect the concept that the widely used 2 ml bilayer agar clonogenic system is essentially a closed system with a finite ability to support growth. Although the system is open with respect to gaseous exchange, the amount of nutrients to support growth is limited to that in the 2 ml of media and agar. Thus the extent of individual cellular proliferation may depend on how many other cells are present that compete for the limited amount of nutrients.

To further test this hypothesis, we plated different concentrations of murine melanoma cell line cells that have extensive proliferative capacity in agar, allowed them to proliferate until they stopped, and then counted the colonies which formed by size. We found that the extent of individual cellular proliferation, as measured by colony size, was inversely related to the number of cells plated (Fig 2). Note that at the higher cell concentrations, few cells proliferated enough to form 100μm diameter colonies while at lower cell concentrations many or all of the clonogenic cells proliferated enough to form 100μm diameter or larger colonies.

We have made a detailed analysis of these colony size distributions (Fig. 2) using a formula we developed that relates colony diameter, colony cell diameter and number of cells per colony[6]. We found that the number and volume of cells within the colonies were similar for the different concentrations of murine melanoma cell line plated[7]. We have also quantitated the extent of proliferation of cells from melanoma biopsies of 7 patients and 11 cell lines derived from various tissues after they were plated and allowed to proliferate in agar until they stopped. We found that the total cellular volume within the colonies at the end of proliferation was similar, approximately 10^9 μm^3 , while the total numbers of cells were different because different samples had average colony cell sizes which were different[8].

These data suggested that the proliferative characteristics of clonogenic cells were significantly determined by the number of cells plated. Furthermore, there may be limit or ceiling of approximately $10^9 \mu m^3$ on the volume of cells that can form within the "closed" 2 ml bilayer agar system.

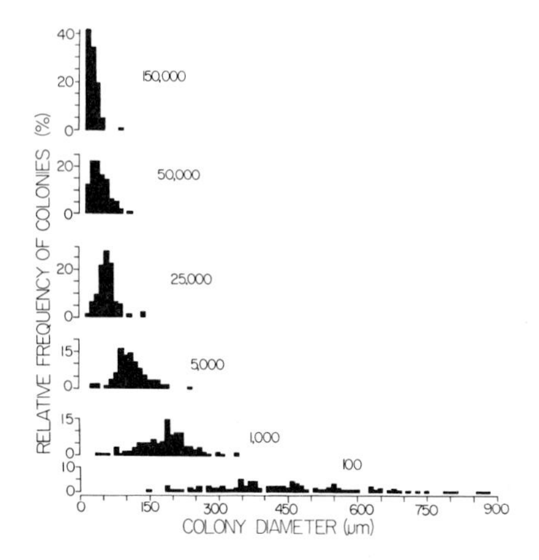

FIGURE 2. Colony size is inversely related to the number of cells plated. Murine melanoma cells (CCL 53.1) were plated in the 2 ml bilayer agar system at concentrations from 150,000 to 100 cells per 35 x 10μm petri dish. The frequency and size of multicellular colonies were estimated after proliferation stopped. Plating efficiencies, using the smallest proliferative units (2 cells), were approximately 80% for each concentration, see reference 7.

ASSAY LINEARITY IS AFFECTED BY THE LIMIT ON PROLIFERATION.

The limit on proliferation within the bilayer agar clonogenic system is consistent with the general curve shape obtained for the number of cells plated versus colony formation (Fig. 3). In general the relationship is linear only up to a certain cell concentration. At higher cell concentrations some cells are not allowed to proliferate enough to reach colony size, which reflects the inverse relationship between colony size and the number of cells plated. Indeed at very high concentrations the cells consume all the nutrients before any colonies can form. At lower cell concentrations two patterns are commonly observed. One is a linear decrease in colony formation, the other is a greater decline in colony formation usually ascribed to a "feeder effect", where a critical number of neighboring cells are required for colony formation.

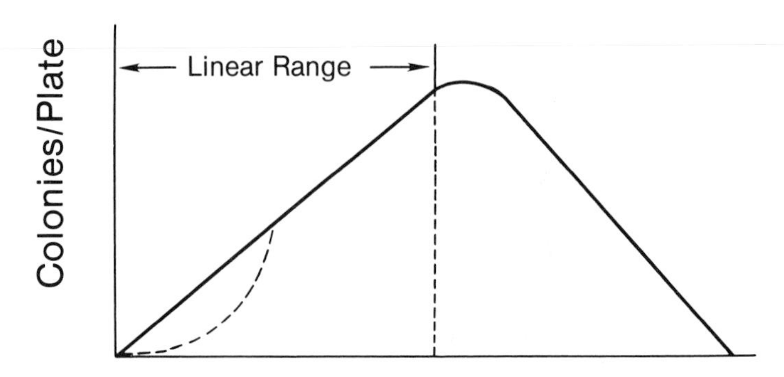

FIGURE 3. The limit on growth affects assay linearity. This is a general curve showing the relation between number of cells plated and colony formation. Note that colony formation is only linear up to a particular cell concentration and that low cell concentrations may be associated with a steep decline in colony formation, a "feeder affect". At high concentrations, cells can not proliferate enough to form colonies.

Our experience with plating different concentrations of cells has confirmed the general curve shape shown in figure 3. We found lower plating efficiencies with increasing cell concentrations for several cell lines of human and animal sources and for cells from human melanoma biopsies[7]. However, the cell concentration at which colony formation declined was highly variable so that the range of linearity varied between the different sources of clonogenic cells. Furthermore decreases in plating efficiencies were observed at both high and low cell concentrations in 4 of 5 melanoma biopsy cell samples, confirming the presence of a "feeder effect". Other investigators have reported data consistent with the general curve shape shown in figure 3[9,10,11] but in general complete number of cells plated versus colony formation curves are not described for many samples used in clonogenic assays.

Another general finding is that the linear range is less for large colonies than for small ones (Fig. 4). The experiments that defined the relationship between the number of cells plated and colony formation for cell lines from several tumor types and cells from melanoma biopsies

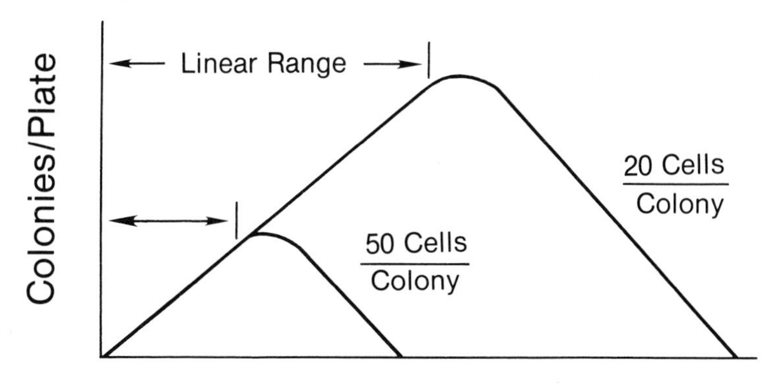

FIGURE 4. The linear range is more restricted for larger colonies. This is a general curve showing relation between number of cells plated and formation of colonies containing 20 or 50 cells. Note that larger colonies do not form at higher cell concentrations, making the linear range for colonies of 50 cells smaller.

have consistently shown a more restricted linear range for larger colonies[7]. This is consistent with the limit on proliferation within the bilayer agar system and simply reflects the inverse relationship between the number of cells plated and colony size. For example, at the end of the linear range for 20 cells per colony (Fig. 4), the system contains so many cells that the total cellular volume is near the limit of the system. Thus few, if any, of the cells can continue to proliferate to form colonies with 50 cells. Therefore the concentration of cells at the end of the linear range for 20 cells per colony is beyond the linear range for 50 cell colonies.

COLONY SIZE AFFECTS SURVIVAL CURVES

A direct implication of changes of linearity for colonies of different sizes is that different survival curves may be obtained for colonies of different sizes. Consider the general case shown in figure 5, where the concentraion of cells at point A were plated. Note that concentration A is in the linear range for 20 cells per colony but is beyond

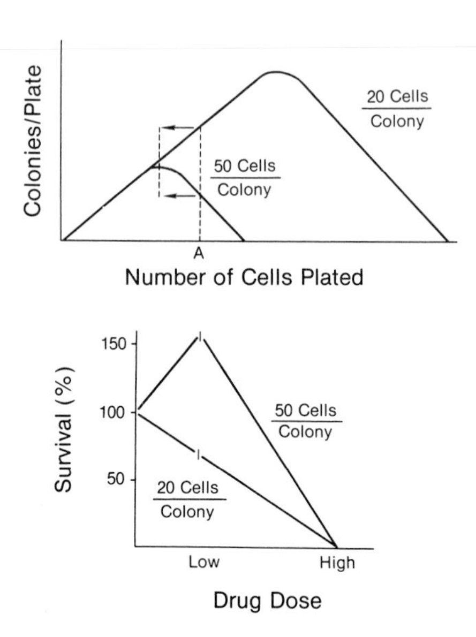

FIGURE 5. Colony survival curves depend on the size used to define a "colony". Note (top panel) that if cell concentration A is plated and a low drug dose lowers cell concentration , as shown by arrows, then a linear kill of colonies containing 20 cells and an apparent "stimulation" of colonies containing 50 cells would occur (bottom panel).

the linear range for 50 cells per colony. If a cytotoxic drug is used at low concentration to lower the cell number, as shown by the arrows, two opposite affects occur. There is a decrease in survival of colonies containing 20 cells while the survival of colonies with 50 cells increases, an apparent "stimulation" by a cytotoxic agent (Fig. 5). Both of these survival curves are consistent with the relation between number of cells plated and formation of colonies containing 20 or 50 cells (Fig. 5). Note that high doses of cytotoxic agents would reduce cell numbers into the linear range for large and small colonies and induce very little survival. We recently reported data that confirmed these effects, with both melphalan and actinomycin D inducing linear kill of smaller colonies and apparent "stimulation" of larger colonies. The apparent stimulation of colony formation by low doses of cytotoxic agents may be due to the lowering of the effective cell concentration which decreases consumption of the limited amount of nutrients and allows cells to proliferate into larger colonies[7].

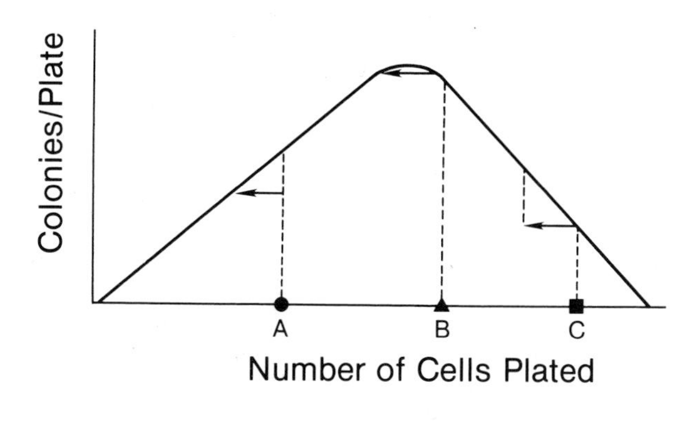

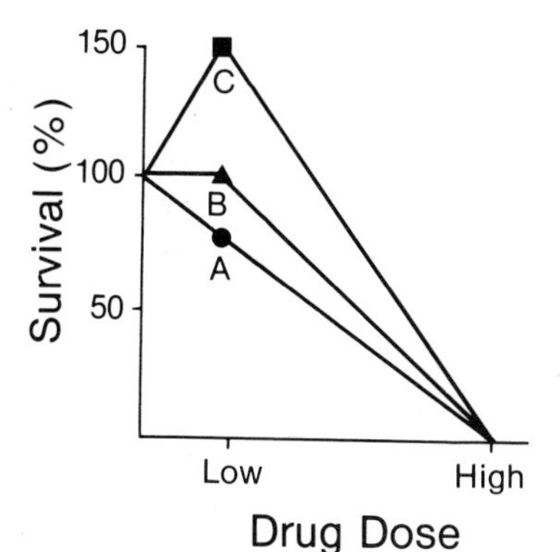

FIGURE 6. Number of cells plated affects survival curves. The general relation between number of cells plated and colony formation (top panel) predicts different survival curves using cell concentration A,B or C. Note that if a low dose of a cytotoxic agent is used, as shown by the arrows, the reduction in effective cell concentration would produce a decrease (A), no change (B) or increase (C) in colony survival.

NUMBER OF CELLS PLATED AFFECTS SURVIVAL CURVES

A direct implication for the general relation between number of cells plated and colony formation (Fig. 2) is that different survival curves may be obtained for different numbers of cells plated. Consider the effect of a low dose of a cytotoxic agent in assays using one of three different cell concentrations (A,B,C), shown in figure 6. Cell concentration A is in the linear range would give a decrease in survival, curve A, as expected for a cytotoxic agent. However, cell concentrations B and C are beyond the linear range so that low doses of a cytotoxic agent would give survival curve B with a plateau and survival curve C with an apparent stimulation. Note that at high doses of a cytotoxic agent colony formation would decrease at cell concentrations A,B, and C. Furthermore, if a stimulator of growth was added to the assay then an increase in colony formation would occur only at cell concentration A, cell concentrations B and C would give an apparent inhibition of colony formation. Our laboratory has had relatively little experience with comparison of survival curves at different cell concentrations since we define the relationship between number of cells plated and colony formation and then plate at a concentration in the linear range, similar to concentration A. However, experiments using different cell concentrations have confirmed the dependence of survival curves on the number of cells plated (Fig. 7). Survival curves similar to A,B,C (Fig. 6) were obtained using 1,000, 5,000 and 25,000 murine melanoma cells respectively and irradiation as the cytotoxic agent. Note that there was a dose dependent decrease with 1,000 cells per plate and a "plateau" and an apparent stimulation with 5,000 and 25,000 cells/plate.

MARKED VARIATION OF CLONOGENIC GROWTH BETWEEN INCUBATORS

We noticed considerable variation between control plates of the same experiment placed in different incubators. Replicate control plates of a murine (Cloudman S91-CCL 53.1) and human melanoma cell line (81-46A) were placed in several different CO_2 incubators which were in routine use for clonogenic assays. Variation of clonogenic growth within a single incubator for the murine and human cell lines was low, with coefficients of variation of 9.3 and 11% respectively. Variation between incubators was significantly higher with C.V. of 25.9 and 70.3%

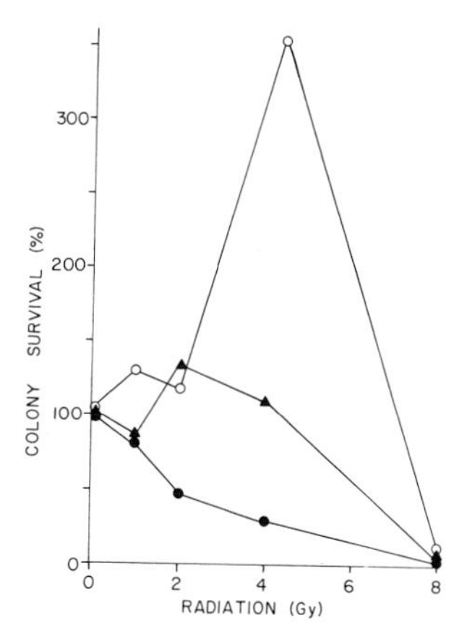

FIGURE 7. Colony survival curves for three different cell concentrations. Murine melanoma cells (CCL 53.1) were plated in bilayer agar at (●) 1000, (▲) 5,000 and (O) 25,000 cells per 35 x 10mm petri dish and irradiated at low doses. Colony (> 150µm diameter) formation decreased, "plateaued", and increased for 1000, 5000 and 25,000 cells/plate respectively. Note that 8 Gy decreased colony formation at all cell concentrations.

respectively (Fischer ratio $p < 0.025$). A representitive experiment is shown in figure 8. Note the large variation between incubators compared to the variation between replicates in the same incubator. Control plates carried to other incubators but returned to the original incubator showed similar colony formation, suggesting that mere distribution of the plates did not account for the variation between incubators. The CO_2 incubators had similar temperature, humidity and CO_2 levels and supported cell lines in monolayer culture adequately. These results suggest that colony formation in agar is sensitive to slight differences in incubator conditions. Furthermore, differences in colony formation between incubators make comparisons of absolute colony formation between incubators or laboratories difficult.

LOSS OF WATER DURING INCUBATION

We detected a variable and significant water loss of 5 to 40% by weighing culture plates prior to and following 12

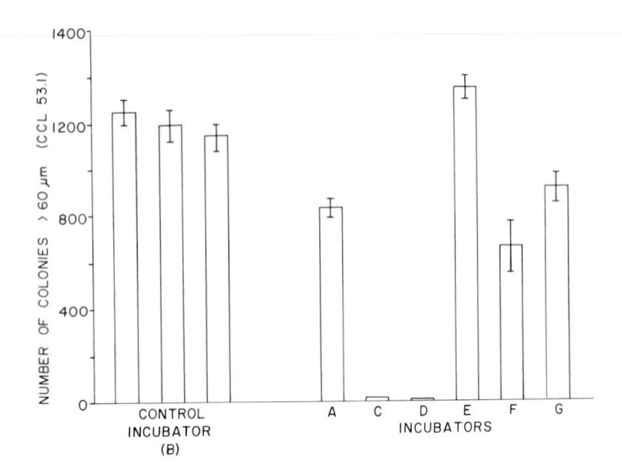

FIGURE 8. Marked variations of clonogenic growth between incubators. Replicate control plates were incubated in several incubators, A-G. Note that variation between incubators is greater than within the same incubator. Mean ± SE are shown.

days of incubation in 12 different CO_2 incubators. We measured the kinetics of the water loss under three different incubation conditions and found significant losses under standard conditions (Fig. 9). Incubation in an unhumidified incubator resulted in complete water loss over a few days. Incubation in the standard incubator, humidified by flooding the bottom, placing a tray of water on the top shelf and turning the door heater off, was associated with a 20% loss over 14 days and more than 30% over 21 days. Macroscopic changes in agar appearance were not seen until 35-40% water loss had occured. Incubation within sealed chambers with water in the bottom reduced the water loss. These results suggest that large water losses occur during standard incubation conditions. The resultant increase in the agar systems osmolarity may inhibit cellular proliferation.

SUGGESTIONS TO IMPROVE THE TECHNIQUE AND INTERPRETATION OF CLONOGENIC ASSAYS

Ideally one should define the relation between number of cells plated and colony formation for each sample because

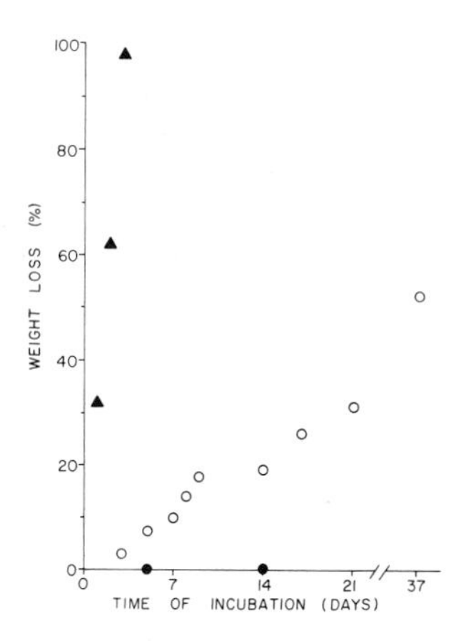

FIGURE 9. Rate of agar water loss during incubation. Plates were weighed prior to and after indicated days of incubation under three conditions, (△) unhumidified incubator, (O) standard humidified incubator and (●) sealed chamber containing water placed in humidified incubator.

the linear range is often different for different cellsamples. We do this routinely for cells from cell lines. For cells from tumor biopsies, we cryopreserve several vials of cells if possible, and plate several cell concentrations to determine optimal concentrations in the linear range for use in further experiments. When time and cell numbers are limited so that the above approach is impractical the careful evaluation of the clonogenic growth pattern is important because it can indicate when too many cells have been plated. For example, if control plates have large numbers of clusters then cell concentration may be too high. Additionally if the media in control plates turns yellow and acidic after a couple days or if stimulation of colony formation occurs by known cytotoxic agents then again cell concentrations may have been too high. Under high cell conditions, artefactual survival curves may be obtained, as described above, so we suggest a replating at lower cell concentrations if possible.

Another consideration is that the initial proliferation usually is not affected by cell crowding because the cells have not proliferated enough to reach the limit on cellular volume within the system. This is why smaller colonies have a greater linear range and their survival curves reflect the true effect of the agent tested in the clonogenic assay. These data, combined with our evidence that, at least for melanoma, small colonies have self-renewal capacities similar to larger colonies (Meyskens and Thomson this volume), suggest that one should use small colonies to define survival curves.

REFERENCES

1. Persky B, Thomson SP, Meyskens FL Jr, Hendrix MJC: Methods for evaluating the morphological and immunohistochemical properties of human tumor colonies grown in soft agar. In Vitro 18:929-936, 1982.
2. Trent JM, Rosenfeld SB, Meyskens FL Jr: Chromosome 6q involvement in human malignant melanoma. Cancer Genet Cytogen 9:177-180, 1983.
3. Thomson SP, Wright MD, Meyskens FL Jr: Improvement of human melanoma colony formation in soft agar using boiled instead of autoclaved agar. Int J Cell Cloning 1:85-91, 1983.
4. Thomson SP, Meyskens FL Jr: Method for measurement of self-renewal capacity of clonogenic cells from biopsies of metastatic human malignant melanoma. Cancer Res 42:4606-4613, 1982.
5. Meyskens FL Jr, Thomson SP: Biology and chemosensitivity of clonogenic human melanoma tumor cells, in Costanzi J (ed), Melanoma II. New York: Raven Press, 1984, in press.
6. Meyskens FL Jr, Thomson SP, Moon TE: Quantitation of the number of cells within tumor colonies in semisolid medium and their growth as oblate spheroids. Cancer Res, 1984, in press.
7. Meyskens FL Jr, Thomson SP, Hickie RA, Sipes NJ: The size of tumor colonies in semisolid medium is inversely related to the number of cells plated. Brit J Cancer, 1983, in press.
8. Thomson SP, Meyskens FL Jr: The extent and kinetics of proliferation of melanoma clonogenic cells. J Cell Physiol (submitted), 1984.

9. Thomson JE, Rauth AM: An in vitro assay to measure the viability of KHT tumor cells not previously exposed to culture conditions. Radiation Res 58:262, 1974.
10. Tveit KM, Fodstad O, Lotsberg J, Vaages, Phil A: Colony growth and chemosensitivity in vitro of human melanoma biopsies. Relationship to clinical parameters. Int J Cancer 29:533, 1982.
11. Page R, Tilchen E, Davis H, Talle Y: Cloning of human tumor cells in soft agar-relationship of specimen processing and inoculum density to evaluable growth for in vitro chemosensitivity. Proc Am Assoc Cancer Res 24, abs 1241, 1983.

Modulation of Tumor Colony Growth by Irradiated Accessory Cells

Anne W. Hamburger

Cell Culture Department
American Type Culture Collection
12301 Parklawn Drive
Rockville, MD. 20852

INTRODUCTION

Despite the expanding clinical use of the human tumor clonogenic assay (HTCA) (8), the factors controlling the ability of human tumor cells to form colonies in soft agar are poorly understood. Clonogenic cells may respond to growth promoting or inhibiting substances produced by accessory cells. Selective manipulation and characterization of the different regulatory cell subsets are necessary to understand these complex interrelationships.

In 1978, we described the effect of phagocytic and adherent cells on the gowth of human tumor colonies. We found that depletion of phagocytic macrophages from ovarian carcinoma effusions resulted in a decrease of tumor colonies (9). In this case, tumor growth appeared to be macrophage-dependent. These observations were subsequently confirmed by Buick et al. (3) using human malignant effusions derived from patients with a variety of adenocarcinomas and Schultz et al. (18) using tumor cell lines and transplantable tumor systems. Similarly, Mantovani et al. (14) have demonstrated that macrophages from patients with ovarian cancer enhance the growth of ovarian tumor cells under defined conditions.

However, mononuclear phagocytes isolated from murine and human neoplasms are cytotoxic for tumor cells both in vitro and in vivo (7). Thus, macrophages appear to be heterogenous in their interaction with tumor cells. It is not known whether the diverse effects of macrophages towards tumor cells are due to different subsets of macrophages, or different activities of the same

HUMAN TUMOR CLONING
ISBN 0-8089-1671-8

population. Macrophage heterogeneity has been amply demonstrated and cells may vary in their velocity sedimentation, buoyant density, enzyme production, and the presence of surface molecules such as receptors for complement and immunoglobulins, and Ia antigens (11). However, little is known about the antigenic profile of macrophages performing different functions.

In an attempt to define the macrophage subpopulations responsible for enhancing tumor growth, we depleted adherent cells of macrophages bearing surface Ia antigens and assessed the ability of the residual population to support growth (10). The results indicated that Ia negative macrophages enhanced the growth of tumor cells in soft agar and that Ia positive macrophages may have limited cell growth.

To further investigate the nature of potentiating cells, we determined the effect of irradiation on the ability of adherent cells to support tumor colony growth. Other investigators have demonstrated irradiation does not alter the ability of adherent cells to support growth of hematopoietic cells (22). However, the following results indicate irradiation of adherent cells increased their ability to enhance the growth of human tumor colonies.

MATERIALS AND METHODS

Cell Preparation Pleural or ascitic fluids (200-4,000 ml) were obtained aseptically in heparinized (10 units/ml) vacuum bottles from patients with histologically proven epithelial neoplasms. Fluid was passed through sterile gauze, centrifuged at 600 g for 10 minutes and the cell pellet resuspended in McCoy's 5A medium containing 10% fetal bovine serum (FBS). Cells were washed twice in this medium and counted in a hemocytometer. Viable nucleated cell counts (determined using erythrosin-B) were routinely more than 90%. Differential counts were performed on slides prepared with a cytocentrifuge.

Culture Assays for Colony Forming Cells (CFCs) Cells were cultured as described (8). One ml underlayers, containing enriched McCoy's 5A medium in 0.5% agar were prepared in 35 mm plastic Petri dishes. Cells to be tested were suspended in 0.3% agar in enriched CMRL 1066 medium (GIBCO, Grand Island, N.Y.) with 15% horse serum (Sterile Systems, Logan, Utah). Each culture received $5x10^5$ cells in 1 ml of agar-medium mixture. Cultures were incubated at 37°C in a 5% CO_2 humidified incubator. Cultures were scored using an inverted phase microscope as previously described (8).

Depletion of Adherent Cells Cells were incubated overnight at 37° C in a humidified atmosphere of 5% CO_2 in air in 100 mm plastic tissue culture dishes at a concentration of $2x10^6$ cells/ml in McCoy's 5A medium containing 10% autologous effusion fluid. Nonadherent cells were removed and the adherent cell layer washed twice with 5 ml of .002% EDTA-saline. The washings were pooled with the nonadherent fraction. Nonadherent cells contained 5 ± 3% macrophages based on morphology and NSE stains. The washed adherent layer was then removed with a rubber policeman. Cell yields were 60-85% of input values. Viabilities were 85% of that observed before overnight culture. In cases where adherent cells were used as a feeder layer, a known number of adherent cells was suspended in the bottom layer of enriched McCoy's 5A media in 0.5% agar. The nonadherent cell suspensions were then overlaid in 0.3% agar.

Irradiation Cells were irradiated in suspension with 1000 rads (unless otherwise specified) using a ^{137}Cs irradiation source (500 rads/min) (Gammator B, Parsnippany, NJ).

Treatment of Adherent Cells with Monoclonal Antibodies Adherent cells were resuspended at $1x10^6$ cells in 1.0 ml McCoy's 5A medium with the appropriate monoclonal antibody and complement (C) (10). Cells were treated with OKT3 antibodies obtained from Ortho Pharmacentical (Raritan, NJ.).

Depletion of E-rosetting T lymphocytes T lymphocytes were isolated by the method of Pellegrino (15). Cells ($5x10^6$ ml) were mixed with AET (2-aminoethyliso-thiouronium bromide, Sigma, St. Louis, MO) treated sheep erythrocytes (SRBC). The rosetted cells were then subjected to sedimentation through Ficoll Hypaque. E negative cells were recovered at the interface and E positive cells in the pellet.

Statistical Analysis The probability of differences between samples being statistically significant was determined by the use of the two-tailed Student's t-test. Four or five plates were scored per point. The results are expressed as mean ± SE.

RESULTS

Effect of Irradiation on the Ability of Adherent Cells to Support Tumor Colony Growth We examined the effect of irradiation on the ability of adherent cells to support the growth of tumor colonies. Sufficient numbers of adherent cells were successfully isolated from effusions of 19/35 patients with adenocarcinoma of the ovary, breast or colon, or melanoma. In all successful cases, adherent cells were composed or macrophages, lymphocytes and fewer than 5% tumor cells (10). This was confirmed by morphology, and by staining for non-specific esterase and macrophage, lymphocyte and epithelial cell surface antigens. Depletion of adherent cells resulted in a loss of colony forming capacity by the residual nonadherent tumor cells (Figure 1). In 18/19 cases, addition of $2x10^5$ adherent cells to underlayers of cultures containing only nonadherent cells significantly increased the number of colonies observed.

Effect of Irradiating Adherent Cells

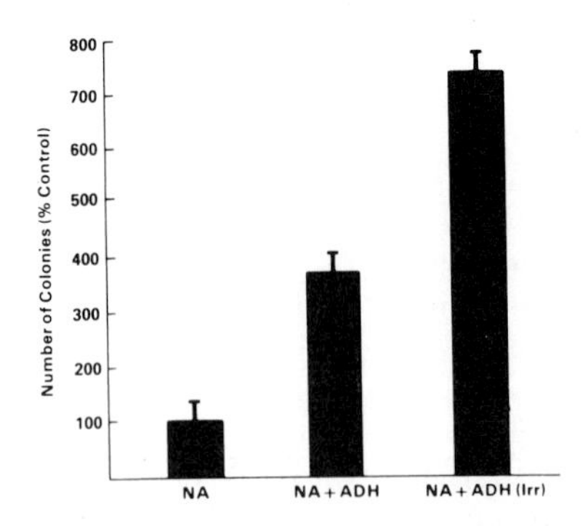

Figure 1: The effect of irradiation on the ability of adherent cells to support tumor colony growth. Bars represent mean ± S.E. (15 experiments). Adherent cells were isolated as described and one aliquot irradiated. Adherent cells ($2x\ 10^5$) were then added to cultures of $5x10^5$ nonadherent tumor cells.

One group of adherent cells from each evaluable sample was exposed to 1000 rads and the ability of irradiated cells to support growth of NA tumor cells compared to controls. The results (Figure 1) indicate that in 17/19 cases irradiated adherent cells supported the growth of more tumor colonies than untreated adherent cells. In 2 cases, there were no significant differences between the number of colonies observed in the presence of irradiated or untreated adherent cells.

Effect of Increasing Doses of Irradiation Adherent cells were treated with 250-4000 rads to define the dose-dependence of the enhancing effect. Figure 2 demonstrates that as few as 250 rads significantly increased the ability of adherent cells to support tumor colony growth. A maximal enhancement was reached between 500-1000 rads.

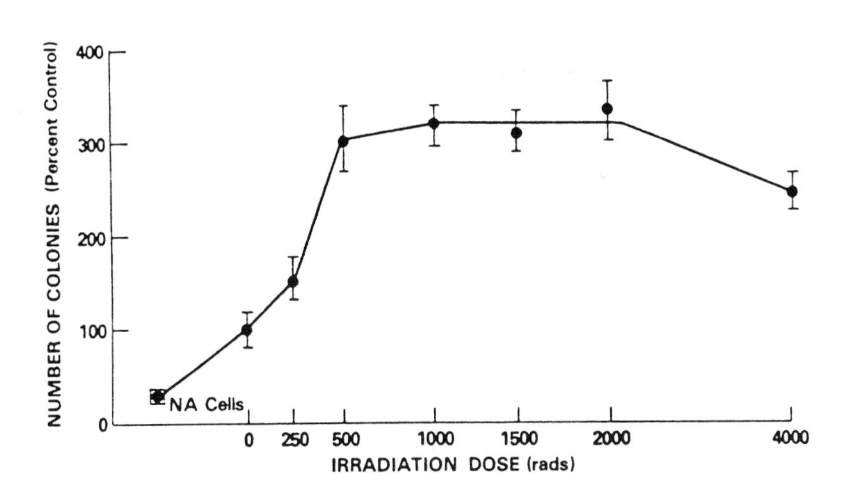

Figure 2: Effect of increasing doses of irradiation on the ability of adherent cells to support tumor colony growth. Adherent cells ($2x10^5$) from eight patients were treated with increasing doses of irradiation and then added to cultures containing nonadherent tumor cells. The results from each specimen were normalized to percent control. The number of colonies observed in the presence of unirradiated adherent cells was equal to 100%.

Effect of Depletion of Adherent T-Cells Although adherent cells in this study consisted of greater than 90% macrophages, (3,5,10), variable numbers of lymphocytes were present. The majority of the lymphocytes in malignant effusions were T-cells. Radiosensitive suppressor T-cells have been demonstrated to interact with macrophage-monocytes in a variety of immune responses (19) and in the control of hematopoietic colony growth (21). Therefore, we interpreted our initial results as indicating that the enhanced ability of irradiated adherent cells to support tumor growth was due to inactivation of T cells that may participate in the arrest of tumor colony growth.

To test this hypothesis directly, adherent cells were depleted of T-cells by the use of a pan-T monoclonal antibody (OKT3) and C. Antibody-treated adherent cells or untreated adherent cells ($2x10^5$) were then added to the agar underlayers as described. The results from the average of studies of 9 patients are shown in Figure 3.

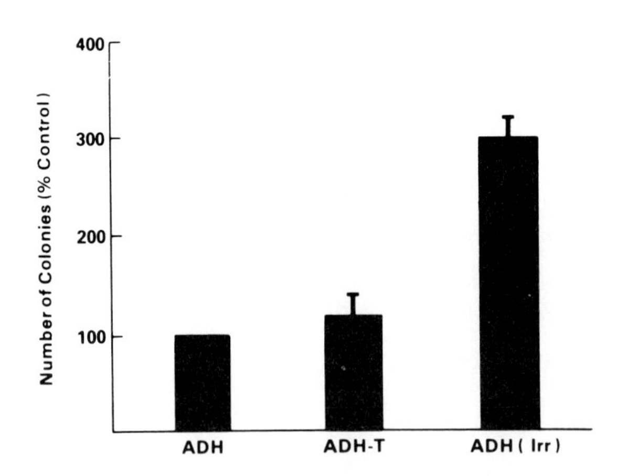

Figure 3: The effects of adherent cells depleted of T lymphocytes on the growth of tumor colonies from nonadherent tumor cells. Bars represent mean ± S.E. (eight experiments). Adherent cells were isolated from effusions and pretreated with C' and OKT3 monoclonal antibody. Adherent cells ($2x10^5$) were then added to cultures of $5x10^5$ nonadherent cells as described.

In 8/9 cases, the number of tumor colonies observed in the presence of T depleted-unirradiated adherent cells was significantly less than the number observed in the presence of irradiated adherent cells. Therefore, depletion of T cells by antibody-mediated cytolysis was not as effective as irradiation in enhancing tumor colony growth.

We also investigated whether the irradiation induced enhancement of tumor colony growth was dependent on the presence of T cells. Adherent cells were isolated as described and divided into two groups. One group was treated with the pan-T OKT3 antibody. Half the cells in each group were subsequently irradiated. Table 1 summarizes data from ten patients (7 ovarian, 3 breast). The results of each experiment were normalized to percent of control observed in the presence of non-adherent cells. The average increase in growth over that observed with nonadherent cells only was calculated for each group. Depletion of T cells by antibody from control adherent cells resulted in variable increases in tumor growth. In contrast, the increased stimulating activity of the irradiated macrophages was found only when T cells were present. This indicates that the increased tumor colony enhancing activity of the irradiated macrophages was T cell dependent.

Table 1: Effect of T cells on the Ability of Irradiated Macrophages to Support Tumor Colony Growth.

	% Control Number of Colonies per 5×10^5 Cells	
	T cells Present	T cells Removed
Untreated Adherent Cells	352 ± 41	407 ± 80
Irradiated Adherent Cells	735 ± 42	182 ± 30

Results are expressed as % of control colony growth (non-adherent cells only) and represent mean ± S.E. of 10 experiments. Adherent cells were isolated as described. Groups were either irradiated and/or depleted of T cells by treatment with a monoclonal antibody, as indicated in the text. Adherent cells were then added to underlayer and 5×10^5 autologous non-adherent tumor cells added to the agar overlayer.

We interpreted the above findings as indicating that the effect of eliminating radiosensitive T cell populations that may suppress tumor CFC growth was obscured by removing the entire population of T lymphocytes. Another subpopulation of T cells that enhance tumor colony growth may have been eliminated.

Selective Effect of Irradiation on the Ability of Adherent Cells to Enhance Colony Formation Finally, we wished to demonstrate directly that the irradiation induced augmentation of the ability of adherent cells to support tumor colony formation was due to a selective effect on T cells. We therefore isolated and selectively irradiated macrophages and lymphocytes. Adherent cells were isolated as described and depleted of T cells by E-rosetting prior to irradiation. The cell populations were then reconstituted and assayed for their ability to support tumor colony formation. The results of one representative experiment (out of 4), using cells from a patient with ovarian cancer, are shown in Table 2.

Table 2: Effect of Irradiation of Isolated Macrophages and Lymphocytes on Tumor Colony Formation.

Cell Combination	Cell Irradiated	No. of Colonies (% Control)
M + L	None	290 ± 25
M + L	Both	509 ± 12
M (1000 rads) + L	Macrophage	330 ± 10
M + L (1000 rads)	Lymphocyte	569 ± 42

Adherent cells were separated into macrophages (M) and T lymphocytes (L) as described. The isolated populations were irradiated, recombined and added to cultures of NA tumor cells. Results are expressed as percent of colonies observed in the presence of nonadherent cells only.

Irradiation of macrophages alone did not increase the ability of adherent cells (macrophages and lymphocytes) to support tumor colony formation. By contrast, irradiation of lymphocytes alone increased the number of colonies to values observed when the reconstituted (irradiated macrophages and irradiated lymphocytes) were used. While T-cell depleted adherent cells could support growth, the enhanced ability of the irradiated macrophages to support growth was T cell dependent.

DISCUSSION

The results of this study confirm our previous findings indicating that adherent and phagocytic cells are required for optimal in vitro tumor colony growth (9). We now report that in 17/19 cases, irradiation enhanced the ability of adherent cells to support tumor colony formation. The irradiation induced enhancement may have been due to several factors. However, one of the more plausible explanations is that cells that suppress colony growth, either directly or indirectly, may have been inactivated.

We felt it was unlikely irradiation affected macrophage secretion of tumor colony stimulating factors directly (22). The fact that the majority of adherent lymphoid cells in effusions are T lymphocytes, (5,12), led us to hypothesize that a major effect of irradiation was to inactivate a T lymphocyte that inhibits tumor colony formation. There is evidence that T cells may either directly suppress tumor growth (6) or modulate secretion of cytostatic products of macrophages (4).

Further experiments in which either isolated T cells or macrophages were irradiated supported the hypothesis that the irradiation induced effect was via a T-lymphocyte. Irradiation of isolated macrophages alone did not enhance the ability of adherent cells (irradiated macrophages and untreated T lymphocytes in combination) to support tumor colony growth. In contrast, when irradiated T cells were added to untreated macrophages, increased colony formation over that observed in the presence of the reconstituted untreated macrophage-lymphocyte combination was observed.

In contrast, the existence of T lymphocytes that may augment secretion of CSFs for tumor cells was supported by the finding that the increased stimulating activity of irradiated macrophages was found only when irradiated T cells were present. This indicates that the increased tumor colony enhancing activity of the macrophages was T cell dependent. This is reminiscent of the modulation of erythroid and granulocyte (21) colony formation in soft agar and tumor growth in vivo (6) that is dependent on T cell-macrophage collaboration.

The results of these studies demonstrate that both adherent lymphocytes and macrophages can affect the growth of human tumor colonies in soft agar. Macrophages have been demonstrated to either enhance or inhibit colony formation. Interacting subsets of T cells may modulate the ability of macrophages to produce colony stimulating or colony inhibiting factors. It is likely that individual patients can be characterized by a unique balance of tumor cell-accessory cell interactions at any time in the natural history of their disease.

The demonstration that manipulation of tumor-accessory cell interactions in vitro influences tumor cell growth may have implications for clinical treatment. The studies presented here are consistent with the hypothesis that many human tumors are to varying extents dependent on immune cells (16). Different subsets of macrophages and lymphocytes may have opposing effects on tumor cell survival and proliferation. In addition, the same populations that are stimulatory at low concentrations may be inhibitory at higher concentrations. Thus, alterations in either subset type, number, or activation state of immune cells by biological response modifiers or monoclonal antibodies may prove beneficial. Further proof that many human tumors are influenced by macrophage and lymphocytes may facilitate development of effective immunological approaches to therapy.

ACKNOWLEDGEMENTS

The author thanks Marc Citron, M.D., Washington Veteran's Administration Hospital and Sherilyn Hummel, M.D., Georgetown University School of Medicine for obtaining the human cells used in these studies. We also thank Shirley Mazur and Penny Lavery for preparation of the manuscript.

This work was supported in part by Grants CA 28669 from the National Cancer Institute and PDT 214 from the American Cancer Society.

REFERENCES

1) Blazar, B. A., Milller G. and Heppner, G. (1978) In situ lymphoid cells from mouse mammary tumors IV. In vitro stimulation of tumor cell survival by lymphoid cells separated from mammary tumors. J. Immunol. 120: 1887.

2) Broxmeyer, H. E. (1981) The association between Ia antigen and regulation of myelopoiesis in vitro by iron binding proteins. In Comparative Research on Leukemia and Related Disorders. Yohn, D. (Ed) Elsevier, Holland, 1981.

3) Buick, R. B., Fry, S. E. and Salmon, S. E. (1980) Effect of host-cell interactions on clonogenic carcinoma cells in human malignant effusions. Br. J. Cancer 41: 695.

4) Cameron, R. J. and Churchill, W. H. (1979) Cytotoxicity of human macrophages for tumor cells. Enhancement by human lymphocyte mediators. J. Clin. Invest. 63: 977.

5) Domagala, W., Emeson, E. and Kass, L. G. (1978) Distribution of T-lymphocytes in peripheral blood and effusions of patients with cancer. J. Natl. Cancer Inst. 61: 295.

6) Gabizon, A., Leibonich, S. J. and Goldman, R. (1980) Contrasting effects of activated and non-activated macrophages and macrophages from tumor-bearing mice on tumor growth in vivo. J. Natl. Cancer Inst. 65: 913.

7) Evans, R. Tumor macrophages in host immunity to malignancies; in Finck, The Macrophage in Neoplasia, pp. 27-42 (Academic Press, New York 1976).

8) Hamburger, A. W. and Salmon, S. E. (1977) Primary bioassay of human tumor stem cells. Science 197: 461.

9) Hamburger, A. W., Salmon, S. E., Kim, M. B., Trent, J. M., Soehnlen, B., Alberts, D. S. and Schmidt, H. J. (1978) Direct cloning of human ovarian carcinoma cells in agar. Cancer Res. 38: 3438-3444.

10) Hamburger, A. W., and White, C. P. (1982) Interactions between macrophages and human tumor clonogenic cells. Stem Cells. 1: 209.

11) Hopper, K. E., Wood, P. R., Nelson, D. S. Macrophage heterogenicity. Vox Sang 36: 257-274 (1979).

12) Haskill, S., Becker, S., Fowler, W. and Walton, L. Mononuclear cell infiltration in ovarian cancer. I. Inflammatory cell infiltrates from tumor and ascites material. Brit. J. Cancer 45: 728.

13) Luna, A. L. (1968) Manual of Histologic Staining Methods, pp. 162, McGraw-Hill, New York.

14) Mantovani, A., Peri, G., Polentarutti, N., Bolis, G., Mangioni, C. and Spreafico, F. (1979) Effects on in vitro tumor growth of macrophages isolated from human ascitic ovarian tumors. Int. J. Cancer 23: 157.

15) Pelligrino, M. A., Ferrone, S., Theofilopoulos, A. W. (1976) Isolation of human T and B lymphocytes with 2-amino-ethylisothiouronium bromide (AET) treated sheep red blood cells. J. Immunol. Methods 11: 273.

16) Prehn, R. T. (1977) Immunostimulation of the lymphodependent phase of neoplastic growth. J. Natl. Cancer Inst. 59: 1043.

17) Raff, H. V., Picker, L. J., Stabo, J. D. Macrophage heterogeneity in man. J. exp. Med. 152: 581-593 (1980).

18) Schultz, R. M., Chirigos, M. A. and Olkowski, Z. L. (1980) Stimulation and inhibition of neoplastic cell growth by tumor-promoter-treated macrophages. Cell. Immunol. 54: 98.

19) Siegel, F. P. and Siegel, M. (1977) Enhancement by irradiated T cells of human plasma cell production: dissection of helper and suppression function. J. Immunol. 118: 642.

20) Small, M. and Trainin, M. (1976) Separation of populations of sensitized lymphoid cells into fractions inhibiting and fractions enhancing syngeneic tumor growth in vivo. J. Immunol. 117: 292.

21) Torok-Storb, B. and Hansen, J. A. (1982) Modulation of in vitro BFU-e growth by normal Ia positive T cells is restricted by HLA-Dr. Nature 298: 473.

22) Williams, N., Jackson, H., Ralph, P. and Nakoinz, I. (1981) Cell interactions influencing murine marrow megakaryocytes. Nature of the potentiating cell in the bone marrow. Blood 57: 157.

23) Williams, W., Beutler, E., Erslev, A. and Rundles, R. W. (1977) Hematology, p. 1627, McGraw-Hill, New York.

VARIABILITY OF CELL LINES FROM PATIENTS WITH SMALL CELL LUNG CANCER

Desmond N. Carney, M.D., Ph.D.;
Marion M. Nau;
John D. Minna, M.D.

NCI-Navy Medical Oncology Branch, DCT, NCI, NIH, and National Naval Medical Center Bethesda, MD 20814

Introduction

In 1984, in the USA, it is estimated that there will be approximately 135,000 new cases of lung cancer and approximately 100,000 deaths from the disease (1). Of the new cases of lung cancer, small cell lung cancer (SCLC) will account for 20-25% of these. In the other major forms of lung cancer including adenocarcinoma, squamous cell carcinoma and large cell carcinoma (collectively referred to as non-SCLC lung cancer), early diagnosis with surgical resection offers the main hope for cure. However SCLC has a propensity for early, widespread dissemination at diagnosis, hope for cure and prolonged survival in these patients is related to the responsiveness of the patient's tumor to intensive combination chemotherapy with or without radiation therapy (2,3).

While many factors may influence the response rate to cytotoxic therapy and ultimate survival of patients with SCLC, including performance status and extent of disease, recent data suggest that the biological behavior of the tumor may be of prognostic importance. Although histologic subtypes of SCLC have long been recognized including the oat cell (lymphocytelike) variety, intermediate cell type and others (4), studies of large numbers of patients aggressively staged and treated with intensive combination chemotherapy, show no differences in clinical presentation, response rate or survival among these histological subtypes (5,6).

HUMAN TUMOR CLONING
ISBN 0-8089-1671-8

However, in those patients with a "mixed" small cell-large cell morphology, response rate and survival were significantly less than patients with pure SCLC (Table 1) suggesting that these tumors had a more "aggressive" behavior in vivo (7).

Over the past several years we and others have reported on our attempts to establish continuous cultures of cell lines of human small cell lung cancer (8,9). Although initial attempts were associated with limited success, the use of improved culture conditions, including serum-free medium, and experience, has increased our ability to establish cell lines from approximately three-quarters of clinical specimens received (10). In this chapter the in vitro biological characteristics of 52 SCLC cell lines will be described and their properties contrasted with non-SCLC cell lines. Of importance is the recognition that among established cell lines of SCLC considerable heterogeneity exists such that these cell lines fall into 3 different categories: classic, variant and multipotent. The variant cell lines may represent the in vitro model of the small cell/large cell histologic subtype which in vivo has a much worse prognosis than other types of SCLC.

TABLE 1

CLINICAL PRESENTATION, RESPONSE RATE TO COMBINATION CHEMOTHERAPY, AND SURVIVAL IN PATIENTS WITH "PURE" SCLC AND MIXED LARGE CELL/SMALL CELL LUNG CANGER.*

	Histology		
	"Pure" SCLC	SCLC/LC	P Value
No. Patients	103	19	
Disease Stage:			
Limited	31%	21%	0.54
Extensive	69%	79%	
Complete Response	45%	16%	< 0.05
Partial Response	45%	42%	
			< 0.005
No Response	10%	42%	
Median Survival (months)	10.5	6	0.008

*Adapted from Carney et al. (5), and Radice et al. (7).

MATERIALS AND METHODS

Specimens for culture were obtained at the time patients were undergoing approved protocol staging procedures and included bone marrow aspirates, lymph node aspirates and biopsies, pleural effusions, and surgically resected tumor masses. Techniques for handling specimens for culture were as previously described (10, 11). Although initial culture conditions utilized serum-supplemented medium alone (SSM) (RPMI-1640 medium, (GIBCO, New York) supplemented with 10% fetal bovine serum), for the past 2-3 years all specimens have been cultured either in SSM or in a serum-free chemically defined medium (HITES medium: RPMI 1640 supplemented with hydrocortisone, insulin, transferrin, 17β-estradiol and selenium) which we have previously shown would support the continual growth of established SCLC cell lines (12), and the selective growth of SCLC tumor cells from fresh biopsy specimens (10). For longterm culture tumor cells were maintained in the culture medium (HITES or SSM) in which best initial and sustained proliferation was observed.

Once established (usually after 2-4 months culture) cell lines were characterized by gross morphology, cytology, electron microscopy, tumorigenicity in athymic nude mice, colony-forming efficiency (CFE) in agarose, and doubling times as previously described (8,11). In addition, all cell lines were assayed for the expression of biomarkers including L-dopa decarboxylase (DDC), (EC 41.1.28) neuron-specific enolase (NSE), (EC 4.2.1.11), bombesin-like immunoreactivity (BLI), and creatine kinase BB (CK.BB) (EC2.7.3.2) activity as previously described (14-17). Some cell lines were tested for their in vitro sensitivity to x-rays, and for their expression of the c-myc oncogene (18,19).

RESULTS

Fifty-two independent cell lines have been established from SCLC patients using the culture conditions described. The source of these cell lines, the sex of patients from whom they were derived, and their treatment status at time biopsy was obtained, are indicated in Table 2. As can be observed cell lines were established from all sites assayed including bone marrow, pleural effusions, lymph nodes and other surgically resected masses. Of the 52 cell lines, all but 10 were established directly from patient biopsy specimens, while the remainder were established from nude mouse heterotransplants of patient biopsies.

TABLE 2

ORIGIN AND GROWTH MEDIA FOR 52 INDEPENDENT CELL LINES DERIVED FROM PATIENTS WITH SMALL CELL LUNG CANCER.

Source	Number cell lines
Bone marrow	15
Pleural effusion	15
Lymph node	13
Lung	6
Other*	3
Sex of Patients	**Number cell lines**
Males	41
Females	11
Treatment Status	**Number cell lines**
Prior Therapy	32
No Therapy	20

*Includes one specimen from liver, brain and adrenal gland.

The success rate of establishing cell lines from SCLC nude mouse heterotransplants is approximately 85% (20,21). Forty-one of these 52 cell lines were established from males and 11 from females; 32 were established from heavily pretreated patients and 20 from newly diagnosed patients.

As some cell lines were established in SSM, and others in serum-free HITES medium, and in order to demonstrate that these different culture conditions did not select out specific populations of cells we first compared and contrasted the biological properties of cell lines cultured in parallel from the same biopsy specimen under both culture conditions. Results for one cell line are shown in Table 3. For this cell line NCI-H123, as for other cell lines, no differences were observed in morphology, growth characteristics, or expression of biomarkers in the cells cultured in either SSM or HITES medium. Although there are some data to suggest that different culture conditions may be selective for one population of cells compared to another (22,23), in the majority of SCLC specimens no differences can be observed for cells cultured in different media conditions and, thus, data for cell lines, whether cultured in HITES or SSM, are combined for analysis

TABLE 3

BIOLOGICAL CHARACTERISTICS OF SCLC CELL LINE NCI H123, CULTURED FROM A LYMPH NODE ASPIRATE IN BOTH SSM AND SERUM-FREE HITES MEDIUM

	Growth Medium	
Characteristic*	HITES	SSM
Morphology	Floating	Floating
Cytology	SCLC	SCLC
DDC	300	328
BLI	2.5	3.6
NSE	702	833
CK-BB	3024	2721

DDC, units/mg protein, BL1, pmol/mg protein, NSE ng/mg of protein, and CK-BB, ng/mg protein.

Biological Characterization of Cell Lines

Based on their expression of the biomarkers analyzed, and to a lesser degree on their in vitro morphology, the 52 independent cell lines could be subgrouped into 3 categories: 1) Classic SCLC cell lines; 2) Variant SCLC cell lines and 3) Multipotent cell lines.

In general, classic cell lines which account for 70% of all lines grow as tightly packed aggregates of cells with a relatively prolonged doubling time, a low CFE in agarose and form tumors in nude mice with histological characteristics typical of the intermediate cell type of SCLC (8,11). In these classic cell lines, electron microscopy examination usually revealed easily recognizable clusters of dense core granules, typical of SCLC. In these classic SCLC cell lines, all expressed elevated levels of DDC, NSE bombesin-like immunoreactivity and creatine-kinase BB. No significant differences were observed in the expression of these biomarkers among cell lines derived from the different sites assayed, or from treated or untreated patients. While these cell lines have been in continuous culture for periods ranging from 6 months to 6 years, with rare exceptions (vide infra) changes in their biological behavior have not been observed with continual growth in culture.

Variant cell lines, in contrast to the classic cell lines usually grow as looser aggregates of floating cells. They have a faster doubling time, higher CFE and shorter latent period to tumor formation in nude mice than classic SCLC cell lines.

Cytological examinations of these cell lines revealed cells with more cytoplasm and more prominent nucleoli than classic SCLC. In several variant cell lines analyzed, electron microscopy did not reveal typical dense core granules.

All of the variant cell lines were characterized by their lack of expression of DDC and bombesin; their decreased level of NSE compared to classic SCLC, but had elevated levels of CK-BB, similar to that of classic SCLC cell lines. The biomarker expression of the 2 major categories of SCLC cell lines, classic and variant, are compared and contrasted both with each other and with a panel of 12 non-SCLC lung cancer cell lines in Table 4.

TABLE 4

COMPARISON OF THE IN VITRO BIOLOGICAL PROPERTIES OF CLASSIC SCLC CELL LINES, VARIANT SCLC CELL LINES, AND NON-SCLC LUNG CANCER CELL LINES*

	SCLC Cell Lines		Non-SCLC
Characteristic	Classic	Variant	Cell lines
Morphology	Floating	Floating	Adherent
Cytology	SCLC	SCLC/LC	Non-SCLC
Cloning Efficiency	1-5%	10-30%	1-25%
Nude Mouse Tumorigenicity	Yes	Yes	Yes
1-Dopa Decarboxylase	Elevated	Absent	Absent
Bombesin	Elevated	Absent	Absent
Neuron Specific Enolase	Elevated	Low	Absent
Creatine Kinase BB	Elevated	Elevated	Absent
Radiation Sensitivity	Sensitive	Resistant	NT
c-MYC Amplification (DNA)	1/12	5/6	1/5
c-MYC expression (RNA)	1/12	5/6	1/5

*NT = not tested

From the expression of these biomarkers, cell lines of SCLC lineage (classic and variant) can be readily distinguished from those of non-SCLC lung cancer cells by their expression of elevated CK-BB. Analysis of the remaining markers can clearly segregate the 2 major classes of SCLC cell lines.

Two cell lines were established from 2 untreated patients with a confirmed histological diagnosis of SCLC. In culture, both cell lines underwent simultaneous differentiations into multiple phenotypes of lung cancer including SCLC and squamous cell, and SCLC, squamous cell and adenocarcinoma. These two multipotent cell lines, in addition, expressed all the biochemical characteristics of classic SCLC cell lines.

Radiation Biological Properties

The in vitro radiation biological properties of 9 SCLC cell lines including 6 classic SCLC cell lines and 3 variant cell lines were carried out using standard techniques (18). Cells were irradiated over a wide dose range and reproductive integrity was assayed using a soft agarose clonogenic assay (11,24). Results are indicated in Table 5. Classic cell lines were characterized by Do's ranging from 51 to 140 rad and a low extrapolation number $\bar{n}$ ranging from 1-3.3. In contrast, while similar Do's were observed in the variant cell lines (80-91 rad) the $\bar{n}$ values in these lines were much greater (5.6-11.1) indicating a greater ability of variant cell lines to accumulate sublethal radiation damage and, thus, an increased (in a relative sense) radiation resistance in vitro in contrast to classic SCLC cell lines.

TABLE 5

IN VITRO RADIOBIOLOGICAL PROPERTIES OF CLASSIC AND VARIANT SMALL CELL LUNG CANCER CELL LINES

Cell Type	No. Cell Lines	Do's	$\bar{n}$	Fraction Surviving 200 R
Classic	6	51-140	1-3.3	2-32%
Variant	3	80-91	5.6-11.1	56-58%

In addition, the proportion of cells surviving 200 rad fraction was much greater for variant cell lines than classics, again indicating marked radioresistance for variant cell lines.

c-myc Oncogene Analysis

The recognition that patients with a variant histology had a worse prognosis than patients with classic SCLC, and the more aggressive behavior of variant cell lines in vitro (higher CFE, faster doubling time, etc.), in contrast to classic cell lines, prompted a study in these cell lines of the expression of oncogenes including the c-myc oncogene (19). (Table 4) Twelve classic SCLC cell lines, 6 variant cell lines and 5 non-SCLC lung cancer cell lines were evaluated while minimal amplification of c-myc DNA (5-fold) was observed in 1 classic cell line, a much greater degree of c-myc DNA amplification (20- to 75-fold) was observed in 5/6 variant cell lines evaluated. The classic cell line with c-myc amplification, NU-N231, gave rise to the variant line NU-N417 after several months culture (vide intra). With this change from classic to variant c-myc amplification was further increased. c-myc RNA expression was also greatly increased in these cell lines (Table 4). Of interest, cytogenetic analysis of 3/3 variant cell lines revealed one or more homogeneous staining regions (HSRs) in each cell line, while studies of 8 classic cell lines, did not reveal an HSR in any cell line (19).

Discussion

Neoplasms are thought to consist of subpopulations of tumor cells with varying expressions of a wide range of phenotypes, including metastatic potential, antigenic expression, growth rate, radiation sensitivity, and sensitivity to cytotoxic agents (25-27). Heterogeneity among human tumors in respect to sensitivity of cytotoxic agents is felt to account for the frequent relapses of human tumors after initial response to chemotherapy.

In this chapter the biological characteristics of multiple cell lines derived from patients with a single tumor type, i.e., small cell lung cancer are described. Although clonal cell lines from individual tumors have not been described, it is clear that there is significant heterogeneity in the biological characteristics of cell lines established from patients with SCLC.

The establishment of continuous cell lines in culture of human tumors has greatly increased our understanding of the biological behavior of human tumors in general, and of specific tumors including lung cancer. While there have been several reports over the past 2 decades of the establishment of individual cell lines of lung cancer, and because most of the studies were done on individual lines, general statements concerning the properties of both SCLC and non-SCLC cell lines could not be made. In fact, in review ing the initial reports of cell lines of SCLC (8,9,28-30) it became clear that considerable heterogeneity existed within these cell lines. Ohara et al. (29) in 1975 reported on a cell line (OAT-1975) established from a patient with SCLC. Although the tumor cell initially grew as floating aggregates, after some time in culture it became attached and was established as a monolayer culture. This cell line had a modal chromosome number of 74, a doubling time of 24 hr and a CFE of 60%. Neurosecretory granules were not evident on electron microscopy examinations. Obashi et al. in 1971 (28) reported on the establishment of another SCLC cell line OAT. This cell cell line initially grew as tightly packed floating cell aggregates with a relatively prolonged doubling time. Subsequently, its growth rate improved reaching a doubling time of 24 hr. In addition, the cells became "more loose" in culture, demonstrated prominent nucleoli, and cytogeneticaly an HSR was demonstrated; neurosecretory granules were evident on electron microscopy examinations. It is likely that both these cell lines belong to the variant class of SCLC cell lines. Gazdar et al. in 1980 (8) reported on the establishment and characterization of 9 SCLC cell lines. All grew as floating cell aggregates, with a relatively long doubling time, had high levels of dopa-decarboxylase, and on electron microscopy examination demonstrated neurosecretory granules. Several other papers on SCLC add to the evidence that, indeed, there is significant heterogeneity in cell lines of SCLC (8,9, 28-30).

The use of defined culture conditions has greatly improved our ability to establish continuous SCLC cell lines from both a variety of different biopsy sites and from nontreated and previously treated patients (31). Analysis of the 52 cell lines established and characterized at a single institution has permitted the definition of the biologically different phenotypes that may be observed in cultured cell lines of SCLC. The 3 categories include classic SCLC cell lines, variant SCLC cell lines, and multipotent cell lines. While this separation is based on the analysis of in vitro cell lines, it is likely that clinical correlates of these cell lines, in particular variant cell lines, exist.

Radice et al. (7) have previously reported on the clinical presentations, reponse rate and survival of 19 patients with a mixed SCLC morphology, and contrasted these data with 103 patients with "pure" SCLC and who were treated in a similar manner. As previously discussed (Table 1), patients with mixed SCLC morphology had a significantly poorer response to therapy, and a shorter survival.

More recently, Hirsch et al. have reported on the prognostic significance of histopathologic subtyping of SCLC in 200 patients (32). As was observed by Carney et al. (5) and Radice et al. (7), "pure" SCLC including the oat-cell variety and the intermediate cell types accounted for 86% of all patients, with 14% of patients having a mixed small cell/large cell morphology. No differences were observed in survival among the groups of patients with "pure" SCLC. However, in patients with mixed small cell/large cell morphology, their survival was significantly shorter than patients with "pure" SCLC, confirming the data of Radice et al. (7). Based upon the in vitro knowledge generated since that time, and the recognition that several of the varient cell lines were established from patients with such a mixed SCLC/LC morphology, it is likely that this histological subtype is the in vivo correlate of the in vitro variant cell lines. Clearly, the in vitro data including the rapid doubling time, higher cloning efficiency, and the c-myc oncogene amplification observed in several of these cell lines would suggest that patients with this phenotype would have a worse prognosis.

Not yet understood is the "origin" of these variant SCLC cells. Whether these cells represent a genetic drift from classic SCLC or indeed represent, de novo, a subtype of SCLC remains unclear. Indeed, there are data to suggest that both, if not other, possibilities exist. Among newly diagnosed patients with SCLC, another histological type, usually large cell, is observed in approximately 6% of patients (33). However, in autopsy studies of patients treated for SCLC, this figure rises to 35-40%, suggesting either that the emergence of second tumors in these patients is very common, or that there is a drift in lung cancer from one cell type to another; this change may be hastened by cytotoxic therapy. The "drift" would suggest the presense of common stem cells for lung cancer. The growth of cell lines with multipotent capacities would also support the hypothesis that, at least for some lung cancers, a common stem cell exists.

Of the variant cell lines described here, 2 cell lines "arose" in vitro after long-term passage of 2 classic SCLC cell lines. With passage, the cells became more loose in morphology, and gradually lost their APUD cell characteristics including DDC and bombesin (17,34). However, CK-BB expression was retained. In addition, in one cell line, with change from classic to variant, amplification of the c-myc oncogene was observed (19). These data would support the concept of a genetic drift in SCLC from classic to variant. If this is so, at least in vitro, it is an uncommon event, occuring in only 10% or less of established SCLC cell lines.

The possibility that variant cell lines may arise de novo is supported by the fact that establishment of several such lines from newly diagnosed, previously untreated patients with a histologically confirmed diagnosis of SCLC. At least for these patients no selective therapy could account for the development of the varient phenotype.

Over the past decades many prognostic factors have been recognised in patients with SCLC. The data presented here on cell lines further expands our knowledge regarding the biology of this tumor type. As has been observed for other tumor types, e.g., breast carcinoma, SCLC is not one disease entity, but a spectrum of diseases with different behavior and prognosis. While many different studies and tests, including in vitro chemosensitivity studies, have been carried out to improve therapeutic responses in patients with this tumor (35-37), determination of the biological characteristics of individual tumors may become an important requirement to better define the prognosis of patients with SCLC.

References

1. Cancer Statistics; 1983; CA-A; 9-25. 1983.
2. Minna, J.D., Higgins, G.A., and Glatstein, E.J.,: Cancer of the Lung, in DeVita., Hellman S., and Rosenberg S.A., (eds). Cancer: Principles and Practice of Oncology, Philadelphia, J.B. Lippincott Company. pps 396-474,1982
3. Carney D.N., Minna J.D.,: Small cell lung cancer. Clinics in Chest Medicine 3:389-398, 1982.
4. Kreyberg L. Histological typing of lung tumors, International classification of tumors. Geneva:World Health Organization, 1967.
5. Carney D.N., Matthews M.J., Ihde, D.C., Bunn, P.A. Jr., Cohen, M.H., Makuch R.W., Gazdar, A.F., Minna, J.D.: Influences of histologic subtype of small cell carcinoma of the lung on clinical presentation, response to therapy and survival. J. Natl. Cancer Inst. 65:1225-1229, 1980.
6. Hansen H.H., Domernowsky, P. Hansen M, et al. Chemotherapy of advanced small cell anaplastic carcinoma. Superiority of a four-drug combination to a three drug combination to the three drug combination. Ann. Int. Med. 89:177-181, 1978.
7. Radice, P.A., Matthews, M.J., Ihde, D.C., Gazdar, A.F., Carney, D.N., Bunn, P.A., Cohen, M.H., Fossieck, B.E., Makuch, R.W., Minna, J.D.: The clinical behavior of mixed small cell/large cell bronchogenic carcinoma compared to pure small cell subtypes. Cancer 50:2894-2902, 1982.
8. Gazdar, A.F., Carney, D.N., Russel, E.K., Simms, H.L., Baylin, S.B., Bunn, P.A. Jr., Guccion, J.G., Minna, J.D.: Small cell carcinoma of the lung: establishment of continuous clonable cell lines having APUD properties. Cancer Res. 40:3502-3507, 1980.
9. Pettengill O.S.; Sorenson G.D., Wurster-Hill, D.H. et al. Isolation and growth characteristics of continuous cell lines from small cell carcinoma of the lung. Cancer 45; 906-918, 1980.
10. Carney, D.N., Bunn, P.A., Gazdar, A.F., Pagan, J.A., Minna, J.D.: Selective growth in serum-free hormone supplemented medium of tumor cells obtained by biopsy from patients with small cell carcinoma of the lung. Proc. Natl. Acad. Sci. USA 78:3185-3189, 1981.
11. Carney, D.N., Gazdar, A.F., Bunn, P.A., Guccion, J.G.: Demonstration of the stem cell nature of clonogenic cells in lung cancer specimens. Stem Cells 1:149-164, 1981.

12. Simms E., Gazdar A.F., Abrams P.A., and Minna J.D. Growth of human small cell (oat-cell) carcinoma of the lung in serum-free growth factor supplemented medium. Cancer Res. 40:4356-4361, 1980.
13. Carney, D.N. Brower, M., Bertness, V., Oie, H.K.: The selective growth of human small cell lung cancer cell lines and clinical specimens in serum-free medium. In Methods in Molecular and Cell Biology. Sato, Barnes (eds.), Alan R. Liss, Inc., N.Y. In press, 1983.
14. Baylin, S.B., Abeloff, M.D., Goodwin, G., Carney, D.N., Gazdar, A.F.: Activities of L-dopa decarboxylase and diamine oxidase (histaminase) in human lung cancers: the decarboxylase as a marker for small (oat) cell cancer in tissue culture. Cancer Res. 40:1990-1996, 1980.
15. Marangos, P.J. Gazdar, A.F., Carney, D.N.: Neuron specific enolase in human small cell carcinoma cultures. Cancer Let. 15:67-71, 1982.
16. Moody, T.W., Pert, C.B., Gazdar A.F., Carney, D.N., Minna, J.D.: High levels of intracellular characterize bombasin human small cell lung cancer. Science 214:1246-1248, 1981.
17. Gazdar, A.F., Zweig, M.H., Carney, D.N., Van Stierteghen, A.C., Baylin, S.B., Minna, J.D.: Levels of creatine kinase and its isozyme in lung cancer tumors and cultures. Cancer Res. 41:2773-2777, 1981.
18. Carney, D.N., Mitchell, J.B., Kinsella, T.J.: In vitro radiation and chemosensitivity of established cell lines of human small cell lung cancer and its large cell variants. Cancer Res. 43:2806-2811, 1983.
19. Little C.D., Nau M.M. Carney D.N., Gazdar A.F., and Minna J.D.: Amplification and expression of the c-myc Oncogene in human lung cancer cell lines. Nature 306:194-196,1983.
20. Gazdar, A.F., Carney, D.N., Sims, H.L., Simmons, A.: Heterotransplantation of small cell carcinoma of the lung into nude mice. I. Comparison of intracranial and subcutaneous routes. Int. J. Cancer 28:773-783, 1981.
21. Chambers W.F., Pettengill O.S., and Sorenson G.D.: Intracranial growth of pulmonary small cell carcinoma cells in nude athymic mice. Experimental cell biology. 49: 90-97, 1981
22. Raff M.C., Miller R.H., and Noble M.,: A glial progenitor cell that develops in vitro into an astrocyte or an oligoderdrocyte depending on culture medium. Nature: 303, 390-396, 1983

23. Ossowski L., and Reich E.,: Changes in malignant phenotype of human carcinoma conditioned by growth environment. Cell 33: 323-333, 1983.
24. Carney, D., Gazdar A.F., Minna J.D.,: Positive correlation between histologic tumor involvement and generation of tumor cell colonies in agarose in specimens taken directly from patients with small cell carcinoma of the lung. Cancer Res. 40: 1820-1823, 1980.
25. Fidler I.J., and Hart I.R.,: Biological diversity in metastatic neoplasms: origins and implications. Science. 217: 998-1003, 1982.
26. Weiss L., Holmes J.C., and Ward P.M.,: Do metastases arise from pre-existing subpopulations of cancer cells? Brit. J. Cancer 47:081-089, 1983.
27. Heppner G.H., Dexter D.L., DeNucci T., Miler F.R., and Calabresi P.,: Heterogeneity in drug sensitivity among tumor cell subpopulations of a single mammary tumor. Cancer Res. 38: 3758-3763, 1978.
28. Oboshi S., Tsugawa S., Seido T., Shimosato, Koide T., and Ishikawa S. A new floating cell line derized from human pulmonary carcinoma of the oat cell type. Gann. 62:505-514, 1971.
29. Ohara H., and Okamoto T.,: A new in vitro cell line established from human oat cell carcinoma of the lung. Cancer Res. 37: 3088-3095, 1977.
30. Bergh J., Larson E., Zech L., and Nilson K.,: Establishment characterization of two neoplastic cell lines (U-1285 and U-1568) derived from small cell carcinoma of the lung. Acta Path Microbiol. Immunol Scand. Sect A,90: 149-158, 1982.
31. Carney D.N., Broader L., Edelstein, M., Gazdar, A.F., Hansen, M., Havemann, K., Matthews, M.J., Sorenson, G.D., Vindelov, L.: Experimental studies of the biology of human small cell lung cancer. Cancer Treat. Rep. 57:27-36, 1983.
32. Hirsch F.R., Osterlind K., and Hansen H.H. The prognostic significance of histopathologic subtyping of small cell carcinoma of the lung according to the world health organization. Cancer 52:2144-2150, 1983.
33. Gazdar, A.F., Carney, D.N., Guccion, J.G., Baylin, S.B.,: Small cell carcinoma of the lung: cellular origin and relationship to other pulmonary tumors. In Small Cell Lung Cancer. Greco, Bunn, Oldham, (eds.), Grune and Stratton, NY, pp. 145-175, 1981.
34. Salmon S.E., Hamburger A.W., Soehnlen B., et al. Quantitation of differential sensitivity of human tumor stem cells to anticancer drugs New Engl J. Med 298: 1221-1327, 1978.

35. Von Hoff D., Casper J., Bradley E., et al Association between human humor colony forming assay results and response of an individual patients tumor to chemotherapy. Am. J. Med. 70:1027-1032, 1981.
36. Carney D.N., Gazdar A.F., and Minna J.D.,: In vitro chemosensitivity of clinical specimens and cell lines of small cell lung cancer Proc. Amer. Soc. Clinical Oncology. 1: C-37, 1982.

TUMOR INHIBITORY FACTOR

MICHAEL G. BRATTAIN

Director, Bristol-Baylor Laboratory
Department of Pharmacology

ALAN E. LEVINE
Bristol-Baylor Laboratory
Department of Pharmacology
Baylor College of Medicine
Houston, Texas 77030

INTRODUCTION

The short term growth of malignant cells from tumor biopsies in semi-solid medium has shown good potential for application in studies of tumor biology and pharmacology (Salmon, 1980). However, a frequent difficulty encountered in attempts to culture tumors has been the low colony-forming ability of many specimens (Pathak *et al*., 1982; Hamburger *et al*., 1981; Brattain *et al*., 1982). Consequently, the development of methodology for improved cloning efficiencies has received a great deal of attention (Brattain, 1983). A number of approaches utilizing the addition of exogenous growth factors or feeder cells to assays have been tested with modest success (Pathak *et al*., 1982; Hamburger *et al*., 1981; Brattain *et al*., 1982, 1983; Hill and Whelan, 1983; Laboisse *et al*., 1981).

The work described in this report was initiated in an effort to identify polypeptide factors which stimulate anchorage independent growth of human malignant cells. Early studies from our laboratory had described the anchorage independent growth of malignant cells from primary human colon carcinomas after purification of the malignant cells by isokinetic sedi-

Supported by Grants PDT-109 from the American Cancer Society, CA 34432 from the National Institutes of Health.

HUMAN TUMOR CLONING
ISBN 0-8089-1671-8

mentation (Brattain _et al_., 1977a,b). When equal numbers of cells from starting samples and from purified samples were plated in soft agar, the samples of purified malignant cells showed a 4-fold higher cloning efficiency than predicted by the degree of purification obtained (Kimball _et al_., 1978). Others have shown the concentration dependence of growth in semi-solid medium by cell lines of malignant cells (Todaro _et al_., 1980; Halper and Moses, 1983; Tucker _et al_., 1983). These studies suggested that malignant cells secreted factors which were capable of enhancing anchorage independent growth. Sporn and Todaro (1980) hypothesized that the endogenous factor responsible for concentration effects of anchorage independent growth was transforming growth factor (TGF). TGF's are a family of polypeptides which reversibly confer to normal fibroblasts the ability to grow in semi-solid medium (Todaro _et al_., 1982). Later work by Moses and his colleagues led to the isolation of endogenous factors from mouse (Tucker _et al_., 1983) and human (Halper and Moses, 1983) malignant cells which were autostimulatory with respect to anchorage independent growth. In our own work we found that a TGF secreting mouse embryo fibroblast cell line, designated C3H 10T½, was effective in promoting anchorage independent growth of some, but not all, malignant cell lines. Consequently, we tested the effects of human TGF in the clonogenic assay. Crude extracts containing TGF activity (as assessed by anchorage independent growth of normal AKR-2B target fibroblasts) were isolated from serum-free conditioned medium of human colonic carcinoma cells as previously described (Marks and Brattain, 1984). The crude extracts were tested for their effects on the cloning ability of tumor biopsy material in the laboratories of Drs. J.K.V. Willson (Univ. of Wisconsin) and J.J. Catino (Bristol-Myers). When directly compared with untreated controls both investigators found that colony formation was inhibited in more than 33% of the samples tested (7 of 18 biopsies). Enhancement of colony formation occurred in less than 20% of the samples, thus suggesting that crude extracts of TGF contained an inhibitory factor for anchorage independent growth which was more potent than the stimulatory activity. This report describes the identification and partial characterization of a polypeptide factor (tumor inhibitory factor, TIF) which inhibits anchorage independent growth of some types of malignant cells.

METHODS

Extraction of TIF and TGF. Both polypeptides were isolated from previously characterized MOSER human colon carcinoma cells (Brattain *et al.*, 1981) and from rat Novikoff cells and ascites fluid. Serum-free conditioned medium from spinner cultures of MOSER cells was also utilized as a source of material and gave essentially the same results as MOSER cells. Cells and ascites fluid were extracted by acid/ethanol and crude extracts were obtained by lyophilization after dialysis against 1% acetic acid (Roberts *et al.*, 1981). Serum-free conditioned medium was extracted as previously described (Marks and Brattain, 1984).

Purification. Lyophilized crude extracts were resuspended and fractionated on Bio-Gel columns of either P-10, P-30 or P-100 in 1M acetic acid. Fractions containing both TGF and TIF activity were pooled, lyophilized, resuspended in 0.1% trifluoroacetic acid (TFA) and fractionated by HPLC on a C_{18} μBondapak column eluted with an acetonitrile gradient in TFA as described by others for purification of TGF (Anzano *et al.*, 1982).

TGF Assay. TGF was assayed as previously described (Marks and Brattain, 1984) utilizing either NRK cells (obtained from Dr. G. Todaro) or AKR-2B cells (obtained from Dr. H.L. Moses). Briefly, lyophilized samples were resuspended in McCoy's tissue culture medium and mixed with agarose in tissue culture medium containing 5×10^3 target cells. The 0.4% agarose (1ml) which contained cells and the samples to be assayed were plated onto 1ml underlayers of 0.8% agarose in 35mm tissue culture plates. Colonies (>10 cells) were visually counted after 10-15 days.

TIF Assay. The standard assay for TIF was performed in an identical manner as the TGF assay described above except that MOSER cells were substituted for the normal NRK or AKR-2B cells utilized in the TGF assay. While the normal fibroblasts will not grow, MOSER cells grow well under anchorage independent conditions with a cloning efficiency of approximately 10% for an inoculum of 5×10^3 cells. TIF activity was expressed as the percent inhibition of colony formation by MOSER cells. Several other cell lines were utilized as potential targets for TIF activity. These included several human colon carcinoma cell lines which have been described in detail (Brattain *et al.*, 1981a,b; 1983) and the A431 epidermoid carcinoma cell line obtained from Dr. G. Todaro.

RESULTS

Identification of TIF Activity. Results from 2 other laboratories (Drs. J.K.V. Willson and J.J. Catino) had indicated that lyophilized extracts containing TGF activity (as assessed by stimulation of anchorage independent growth of normal fibroblasts) from serum free conditioned medium of a human colon carcinoma cell line (MOSER) were significantly inhibitory toward the anchorage independent growth of a substantial percentage of primary tumor specimens (7/18). This suggested the existence of an inhibitory factor in the crude extracts which we examined further. Lyophilized extracts of serum free conditioned medium from MOSER cultures and acid/ethanol extracts of MOSER cells, Novikoff hepatoma ascites fluid and Novikoff hepatoma cells were all tested for TIF activity as assessed by inhibition of the growth of MOSER cells in semisolid medium. All of the extracts contained both TGF and TIF activity (Table 1). In both ascites fluid and in serum free conditioned medium the IC_{50} protein concentration for TIF activity was 50-100-fold lower than the EC_{50} for TGF activity. By contrast, the IC_{50} for TIF from cellular extracts of Novikoff and MOSER cells were only 2 and 14-fold lower than the EC_{50} for TGF.

Like TGF, TIF activity was apparently stable to acid treatment. TGF has been shown to be sensitive to treatment with dithiothreitol (DTT) and trypsin. A comparison of the stabilities of TGF and TIF from MOSER cellular extracts indicated that the inhibitory factor was also sensitive to trypsin and DTT (Table 2). Similar results were obtained from stability studies of the extracts from the other sources of TIF. Consequently, we concluded that TIF like TGF, is an acid stable polypeptide with hormonal-like activity which is dependent upon disulfide linkages. Further characterization would be dependent upon resolution of the two activities.

Partial Purification of TIF. TIF was purified from MOSER cell extracts by chromatography on Bio-Gel P-10 in 1M acetic acid followed by HPLC using a C_{18} μBondapak column and an acetonitrile gradient. Bio-Gel P-10 chromatography resulted in 2 distinct peaks of TGF activity at ~15kd and 7kd as previously noted for MOSER conditioned medium (Marks and Brattain, 1984). TIF activity occurred as a single broad peak which overlapped both the 15 and 7kd peaks. The fractions comprising the 7kd TGF peak were pooled, lyophilized and resuspended in 0.1% TFA and chromatographed by reverse phase HPLC to separate TGF activity from TIF activity. This procedure resolved the two activities as TGF was eluted as a single peak of activity at 26% acetonitrile while TIF was eluted as a single peak at 35% acetonitrile.

A similar scheme of purification was employed for the purification of TIF from Novikoff ascites fluid except that chromatography was performed on Bio-Gel P-30 rather than Bio-Gel

TABLE 1
Comparison of TIF and TGF Levels from Different Sources

	TIF, IC_{50} (μg/ml)	TGF, EC_{50} (μg/ml)
Novikoff Cells	38	55
Ascites Fluid	12	600
MOSER Cells	5	70
MOSER Conditioned Medium	0.1	15

TABLE 2
Stability of TIF and TGF from MOSER Cells

	TIF Activity (MOSER Colonies)	TGF Activity (NRK Colonies)
No Addition	693	0
Extract	37	300
Extract + DTT	677	0
Extract + Trypsin	671	0
Extract, 56°, 30 min.	0	285
Extract, 100°, 3 min.	0	219

P-10 in an attempt to obtain better initial resolution of TIF and TGF activities. However, this was not successful as both activities were eluted in a single broad peak from Bio-Gel P-30. As with the MOSER cell TIF, Novikoff ascites TIF was completely resolved from TGF activity by reverse phase HPLC and eluted at 36% acetonitrile thus suggesting a high degree of structural similarity between rat and human TIF.

In a final purification, Novikoff hepatoma cells obtained from rat ascites fluid were utilized as a starting source for TIF activity and Bio-Gel P-100 was employed in an attempt to obtain initial resolution of TIF activity from TGF activity. TGF activity at 15kd was well resolved from 7kd TGF activity on Bio-Gel P-100. The fractions comprising each of the TGF peaks were pooled and tested for TIF activity to determine whether separation of the two activities could be achieved by chromatography on Bio-Gel P-100. The 7kd peak of TGF activity also contained potent TIF activity which was purified 10-fold over the starting material. The 15kd peak of TGF activity did not contain TIF activity. These results indicated that, although different MW classes of TGF are identified by molecular sieve chromatography, TIF activity could not be resolved from TGF activity with this procedure. Characterization of TIF activity free from TGF activity required additional purification by other procedures such as HPLC.

Characterization of the Biological Activity of TIF. In previous work we described the establishment and characterization of a large bank of human colon carcinoma cell lines with diverse biological properties (Brattain *et al.*, 1981a). The lines were assigned to arbitrary groups according to various criteria for aggressiveness. Group I lines were highly aggressive as characterized by tumorigenicity in athymic mice, high plating efficiencies in semi-solid medium and lack of differentiation. Group III lines were relatively indolent as characterized by poor tumorigenicity, poor anchorage independent growth and a high degree of differentiation. Group II lines had intermediate properties. Although the original classification of the lines was based on their biological properties subsequent work has shown that the system is supported by cell surface protein and glycoprotein markers (Marks *et al.*, 1983), as well as cytoplasmic immunochemical markers (Taylor *et al.*, 1984).

MOSER cells were classified as a Group II cell line in this system (Brattain *et al.*, 1981a). Consequently, we characterized the effects of TIF on anchorage independent growth of cell lines from Group I and III. Group I lines included the previously characterized lines HCT 116 (Brattain *et al.*, 1981b) and HCT C (Brattain *et al.*, 1983) while Group III lines included CBS and FET (Brattain et al., 1981a). Due to limited amounts of TIF most of the characterizations of activity were performed with crude extracts from all of the sources of the factor mentioned above and contained TGF activity. However, limited confirmatory characterizations were performed with HPLC purified TIF which was free of TGF activity. The anchorage independent cloning efficiencies of the cell lines in the presence of crude TIF in extracts of MOSER conditioned medium (5 µg/ml) relative to untreated cells is shown in Table 3. Group I cell lines were unaffected by TIF while Group III cells were completely inhibited. Further characterization showed a similar spectrum of activity from ascites fluid (Table 4). HPLC purified TIF from ascites fluid completely inhibited growth of CBS cells but did not affect anchorage independent growth of HCT 116 cells. It is significant to point out that TIF did not affect the doubling time of MOSER or CBS cells under anchorage dependent conditions in standard tissue culture on plastic thus suggesting that the effects of the factor are specific for anchorage independent growth.

Those cell lines which were differentiated responded to TIF while undifferentiated lines showed no response to the factor thus suggesting a positive correlation between differentiation and response to TIF. In previous work, we had observed that differentiation of malignant cells was correlated to EGF response (Chakrabarty et al., in press). This, and work from other laboratories showing that anchorage independent growth of some, but not all, malignant cells was stimulated by EGF (Hamburger et al., 1981; Pathak et al., 1982), suggested that there may be a relationship between EGF and TIF responsiveness. Consequently, we investigated the effects of EGF on the anchorage independent growth of the colon carcinoma cell lines described above (Table 3). In addition, A431 human epidermoid carcinoma cells were included as a control since these cells have been shown to be inhibited by EGF. EGF increased the cloning efficiency of Group III cell lines, but did not affect Group I lines or MOSER cells. A431 cells were inhibited by both EGF and TIF. Consequently, although the number of samples is small it appears as though response to TIF is broader than EGF response. Table 5 summarizes the relationships between biological properties of colon carcinoma cells and the effects of EGF and TIF on anchorage independent growth.

TABLE 3

Effect of EGF and TIF on Growth of Various Cell Lines

	Cells plated x 10^3	P.E. (%) No addition	EGF	TIF
MOSER	5	10.3	11.9	0
FET	50	0.9	2.0	0
	25	1.0	2.0	0
	5	0	0.4	0
CBS	50	1.4	2.1	0
	25	1.1	1.9	0
	5	0	1.5	0
HCT 116	5	29	25	26
HCT C	5	11	10	11
A431	100	2.0	0.9	0.4
	25	1.6	0.3	0.7
	5	0.4	0	0.3

Cells were plated in soft agarose at the indicated concentrations with no additions or the addition of mouse EGF (20 ng) or of MOSER conditioned medium (5 μg protein).

TABLE 4

Effect of Ascites TIF on Anchorage Independent Growth

	Indicator Cell Line (Colonies/Plate)				
	NRK	MOSER	CBS	HCT 116	HCT C
No addition	0	480	700	568	1205
Crude Extract	82	0	89	525	1035
Bio-Gel	336	72	ND	ND	ND
TIF (HPLC)	0	216	ND	528	1135

TABLE 5

Summary of Growth Factor Responses of Colon Cell Lines

Cell Line	Tumorigenicity	Differentiation	CEA	A.I. Growth	TIF Response	EGF Response
I. HCT 116	+++	-	-	+++	-	-
HCT C	+++	-	-	+++	-	-
II. MOSER	+	+	+	+++	+	-
III. CBS	±	++	++	+	+	+
FET	±	++	++	+	+	+

DISCUSSION

We have described the isolation and partial purification of an hormonal-like polypeptide growth factor which inhibits the anchorage independent growth of some malignant cells. The factor appears to be chemically similar to TGF in that it is heat and acid stable, but is trypsin and DTT sensitive. Consequently, it is not surprising that the factor co-purifies with TGF through Bio-Gel chromatography. We have called this polypeptide a tumor inhibitory factor (TIF) as originated by Todaro and his colleagues (1982). These investigators also described a TIF with similar physical characteristics to TGF (Todaro *et al.*, 1982) which may be similar to the TIF reported here. However, there appear to be significant functional differences between our TIF and that described by Todaro *et al.* (1982). Todaro *et al.* (1982) reported the inhibition of TGF dependent growth of normal NRK fibroblasts in soft agar by the TIF isolated in their laboratory. Inhibition curves for crude TIF from either colon carcinoma or Novikoff ascites fluid are almost 100-fold lower than the EC_{50}'s for TGF stimulated growth of normal fibroblasts suggesting that TIF has no effect on TGF. In an additional experiment, TIF purified by HPLC was added to TGF free of TIF activity and no inhibition of NRK colony formation was observed. Consequently, we conclude that TIF from Novikoff ascites fluid and human colon carcinoma is not antagonistic to TGF activity involving normal fibroblasts.

Comparison of TIF from ascites and human colon carcinoma suggests that there is little interspecies difference associated with the factor. Both types of activity showed the same selectivity for the types of malignant cells which they inhibited in these studies. Structural similarities were indicated by the behavior of the TIF's from the two sources during purification. Particularly striking was the elution of the two polypeptides at 35 and 36% acetonitrile from HPLC.

Our results concerning the relationship between TGF and TIF raise several issues concerning their activity *in vivo*. One question relates to the seemingly contradictory result that MOSER cells grow quite well under anchorage independent conditions, but the addition of crude extracts from conditioned medium containing both TIF and TGF inhibit growth. There are several possible explanations for this observation. One possibility is that TGF is the major secreted product at low cell densities (e.g. conditions for anchorage independent growth in semi-solid medium) while TIF is the major product at higher cell densities. Conditioned medium was always harvested for TIF and TGF growth factor production at high cell densities in order to maximize the amount of growth factor present. A second possibility is that MOSER cells have a higher sensitivity to TGF than to TIF so that maximal stimulation of the malignant cells occurs at a much lower level of TGF than with TIF. Consequently, when TIF was added to soft agarose systems, MOSER cells were already maximally stimulated by TGF, but were still capable of responding to additional increments in the concentration of TIF. This explanation would require that malignant cells are at least 10^2-fold more sensitive to TGF than normal fibroblasts since the EC_{50} for TGF was 10^2-fold higher for NRK cells than the IC_{50} for MOSER cells. Finally, TGF may not be autostimulatory to MOSER cells while TIF is autoinhibitory. Factors responsible for the anchorage independent growth of MOSER cells may have been denatured during acid extraction. It should be noted that Hirai *et al*. (1983) have identified an acid-labile TGF. The selection of any of these explanations is speculative, but current evidence does indicate that the TGF's isolated in these studies are not autostimulatory factors for anchorage independent growth of MOSER cells. TGF separated from TIF by HPLC is not stimulatory to MOSER cells and purified TIF does not interfere with the colony formation of normal fibroblast cells stimulated by purified TGF.

This does not mean, however, that stimulatory factors for the anchorage independent growth of malignant cells do not exist. On the contrary, factors stimulatory to human (Halper and Moses, 1983) and to mouse (Tucker *et al*., 1983) malignant cells have been reported and the density dependence of anchorage independent growth by malignant cells also strongly suggests the existence of such factors. We suggest that MOSER cells produce stimulatory factors, but these may be denatured during isolation of TGF and TIF. The TGF's which we have isolated may have an extracellular function during tumor growth. One possibility for the function of TGF might be the stimulation of the growth of supporting stromal fibroblasts. This kind of function would be consistent with the stimulation of fibroblast growth by TGF during would healing which has been described (Sporn *et al*., 1983).

REFERENCES

Anzano, M.A., Roberts, A.B., Smith, J.M. et al. Purification by reverse-phase high performance liquid chromatography of an epidermal growth factor-dependent transforming growth factor. Anal. Biochem. (1982) 125: 217-224.

Brattain, M.G. Short-term culture of cells from human solid tumors in semi-solid medium. In: Cell Separation: Methods and Selected Applications, Vol II (T.G. Pretlow II and T.P. Pretlow, eds.) Academic Press, New York (1983) pgs. 235-249.

Brattain, M.G., Brattain, D.E., Fine, W.D. et al. Initiation of cultures of human colonic carcinoma with different biological characteristics utilizing feeder layers of confluent fibroblasts. Oncodevel. Biol. and Med. (1981a) 2: 355-366.

Brattain, M.G., Brattain, D.E., Sarrif, A.M. et al. Enhancement of growth of human colon tumor cell lines by feeder layers of murine fibroblasts. J. Nat'l. Cancer Inst. (1982) 69: 767-771.

Brattain, M.G., Fine, W.D., Khaled, F.M. et al. Heterogeneity of malignant cells from a human colonic carcinoma. Cancer Res. (1981b) 41: 1751-1756.

Brattain, M.G., Kimball, P.M., Pretlow, T.G. and Pitts, A.M. Partial purification of human colonic carcinoma cells by sedimentation. Brit. J. Cancer (1977) 35: 850-857.

Brattain, M.G., Marks, M.E., McCombs, J. et al. Characterization of human colon carcinoma cell lines isolated from a single primary tumor. Brit. J. Cancer (1983) 47: 373-381.

Chakrabarty, S., McRae, L.J., Levine, A.E. and Brattain, M.G. Restoration of normal growth control and membrane antigen composition in malignant cells by N,N-dimethylformamide. Cancer Res. (in press).

Halper, J. and Moses, H.L. Epithelial tissue derived growth factor-like polypeptides. Cancer Res. (1983) 43: 1972-1979.

Hamburger, A.W., White, C.P. and Brown, R.W. Effect of epidermal growth factor on proliferation of human tumor cells in soft agar. J. Nat'l. Cancer Inst. (1981) 67: 825-830.

Hill, B.T. and Whelan, R.D.H. Attempts to optimise colony-forming efficiencies using three different survival assays and a range of human tumour continuous cell lines. Cell Biol. Int'l. Rpts. (1983) 7: 617-624.

Hirai, R., Yamaoka, K., and Mitsui, H. Isolation and partial purification of a new class of transforming growth factors from an avian sarcoma virus-transformed rat cell line. Cancer Res. (1983) 43: 5742-5746.

Kimball, P.M., Brattain, M.G. and Pitts, A.M. A soft agar procedure for measuring the growth of human colonic carcinomas. Brit. J. Cancer (1978) 37: 1015-1019.

Laboisse, C.L., Augeron, C. and Potet, F. Growth and differentiation of human gastrointestinal adenocarcinoma stem cells in soft agarose. Cancer Res. (1981) 41: 310-315.

Marks, M.E. and Brattain, M.G. Induction of plasma membrane alterations in AKR-2B mouse embryo fibroblasts by endogenous growth factors from malignant cells. Int. J. Cancer (1984) 33: in press.

Marks, M.E., Danbury, B.H., Miller, C.A. and Brattain, M.G. Plasma membrane proteins and glycoproteins from colonic carcinoma cell lines with different biological properties. J. Nat'l. Cancer Inst. (1983) 71: 663-671.

Pathak, M.A., Matrisian, L.M., Magun, B.E. and Salmon, S.E. Effect of epidermal growth factor on clonogenic growth of primary human tumor cells. Int. J. Cancer (1982) 30: 745-750.

Roberts, A.B., Anzano, M.A., Lamb, L.C. et al. New class of transforming growth factors potentiated by epidermal growth factor: Isolation from non-neoplastic tissues. Proc. Nat'l. Acad. Sci. USA (1981) 78: 5339-5343.

Salmon, S.E. Introduction. In: Human Tumor Cloning (S. Salmon ed.) Liss, New York, 1980, pgs. 3-12.

Sporn, M.B., Roberts, A.B., Shull, J.H. et al. Polypeptide transforming growth factors isolated from bovine sources and used for wound healing in vivo. Science, (1983) 219: 1329-1331.

Sporn, M.B. and Todaro, G.J. Autocrine secretion and malignant transformation of cells. New Eng. J. Med. (1980) 303: 878-880.

Taylor, C.W., Brattain, M.G., and Yeoman, L.C. Cytoplasmic proteins and antigens of human colon tumor cell lines reflect their rate of growth and differentiation. Cancer Res. (1984) 44: in press.

Todaro, G.J., Fryling, C. and DeLarco, J.E. Transforming growth factors produced by certain human tumor cells: polypeptides that interact with epidermal growth factor receptors. Proc. Nat'l. Acad. Sci. USA (1980) 77: 5258-5262.

Todaro, G.J., Marquardt, H., Twardzik, D.R. et al. Transforming growth factors produced by tumour cells. In: Tumor Cell Heterogeneity: Origins and Implications Bristol-Myers Cancer Symposium, Vol. 4, (A.H. Owens, D.S. Coffey and S.B. Baylin, eds.) Academic Press, New York, 1982, pgs. 205-224.

Tucker, R.F., Volkenant, M.E., Branum, E.L. and Moses, H.L. Comparison of intra-and extracellular transforming growth factors from non-transformed and chemically transformed mouse embryo cells. Cancer Res. (1983) 43: 1581-1586.

TUMOR CHARACTERISTICS CORRELATED WITH IN VITRO GROWTH IN THE HUMAN TUMOR CLONOGENIC ASSAY (HTCA)

VICTOR E. HOFMANN, MICHAEL E. BERENS, URSINA FRUEH

Division of Oncology
University Hospital
8091 Zürich, Switzerland

INTRODUCTION

Epithelial tumors clone with varying and unpredictable degrees of success in the HTCA, thus limiting its applicability for routine clinical use. In fact, little is known of tumor characteristics related to successful in vitro growth (≥5 colonies). Some tumor types, such as neuroblastomas and ovarian adenocarcinomas, appear to grow with higher frequency than other tumors, e.g. colorectal cancers and sarcomas. In addition, tumor cells from effusions more readily clone in vitro than cells isolated from solid tumors. Taken together, these findings suggest that there may be some characteristic(s) of the tumor sample that are associated with successful cloning.

This study investigates whether cellular composition and tumor cell kinetics of human malignant effusions are correlated with tumor growth in the soft agar clonogenic assay.

MATERIAL AND METHODS

Patients: Ascites and pleural effusions from 84 patients with various epithelial neoplasias were investigated. Patients were untreated or had received a treatment course more than 4 weeks prior to paracentesis.

HUMAN TUMOR CLONING
ISBN 0-8089-1671-8

Tumor sample characteristics: Cytospin preparations from each sample were stained with May-Grünwald Giemsa or according to Papanicolaou to determine the % of granulocytes, lymphocytes, monocytes and tumor cells. The labeling index (L.I.) was scored by high speed scintillation autoradiography (1). Briefly, $2x10^6$ cells were exposed to 10 uCi ^{3}(H)-thymidine (specific activity 77 Ci/mmol) for 1 hour; unincorporated nucleotide was removed by two washes. Cells were cytocentrifuged, dipped in emulsion (Kodak, NTB 3) and developed. Labeling index was calculated as the percent tumor cells with 5 granules located over the nucleus in relation to total tumor cell number.

Tumor cloning assay: Cells were isolated by centrifugation from effusions anticoagulated with 10 U/ml preservative-free heparin (Novo). Thereafter, cells were washed twice, adjusted to appropriate concentrations and aliquoted for differentials, labeling index and tumor cloning. The double layer agar system was prepared exactly as described by Hamburger and Salmon (2) with the exception that no conditioned medium was included. Growth was monitored every other day using an inverted microscope and the final scoring was usually performed between days 10 and 28. A colony was defined as any new, round, cell aggregate >40 cells and/or >80 u in diameter. Successful growth was defined as 5 colonies arising from 0.5×10^6 cells plated.

RESULTS

Positive cytology was found in 72 of 84 effusions (86%). Granulocytes were rarely found and, when present, comprised only a small portion of the cellular make-up. Overall, lymphocytes were the principal cell type with a median of 38.5% (range: 3-98%). Monocyte/macrophage content had a median of 24% (range: 1-87%). Tumor cells in the malignant samples ranged from 0.1 - 93% with a median of 7.9% (Table I).

TABLE I

Characteristics of 72 Malignant Effusions

	Median (%)	Range (%)
Cellular composition		
Tumor cells	7.9	0.1 - 93.4
Lymphocytes	38.5	2.8 - 97.8
Monocytes/Macrophages	24.0	1.0 - 86.8
Labeling Index (L.I.)	3.3	0.0 - 20.8
Cloning Success (Colonies)	47	5 - 307

Successful clonal growth occurred in 34/72 cytologically positive samples (47%), while no colony growth was observed in the cytologically negative specimens. Individually, tumor cell number, lymphocyte or macropahge infiltration were not found to separate growers from non-growers (Table II). For those samples which grew, there was no relationship between the colony number and the content of any individual cell type.

For all malignant effusions, the L.I. ranged from 0 to 21% with a median of 3.3%. In the samples which successfully grew, the median L.I. was 6.8% which was statistically different ($p < 0.008$) from non-growing samples (Table II).

TABLE II

Characteristics of Growers vs. Non-Growers

	growth		
	<5 colonies	>5	
Tumor Cells % (mean/median)	13/ 3	21/10	n.s.*
Lymphocytes % (mean/median)	44/56	41/37	n.s.*
Monocytes/Macrophages % (mean/median)	34/19	27/21	n.s.*
Labeling Index % (mean/median)	4/ 2	7/ 6	$p < 0.008$

* n.s. = not significant

Combining cell composition and the L.I., it was determined that a subset of samples with at least three of the next four characteristics had a higher frequency of growth ($p < 0.05$): tumor cells >5%, lymphocytes <25%, monocytes/macrophages > 20%, and L.I. > 7% (Table III).

TABLE III

Association of Growth Success with Number of Characteristics

In vitro Growth (colonies)	No. of Characteristics* 0	1	2	3	4	Total
< 5	2	11	9	5	0	27
≥ 5	4	3	6	10	2	25

*Characteristics: > 5% tumor cells, < 25% lymphocytes, > 20% macrophages, > 7% L.I.

DISCUSSION

It is well known, that establishing the optimal growth conditions of a tumor cell line is an intense laboratory procedure. In general, optimal conditions vary from one cell line to another Therefore, it would be surprising if one culture milieu such as the one described for the HTCA could be satisfactory for all malignant tissues. In order to make the HTCA a routine test, one culture system was defined for all tumor types. However, because each specimen is unique it is likely that at best, these culture conditions are suboptimal. Additionally, a variable number of non-malignant cells may modulate in vitro cloning. For instance, it has been described that endogenous macrophages can greatly improve ovarian colony formation (3).

In this study, we have looked at four parameters of 72 cytologically proven malignant effusions. We found, that neither the number of tumor cells, lymphocytes or macrophages indicated which samples would produce tumor clones in the HTCA. The L.I. proved to be of some usefulness in that samples which grew had a statistically higher L.I. than the non-growers. For this reason, this simple factor could be used to discriminate which samples are unlikely to grow in the presently defined system. For such samples alternate growth conditions should be tested.

In an effort to delineate a subgroup of samples with higher likelihood of growth, we found that samples with $>$5% tumor cells, $<$25% lymphocytes, $>$20% macrophages and $>$7% L.I. had a growth advantage. This finding, and previously described reports using macrophage enriching techniques, would suggest that in addition to changing the culture media and growth factors, manipulation of cellular composition could improve clonogenicity.

REFERENCES

1. Durie BGM, Salmon SE. High speed scintillation autoradiography. Science 190: 1093-1095, 1975.

2. Salmon SE, Hamburger AW, Soehnlen BJ, et al. Quantitation of differential sensitivity of human tumor stem cells to anticancer drugs. N. Engl. J. Med. 298: 1321-1327, 1978.

3. Hamburger AW, Salmon SE, Kim MB, et al. Direct cloning of human ovarian carcinoma cells in agar. Cancer Res. 38: 3438-3444, 1978.

The expert assistance of Rosmarie Fringeli in preparation of the manuscript is gratefully acknowledged.

CYTOGENETIC AND CYTOKINETIC ANALYSIS OF HUMAN TUMOR COLONY FORMING CELLS (TCFUs).

J. M. TRENT, Ph.D.
Associate Professor of Medicine
University of Arizona Cancer Center
Tucson, Arizona 85724

I. INTRODUCTION

The blending of two seemingly divergent disciplines, molecular biology and cytogenetics has produced a renaissance in the field of cancer genetics. Specifically, a remarkable concordance has recently been observed between chromosome breakpoint sites specific to certain human tumors and the chromosomal locus of several human cellular oncogenes. These recent advances in the molecular biology of oncogenic viruses (and their cellular equivalents in man) have provided a renewed stimulus for research in the field of cancer cytogenetics. Additionally, these recent discoveries may point toward a common molecular basis for the previously recognized, but ill-understood, chromosome alterations specific to many human cancers. The viewpoint then to be addressed by this manuscript is that chromosomal alterations may represent a by-product of those molecular events which may be of germinal importance in the genesis and clinical progression of human cancer.

Three specific areas will be addressed which represent this author's view of the "cutting edge" in this field. First, a brief description of recent methodologic advances in both hematopoietic and solid tumor cytogenetics will be discussed as they have been instrumental in our recognition of tumor-associated chromosome change. Secondly, the use of chromosomal analysis for cytokinetic measurements of cell cycle transit time (T_C), as well as individual cycle phases may prove of significance in studies of the cellular proliferation of human tumors. Thirdly, the application of molecular cytogenetic techniques to study the localization and amplification of cellular oncogenes is hoped to further improve our knowledge of the mechanisms of oncogenesis.

HUMAN TUMOR CLONING
ISBN 0-8089-1671-8

II. TUMOR CYTOGENETICS

A. Methods of Solid Tumor Karyology

Cytogenetic analysis has provided valuable insights into the association of human leukemias and solid tumors with specific chromosomal alterations (for review see 1). An abbreviated summary of those human cancers displaying non-random chromosome change is presented in Table 1A,1B.

Table 1A. Association of Human Cancers with Specific Chromosomal Defects*

Tumor	Chromosome Aberration
ALL General	t(4;11)(q21;q23) t(9;22)(q34;q11)
T-cell	t(11;14)(p11;q13)
Pre-B	t(1;19)(q23;p13.3)
ANLL	
M2	t(8;21)(q22;q22). Often with loss of X or Y
M3	t(15;17)(q22;q12-21) 17q12>q21
M4-M5	t(9;11)(p21;q23) 11q13>q25
M5	t(11;19)(p23;p12 or q12)
Burkitt's Leukemia/Lymphoma	t(8;14)(q24;q32) t(2;8)(p11-13:q24) t(8;22)(q24;q11)
CLL	12q13>q22
CML	t(9;22)(q34;q11)
Blast crisis	i(17q)
Lymphoma	
Non-Burkitt's	t(8;14)(q24;q32)
Prolymphocytic	t(6;12)(q15;p13)

[Reference 1]

Table 1B. Association of Human Cancers with Specific Chromosomal Defects*

Tumor	Chromosome Aberration
Ewing's Sarcoma	t(11;22)(q24;q12)
Melanoma	6q11>q31. Translocations, deletions
Meningioma	del(22)(q11>qter) -22
Neuroblastoma	1p32>pter
Ovarian Carcinoma	del(6)(q15>q21) t(6;14)(p21;q24)
Renal Carcinoma	t(3;8)(p21;q24)
Retinoblastoma	13q14. Deletions and translocations, constitutional and tumor -13
Salivary Gland Tumors	t(3;8)(p25;q21) 8q12>q22. Other rearrangements 12q13>q15. Translocations, deletions
Wilm's Tumor	11p13. Deletions and translocations, constitutional and tumor
**Rhabdomyosarcoma	3p21. Deletions and translocations

*Reference 1
**Unpublished results of author

As can be noted, most information on tumor cytogenetics comes from the study of hematopoietic malignancies, with technical difficulties severely restricting the study of human solid tumors. However, in both hematopoietic and non-hematopoietic malignancies the major obstacle to be overcome before successful cytogenetic analysis can occur remains the acquisition of sufficient quantity and quality of mitotic figures.

Recently, culture methods developed to select and study tumor progenitor/stem cells have been used in cytogenetic studies in order to enhance the number of tumor mitoses and (especially in studies of hemopoiesis) to select cells from a specific cell lineage. New cytogenetic methods for analysis of colony forming cells (CFUs) from acute and chronic myelogenous leukemia [AML, CML] (2,3) have been reported and demonstrate conclusively that chromosomal analysis of abnormal hematopoietic progenitors is now routinely possible. In related _in vitro_ studies of CML progenitor populations, recent results suggest that karyotypically normal progenitors are maintained in the early stages of this disease although they can usually only be demonstrated following long term _in vitro_ culture (4). This method of long term culture represents a fascinating new approach to compare abnormal with normal progenitor cell function in this important disease.

In addition to the discovery that culture conditions can affect the proportion of normal versus abnormal progenitors in CML cultures (4), numerous recent reports have described the selection of different karyotypically abnormal populations as a consequence of altering culture harvest times or culture methods (5,6). Clearly, this finding demonstrates the importance of utilizing an appropriate culture method for obtaining cell growth of a clinically relevant tumor population. Using colony assay techniques, our laboratory has recently described the presence of multiple karyotypically unique sub-populations within a single tumor in both malignant melanoma and ovarian carcinoma (7,8), and while using a monolayer method similar results have been observed in malignant gliomas by Shapiro et al. (9). The former studies compared CFUs versus established cell lines developed from the clonogenic population and demonstrated significant karyotypic heterogeneity in both "direct" and _in vitro_ established cultures with progressive selection of karyotypically distinct subclones occuring with increasing time _in vitro_. In the work of Shapiro and colleagues (9), the demonstration of significant karyotypic heterogeneity was also shown to be associated with significant heterogeneity of chemo-sensitivity response in malignant glioma cultures (10).

Despite remaining technical difficulties in obtaining growth and subsequent cytogenetic analysis of many tumor types (especially human solid tumors) tremendous advances have occurred in recent years as evidences by the number of tumors with recognized chromosome change (Table 1). Exploitation of clonal assays as a means to select cells with both high proliferative potential and a documented cell lineage appears among the most important new methods in cancer cytogenetics.

III. CYTOKINETIC ANALYSIS OF CFUs USING BROMODYOXYURODINE

BrdU incorporation followed by sister chromatid differential staining is now a well established methodology for both the detection of chromosome alterations associated with mutagens and carcinogens, and for analyzing cellular proliferation (10-13). In human cancers, cell proliferation using BrdU incorporation followed by sister chromatid differential staining has principally been performed upon leukemic whole marrow populations. These results suggest that leukemic cells divide more slowly than their normal cellular counterparts (14-16). However, the majority of proliferative studies in human leukemia have used short term culture and routine cytogenetic techniques. With these techniques it is not possible to determine from which hemopoietic series the observed metaphase was derived. Moreover, as previously mentioned, recent reports indicate that differences in culture techniques may select different subpopulations with different karyotypes and characteristics (5,6). Accordingly, we have used the leukemic blast cell assay of Buick et al. (17) for initial studies of cellular proliferation in human leukemia. Cells grown in this assay have been shown by both light and electron microscopy to retain their leukemic blast cell morphology (18,19). In addition, the leukemic nature of cells grown in this assay has been confirmed by cytogenetic studies (3). The combination of the BrdU/Giemsa technique with the blast cell assay allows rapid and reproducible measurements of cellular proliferation in the key leukemic clonogenic population.

Briefly, the method of colony growth involves obtaining mononuclear cells for culture by Ficoll Hypaque density centrifugation followed by rosetting with neuraminidase treated sheep red-blood cells to remove rosette forming T-lymphocytes from the leukemic precursors. Cells are plated in 0.8 percent (v/v) methyl cellulose in Alpha medium supplemented with 10 percent fetal bovine serum and 20 percent (v/v) PHA stimulated leukocyte conditioned medium (PHA-LCM) (17). Cultures were then exposed to BrdU at a final concentration of 5 μg/ml 72 hours after plating followed by chromosome harvest (3) at 24 hour intervals up to 96 hours.

As an example, Figure 1 present the results from serial cytogenetic harvests of cultures of CFUs from one patient with AML analyzed at 12 hour intervals up to 108 hours (20). As expected, all metaphases observed at 12 hours were in their first division, with the number of 1st division cells declining steadily until zero by 60 hours of culture and subsequently were not observed. The drop in the number of 1st division cells was accompanied by a corresponding rise in the number of 2nd, and then 3rd division metaphases. The number of 2nd division cells subsequently dropped to zero while the number of third division cells reach 100 percent. Using the recently described method of Trent et al. for estimating mean cell cycle time (T_C) (21), the time interval between the appearance of 2nd and 3rd division metaphases measured at the 50 percent level resulted in an estimate of T_C in this patient of 19.5 hours (Figure 1).

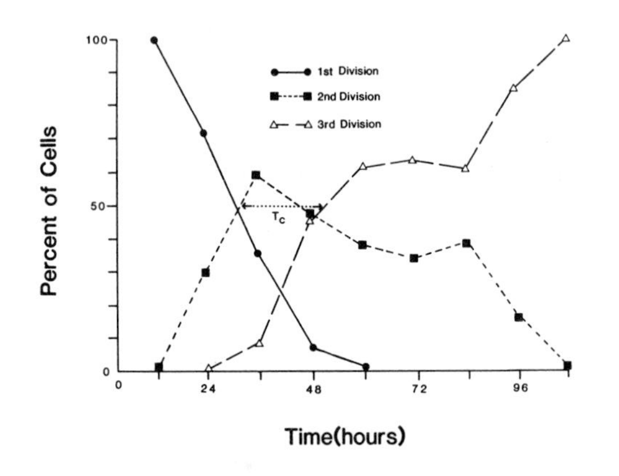

Figure 1. Distribution of 1st, 2nd, and 3rd division metaphases in serial cultures from a patient with AML. (20).

Based upon these results, we have suggested that the previously reported lengthy cell cycle times observed for leukemic marrow cells might, in fact, represent an artifact of the BrdU measurement technique utilized. Specifically, early BrdU techniques have derived their estimates of cell cycle times by dividing the total culture time by the number of cellular divisions completed (22). These techniques have been shown conclusively to overestimate cell cycle times and, accordingly, are in error (21). Further, our data suggests (as supported by earlier work of Steel [23]) that the initial phase of CFU growth in AML cultures may be a quiescent state (G_0) with cells requiring 48-72 hours before entering active cell division. At present studies are underway to further study this phenomenon.

Finally, as mentioned the mean T_C of CFUs in this patient measured by the 50 percent frequency distribution method (21) is 19.5 hours. This is considerably less than estimation of T_C from other published kinetic studies of leukemic marrows (range 34.3 - 67.3 hours) (24-26). However, it must be emphasized that in all previous studies whole marrow samples were studied whereas in our study only the clonogenic subpopulation was studied. Moreover, the culture conditions used in the above studies and ours are completely different. Nevertheless, our data suggest that some leukemic cells do have the potential to divide quite rapidly. Whether the clonogenic leukemic cells divide at this rate *in vivo* remains to be determined.

Our studies indicate that BrdU incorporation followed by differential chromatid staining can readily be applied to the cytokinetic study of CFUs from leukemia patients. Further studies utilizing this technique are underway to investigate leukemic cell proliferation *in vitro*, as well as to expand these studies to CFUs from human solid tumors.

IV. RELATIONSHIP OF CHROMOSOME ALTERATIONS TO ONCOGENESIS

It is now well established that certain retroviruses (termed v-*onc* genes) can cause rapid transformation of vertebrate cells (27). In vertebrate cells those genes which exhibit DNA sequence homology to v-*onc* genes have been termed cellular or c-*onc* genes. Various lines of evidence suggest that these genes may be capable of producing malignant transformation under certain conditions (i.e. through mutation or altered gene expression) (27).

As previously mentioned, cytogenetic analysis of a variety of human cancers has demonstrated tumor-associated chromosome changes perhaps suggestive of an underlying molecular-genetic mechanism. Recently, this notion has been strengthened by the localization of several cellular oncogenes at chromosomal regions nonrandomly altered in human cancers. The most extensively studied example of a cellular oncogene located at a specific site of nonrandom chromosome change is the association of c-*myc* with chromosome 8q24, a region in Burkitt's lymphoma often reciprocally translocated to the immunoglobulin heavy chain locus on chromosome 14 (28,29). In addition to c-myc, other oncogenes have been regionally mapped to sites of frequent chromosome change in human tumors including: c-abl (30) and c-sis (31,32) (chronic myelogenous leukemia); c-mos (33) (acute non-lymphocytic leukemia - M2 category); and c-myb (34) (ovarian adenocarcinoma and malignant melanoma) (35).

However, although several examples have appeared suggesting chromosome alterations may usually be associated with an oncogenic sequence this may not always be true. A case in point is human neuroblastoma, a tumor where three major cytogenetic abnormalities have been characterized: 1) double minutes [DMs] (36-38), 2) Homogeneously staining regions [HSRs] (39-41), and 3) structural alteration (usually deletion) of chromosome 1p at bands 1p32>pter (41-43). DMs and HSRs have been demonstrated to contain amplified copies of cellular genes (46,45) including cellular oncogenes (46,47). Recently, we described an amplified DNA segment of limited homology to c-myb (termed N-myc) which was isolated from a human neuroblastoma cell line (48). Although N-myc was shown to be present in single copy in normal cells, it was also shown to be selectively amplified in tumor cells from human neuroblastomas but not in other HSR bearing tumors (48). Using in situ hybridization (49) of ^{3}H-labelled N-myc probe DNA we have now mapped N-myc to 2p23 or 24 normal human mitotic cells (50). We have also localized N-myc to homogeneously staining regions (HSRs) on three different chromosomes from 3 HSR-bearing neuroblastoma cell lines (50). Our results demonstrated that the locus of N-myc in normal cells (2p23 or 24) bears no obvious relationship to either chromosomal sites of amplified N-myc, or to recognized sites of non-random chromosome alteration in human neuroblastoma. Therefore, in contrast to cellular oncogenes specifically related to chromosomal breakpoints in certain tumors (i.e. Brukitt's lymphoma), in human neuroblastoma numeric or structural chromosome alterations involving the resident site on chromosome 2 of N-myc are extremely rare (42,51). For example, whereas a recent report identified deletions or rearrangements involving 1p32>pter in >70 percent of all neuroblastomas (42), this same report revealed structural rearrangements of 2p in only one out of 20 (<5 percent) cases analyzed.

Both enhanced expression of cellular oncogenes (possible through gene amplification) and mutational change of the proteins encoded by these genes have both been implicated in tumorigenesis (52). Translocations between chromosomes may offer examples of both mechanisms (53). Amplification of

DNA is another means by which the expression of cellular oncogenes might be enhanced, as exemplified by our findings with N-*myc* (reference 50 and unpublished data). It is furthermore possible that genes become mutated during the course of amplification. However it might affect cellular oncogenes, gene amplification may contribute to the genesis of human cancers.

The aforementioned findings appear to provide strong support for models of oncogenesis which suggest that a specific chromosomal alteration may result from changes in a "normal" cellular gene (i.e. c-*onc* gene), then leading to expression of the transformed or malignant phenotype (54). The possibility that these findings will prove of importance to spontaneous human cancers appears extremely high.

V. SUMMARY AND FUTURE PROSPECTS

This chapter has attempted to address several important new developments related to cytogenetic analysis of human malignancy. Tumor associated chromosome change has been documented to frequently occur in virtually all human cancers receiving sufficient study. Additionally, utilization of cytogenetic techniques for analysis of cellular proliferation, particularly in clonogenic populations, may be of substantial benefit in understanding the kinetics of tumor cell growth. Finally, the association of c-*onc* genes with sites of specific chromosomal change appears to further support mechanisms of oncogenesis resulting from a basic molecular defect. Perhaps within the next decade the discoveries now being made on the basic molecular mechanisms of oncogenesis might translate into novel treatment modalities allowing a correction of those genetic changes responsible for the malignant state. Our studies of the mechanisms and consequences of chromosome alterations is only now beginning to bear significant fruit. Without doubt, cytogenetic analysis of human cancers will continue to receive increasing attention in the coming years.

REFERENCES

1. de la Chapelle A, Berger R: Human gene mapping workshop 7. Cytogene and Cell Gene 37:274-311, 1984.

2. Dube IO, Eaves CJ, Kalousek DK, Eaves AC. A method for obtaining high quality chromosome preparations from single hemopoietic colonies on a routine basis. Cancer Gene and Cytogene 4:157-168, 1981.

3. Trent JM, Davis JR, Durie BGM: Cytogenetic analysis of leukemia colonies from acute and chronic myelogenous leukemia. Brit J Cancer 47:103-109, 1983.

4. Coulombel L, Kalousek D, Eaves C, Gupta C, Eaves AC. Long term marrow culture reveales chromosomally normal hematopoietic progenitor cells in patients with Philadelphia chromosome - positive chronic myelogenous leukemia. N Eng J Med 308:1493-1498, 1983

5. Carbonell F, Grilli G, Fliedner TM: Cytogenetic evidence for a clonal selection of leukemic cells in culture. Leukemia Res 5:395-398, 1981.

6. Lowenberg B, Hagemeijer A, Swart K: Karyotypically distinct subpopulations in acute leukemia with specific growth requirements. Blood 59:641-645, 1982

7. Trent JM, Thompson FH, Ludwig C. Homogeneously staining regions: Evidence for selection in a human melanoma cell line. In press, Cancer Res.

8. Trent JM, Thompson FH, Buick RNM. Generation of clonal varients in a human ovarian carcinoma studied by chromosome banding analysis. In press, Cancer Gene and Cytogene.

9. Shapiro JR, Yung W-KA, Shapiro WR. Isolation, karyotype, and clonal growth of heterogeneous subpopulations of human malignant gliomas. Cancer Res 41:2349-2359, 1981.

10. Yung W-KA, Shapiro JR, Shapiro WR. Heterogeneous chemosensitivities of subpopulations of human glioma cells in culture. Cancer Res 42:992-998, 1982.

11. Wolff S: Sister chromatid exchange: The most sensitive mammalian system for determining the effects of mutagenic carcinogens. In: Genetic Damage in Man caused by Environmental Agents, K Berg, ed. Academic Press, New York, pp. 229-246, 1979.

12. Tice R, Schneider EL, Rary JM: The utilization of bromodeoxyuridine incorporation into DNA for the analysis of cellular kinetics. Exp Cell Res 102:232-236, 1976.

13. Crossen PE, Morgan WF: Analysis of lymphocyte cell cycle time in culture measured by sister chromatid differential staining. Exp Cell Res 104:453-457, 1977.

14. Abe S, Kakati S, Sandberg AA: Growth rate and sister chromatid exchange (SCE) incidence of bone marrow cells in acute myeloblastic leukemia (AML). Cancer Genet Cytogenet 1:115-130, 1979.

15. Abe S, Sandberg AA: Sister chromatid exchange and growth kinetics of marrow cells in aneuploid acute nonlymphocytic leukemias. Cancer Res 40:1292-1299, 1980.

16. Becher R, Zimmer G, Schmidt CG: Sister chromatid exchange and growth kinetics in untreated acute leukemia. Int J Cancer 27:199-204, 1981.

17. Buick R, Till J, McCulloch EA: Colony assay for proliferative blast cells circulating in myeloblastic leukemia. Lancet 1:862-863, 1977.

18. Minden MD, Till JE, McCulloch EA: Proliferative state of blast cell progenitors in acute myeloblastic leukemia. Blood 52:592-600, 1978.

19. McCulloch EA, Howatson AF, Buick RN, Minden MD, Izaguirre CA: Acute myeloblastic leukemia considered as a clonal homopathy. Blood Cells 5:261-282, 1979.

20. Crossen PE, Broderick R, Durie BGM, Trent JM. Proliferative characteristics and sister chromatid exchange (SCE) of colony forming cells (CFUs) in acute myelogenous leukemia. In press, Cancer Gene and Cytogene.

21. Trent J, Gerner E, Broderick RD, Crossen PE: Cell cycle analysis using bromodeoxyuridine: Comparison of methods for measurement of total cell transit time. Cell and Tissue Kinetics (In Press), 1983.

22. Crossen PE, Morgan WF: Proliferation of PHA and PWM stimulated lymphocytes measured by sister chromatid differential staining. Cell Immunol 32:432-438, 1977.

23. Steel, GG. Growth kinetics of tumours. Oxford, Clarendon Press, 1977.

24. Clarkson B, Ohkita T, Ota K, Fried J: Studies of cellular proliferation in human leukemia: Estimation of growth rates of leukemic and normal hematopoietic cells in two adults with acute leukemia given single injections of tritiated thymidine. J Clin Invest 46:506-529, 1967.

25. Saunders EF, Lampkin BC, Mauer A: Variation of proliferative activity in leukemic cell populations of patients with acute leukemia. J Clin Invest 46, 1356-1363:1967.

26. Greenberg M: The generation time of human leukemic myeloblasts. Lab Invest 46:245-252, 1972.

27. Cooper GM: Cellular transforming genes. Science 218:801-806.

28. Dalla-Favera, R. et al.: Human c-myc onc gene is located on the region of chromosome 8 that is translocated in Burkitt lymphoma cells. Proc Natl Acad Sci USA 79:7824 -7827, 1982.

29. Taub, R. et al.: Translocation of the c-myc gene into the immunoglobulin heavy chain locus in human Burkitt lymphoma and murine plasmocytoma cells. Proc Natl Acad Sci USA 79:7837-7841, 1982.

30. Heisterkamp, N. et al.: Chromosomal localization of human cellular homologues of two viral oncogenes. Nature 299:747-749, 1982.

31. Swan DC et al.: Chromosomal mapping of the simian sarcoma virus onc gene analogue in human cells. Proc Natl Acad Sci USA 79:4691-4695, 1982.

32. Dalla-Favera R, Gallo RC, Giallongo A, Croce CM: Chromosomal localization of the human homolog (c-sis) of the simian sarcoma virus onc gene. Science 218:686-688, 1982.

33. Neil BG, Jhanwar SC, Chiganti RSK, Hayward WS: Two human c-onc genes are located on the long arm of chromosome 8. Proc Natl Acad Sci USA 79:7842-7846, 1982.

34. Harper ME et al.: Chromosomal sublocalization of human c-myb and c-fes cellular onc genes. Nature 304:169-171, 1983.

35. Trent JM, Rosenfeld SB, Meyskens FL: Chromosome 6q involvement in human malignant melanoma. Cancer Gene and Cytogene 9:177-180, 1983.

36. Biedler JL, Helson L and Spengler BA: Morphology and growth, tumorigenicity and cytogenetics of human neuroblastoma cells in continuous culture. Cancer Res 33:2643-2652, 1973.

37. Balaban-Malenbaum G, Gilbert F: Double minute chromosomes and the homogeneously staining regions in chromosomes of a human neuroblastoma cell line. Science 198:739-741, 1977.

38. Barker PE: Double minutes in human tumor cells. Cancer Gene and Cytogene 5:81-84, 1982.

39. Biedler JL, Spengler BA: Metaphase chromosome anamoly: Association with drug resistance and cell specific products. Science 191:185-187, 1976.

40. Biedler JL, Spengler BA: A novel chromosome abnormality in human neuroblastoma and antifolate resistant chinese hamster cell lines in culture. J. Natl Cancer Inst 57:683-695, 1976.

41. Brodeur GM et al.: Cytogenetic features of human neuroblastomas and cell lines. Cancer Res 41: 4678-4686, 1981.

42. Brodeur GM, Sekhon GS, Goldstein MN: Chromosomal aberrations in human neuroblastomas. Cancer 40:2256-2263, 1977.

43. Gilbert F et al.: Abnormalities of chromosome 1p in human neuroblastoma tumors and cell lines. Cancer Gene and Cytogene 7: 33-42, 1982.

44. Schimke RT, Kaufman RJ, Alt FW and Kellems RF: Gene amplification and drug resistance in cultured murine cells. Science 202:1051-1055, 1978.

45. Cowell JK: Double minutes and homogeneously staining regions: Gene amplification in mammalian cells. Ann Rev Genet 16:21-59, 1982.

46. Alitalo K, Schwab M, Lin CC, et al.: Homogeneously staining chromosomal regions contain amplified copies of an abundantly expressed cellular oncogene (c-myc) in malignant neuroendocrine cells from a human colon carcinoma. Proc Natl Acad Sci USA 80:1707-1711, 1983.

47. Tannock I, Buick R, Kopelyan I, et al.: Amplification of the HA-RAS Oncogene in a Human Bladder Cancer Cell Line is Associated with Several Indices of Malignancy. In press, Nature, 1983.

48. Schwab, M et al.: Amplified DNA with limited homology to myc cellular oncogene is shared by human neuroblastoma cell lines and a neuroblastoma tumor. Nature 305:245-248, 1983.

49. Trent JM, Olson S and Lawn RM: Chromosomal localization of human leukocyte, fibroblast and immune interferon genes by means of in situ hybridization. Proc Natl Acad Sci USA 79:7809-7813, 1982.

50. Schwab M, Varmus HE, Bishop JM, et al.: Human N-myc maps to chromosome 2p23 or 24 in normal cells and to homogeneously staining chromosomal regions (HSRs) in neuroblastomas. In press, Nature, 1984.

51. Biedler JL, Meyers MB, Spengler BA: Homogeneously staining regions and double minute chromosomes: Prevalent cytogenetic abnormalities of human neuroblastoma cells. In: Advances in Cellular Neurobiology Vol 4. New York, Academic Press, 1983, pp 267-307.

52. Bishop, J.M.: Cellular oncogenes and retroviruses. Ann Rev Biochem 52:301-354, 1983.

53. ar-Rushdi A, Nishikura K, Erikson J, et al: Differential expression of the translocated and the untranslocated c-*myc* oncogene in Burkitt lymphoma. Science 222:390-393, 1983.

54. Klein G: The role of gene dosage and genetic transpositions in carcinogenesis. Nature 294:313-318, 1981.

Research supported in part by PHHS Grant CA-29476 and CA-17094 awarded by the National Cancer Institute. Dr. Trent is a Scholar of the Leukemia Society of America.

In Vitro Cytogenetic Studies in CML

IAN D. DUBÉ, DAGMAR K. KALOUSEK, LAURE COULOMBEL,
CONNIE J. EAVES, AND ALLEN C. EAVES

Terry Fox Laboratory
British Columbia Cancer Research Centre
Vancouver, B.C., Canada

Chronic myeloid leukemia (CML), like most malignancies, is a clonal neoplasm, tracing its origin to a single transformed cell (1). In it is exemplified the hierarchical structure typical of many malignancies in which cells of limited proliferative potential are produced by an expanding stem cell population (2).

Considerable evidence indicates that in the case of CML, the cell that is initially transformed is a member of the normal pluripotent stem cell compartment of the blood-forming system. During the chronic phase of the disease this cell proliferates extensively, while retaining its capacity for differentiation into functionally and morphologically diverse blood cells. As a result, by the time of diagnosis all of the circulating red cells, granulocytes, monocytes, and platelets are progeny of the neoplastic clone (1), as may also be the case for some B lymphocytes (3). The morphologically unmarked neoplastic members of the stem cell compartment, like their normal counterparts, are thus greatly diluted by their terminally differentiating progeny.

Information about changes in the size and composition of

HUMAN TUMOR CLONING
ISBN 0-8089-1671-8

the stem cell compartment is important, both for the development of new treatment strategies, and as an approach to the analysis of mechanisms leading to clonal dominance. However, the acquisition of such information requires the use of methods that allow both normal and neoplastic stem cells to be detected and distinguished. Methods for quantitating normal primitive hemopoietic cells based on their ability to generate colonies of recognizable blood cells are available (4), and a long-term marrow culture system in which such primitive cells can be maintained for many weeks has also been recently developed (5). The Philadelphia (Ph^1) chromosome provides a consistent marker of the neoplastic clone in most patients (6). We therefore focused our efforts on the use of cytogenetic analysis of hemopoietic colonies to evaluate the stem cell compartment in this disease. A technique for obtaining analyzable metaphases in high yield from individual colonies was developed, and then used to determine the number of Ph^1-negative and Ph^1-positive colony-forming progenitors present in fresh marrow or blood samples, or in long-term marrow cultures maintained for various periods of time.

In this paper, we will first review briefly the techniques we have used and then summarize the findings obtained.

Methylcellulose Assays for Hemopoietic Progenitors

Hemopoietic progenitors exist at low incidence in the marrow, and to an even lesser degree, in the blood. They are routinely assayed by plating the red cell depleted fraction of marrow (i.e., buffy coat cells) or a light density (<1.077 gm/ml) fraction of blood in a nutrient medium (e.g., alpha-medium) made viscous with 0.8% methylcellulose, and supplemented with appropriate known and undefined growth factors. Additives include fetal calf serum (FCS - 30%), deionized BSA (1%), 2-mercaptoethanol (10^{-4} M), erythropoietin (2-5 units per ml), and a source of other factors present in human leukocyte conditioned media.

For additional details, see Refs. 7,8. All reagents should be prescreened and titrated for optimal erythroid and granulocyte colony growth-supporting capacity against laboratory standards. Peripheral blood preparations from normal individuals are plated at a final cell concentration of 4×10^5 cells per ml, while marrow preparations are plated at a final concentration of 2×10^5 cells per ml. In patients with CML, particularly those with high WBC counts, progenitor levels are often elevated in a relative as well as absolute fashion (9,10). Therefore, lower plating concentrations, e.g., 5×10^4-10^5 cells per ml are required in these cases. Cultures are incubated at 37°C in a 5% CO_2-air environment and maintained at high humidity.

During the first week of incubation, the vast majority of the cells plated die. However, if conditions are optimal, various myeloid progenitor cells proliferate and differentiate, eventually giving rise to terminally differentiated, nondividing progeny. Erythroid colonies are readily identified by their red color (indicative of ongoing hemoglobin synthesis) and their characteristic arrangements in subcolonies or clusters of approximately 50-100 erythroblasts each. The smallest colonies derived from the most differentiated progenitors are the first to contain hemoglobin-synthesizing erythroblasts. By 9-12 days, larger colonies, already containing a minimum of 500-1,000 cells, begin to be detectable as erythroid. Although the number of large colonies obtained varies considerably between individuals (8), the timing of their formation does not. The delay in the maturation of large colonies is believed to be related to the primitive state of differentiation of the progenitors from which they arise (11).

During the second week of incubation, large granulopoietic colonies containing 200-1,000 cells can be reliably identified by virtue of their characteristic diffuse morphology and lack of red color. When cells are plated at a concentration sufficiently low that colony overlap is minimal (<30 large colonies per 1 ml culture), single colonies containing both erythroid and granulocytic cells can also be reliably identified. These occur, however, at a much lower frequency, and represent the clonal progeny of a more primitive pluripotent progenitor cell (12).

Long-Term Marrow Cultures

In normal long-term marrow cultures, an environment is reproducibly attained that allows hemopoiesis to continue for periods of at least 8 weeks. The maintenance of hemopoiesis appears to be associated with the development of a complex adherent cell layer containing a variety of cell types including fibroblasts, fat-laden cells, cells with endothelial characteristics, and macrophages (13,14), as well as the majority of the primitive hemopoietic progenitor cells (5). Although the mechanisms by which non-hemopoietic and hemopoietic components of the adherent layer may influence one another are unknown, several lines of evidence suggest that short-range interactions, possibly similar to those regulating stem cell activity in vivo, may be involved (15).

Long-term cultures are routinely initiated by incubating 2.5×10^7 unwashed marrow buffy coat cells in 8 ml of growth medium, as previously described (5). Cultures are incubated for the first 3-5 days at 37^oC. The medium is then completely changed and the cultures subsequently transferred to 33^oC. Cultures are fed again on the 7th day after initiation and at weekly intervals thereafter by the removal of half of the liquid volume, including half of the nonadherent cell fraction, and replacement with fresh growth medium. The nonadherent cells removed are counted and plated in methylcellulose assays to provide a weekly monitor of culture viability. However, since the majority of the most primitive progenitors remain in the adherent layer, cytogenetic analysis of progenitors usually requires the evaluation of colonies generated from assays of the adherent layer.

Evaluation of adherent layer progenitors involves

sacrificing an entire culture. Nonadherent cells and all of the growth medium are first removed, and the remaining cells then washed twice with calcium- and magnesium-free Hank's balanced salt solution (CaMg-free HBSS). 10 ml of 20% (FCS) in CaMg-free HBSS containing 0.1% collagenase is then added and the cultures incubated for 3 hours at 37°C. At the end of this incubation, most of the adherent layer cells (including the vast majority of the hemopoietic cells) can be readily detached by vigorous pipetting. After centrifugation and 2 washes in 2% FCS in CaMg-free HBSS, the cell pellet is carefully resuspended (5). Colony assays for myeloid progenitors are then carried out by plating these cells at a final concentration of 10^{5} cells per ml in standard 0.8% methylcellulose assays as described above. Adherent layers can also be trypsinized, although this removes all cells and the progenitor concentration is therefore lowered (5).

Cytogenetic Studies of Hemopoietic Colonies

Cytogenetic analysis of individual hemopoietic colonies offers a powerful approach to the study of primitive cell types that are morphologically indistinguishable and present at low concentrations in vivo. However, most hemopoietic colonies in which the number of dividing cells is still significant contain less than 1,000 cells, and cell loss poses a potentially major technical stumbling block. This problem is overcome by the use of a technique in which colonies are processed on polylysine coated slides (16). Individual large colonies are selected when they can be identified without ambiguity, but before proliferation has stopped. For large erythroid colonies, this is usually between days 9 and 12 of incubation, and for granulocyte colonies, between days 12 and 16. Cell divisions are arrested at metaphase by the addition of 0.1 ug/ml of colcemid. Single colonies are removed individually into a finely drawn out Pasteur pipette (internal diameter approx. 0.5 mm) using an inverted microscope. The colony, suspended in <0.01 ml of the surrounding culture media is then transferred into a microtiter well containing 0.1 ml of 0.075 M hypotonic KCL, and dispersed by gently pipetting up and down 3-4 times. After 20 minutes at room temperature, the entire contents of each microtiter well are transferred onto a microscope slide that has been pretreated with polylysine (Sigma p1886) for 90 minutes. The polylysine solution is made in batches of 250 ml, each containing 25 mg of polylysine in distilled water. Aliquots of 1 ml are frozen and kept for several months. The pretreatment is carried out by placing one drop of freshly thawed polylysine solution onto each microscope slide and then coverslipping with a 22 mm^2 coverslip. This facilitates the application of an adhesive coat of polylysine on the glass surface. Immediately prior to use, the coverslip is washed off and the slide gently blotted dry.

The colony, still suspended in the hypotonic solution, is allowed to sit on the polylysine treated area of the slide in a humid environment for 10 minutes. Then the excess hypotonic solution is gently removed using the corner of a piece of absorbent paper, and 2 drops of 3:1 methanol:acetic acid fixative are gently dropped with a Pasteur pipette onto the area of the slide where the colony had been placed. After 15-30 seconds, 2 more drops of the fixative are added. After about 1 minute, the slide is gently blown and quickly dried high over an open flame. The slide, with the colony firmly fixed to it, is then immersed in fresh fixative for 15 minutes before airdrying and scanning. Cell recovery using this procedure is consistently 80-90%, and 50-75% of selected colonies give 2 or more analyzable metaphases.

RESULTS AND DISCUSSION

In CML, residual normal hemopoiesis is diluted and/or suppressed to undetectable levels in most patients by the time of diagnosis. This has been previously documented in a large number of cytogenetic studies of direct preparations of CML marrow cells where the Ph^1 chromosome is typically found in all metaphases examined throughout the course of the disease (6). However, the Sloan-Kettering group (17) and others (18) have recently demonstrated that aggressive treatment regimens can lead to re-establishment of Ph^1-negative hemopoiesis, indicating the persistence of a clinically significant residual normal stem cell population in at least some patients. Unfortunately, the toxicity associated with such treatment has been high, and permanent eradication of the leukemic clone has not been possible. Demonstration of the presence or absence of the Ph^1-chromosome among primitive precursor cell populations could theoretically be used to provide a relatively noninvasive, predictive approach to evaluate the extent to which the Ph^1-positive clone dominates the stem cell compartment in individual patients with CML.

In the present studies, hemopoietic progenitors from 27 Ph^1-positive patients in typical chronic phase were cultured in methylcellulose and their clonal progeny cytogenetically analyzed. Seven of these patients were studied twice and 3 were studied 3 times, giving a total of 40 separate investigations. Of these, 22 were carried out using specimens from patients studied at diagnosis and prior to the initiation of chemotherapy. In 10 cases, hemopoietic colonies analyzed cytogenetically were derived from progenitors in marrow only, in 11 cases they were cultured from blood progenitors only, and in 19 cases both marrow and blood derived progenitors were assayed.

Fourteen marrow aspirates collected on different occasions from 12 of the 27 patients were used to initiate long-term cultures that were maintained with weekly feeding for 2-8 weeks before hemopoietic progenitors in the adherent cell fraction were removed and stimulated to form colonies in methylcellulose assays. These were, in turn, cytogenetically analyzed. In all cases, only colonies yielding at least 2 G-banded metaphases were included in the data.

The results, presented in detail elsewhere (19-21), are summarized in Table 1. Ph^1-negative metaphases were

TABLE 1

Number of Instances Where Ph[1]-Negative Cells Were Detected

Cell Source	Samples from Untreated Patients	Samples from Treated Patients	Total
Direct marrow and/or unstimulated blood[1]	2/22	0/18	2/40 (5%)
Marrow/blood progenitor assay[2]	8/22	1/18	9/40 (22.5%)
Long-term culture adherent layer progenitor assay[3]	8/9	3/5	11/14 (79%)

Based on the analysis of: (1) 25 or more metaphases; (2) 10 or more colonies in all but 3 cases; (3) 10 or more colonies in all but 2 cases.

detected at low incidence (<5%) in direct preparations in only 2 of the 40 specimens. These were both from the same newly diagnosed patient who had a relatively low white blood cell count (<17,000 cells per mm^3) and was studied twice in 2 months. Ph[1]-negative cells were detected in both the blood and marrow progenitor compartments from these 2 specimens at much higher incidences (13-26%) than seen in the corresponding direct preparations. Chromosomally normal progenitors were detected among blood and marrow progenitors in an additional 7 cases. All but one of these were newly diagnosed, untreated patients. The incidence of Ph^1-negative progenitors ranged from 5% to 100%. As a group, these patients were among the most recently diagnosed with lower than average white blood cell counts (<31,000 cells/mm^3).

Chromosomally normal hemopoietic progenitors were readily detected in the adherent layers of long-term marrow cultures established from 11 of the 14 specimens. In the remaining 3, only Ph^1-positive progenitors were detected. Eight of the 11 specimens showing chromosomally normal progenitors were from patients with newly diagnosed CML, but 3 were from patients studied 5-14 months post-diagnosis and initiation of chemotherapy. In 9 of the 11 cases, chromosomally normal cells were not detected among direct bone marrow metaphases and, in 6 of these, chromosomally normal progenitors were not detected in methylcellulose assays of the marrow used to initiate the long-term cultures. In long-term cultures established from all of these 11 marrow specimens, the Ph^1-positive clone declined rapidly so that, by 4-6 weeks, 100% of progenitors were chromosomally normal.

In 2 of the 3 long-term cultures where Ph^1-negative progenitors were not detected, the Ph^1-positive clone also declined rapidly. However, in the third culture, the number of Ph^1-positive progenitors present after 3 and 6 weeks remained high, showing kinetics similar to that exclusive to the Ph^1-negative population in other cultures.

The combined in vitro-cytogenetic approach used here reveals a number of interesting findings:

(1) Chromosomally normal progenitors were detected in methylcellulose assays only when the specimens were obtained from patients at or soon after diagnosis with relatively low white blood cell counts. These were probably cases in which the leukemic clone had not yet expanded to such a degree so as to preclude the detection of a normal-sized population of non-neoplastic progenitors, based on the analysis of realistic numbers of colonies. This suggests that in most patients in the chronic phase of their disease, the leukemic

clone greatly dilutes the residual normal hemopoietic cell population, not only at the level of the terminally differentiated cells, but also among the cells of the more primitive myeloid progenitor cell compartments.

(2) The long-term culture assays revealed the presence of significant numbers of chromosomally normal progenitors in the majority of specimens from both newly diagnosed patients and patients with well-established, treated CML. This observation supports the hypothesis that non-neoplastic hemopoietic stem cells persist in the marrows of most patients with CML for considerable periods of time.

(3) The rapid decline of the Ph^1-positive clone in most long-term CML marrow cultures suggests that differences exist between Ph^1-positive and Ph^1-negative stem cell responses during the development of the adherent layer. It is also possible that, as a result of certain as yet undefined properties, Ph^1-negative stem cells may preferentially become incorporated into the adherent cell layer. Nevertheless, the failure of the Ph^1-positive progenitor population to decline rapidly in one long-term culture also suggests that the differential behavior usually exhibited by Ph^1-positive cells is not an intrinsically fixed property of the Ph^1-positive phenotype.

(4) The fact that Ph^1-negative progenitors were not detected in long-term cultures from 2 patients (one newly diagnosed), even though in these the Ph^1-positive population did rapidly decline, suggests that heterogeneity exists among patients with respect to the ability of their residual Ph^1-negative stem cells to persist in vivo. On the other hand, the possibility that such cells were simply diluted to undetectable levels in the particular marrow samples used to initiate these 2 cultures cannot be ruled out.

(5) The results of the short-term methylcellulose assays for progenitors in blood and marrow provide evidence for the suppression of normal stem cell proliferation in vivo. In 9 cases, chromosomally normal cells were detected at significantly higher levels among myeloid progenitors than among marrow metaphases. In all cases, sufficient numbers

of marrow metaphases were analyzed to permit the detection of the progeny of these chromosomally normal progenitors if they had been proliferating in vivo (22).

SUMMARY AND CONCLUSIONS

The level and extent to which normal blood cell formation is suppressed in the hemopoietic system of a series of 27 patients with typical Ph^1-positive CML has been investigated. The Ph^1-chromosome was used as a marker of the transformed clone and its presence or absence among myeloid progenitor cell populations before and after long-term culture assessed. Hemopoietic progenitors in fresh marrow and blood specimens were cultured in standard methylcellulose assays, and the resulting colonies harvested individually for cytogenetic analysis using polylysine coated slides to optimize the yield of metaphases. Using this approach, chromosomally normal progenitors were detectable in fresh specimens from most cases where predicted. This prediction was based on the assumption that normal progenitors persist in vivo at normal levels, and are diluted to varying degrees by the leukemic clone. Conversely, in most of the cases where chromosomally normal progenitors were not detected in fresh specimens, their absence was similarly predicted.

Further evidence of persisting normal progenitors was obtained from studies of the genotype of progenitors present in the adherent layer of long-term cultures initiated with CML marrow cells. For most patients studied, the initial Ph^1-positive progenitor population declined rapidly when placed in this culture system. This provided an even more sensitive method for detecting residual Ph^1-negative progenitors, since they were maintained at levels seen in long-term cultures of normal marrow cells. Using this approach, chromosomally normal cells could be readily demonstrated in both untreated patients, and in patients studied several months after diagnosis and initiation of chemotherapy.

These studies indicate that the disappearance of normal circulating blood cells that characterizes patients with CML is not due to a major decrease in the residual chromosomally normal marrow stem cell reserve. Such cells clearly persist in many patients, but fail to compete with their Ph^1-positive counterparts in generating mature progeny.

REFERENCES

1. Fialkow PJ. Cell lineages in hematopoietic neoplasia studied with glucose-6-phosphate dehydrogenase cell markers. J Cell Physiol (Suppl 1): 37-43, 1982.
2. Salmon SE (ed). "Cloning of Human Tumor Stem Cells," Alan R Liss Inc, New York, 367 pages, 1980.
3. Bernheim A, Berger R, Preud'homme JL, Labaume S, Bussel A & Barot-Ciorbaru R. Philadelphia chromosome positive blood B lymphocytes in chronic myelocytic leukemia. Leuk Res 5: 331-339, 1981.
4. Eaves CJ & Eaves AC. Erythropoiesis. In: "Hematopoietic Stem Cells," (ed. DW Golde & F Takaku), M Dekker Inc, New York (in press).
5. Coulombel L, Eaves AC & Eaves CJ. Enzymatic treatment of longterm human marrow cultures reveals the preferential location of primitive hemopoietic progenitors in the adherent layer. Blood 62: 291-297, 1983.
6. Shaw MT (ed). "Chronic Granulocytic Leukemia," Praeger Scientific, New York, 251 pages, 1981.
7. Eaves CJ, Krystal G & Eaves AC. Erythropoietic cells. In: "Bibliotheca Haematologica, No. 48 - Current Methodology in Experimental Hematology," (ed. SJ Baum), Karger, Basel (in press).
8. Eaves CJ & Eaves AC. Erythroid progenitor cell numbers in human marrow - Implication for regulation. Exp Hematol 7 (Suppl 5): 54-64, 1979.

9. Eaves AC & Eaves CJ. Abnormalities in the erythroid progenitor compartments in patients with chronic myelogenous leukemia (CML). Exp Hematol 7 (Suppl 5): 65-75, 1979.

10. Goldman JM, Shiota F, Th'ng KH & Orchard KH. Circulating granulocyte and erythroid progenitor cells in chronic granulocytic leukemia. Br J Haematol 46: 7-13, 1980.

11. Eaves CJ, Humphries RK & Eaves AC. In vitro characterization of erythroid precursor cells and the erythropoietic differentiation process. In: "Cellular and Molecular Regulation of Hemoglobin Switching," (ed. G Stamatoyannopoulos & A Nienhuis), Grune & Stratton, New York, pp 251-273, 1979.

12. Fauser AA & Messner HA. Granuloerythropoietic colonies in human bone marrow, peripheral blood and cord blood. Blood 52: 1243-1248, 1978.

13. Dexter TM, Allen TD, Lajtha LG. Conditions controlling the proliferation of haemopoietic stem cells in vitro. J Cell Physiol 91: 335-344, 1977.

14. Gartner S & Kaplan HS. Long-term culture of human bone marrow cells. Proc Natl Acad Sci USA 77: 4756-4759, 1980.

15. Dexter TM, Spooncer E, Toksoz D & Lajtha LG. The role of cells and their products in the regulation of in vitro stem cell proliferation and granulocyte development. J Supramol Struct 13: 513-524, 1980.

16. Dubé ID, Eaves CJ, Kalousek DK & Eaves AC. A method for obtaining high quality chromosome preparations from single hemopoietic colonies on a routine basis. Cancer Genet Cytogenet 4: 157-168, 1981.

17. Goto T, Nishikori M, Arlin Z, Kempin S, Burchenal J, Strife A, Wisniewski D, Lambek C, Little C, Jhanwar S, Chaganti R & Clarkson B. Growth characteristics of leukemic and normal hematopoietic cells in Ph^1+ chronic myelogenous leukemia and effects of intensive treatment. Blood 59: 793-808, 1982.

18. Smalley RV, Vogel J, Huguley CM & Miller D. Chronic granulocytic leukemia: Cytogenetic conversion of bone marrow with cycle-specific chemotherapy. Blood 50: 107-113, 1977.

19. Dubé ID, Gupta CM, Kalousek DK, Eaves CJ & Eaves AC. Cytogenetic studies of early myeloid progenitor compartments in Ph^1-positive chronic myeloid leukemia (CML). I. Persistence of Ph^1-negative committed progenitors that are suppressed from differentiating in vivo. Br J Haematol (in press).

20. Coulombel L, Kalousek DK, Eaves CJ, Gupta CM & Eaves AC. Long-term marrow culture reveals chromosomally normal hematopoietic progenitor cells in patients with Philadelphia chromosome-positive chronic myelogenous leukemia. N Engl J Med 308: 1493-1498, 1983.

21. Dubé ID, Kalousek DK, Coulombel L, Gupta CM, Eaves CJ & Eaves AC. Cytogenetic studies of early myeloid progenitor compartments in Ph^1-positive chronic myeloid leukemia (CML). II. Long-term culture reveals the persistence of Ph^1-negative progenitors in treated as well as newly diagnosed patients. Blood (in press).

22. Hook EB. Exclusion of chromosomal mosaicism. Tables of 90%, 95% and 99% confidence limits and comments on use. Am J Hum Genet 29: 94-97, 1977.

Resistant Cell Types in Human Gliomas

J.R. SHAPIRO, P-Y. PU and W.R. SHAPIRO

Department of Neurology
Memorial Sloan Kettering Cancer Center
New York, New York 10021

INTRODUCTION

Human gliomas remain a most devastating disease with a median survival time of only 52 weeks.[7] In general, failure of chemotherapy occurs because the tumor is resistant to the drugs being used. This resistance may be related to heterogeneity or chemosensitivity of the cells in the tumor. Our laboratory has defined karyotypic heterogeneity and some of the specific nonrandom changes that occur in these tumors,[4,5] and has demonstrated differential chemosensitivity (or resistance) among these clonal populations.[9] Our protocol permits us to identify and isolate many of the cell types that make up the subpopulations in the sample analyzed. We can therefore compare the phenotype of the heterogenous parental populations to individual clonal subpopulations or any mixture of them.

Our initial studies demonstrated that some clonal subpopulations rapidly generated new mutant cell types while others retained the parental cell type for at least 3-4 months in culture.[6] The stable phenotypes were clones that had chromosome numbers in the near-diploid range (40-54 chromosomes), while those that were more prone to segregational errors had chromosome numbers of 55 or more. The parental cell type of these latter clones could not be identified after 2-3 passages.

We extended the chemosensitivity studies in an effort to relate chromosomal number to cellular resistance in the tumors. This report includes preliminary data on four human gliomas, one astrocytoma (tumor WM) and three glioblastomas multiforma (tumors RM, MK and MB). Parental and clonal populations of each tumor were tested in a colony forming assay (CFA) after a one hour exposure to the chemotherapeutic agent BCNU. Parental lines that were predominately near-diploid in chromosome number were resistant, as were

HUMAN TUMOR CLONING
ISBN 0-8089-1671-8

homogeneous clones with diploid and near-diploid modal numbers. In contrast, hyperploid clones contaning 56 or more chromosomes were sensitive to BCNU.

PROTOCOL

Freshly resected human gliomas were dissociated by mechanical disruption into pools of single cells which were divided into three aliquots. The first aliquot was grown in suspension and short-term cultures which were harvested for chromosome analysis during the first 72 hours following tumor resection. The second aliquot was serially diluted and plated in cloning densities to permit isolation of clones from single cells. The third aliquot was used for drug resistance studies.

Karyotypes prepared from the suspension and short-term cultures permitted us to identify the karyotypic deviations found in the subpopulations and isolated cell types of the tumors.[4] Clones isolated from cloning dishes 7-28 days post-plating included some that were karyotypically identified as part of the original tumor, and many that could not be identified and were presumed to be isolated cell types. These clones were karyotyped at passage 2 or 3. All heterogeneous parental populations from each tumor were combined from several flasks and petri dishes to assure a representative sampling.

Cells in early passage (tumors WM and RM) or freshly dissociated (tumors MK and MB) were treated with BCNU in the CFA using the following techniques.[9] The cells were washed several times with medium containing no fetal calf serum (FCS). They were exposed to different concentrations of BCNU for one hour, washed twice with medium containing FCS, and plated at cell density of 2000 cells per 60 mm. integrid petri dish. After 21-23 days in culture, the cells were fixed wth methanol and stained with Giemsa. The parental cells for tumor WM were tested at passage 2 and those from tumor RM were tested at passages 2 and 10. Drug testing on tumors MK and MB used freshly dissociated or primary cells; tumor MK was also tested at passage 8. All the clones analyzed were early passage cells except those from WM clones which were removed from liquid nitrogen prior to testing. Six 60 mm. petri dishes were routinely set up on each parental or clonal population tested. Three plates from each test group were chosen at random and the clones counted. The number of colonies in the treated plates was compared to that in the controls and the results graphed as percent of control. The ED_{50} was established for each tested parental and clonal population.

RESULTS

Within each of the 4 gliomas, the most resistant cells were from the near-diploid populations within each of the 4 gliomas, WM, RM, MK and MB. The heterogeneous parental populations and clonal populations were resistant if the predomiunant cell type was near-diploid in chromosome number while clonal populations with hyperploid chromosomal numbers were sensitive.

TABLE 1

Correlation of Chromosome Number and Cellular Resistance to BCNU in the Parental and Clonal Populations of Tumor WM

Parental Line	Modal[a] Chromosome Number	Chromosome Range	Passage[b] Number Tested	CFA ED_{50} BCNU
WM	46,XX	38-92	-	-
WM-A (fibroblast-like)	46,XX	36-92	2	>15.0 µg./ml.
WM-B (squamous-like)	46,XX	37-100	2	11.8 µg./ml.
Clones				
WMC-1	45,XO	41-90	5	>15.0 µg./ml.
WMC-2	45,XO	35-99	3	12.9 µg./ml.
WMC-10	45,XO	42-90	2	11.6 µg./ml.
WMC-13	45,XO	41-100	2	>15.0 µg./ml.
WMC-14	46,XX	45-92	2	14.0 µg./ml.
WMC-16	46,XX	40-97	6	10.2 µg./ml.

a. The modal chromosome number for the parental lines of the WM tumor were established from short term cultures only.

b. The clones used in the WM line were frozen in liquid nitrogen.

Table 1 depicts the results with tumor WM, the only low grade tumor in this series. One hundred and seventy cells were analyzed; 90% had a chromosomal complement of 45 or 46 chromosomes. Seventy of the 73 cells with 46 chromosomes had a normal karyotype with standard G-banding while each of the other three karyotypes had a different abnormal chromosomal complement. Cells with 45 chromosomes were missing only an X chromosome in 57 or 63 cells analyzed. All the clones isolated and karyotyped in early passage carried near-diploid chromosome numbers. The parental population began to diverge in morphology and was kept separately as a fibroblast-like culture (WM-A) or as a squamous-like culture (WM-B). No structural rearrangements were seen in the primary cytogenetic analysis. When tested by the CFA, all the parental and clonal populations were resistant to BCNU.

TABLE 2

Correlation of the Chromosome Number and Cellular Resistance to BCNU in the Parental Line and Clonal Populations of Tumor RM

Parental Line	Modal Chromosome Number	Chromosome Range	Passage Number Tested	CFA ED_{50} BCNU
RM(P)[a]	46,XX	36->100	-	-
RM	46,XX	38-99	10	>15.0 µg./ml.
RM	46,XX	35-94	2	>15.0 µg./ml.
Clones				
RMC-1	46,XX	39-92	3	>15.0 µg./ml.
RMC-3	46,XX	37-91	3	>15.0 µg./ml.
RMC-6	46,XX	37-97	6	9.3 µg./ml.
RMC-10	45,XO	38-90	4	>15.0 µg./ml.
RMC-12	44,XO	37-90	4	>15.0 µg./ml.
RMC-5	44,XO	38-77	3	>15.0 µg./ml.
RMC-9	85,XX	54-91	4	5.2 µg./ml.
RMC-18	88,XXXX	66-100	3	5.8 µg./ml.
RMC-2	92,XXXX	43-181	4	5.4 µg./ml.

a. P=primary analyses in which cells were harvested from suspension and short-term cultures.

Table 2 depicts the results with tumor RM, a high grade glioblastoma multiforme in which 239 cells were analyzed for chromosomal complement. Three major populations of cells were identified in the primary analysis. Thirty percent carried a normal 46 XX karyotype, 22% had 45 chromosomes, missing an X chromosome, and 14% contained 92 chromosomes and were tetraploid for each homologue. The remaining karyotypes were clustered around the near-diploid mode (19%) or tetraploid mode (4%). Most of the clones isolated in this series also reflected this chromosomal distribution. However, some clones with near-diploid chromosome numbers were tightly clustered around the clones modal number while others appeared to be very heterogeneous even at this early passage. We therefore selected for chemosensitivity testing homogeneous and heterogeneous clones having the same modal number. For example, RMC-6 had a modal chromosome number of 46 XX, but at passage 6 the majority of the cells (53%) contained triploid or tetraploid chromosome numbers. In the RMC-2 clone, 62% of the cells were 92 XXXX, but at passage 4 only 31% of the total cells had the modal chromosome number; the remaining cell population was distributed among cells containing 43-181 chromosomes.

The results of the CFA on the parental line was similar at an early and late passage number, both tests indicating that the RM tumor line was resistant to BCNU. The cytogenetic analysis also showed no major change in the frequency of the near-diploid cells. The homogeneous clones carrying near- diploid chromosome numbers were also resistant to BCNU. The heterogeneous clone RMC-6 was also resistant, but less so than the other diploid and near-diploid clones. The hyperploid clones RMC-9, -18 and -2 were all sensitive to BCNU at levels ranging from 5.2 to 5.8 ug./ml.

TABLE 3

Correlation of the Chromosome Number and Cellular Resistance to BCNU in the Parental and Clonal Populations of Tumor MK

Parental Line		Modal Chromosome Number	Chromosome Range	Passage Number Tested	CFA ED_{50} BCNU
MK	(P)[a]	45,XO	38-100	P	-[b]
MK	(8)	86,XX	41-138	8	6.1 μg./ml.
MK	(12)	75,XO	42-203	-	N.D.[c]
Clones					
MKC-5		45,XO	38-47	6	>15.0 μg./ml.
MKC-17		45,XO	40-46	4	>15.0 μg./ml.
MKC-31		45,XO	40-90	5	12.0 μg./ml.
MKC-39		44,XO	41-86	8	11.3 μg./ml.
MKC-29		46,XX	39-92	5	>15.0 μg./ml.
MKC-25		84,XX	73-116	7	2.8 μg./ml.
MKC-14		88,XX	71-96	8	6.9 μg./ml.

a. P=primary analysis on suspension & short-term cultures; 8 and 12 refer to passage numbers.
b. Primary data lost to contamination.
c. N.D.=not done.

Table 3 depicts the results with tumor MK, a glioblastoma multiforme. Eighty-two percent of the cells were clustered in three subpopulations in the primary analysis. The predominant population contained 45 chromosomes, missing the X chromosome. A second population had 46 chromosomes and a normal G-banded karyotype. The third and smallest population had 44 chromosomes and was missing a number 19 and an X chromosome. The remaining karyotypes contained chromosome complements seen only once. During a three month interval the modal chromosome number of the parental population shifted from 92% near-diploid cells at passage 4 to 90% tetraploid cells at passages 8-12. The clones selected for the CFA were all homogeneous except the MKC-25 clone which contained extensive heterogeneity; of 39 cells counted in this clone, only six cells had 84 chromosomes (modal number). The remaining cells were distributed over a range of 73-155 chromosomes/cell. The parental line tested at passage 8 was sensitive to BCNU; the predominate cell types in this in this culture were hyperploid. All clones resistant to BCNU were diploid or near-diploid in modal chromosome number. The hyperploid clones, like the parental population were sensitive to BCNU.

TABLE 4

Correlation of the Chromosome Number and Cellular Resistance to BCNU in the Parental and Clonal Populatiuons of Tumor MB

Parental Line	Modal Chromosome Number	Chromosome Range	Passage Number Tested	CFA ED_{50} BCNU
MB	45,XO	43-94	Primary	>15.0 μg./ml.
Clones				
MBC-2	46,XX	42-46	4	>15.0 μg./ml.
MBC-6	46,XX	38-92	4	>15.0 μg./ml.
MBC-9	46,XX	40-92	4	>15.0 μg./ml.
MBC-4	46,XX	38-46	4	>15.0 μg./ml.
MBC-3	46,XX	39-46	4	>15.0 μg./ml.
MBC-10	46,XX	40-92	4	>15.0 μg./ml.
MBC-1	45,XO	40-46	4	>15.0 μg./ml.
MBC-17	45,XO	41-48	4	>15.0 μg./ml.
MBC-12	45,XO	38-920	4	>15.0 μg./ml.
MBC-5	45,XO	40-901	4	>15.0 μg./ml.
MBC-18	45,XX	39-92	4	14.2 μg./ml.
MBC-20	47,XX	43-137	4	5.0 μg./ml.
MBC-11	49,52,XX	43-89	4	6.0 μg./ml.
MBC-19	80,XXX	37-83	4	4.1 μg./ml.

Table 4 depicts the results with the tumor MB, a glioblastoma multforme. Seventy-five percent of the cells were distributed among three major subpopulations, 46 XX, 45 XO, and a 92 XXXX (tetraploid for all homologues). The X chromosome was missing in the 45 population; both the 46 and 92 populations appear to have normal banding patterns with standard Giemsa techniques. Most of the clones isolated contained either 45 or 46 chromosomes. Only a few clones contained triploid to tetraploid modal chromosome numbers; the 92,XXXX population was not isolated. As in the RM tumor, several near-diploid clones appeared to be heterogeneous (MBC-11 and MBC-20). The MBC-11 clone was bimodal at passage 2. Seventy-eight percent of the cells were in the near-diploid range of 43-52 chromosomes/cell and the remaining cells had 71-88 chromosomes/cell. At passage 4, 101 cells were scored and only 52% were in the range of 43-54 chromosomes/cell while 49% had 71-89 chromosomes/cell. A similar shift in chromsome number was seen for the MBC-20 clone in which the tetraploid population went from 2% at passage 2 to 42% at passage 4. The results of the CFA were also similar to tumor RM. Homogeneous diploid and near-diploid clones were resistant. The parental population which was 75% near-diploid was also resistant. The only clones sensitive to BCNU were those that had hyperploid modal chromosome numbers or were clones in which nearly half of the cells were split between the near diploid and triploid-tetraploid modes.

DISCUSSION

Drug resistance is a feature of both low grade and high grade gliomas.[2,3,9] It remains to be established whether all tumor cells are resistant or only a portion of the cells contribute to this phenotype. Our data indicate that those cells most likely to be resistant to BCNU, a drug currently used in the treatment of brain tumor patients, are near-diploid in chromosome number.

The work in animal tumors suggested that clones would be differently sensitive,[1] a feature we confirmed in human gliomas.[9] Although our initial studies did not demonstrate a correlation between chromosome number and drug resistance, this failure can be explained by our not doing a chromosome analysis at the passage the cells were being tested. All clonal subpopulations ware karyotyped at passage 2 or 3 following their isolation. This represents about 15-18 generations of cells. To provide enough cells for the assay, this population of cells is passaged 2-4 times before testing. In very unstable clones this is enough time to lose the parental cell type. In this study we prepared chromosomes on the parental and clonal populations being tested.

Clones with modal chromosome numbers of 44, 45, or 46 were found to retain their modal chromosome number. However, in the higher passage numbers these clones became heterogeneous. Occasionally diplochromosomes were seen in these cultures, suggesting that endoreduplication is an active mechanism in these tumors. In clones with triploid and tetraploid modal numbers the cells were rarely clustered around the modal chromosome number at the time of drug testing, but were frequently distributed over a wide range with only 2-3 cells seen for a given chromosome number. The modal number was therefore determined by only a few cells.

We correlated the chromosome numbers of the clones to the results of the CFA. Cells with near-diploid chromosome numbers were found to be resistant to BCNU. This was particularly noteworthy in clones that carried 44, 45 and 46 chromosomes. The MB tumors appeared at first to be an exception in that the two clones MBC-11 and MBC-20 carried near-diploid modal chromosome numbers and were sensitive to BCNU. However, as pointed out in the results, both clones had a dramatic increase in triploid-tetraploid cells at the time of the CFA that may explain their apparent sensitivity to BCNU.

The parental populations were all resistant at clinically achievable levels of drug when tested in early passage (WM and RM tumor) or as a primary culture (MB tumor). The data from the MK primary culture was lost to contamination and the retesting at passage 8 indicated a sensitivity to BCNU. The chromosome number of this parental population shifted from 45 to 86 chromosomes during this same interval. This suggests that parental populations with predominate cell types in the triploid-tetraploid chromosome range will be more sensitive to BCNU than parental lines with near-diploid numbers. Tumors with hyperploid parental cell types are now being tested.

In summary, these data suggest that near-diploid cells in both parental and clonal populations of a tumor are the most resistant cells to clinically achievable blood plasma levels of BCNU while the triploid and tetraploid populations are more sensitive. The "stem cells" we must therefore be concerned with are the near-diploid cells that remain and most likely repopulate the tumor following clinical intervention.

ACKNOWLEDGEMENTS

We wish to thank George Rezac and Ms. Donna Zeff for technical assistance. This work was supported by Grant CA 25956-094 from the National Cancer Institute and BRSG 3854 from Memorial Sloan Kettering Cancer Center.

REFERENCES

1. Heppner, G.H., Dexter, D.L., De Nucci, T., Miller, F.R. and Calabresi, P.: Hetereogeneity in drug sensitivity among tumor cell subpopulations of a single mammary tumor. Cancer Res. 38:3758-3763, 1978.

2. Kornblith, P.A., Dohan, P.L., Wood, F.C. and Whitman, B.O.: Human astrocytoma: Serum-mediated immunologic response. Cancer 33:1512-1519, 1974.

3. Rosenblum, M.L., Vasquez, D.A., Hoshino, T. and Wilson, C.B.: Development of a clonogenic cell assay for human brain tumors. Cancer 41:2305-2314, 1978.

4. Shapiro, J.R., Yung, W.-K.A., and Shapiro, W.R.: Isolation, karyotype, and clonal growth of heterogeneous subpopulations of human malignant gliomas. Cancer Res. 41:2349-2359, 1981.

5. Shapiro, J.R. and Shapiro, W.R.: Specific karyotypic and tumorogenic changes in cloned subpopulations of human gliomas exposed to sublethal doses of BCNU. Chabner, B.A. (ed.): Rational Basis for Chemotherapy. UCLA Symposia on Molecular and Cellular Biology. New York, Alan R. Liss, 1983, pp. 45-59.

6. Shapiro, J.R. and Shapiro, W.R.: Clonal tumor cell heterogeneity. In: Progress in Experimental Tumor Research, Vol 1: Brain Tumor Biology (eds. M.L. Rosenblum and C.B. Wilson), Basel, S. Karger. In press. 1983.

7. Shapiro, W.R.: Treatment of neuroectodermal brain tumors. Ann. Neurol. 12:231-237, 1982.

8. Shapiro, W.R., Yung, K.-W.A., Basler, G.A., and Shapiro, J.R.: Heterogenous response to chemotherapy of human gliomas grown in nude mouse and as clones in vitro. Cancer Treatment Reports 65: (Suppl. 2) 55-60, 1981.

9. Yung, W.-K.A., Shapiro, J.R. and Shapiro, W.R.: Heterogeneous chemosensitivities of subpopulations of human glioma cells in culture. Cancer Res. 42:992-998, 1982.

DIHYDROFOLATE REDUCTASE GENE AMPLIFICATION IN HUMAN TUMOR CELLS SELECTED FOR RESISTANCE TO METHOTREXATE

PAUL MELTZER, M.D., Ph.D.[1]
Y C. CHENG, Ph.D.[2]
SHARON OLSON, M.T.[1]
J. M. TRENT, Ph.D.[1]

[1]University of Arizona Cancer Center
Tucson, Arizona 85724
[2]Departments of Pharmacology and Medicine
University of North Carolina
Chapel Hill, North Carolina 27514

INTRODUCTION

Resistance to methotrexate (MTX) has been documented to occur by several mechanisms including overproduction of the target enzyme dihydrofolate reductase (dhfr, E.C. 1.5.1.3) by increased dosage of the dhfr gene. Likewise, resistance to several other cytotoxic drugs has been demonstrated to occur via gene amplification in several mammalian cell culture systems (1). Increased copies of the dhfr gene (up to 200 fold) have been observed in MTX resistant (MTX^R) rodent cells (2) and human leukemic cells (3) selected by exposure to incremental concentrations of MTX *in vitro*. Additionally, of more clinical relevance, dhfr gene amplification has recently been reported to occur in patients exposed clinically to MTX (4,5). Although the importance of this phenomenon as the clinical basis for acquired resistance to cytotoxic agents remains indeterminate, it has provided a valuable research approach for the analysis of genetic mechanisms of acquired drug resistance.

Two specific cytogenetic anomalies are often found in MTX^R cells. These are homogeneously staining regions (HSR) and double minute chromosomes (DM). HSRs are chromosomal regions of intermediate staining intensity which lack the normal banding pattern in G-banded preparations. DMs are small paired chromatin fragments which are distributed randomly at mitosis (due to their lack of a centromere) and are, therefore, inherently genetically

HUMAN TUMOR CLONING
ISBN 0-8089-1671-8

unstable. In MTX^R rodent cell lines *in situ* hybridization has localized sequences encoding dhfr to HSRs (6,7). This observation supports the hypothesis that these anomalous structures represent the chromosomal sites of the amplified dhfr gene.

Recently Cheng and colleagues have reported the establishment via incremental exposure to MTX of several MTX^R subclones from the human KB carcinoma cell line (8). These lines have previously been demonstrated to overproduce dhfr and to maintain their resistant phenotype when cultured in MTX free medium. We now report the detailed cytogenetic analysis of these cell lines including documentation of cytologically recognizable evidence of gene amplification (DMs and HSRs), as well as localization of dhfr sequences to HSR regions using *in situ* hybridization. Finally, determination of the number of copies of the dhfr gene in increasingly MTX^R subclones has been elucidated by slot and Southern blot hybridization techniques.

RESULTS AND DISCUSSION

KB parental cells are capable of withstanding exposure to 0.01 µM MTX. Selection at higher MTX concentrations has been performed and yielded MTX^R subclones with relative resistance of 10 (clone 2A), 600 (clone 4A), 6,500 (clone 6B), and 13,000 (clone 7A) times the parental level. Dhfr activity increases progressively in these MTX^R subclones, and the enzyme has been shown to be identical with that of the parental cells (unpublished results - T.C.)

Cytogenetic analysis revealed a near triploid chromosome number in the parental cells with increasing polyploidy associated with increasing relative MTX^R. HSRs or DMs, although absent from the parental cell lines, were present in all MTX^R subclones (Figure 1). Clone 2D contained multiple copies of DMs in approximately 15 percent of all cells examined. Subclone 4A demonstrated a single HSR on chromosome 10q in 97 percent of cells. Three percent of cells in clone 4A also contained an isochromosome (iso) of the long arm (q) of chromosome 10. Both the 6B and 7A subclones displayed the iso10q HSR in all cells examined. The emergence and increasing size of the HSR marker is consistent with the increasing relative resistance to MTX observed in these clones.

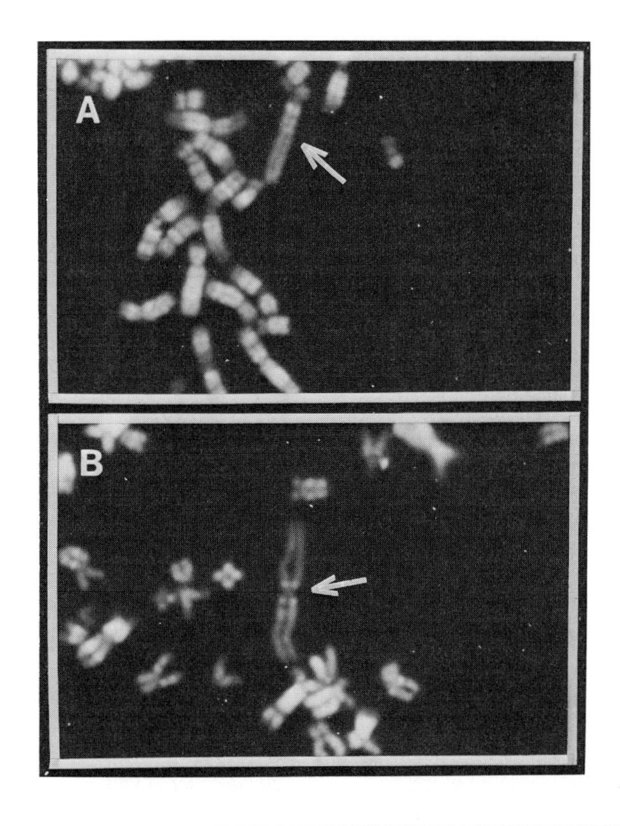

Figure 1.
A) Q-banded metaphase spread from MTX^R subclone 4A. Arrow denotes HSR on chromosome 10q23.
B) Q-banded metaphase spread from MTX^R subclone 7A. Arrow denotes the HSR-bearing isochromosome of 10q marker present in all cells examined from clones 6B and 7A. Additionally, ~3 percent of cells from clone 4A demonstrated the iso10q marker.

Localization of dhfr sequences to the HSR-bearing markers in clone 6B and 7A was confirmed by in situ hybridization (9). For the study a recombinant plasmid (pdhfr 11, generously provided by R.T. Schimke) containing a dhfr sequence derived from a murine cDNA clone was used and provided significant hybridization signal over the iso10q HSR (10).

Dhfr gene amplification in the MTX^R subclones was verified by quantitative slot blot hybridization experiments. High molecular weight DNA was extracted from the MTX^R subclone and the parental cells. Serial dilutions were prepared and applied to nitrocellulose filters. The filters were probed with the pdhfr 11 plasmid and estimates of relative gene copy number were made by linear regression analysis of a densitometric scan of the autoradiograms. By this method we have found that progressively resistant KB subclones contain increasing dhfr gene copy number. In the subclone displaying the highest level of MTX^R (7A) 15 copies of the dhfr gene are present (10). Of interest, dhfr enzyme activity appears proportional to gene copy number in all the MTX^R subclones. These observations are consistent with those in rodent systems which support the concept that dhfr overproduction is a consequence of gene amplification. However, other factors such as dhfr enzyme turnover and mRNA kinetics remain to be investigated and may contribute to MTX^R in these cell lines.

Southern blot hybridizations of restriction endonuclease digested DNA from the parental KB cells and the highly MTX^R subclone 7A confirmed the presence of an increased number of dhfr gene copies (Figure 2). Of interest is the appearance in MTX^R cells of novel restriction fragments not present in the parental cell line. It is evident that the molecular events which underlie the gene amplification event result in translocation of the dhfr gene generating new restriction sites.

Clearly, the large size of the HSRs observed cannot be entirely accounted for by the dhfr gene alone. This conclusion holds even if one considers the large size of the human dhfr gene which spans 28kb of genomic DNA (11), and any underestimation of gene copy number in our studies which may result from the presence of dhfr pseudogenes in the KB parental cells. Evidently, in this as in other systems (6,7,12), the amplified domain includes large segments of DNA extraneous to the dhfr gene. The structure and functions of these segments remain indeterminate.

Investigation of this question may contribute to elucidation of the mechanism of gene amplification.

CONCLUSIONS

The information presented here provides an example of MTX^R as a result of gene amplification in a human carcinoma derived cell line. Evidence is presented that resistance is accompanied by characteristic cytogenetic anomalies (DMs, HSRs) which reflect the process of gene amplification. The emergence of drug resistant cells continues to be a vexing problem for the clinician. Clear-cut examples have recently been reported of dhfr gene amplification in patients exposed to MTX for therapeutic purposes (4,5). The frequency of this occurrence and its contribution to clinical outcome are important points for further study.

Figure 2

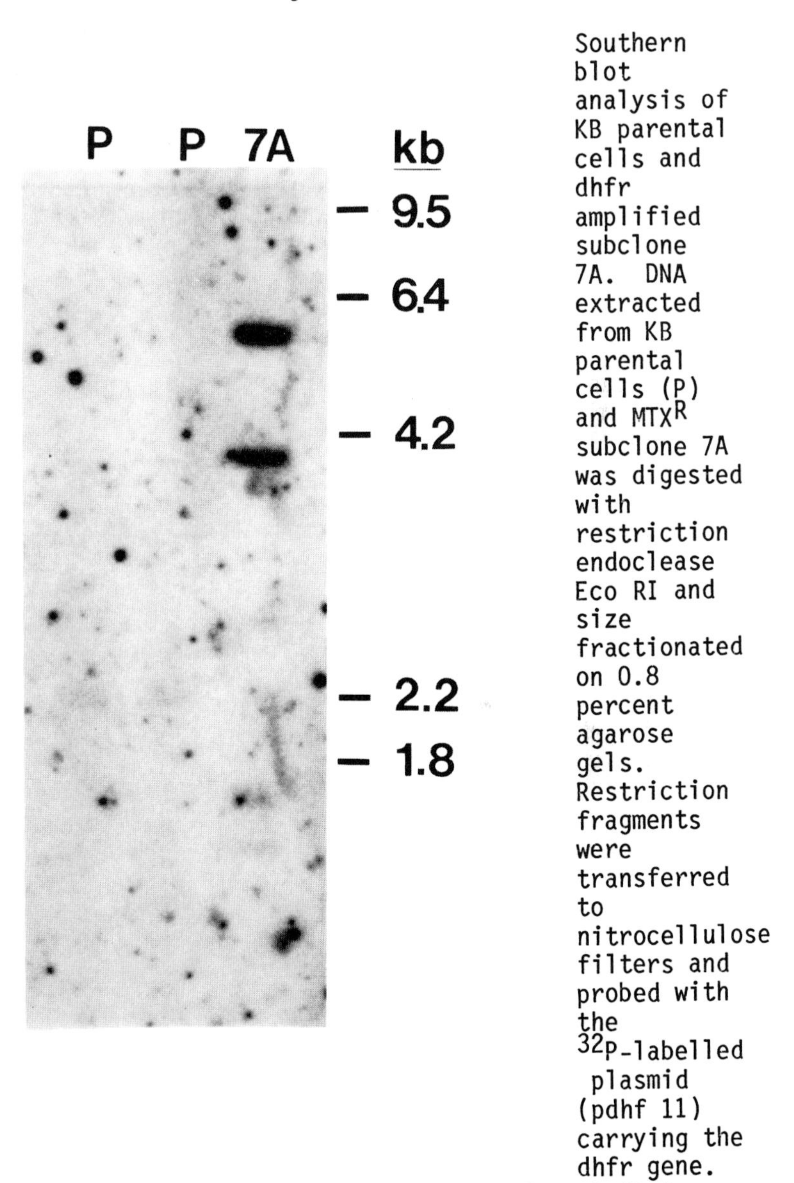

Southern blot analysis of KB parental cells and dhfr amplified subclone 7A. DNA extracted from KB parental cells (P) and MTX^R subclone 7A was digested with restriction endoclease Eco RI and size fractionated on 0.8 percent agarose gels. Restriction fragments were transferred to nitrocellulose filters and probed with the ^{32}P-labelled plasmid (pdhf 11) carrying the dhfr gene. The amounts of DNA applied to each lane are from left to right 8μg, 4μg, and 3μg. The size markers on the right show the positions of the Hind III restriction fragments of λDNA. Results clearly demonstrate amplification of dhfr sequences in MTX^R subclone 7A.

REFERENCES

1. Ling V: Genetic basis of drug resistance in mammalian cells, in Bruchovsky N, Goldie J (eds): Drug and Hormone Resistance in Neoplasia. New York, CRC Press, 1983, p 14.

2. Alt FW, Kellems RE, Bertino JR: Multiplication of dihydrofolate reductase genes in methotrexate-resistant variants of cultured murine cells. J Biol Chem 253:1357-1370, 1978.

3. Bertino JR, Srimatkandada S, Engel D, et al.: Gene amplification in a methotrexate resistant human leukemia line K 562, in Schimke RT (ed): Gene Amplification. Cold Spring Harbor, Cold Spring Harbor laboratory, 1982, pp 23-27.

4. Horns RC Jr, Dower WJ, Schimke RT: Human dihydrofolate reductase gene amplification after methotrexate treatment. Proc Am Assoc Cancer Res 24:280, 1983.

5. Curt GA, Carney DN, Cowand KH, et al.: Unstable methotrexate resistance in human small cell carcinoma associated with double minute chromosomes. N Engl J Med 308:199-202, 1983.

6. Nunberg JH, Kaufman RJ, Schimke RT, et al.: Amplified dihydrofolate reductase genes are localized to a homogeneously staining region of a single chromosome in a methotrexate resistant chinese hamster ovary cell line. Proc Natl Acad Sci, USA 75:5553-5556, A78.

7. Dolnick BJ, Berenson RJ, Bertino JR, et al.: Correlation of dihydrofolate reductase elevation with gene amplification in a homogeneously staining region in L5178Y cells. J Cell Biol 83:394-402, 1979.

8. Wolman SR, Craven ML, Grill SP, et al.: Quantitative correlation of homogeneously stained regions on chromosome 10 with dihydrofolate reductase enzyme in human cells. Proc Natl Acad Sci USA 80:807-809, 1983.

9. Trent JM, Olson S, Lawn RM: Chromosomal localization of human leukocyte, fibroblast and immune interferon genes by means of _in situ_ hybridization. Proc Natl Acad Sci USA 79:7809-7813, 1982.

10. Meltzer PS, Cheng YC, Trent JM: Dihydrofolate reductase gene amplification in a human carcinoma cell line. In press, 1984.

11. Chen M-J, Shimada T, Davis A, et al.: Characterization of the human dihydrofolate reductase gene. Blood 62 Suppl 1:158a, 1983.

12. Bostock CJ, Clark EM, Harding NG, et al.: The development of resistance to methotrexate in a mouse melanoma cell line. I Characterization of the dihydrofolate reductases and chromosomes in sensitive and resistant cells. Chromosoma 74:153-177, 1979.

II. METHODOLOGIC CONSIDERATIONS

Plating Efficiencies of Human Tumors in Capillaries Versus Petri Dishes

DANIEL D. VON HOFF, M.D.

Associate Professor
Department of Medicine
University of Texas Health Science Center at San Antonio
San Antonio, Texas

INTRODUCTION

It is well recognized by the participants in this conference that, as now constituted, the conventional two layer agar technique has several problems. The first problem is that not all patients' tumors form colonies in the system. For example, in our experience, only 35 to 50% of patients' tumors will form enough colonies in control plates to enable drug sensitivity test.[10,11]

Acknowledgment:
Supported by grant NCI RO1 CA27733-04 from the Department of Health and Human Sciences. The expert secretarial assistance of Clair Waggoner is gratefully acknowledged.

HUMAN TUMOR CLONING
ISBN 0-8089-1671-8

A second problem is that the plating efficiencies are very low. This low plating efficiency has led many investigators to fear that the two layer agar system may be too selective and allow only the hearty cells to survive. Finally, a large number of tumor cells is required to perform drug sensitivity testing. Since the tumor biopsy may be small, it is often very difficult to test a large number of drugs on the patients tumor.

The system we are currently examining, to circumvent problems noted with the conventional petri dish technique, is a capillary cloning system. It is important to point out that the growth of tumor cells in capillary tubes is not a new concept. In fact, capillary tubes were initially used to start cell lines from a single cell by Sanford and colleagues as early as 1948.[9] Their rationale for utilizing capillary tubes was their observation that to proliferate, an individual cell requires the presence of other cells. They reasoned that if a capillary tube were utilized (to reduce the volume of a vessel) then a single cell might be able to condition that smaller amount of media by its own metabolic activity (rather than depend on other cells for growth factors). Using this rationale, they successfully utilized capillary tubes to initiate a number of human tumor cell lines.

After Sanford's initiated work, capillary tubes have been used to grow granulocyte colonies,[1, 6] to clone and perform drug sensitivity tests on tumor cell lines[2] and, most recently, by Maurer and Oli-Osman,[5] to clone tumors taken directly from patients with a variety of malignancies. Therefore cloning of tumors in capillaries is not a new technique.

In this study we compare the capillary cloning system with the conventional two layer agar system of Hamburger and Salmon[3, 4, 8] to determine if the capillary system offers any advantages over the two layer 35 mm petri dish method.

MATERIALS AND METHODS

Collection and Preparation of Tumor Cells

After obtaining informed consent in accordance with federal and institutional guidelines, malignant effusions, ascites, bone marrow-containing tumor cells, and solid tumor specimens were collected from patients undergoing procedures done as part of a diagnostic work-up or as part of treatment for their disease. No major surgical procedures were performed solely to obtain specimens for drug sensitivity testing. Solid tumors or lymph nodes were minced into 2 to 5 mm fragments in the operating room and were immediately placed in McCoy's Medium 5A plus 10% heat-inactivated newborn calf serum plus 1% penicillin and streptomycin (all from Grand Island Biological Co., Grand Island, N.Y.) under aseptic conditions. Within 4 hours, these solid tumors were mechanically dissociated with scissors; forced through a No. 50 stainless steel mesh, through 20-, 22-, and 25-gauge needles, and through a nylon mesh; and then washed with Hanks' balanced salt solution as described previously.[3, 4, 8, 10, 11] Ascitic, pleural, and pericardial fluids and bone marrows were obtained by standard techniques. The fluid or marrow was placed in sterile containers containing 10 units of preservative-free heparin (Grand Island Biological Co.) per ml of malignant fluid or marrow. After centrifugation at 150 x g for 10 minutes, the cells were harvested and washed twice in Hanks' balanced salt solution plus 10% heat-inactivated fetal calf serum. The viability of cell suspensions was determined in a hemocytometer with trypan blue. Viability of cells derived from solid tumors ranged from 0 to 100% (median, 26%), while viability of cells from effusions and bone marrows ranged from 0 to 100% (median, 90%).

Culture of Cells in 2 Layer Soft Agar System

The culture system used in this study has been extensively described elsewhere.[3, 4, 8, 10, 11] In brief, cells to be tested were suspended in 0.3% agar in enriched Connaught Medical Research Laboratories Medium 1066 (Grand Island Biological Co.) supplemented with 15% horse serum, penicillin (100 units/ml), streptomycin (2 mg/ml), glutamine (2 mM), $CaCl_2$ (4 mM), and insulin (3 units/ml). Prior to plating, asparagine (0.6 mg/ml), DEAE-dextran (0.5 mg/ml; Pharmacia Fine Chemicals, Inc., Piscataway, N.J.), and freshly prepared 2-mercaptoethanol (final concentration, 50 mM) were added to the cells. One ml of the resultant mixture was pipetted onto 1 ml feeder layers in 35 mm plastic Petri dishes (Falcon Plastics). The final concentration of cells in each culture was 5×10^5 cells in 1 ml of agar medium. The feeder layers used in this study consisted of McCoy's Medium 5A plus 15% heat-inactivated fetal calf serum and a variety of nutrients described by Pike and Robinson.[7] Immediately before use, 10 ml of 3% tryptic soy broth (Grand Island Biological Co.), 0.6 ml of asparagine, and 0.3 ml of DEAE-dextran were added to 40 ml of the enriched underlayer medium. Agar (final concentration, 0.5%) was added to the enriched medium, and underlayers were poured into 35 mm Petri dishes. After preparation of both bottom and top layers, the plates were examined under an inverted microscope and on a feature analysis system (FAS II; Bausch and Lomb, Inc.) to ensure the presence of a good single-cell suspension. To be considered evaluable for subsequent growth and drug sensitivity information, the plates had to have excellent single-cell suspensions which were defined as ≤ 10 features (diameter, ≥ 40 µm) on each of 3 control plates. The plates were then incubated at 37°C in a 7% CO_2 humidified atmosphere.

Colonies (>50 cells) usually appeared by day 14 of culture, and the number of colonies on control and drug-treated plates was determined by counting the colonies on an inverted stage microscope at x 30 magnification.

Culture of Cells in Capillary Tubes

Tumor cells were collected and prepared as described above. Cells to be cloned were suspended in 0.3% agar in enriched Connaught Medical Research Laboratories media 10^6 (Grand Island Biological Co.) supplemented with 15% horse serum, penicillin (100 units/ml), streptomycin (2 mg/ml), glutamine (2 mM), $CaCl_2$ (4 mM), and insulin (3 units/ml). Prior to plating, asparagine (0.6 mg/ml), DEAE-dextran (0.5 mg/ml); Pharmacia Fine Chemicals, Inc., Piscataway, N.J.), and freshly prepared 2-mercaptoethanol (final concentration, 50 mM), and Hepes Buffer (20 mM) were added to the cells. One hundred microliters of the resultant mixture (containing a variable number of tumor cells (see below) was placed into 100 µl capillary tubes (round capillaries, Fisher Scientific Co., or square capillaries , Glass Co. of America) by capillary action. The ends were sealed with clay. A 1 mm gap was present between the clay and the agar to prevent contamination from the clay. Six capillary tubes were prepared for each data point. The 6 tubes were placed in a 12 x 150 mm test tube and laid flat in a 7% CO_2 incubator at 37°C. The tubes were removed from the incubator on day 2 and 14 for counting under an inverted scope (x 30 magnifications). Colonies ($\geq$50 cells) usually appeared by day 14 of culture. As an alternative method for colony counting, on day 14 the agar was extracted from the capillary tube onto a microscope slide by removing the clay (breaking the ends) and applying positive pressure on one end of the tube. The agar was allowed to air dry and the slide was stained with Diffquik Stain (a hematoxylin stain from Difco Co.). _In vitro_ exposure of tumor cells to drugs was accomplished as outlined above except after the drug incubation and washing procedures, the cells were placed in capillary tubes rather than in 35 mm petri dishes.

Study Design

To perform this comparison of capillary versus petri dish system, human tumors were divided and simultaneously cultured in both systems. For the first comparison, an equal number of cells (50,000 in this case) were plated in each vessel (50,000 cells was chosen because it seemed optimal in the capillary tube). Identical tissue culture media were utilized in each system (for the capillaries we used enriched 50:50 McCoys and double enriched CMRL). Sets of 6 capillary tubes were used versus 6 petri dishes.

The percent plating efficiencies were calculated in each system utilizing the usual formula of number of colonies per vessel over number of cells plated multiplied by 100.

For a further comparison of the two systems additional human tumors were divided and the optimal cell numbers were plated in each vessel. The optimal numbers were 50,000 in each capillary tube and 500,000 in each petri dish.

RESULTS

Overall, for 34 human tumors of 7 different histologic types plated in both systems at 50,000 cells/plate, the median percent plating efficiency for the capillary system was 0.12 (with a range of 0 to 0.35) while the percent plating efficiency in the 35 mm petri dish was 0 with a range of 0 to 0.23%. For 33 of the 34 specimens, the percent plating efficiency was higher in the capillary system than it was in the petri dish.

Overall, for 61 tumors cultured at optimal cell numbers in both systems, (500,000 petri dish, 50,000 capillary) the median percent plating efficiency for the capillary system was 0.12 (range 0 to 0.36) while for the same tumors the percent plating efficiency in the 35 mm petri dishes was 0.04 (range 0 to 0.35). For 51 of the 61 tumors cultured, the plating efficiencies were higher in the capillary system than in the petri dish system. For these 61 tumors the mean improvement in plating efficiency was 9.8 fold while the median improvement in plating efficiency was 3.0 fold.

DISCUSSION

From these initial comparative studies, it is clear that the plating efficiencies for a given tumor are higher when the cells are plated in capillary tubes than when they are plated in 35 mm petri dishes. This is the case whether an equal number of cells are plated in each vessel (50,000 cells/vessel), or when optimal cell numbers are plated in each vessel (500,000 per petri dish; 50,000 per capillary tube). The higher plating efficiencies obtained with the capillary system may be helpful for in vitro drug sensitivity testing because a higher percentage of patients' tumors achieve evaluable growth and because fewer tumor cells are utilized per drug test which enables testing a larger number of drugs per specimen. Because of higher evaluability rates with the capillary systems, we have initiated a clinical trial in which patients have their tumor submitted and are randomized between a choice of a drug by the clinician and a choice by the assay. This trial should define the place of the assay in clinical management.

In addition to the clinical trial, a number of other refinements of the capillary system have been pursued. Instead of utilizing round capillary tubes, we have switched to square tubes. These square tubes offer superior properties for viewing and counting colonies.

For colony counting in the capillary tubes, a relatively inexpensive automated capillary scanner called the Cellscan has been developed by Triton Biosciences, Incorporated. This scanner records the features in each capillary. Each deflection of the light beam represents a cluster or colony of cells. The size of the deflection is related to the size of the colony or cluster. Initial measurements with this instrument have shown very good agreement between technician and machine counts.

Development of a Hepes buffer system for the sealed capillaries has allowed use of a plain incubator without CO_2 and the sealed capillaries have a lower rate of contamination than the open petri dishes. A method has also been devised to precoat the inside of the capillary tube with the antineoplastic agent. When these precoated tubes are filled with cells in agar, the drug comes off the wall of the capillary tube and diffuses into the agar. These precoated tubes will allow more automated in vitro sensitivity testing.

And finally, a perfusion system has also been developed whereby the tumor cells are placed in porous glass capillaries. These porous glass capillaries, containing the colonies are perfused with media or media plus drug. The media or drug can freely exchange across the capillary wall. This modification will potentially allow flexibility of drug scheduling as well as testing of combinations of drugs.

Overall, it is clear that in addition to improving plating efficiency, cloning human tumors in capillary tubes will allow a number of useful modifications to the current petri dish technique.

REFERENCES

1. Abrams L, Carmeci P, Bull JM, Carbone PP: Capillary tube scanning applied to in vitro mouse marrow granulocyte growth. J Natl Cancer Inst 50: 267-270, 1973,
2. Braslow NM, Bowman RL: Capillary tube scanning applied to cell growth kinetics. Science 175: 1436-1440, 1972.

3. Hamburger AW, Salmon SE: Primary bioassay of human myeloma stem cells. J Clin Invest 60: 846-854, 1977.
4. Hamburger AW, Salmon SE: Primary bioassay of human tumor stem cells. Science 197: 461-463, 1977.
5. Maurer HR, Oli-Osman F: Tumor stem cell cloning in agar-containing capillaries. Naturwissenschaften 67: 381-383, 1981.
6. Maurer HR, Henry R: Automated scanning of bone marrow cell colonies growing in agar-containing glass capillaries. Experimental Cell Research 103: 271-277, 1976.
7. Pike B, Robinson W: Human bone marrow colony growth in vitro. J Cell Physiol 76: 77-81, 1970.
8. Salmon SE, Hamburger AW, Soehnlen, et al.: Quantitation of differential sensitivity of human tumor stem cells to anticancer drugs. N Engl J Med 298: 1321-1327, 1978.
9. Sanford K, Earle WR, Likelag D: The growth in vitro of single isolated tissue cells. J Natl Cancer Inst 9: 229-246, 1948.
10. Von Hoff DD, Casper J, Bradley E, et al.: Association between human tumor colony-forming assay results and response of an individual patient's tumor to chemotherapy. Am J Med 70: 1027-1032, 1981.
11. Von Hoff DD, Clark GM, Stogdill BJ, et al.: Prospective clinical trial of a human tumor cloning system. Cancer Res 43: 1926-1931, 1983.

Evaluation of the Omnicon Image Analysis System for Counting Human Tumor Colonies

Sydney E. Salmon
Professor of Medicine, Department of Internal Medicine
Director, Cancer Center

Laurie Young
Joyce Lebowitz
Stephen Thomson
Janine Einsphar
Tony Tong
Research Associates, Cancer Center

Thomas E. Moon
Research Professor of Medicine
Department of Internal Medicine
Assistant Director, Cancer Center

University of Arizona College of Medicine
Tucson, Arizona 85724

ABSTRACT

The Omnicon FAS II image analysis system was applied as a "colony counter" for the soft agar human tumor clonogenic assay. A detailed protocol was used to assess the instrument's sensitivity, specificity, precision and accuracy. We found that the colony counter provided sufficient reliability to be applied to counting human tumor colonies grown in vitro. Additionally, the colony counter performed the petri dish counts ten times faster than experienced technicians and without associated operator fatigue.

INTRODUCTION AND METHODS

The soft agar human tumor clonogenic assay (HTCA) system developed in our laboratory (1) has proven applicable to study fresh tumor biopsy specimens and permits quantitation of anticancer drug sensitivity (2). Clinical trials from a

HUMAN TUMOR CLONING
ISBN 0-8089-1671-8

number of centers have shown that, for a variety of tumor types, in vitro sensitivity or resistance in HTCA correlates well with clinical response or resistance to anticancer drugs (3-10). Other applications of HTCA have been to preclinical new drug screening (11,12), in vitro phase II trials (13), and to various diagnostic applications (14,15).

After development of HTCA, it became apparent that an automated tumor colony counter would be desirable, as counting by technicians proved to be very time-consuming. An automated tumor colony counter using image analysis techniques was therefore developed for HTCA (16). In the current investigation, we assessed the automated tumor colony counter's applicability to routine counting of tumor colonies grown in vitro in HTCA. This brief report provides a synopsis of our findings which are detailed elsewhere (17) along with the detailed protocol and materials and methods. The assessment was carried out on a blinded basis with counts carried out by inverted microscopy by at least two technicians as well as by the colony counter. Metal microspheres (60-160 μ in diameter), four tumor cell lines, and 30 fresh tumor specimens were all employed in the assessment of the colony counter, with the cells grown in HTCA with modifications of the method of Hamburger and Salmon (1). At least two concentrations of five cytotoxic drugs were used with the cell lines and fresh tumors.

A microscope stage mask was prepared so that the total counting area viewed by inverted microscopy had similar field margins and area as that of the total field area of the automated colony counter. Samples with <15 or >2000 colonies/plate were excluded from this study.

The Omnicon model "FAS II" (Bausch and Lomb, Rochester, New York) was utilized as the automated colony counter ("colony counter") in these experiments. Background concerning underlying theory and application to colony counting have been published previously (16). As equipped for this study, the instrument incorporated a keyboard interface and printer, 36 position automated mechanical stage designed to hold 35 mm petri dishes, an inverted microscope with a binocular port for technician observation, a television camera and video monitor and a computer programmed for image analysis. The counting procedure for 36 dishes takes from 35-40 minutes and can be accomplished unattended. As carried out in this study, colonies were counted without any form of staining.

RESULTS

Sensitivity Studies

In these studies, metal microspheres in the size range from 60μ to 160μ were embedded in matrix in the petri dishes and counted by three technicians and the colony counter. There was close agreement between technician counts as well as in the comparison of the mean of the technicians versus the colony counter, r=.999.

Specificity Studies

In these studies, the field-by-field comparisons of technician and machine classification and colony counts were made for ten fresh tumors and one cell line, using a binocular microscope port attached to the Omnicon. There was close agreement between the number of colonies counted by the technicians and the number counted by the colony counter (correlation coefficient .95).

Precision Analysis

Precision of the colony counter was evaluated with colonies from a colon cancer cell line counted 20 times on one day or twice daily for five days. The comparative colony counts between the ten runs over five days were quite reproducible. The coefficient of variation was <10% in all of the precision studies.

Accuracy Studies

A total of 30 fresh tumors (11 ovarian, 9 melanoma, and 10 unknown primary carcinomas) were included in this study with two drug concentrations tested for each of five drugs plus untreated controls. Each was counted independently by two technicians and the colony counter. Thus, there were a total of 300 data points for the fresh tumor comparisons. There were a total of 90 data points available for analogous comparisons of counts on cell lines.

Technician Count Comparisons

The comparison of technician counts of fresh tumor colonies yielded a regression line with a correlation coefficient of 0.93 and the association was linear throughout the entire size range of colonies.

Analogous observations on comparative technician counts were made with the cell line studies. The regression line (technician 1 vs technician 2) yielded a correlation coefficient of .95. Subdividing the studies to assess just high or low drug concentration exposure, the correlation coefficient for technician counts was .95 for each subset.

Technician versus Colony Counter Comparisons

The comparison of mean technician counts versus the colony counter for the 30 fresh tumors studied is depicted in figure 1A.

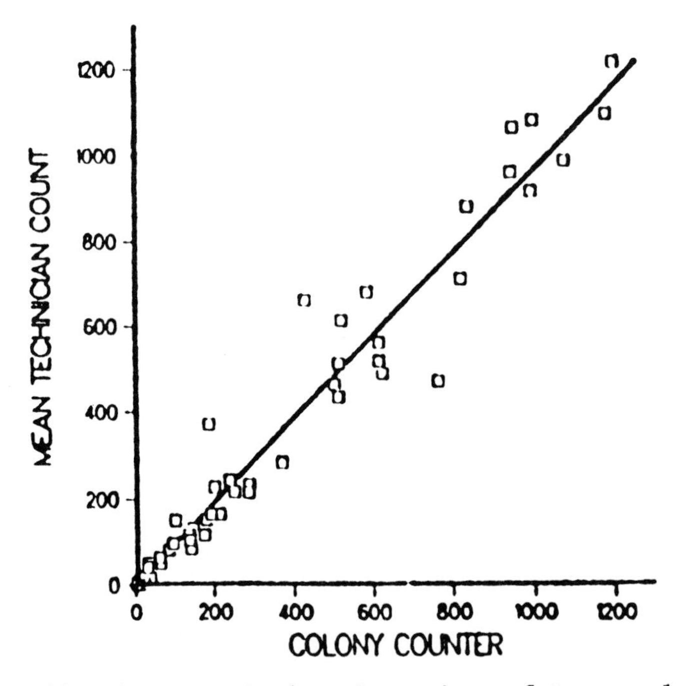

Figure 1A. Accuracy Study: Comparison of tumor colony counts from 30 cancer specimens obtained by each of two technicians (mean value) and the Omnicon colony counter. Data are plotted as absolute counts and include 150 data points from both drug treated and control cultures. Analysis yielded a regression line with a correlation coefficient of .91.

The results of the regression analysis indicate the 95% conf. int. for the slope includes the value of 1.0 and for the intercept includes 0.0, permitting the conclusion that there was not a consistent difference between the enumeration of colonies by technician as compared to the colony counter. The comparison of percent survival of TCFU's, as calculated from the mean technician counts versus the colony counter for the 30 fresh tumors analyzed, is shown in figure 1B.

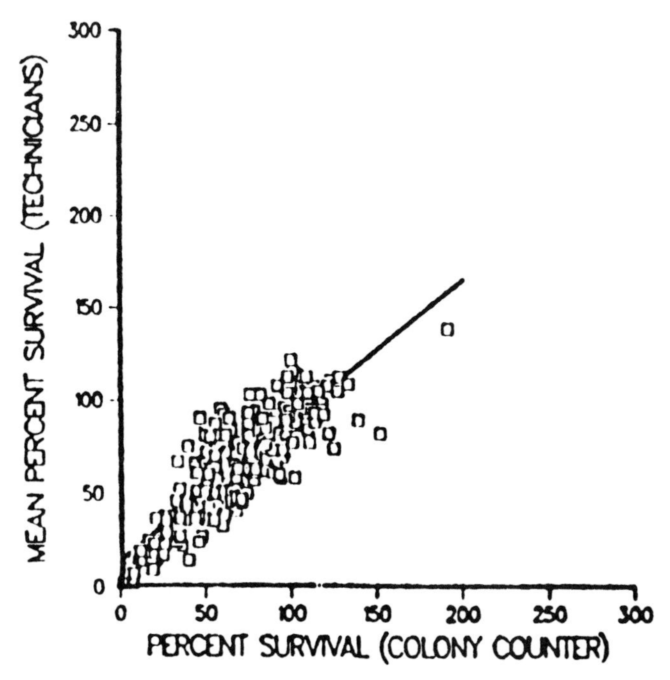

Figure 1B. Accuracy Study: Comparison of technician and Omnicon data expressed as percent survival of tumor colonies after drug exposure. Points on this plot are calculated from the colony counts shown in figure 1A. This analysis normalizes the data from low and high colony count tumors with most data points falling between 1% and 100% of control. The analysis yielded a regression line with a correlation coefficient of .85.

Accuracy data on the tumor cell lines were also compared for the mean technician counts versus the colony counter. An excellent correlation was obtained (r=.98) using tumor cell lines. Under the various conditions and lines tested, there was very close linear association between the mean technician count and Omnicon colony counts of cell lines throughout the entire range of colony sizes studied.

DISCUSSION

These systematic studies provide excellent evidence that the Omnicon image analysis system is a sensitive, specific, precise and accurate means to count tumor colonies in soft agar. Based on these data, we believe it is a highly desirable alternative to manual counting by technicians. Our results markedly extend the initial report by Kressner et al. (16) and the recent paper by Herman et al. (18) who conducted an in-use evaluation of this colony counter. We found the colony counter to be as accurate as technician counts when either gross colony counts or percent survival after drug exposure was compared. In our experience, most technicians have substantial difficulty doing colony counts for more than 3-4 hours per day because of fatigue. The machine's ability to perform colony counts automatically for as long a period of time as desired, as well as its intrinsically greater speed (averaging 1 minute per plate as compared to 10 minutes per plate by the technician), represent two of its highly desirable characteristics. In the current studies, the instrument proved accurate in the assessment of surviving numbers of colonies after exposure to high or low concentrations of a variety of anticancer drugs. The studies of accuracy clearly established that within the count range of up to 1000 colonies per plate, colonies from both fresh tumors and tumor cell lines could be reliably counted with automated methods. When significantly larger numbers of colonies are present in the petri dishes (e.g., more than 2000 per plate), the ability of the automated counter falls off substantially, and for that reason such high count samples were not included within the specimens considered evaluable for this analysis.

Based on these studies, we believe that the automated colony counter provides a useful and practical alternative to manual counting of tumors by technicians. It likely will have a wide variety of preclinical and clinical applications in oncology

ACKNOWLEDGEMENTS

These studies were supported in part by grants CA-21839, CA-17094 and CA-27502 from the National Cancer Institute. The authors thank Bausch and Lomb Corporation, Rochester, New

York, for providing the automated colony counter for evaluation; Barbara Soehnlen, Rebecca Owens, Judi Christensen, Marti McFadden, Chris Hayes, Judie Persaud and Julie Buckmeier for technical assistance.

REFERENCES

1. Hamburger AW, Salmon SE: Primary bioassay of human tumor stem cells. Science 197:461-463, 1977.

2. Salmon SE, Hamburger AW, Soehnlen B, Durie BGM, Alberts DS, Moon TE: Quantitation of differential sensitivity of human tumor stem cells to anticancer drugs. New Engl J Med 298:1321-1327, 1978.

3. Salmon SE, Alberts DS, Meyskens FL Jr, Durie BGM, Jones SE, Soehnlen B, Young L, Chen HSG, Moon TE: Clinical correlations of in vitro drug sensitivity. In: Cloning of Human Tumor Stem Cells, Salmon SE (ed), New York, Alan Liss, 1980, p 223-245.

4. Von Hoff DD, Casper J, Bradley E, Sandbach J, Jones D, Makuch R: Association between human tumor colony forming assay results and response of an individual patient's tumor to chemotherapy. Amer J Med 70:1027-1032, 1981.

5. Von Hoff DD, Clark GM, Stogdill BJ, Sarosdy MF, O'Brien MT, Casper JT, Mattox DE, et al.: Prospective clinical trial of a human tumor cloning system. Cancer Res 43: 1926-1931, 1983.

6. Alberts DS, Salmon SE, Chen HSG, Moon TE, Young L, Surwit EA: Pharmacologic studies of anticancer drugs using the human tumor stem cell assay. Cancer Chemother and Pharmacol 6:253-264, 1981.

7. Meyskens FL Jr, Moon, TE, Dana B, Gilmartin E, Casey WJ, et al.: Quantitation of drug sensitivity by human metastatic melanoma colony forming units. Br J Cancer 44: 787-797, 1981.

8. Carney DN, Broder L, Edelstein M, Gazdar AF, Hansen M, et al.: Experimental studies of the biology of human small cell lung cancer. Cancer Treat Rep 67(1):21-26, 1983.

9. Ozols RF, Young RC, Speyer JL, Sugarbaker PH, Greene R, Jenkins J, Myers CE: Phase I and pharmacological studies of adriamycin administered intraperitoneally to patients with ovarian cancer. Cancer Res 42:4265-4269, 1982.

10. Mann BD, Kern DH, Giuliano AE, Burk MW, Campbell MA, Kaiser LR, Morton DL: Clinical correlations with drug sensitivities in the clonogenic assay. Archives of Surgery 117:33-36, 1982.

11. Salmon SE: Applications of the human tumor stem cell assay to new drug evaluation and screening. In: Cloning of Human Tumor Stem Cells, Salmon SE (ed), New York, Alan Liss, 1980, p 291-312.

12. Shoemaker RH, Wolpert-DeFilippes MK, Makuch RW, Venditti JM: Use of the human tumor clonogenic assay for new drug screening. In: Proceedings, Amer Assoc Cancer Res, San Diego, 1980, 24:311, abstract #1231.

13. Salmon SE, Meyskens FL Jr, Alberts DS, Soehnlen B, Young L: New drugs in ovarian cancer and malignant melanoma: in vitro phase II screening with the human tumor stem cell assay. Cancer Treat Rep 65(1-2):1-12, 1981.

14. Herman CJ, Pelgrim O, Kirkels WJ, Debruyne FMJ, Vooys GP: "Viable" tumor cells in post-therapy biopsies: a potential application of human tumor clonogenic cell culture. Arch Pathol Lab Med 107:81, 1981.

15. Kirkels WJ, Pelgrim O, Debruyne FMJ, Vooys GP, Herman CJ: Soft agar culture of human transitional cell carcinoma colonies from urine. Amer J Clin Pathol 78:690, 1982.

16. Kressner BE, Morton RRA, Martens AE, Salmon SE, Von Hoff DD, Soehnlen B: Use of an image analysis system to count colonies in stem cell assays of human tumors. In: Cloning of Human Tumor Stem Cells, Salmon SE (ed), New York, Alan Liss, 1980, p 179-193.

17. Salmon SE, Young L, Lebowitz J, Thomson S, Einsphar J, et al.: Evaluation of an automated image analysis system for counting human tumor colonies. Int J Cell Cloning, 2:142-159, 1984.

18. Herman CJ, Pelgrim O, Kirkels WJ, Verheigen R, Debruyne FMJ, Kenemans P, Vooijs GP: In-use evaluation of the Omnicon automated tumor colony counter. Cytometry 3(6): 439-442, 1983.

HETEROGENEITY OF CHEMOSENSITIVITY RESPONSE OF HUMAN TUMORS

D.H. KERN

Surgical Service, VA Medical Center
Sepulveda, CA 91343

N. TANIGAWA

Second Department of Surgery
Fukui Medical College, Fukui, Japan

C.A. BERTELSEN

Division of Surgical Oncology
John Wayne Clinic

V.K. SONDAK

Division of Surgical Oncology
John Wayne Clinic

D.L. MORTON

Division of Surgical Oncology
John Wayne Clinic
Jonsson Comprehensive Cancer Center
UCLA School of Medicine
Los Angeles, CA 90024

Supported by VA Medical Research Service and by Grant CA 29605 of the National Cancer Institute.

HUMAN TUMOR CLONING
ISBN 0-8089-1671-8

INTRODUCTION

Tumor cell heterogeneity poses significant problems for the clinical use of chemosensitivity assays. The question arises as to whether a single biopsy specimen is representative of a patient's disease. Tumors are known to consist of heterogenous subpopulations of cells with varying biologic phenotypes. Recent analytic techniques have demonstrated that cells from individual tumors can vary in many biologic properties, such as growth rates, (Schabel, 1975; Citone and Fidler, 1980) karyotypes, (Siracky, 1979; Vindeln, et al., 1980) metastatic potential, (Fidler and Kripka, 1977) surface antigens, (Pimm and Baldwin, 1977; Kerbel, 1979) and response to irradiation, (Leith, et al., 1982). A major focus of experimental work on tumor heterogeneity has been directed to response of tumors to chemotherapeutic agents. Varying sensitivities to drugs have been observed among subpopulations of cells within a single tumor, (Heppner, et al., 1978; Dexter et al., 1981), and among different metastases from the same primary, (Shari and Wolf, 1977). The experimental evidence, however, has been derived almost exclusively from animal tumor models or from human cell lines used after many passages.

We focused our studies of tumor heterogeneity on human tumors assayed for chemosensitivity to anticancer agents.

Chemosensitivity assays measure the inhibition of a proliferating cell population by anticancer drugs. Predictions of chemosensitivity have been obtained for patients with a variety of tumor types. The accuracy for these predictions of sensitivity has been reported to vary between 57% (Bertelsen, et al., 1984; Von Hoff, et al, 1983) and 82% (Mann, et al., 1982). Prediction of resistance has been somewhat better, varying between 85% (Von Hoff, et al., 1983) and 99% (Alberts, et al., 1980). In this study, we sought evidence of tumor heterogeneity by comparing chemosensitivity responses between: 1) different portions of a single tumor, 2) a primary and a metastatic biopsy taken from a patient on the same day, and 3) different metastases from a patient taken either on the same day (synchronous) or at an interval of several months (metachronous).

MATERIALS AND METHODS

Preparation of tumors

Connective and adipose tissues and necrotic areas were removed from fresh surgical specimens. The tumor was minced into pieces less than 2 mm in diameter in the presence of CEM (MA Bioproducts, Walkersville, MD) containing 15% heat-inactivated FCS (Flow Laboratories, McLean, VA), 100 U/ml penicillin (Grand Island Biological Company, Grand Island, NY), 100 μg/ml streptomycin (Grand Island) and 50 μg/ml Fungizone (Grand Island).

Enzymatic dissociation of solid tumors

Ten ml of CEM containing 0.03% DNase (500 Kunitz units/ml) and 0.14% collagenase Type I (Sigma Chemical Company, St. Louis, MO) were used for each gram of tissue. The tumor fragments were added to the enzyme medium, and mixtures were stirred for 90 min at 37° in the presence of 6% CO_2. After enzymatic digestion, the free cells were decanted through 12 layers of sterile gauze and centrifuged at 200 x g for 10 min. The supernatant was removed and the cells were centrifuged again. The pellet was resuspended in 10 ml of supplemented CEM (CEM with 15% heat-inactivated FCS and 100 μg/ml streptomycin and 100 U/ml penicillin). Cell yield and viability were determined by trypan blue exclusion.

Preparation of agar plates

An underlayer of agar was prepared as follows: 16.5 ml of supplemented CEM was warmed to 50° and added to 3.5 ml of 3% agar, also prewarmed to 50°. The solution was rapidly mixed and 1 ml of the solution at 50° was dispersed to each 35 x 10 mm well of a 6-well plate (Linbro 76-247-05, Flow Laboratories). The agar was allowed to set at room temperature. Next, 1.2×10^6 viable tumor cells were suspended in 1.0 ml of supplemented CEM at 37°. After 2.0 ml of 0.6% agar in supplemented CEM was added, the cells were rapidly and evenly suspended. One ml of cell suspension was added to each of three wells over the feeder layer, yielding 4×10^5 viable tumor cells per well. The plates then were placed in a humidified incubator at 37° in the presence of 6% CO_2.

Colony counting

The cells in the agar plates were examined weekly under

an inverted microscope at 4X and 10X. The number of colonies (>20 cells) usually reached a maximum within 2-4 weeks. Colony counts were made with an Omnicon FAS II (Bausch and Lomb, Rochester, NY). Minimum colony size was 60μ.

Inhibition of colony formation by anticancer drugs

Drug solutions were prepared from standard intravenous formulations in continuous exposure. Concentrations used are shown in Table I. Drugs were prepared as 60 X stock solutions and were aliquoted and stored at -70°. Drugs were added to the tumor cell suspension (the upper agar layer) immediately prior to plating; 100 μl of drug were added for each 1.2×10^6 cells. For no-treatment controls, 100 μl of CEM was added to the tumor cell suspension. Drug effects were calculated as percent inhibition of colony formation relative to the controls (no drug). Control wells with an average count of less than 30 colonies were considered to be "no growth" and were not evaluated. All drug tests were performed in triplicate. Six control (no drug) wells were included in each experiment.

TABLE I

CONCENTRATIONS OF ANTICANCER AGENTS USED FOR CHEMOSENSITIVITY TESTING

Agent	Concentration
Adriamycin	0.4 μg/ml
BCNU	2.0
Melphalan	1.0
Mitomycin C	3.0
Methotrexate	4.0
Vincristine	0.5
5-Flourouracil	10.0
DTIC	10.0
cis-Platinum	2.0
Vinblastine	5.0

RESULTS

Comparison of different sections of a tumor

Two different sections from the same tumor were assayed for the effects of anticancer drugs on colony formation. Tumor types were: 5 renal cell carcinomas, 2 germ cell tumors, 1 liposarcoma and 1 neuroblastoma. In these 9 paired samples, no significant differences in cell yield, tumor cell viability or cloning efficiency were observed. A total of 28 paired chemosensitivity tests were performed. Sensitivity to a drug was defined as a 50% or greater inhibition of colony formation. In 14/28 (50%) drug tests, both parts of a tumor were either sensitive or resistant to the drug. However, in 14 instances, one part of the tumor was sensitive to a drug, while the other part from the same tumor was resistant. This 50% discordance rate indicated that chemosensitivity results obtained from a biopsied section of a tumor was not necessarily representative of the whole tumor.

Comparison of primary tumors with its metastasis

Primary tumors and one of its metastases were obtained simultaneously during the operative procedure from 14 patients (2 breast, 6 stomach, 5 colorectal and 1 melanoma). Cloning efficiences varied significantly between primary and metastatic tumors from 10 patients. A total of 47 paired drug tests were performed, and agreement was recorded in 32 instances. However, discordance was noted in 15/47 (32%) tests. These data suggest that chemosensitivity patterns of metastatic tumors may not reflect those of tumors of primary origin.

Comparison of different metastases

A total of 49 paired assays were performed on two separate metastatic tumors from the same patient; 22 were from patients with synchronous biopsies, and 27 were from patients with metachronous biopsies. The mean interval for metachronous assays was 6 months (minimum 1 month). Tumor types were: 17 melanoma, 8 sarcoma, 7 ovarian, 6 breast, 5 colon and 6 miscellaneous. A total of 272 drug comparisons were possible, 151 in synchronous and 121 in metachronous assays.An observed difference in chemosensitivity to a given drug was noted if the drug had greater than 50% inhibition of colony formation against one metastasis but less than 50% inhibition against the other. Differences in chemo-

TABLE II

Demonstration of heterogeneity of chemosensitivity in human tumors. A tumor was considered sensitive to a particular drug if there was 50% or greater inhibition colony formation.

Paired Comparison	Number of Patients	Discordant Drug Tests	(%)
Two sections, same primary	9	14/28	50
Primary vs. metastasis	14	15/47	32
Different metastases:			
Synchronous	22	42/151	28
Metachronous	27	49/121	40

sensitivity among drug comparisons were evident in 42/151 (28%) of synchronous comparisons and 49/121 (40%) of metachronous comparisons, overall 91/272 (33%). Differences in drug sensitivity were more likely to occur in metachronous assays (40%) than in synchronous assays (28%) ($p<0.05$). Discordance in chemosensitivities between two metastases indicated that considerable heterogeneity exists between multiple metastases from the same patient.

In Table II is a summary of these results. Discordance in chemosensitivity tests was observed for 120/347 (34.5%) paired tests.

Determination of inherent assay variability

Variability inherent within the assay system was measured by two different methods: 1) chemosensitivity assays were repeated using cryopreserved cells, and 2) drugs were tested 2 or 3 times within a single assay. No differences in these two approaches was evident. For 385 replicate drug tests, there was discordance in 38 (1.9%)instances. Thus, assay variability may have accounted for up to 10% of the observed discordance in chemosensitivity tests between tumors from the same patient.

DISCUSSION

These data demonstrate the presence of considerable heterogeneity of response to chemotherapy among different tumors from the same patient, and even within the same tumor.

Several theories have been proposed to explain the origin of tumor cell heterogeneity. One of the most widely held theories has been reviewed and expounded upon by Nowell (Nowell, 1965). He argues that tumors arise by neoplastic transformation of a single cell and that early tumor growth is the result of clonal expansion of that cell. Because of the greater genetic instability of tumor cells, mutants are frequently produced. These mutant cells, which differ both genetically and biologically from the original clone, can survive if selection pressures favor their proliferation over the original clonal cells. Such phenotypic diversity is magnified as the malignancy evolves. Thus subpopulations of cells with diverse biologic properties may exist by the time the tumor becomes clinically manifest. Because metastic deposits undergo evolutionary processes similar to the primary tumor, it is not surprising that even simultaneous biopsies of different tumors from the same patient display varying biologic properties such as drug sensitivity. Tumor evolution is a dynamic process, and it is possible that the chemosensitivity of a tumor could change with time.

These differences in drug sensitivity among multiple tumors or within a single tumor have important implications for the clinical use of chemosensitivity assays. First is the problem of representation. Assays are often performed on readily accessible tissues (subcutaneous metastases, soft-tissue lesions, lymph nodes, effusions), but chemosensitivity data derived from them may not reflect the chemosensitivity of more inaccessible, visceral metastases. Second, the data obtained from primary tumors may not accurately reflect the chemosensitivities of metastases should they develop. Furthermore, data derived on a metastasis at one point in a patient's course may not reflect the chemosensitivities of a similar metastasis at a later time. Clinicians must appreciate the limitations that tumor heterogeneity imposes upon application of chemosensitivity assays to drug selection.

However, the problem of tumor heterogeneity does not invalidate use of such assays. Our findings indicate that in over two-thirds of patients, chemosensitivity data derived from one anatomic source will reflect the responsiveness of a cell population at a different site. The existence of tumor heterogeneity may account for the less than perfect accuracy rate for prediction of sensitivity observed for most chemosensitivity tests.

REFERENCES

Alberts, D.S. et al., In vitro clonogenic assay for predicting response of ovarian cancer to chemotherapy, Lancet, 1980, 2, 340-342.

Bertelsen, C.A., Sondak, V.K., Mann, B.D., Korn, E.L., and Kern, D.H., Chemosensitivity testing of human solid tumors, Cancer, 1984, in press.

Citone, M.A., and Fidler, I.J., Correlation of patterns of anchorage-independent growth with in vivo behavior of cells from a murine fibrosarcoma. Proc. Natl. Acad. Sci. USA, 1980, 77, 1039-1043.

Dexter, D.L., Spremulli, E.N., Fligiel, Z., et al, Heterogeneity of cancer cells from a single human colon carcinoma. Am. J. Med., 1981, 71, 949-956.

Fidler, I.J., and Kripka, M.L., Metastasis results from pre-existing variant cells within a malignant tumor. Science, 1977, 197, 893-895.

Heppner, G.H., Dexter, D.L., DeNucci, T., Miller, F.R., and Calabresi, P., Heterogeneity in drug sensitivity among tumor cell subpopulations of a single mammary tumor. Cancer Res., 1978, 35, 3758-3763.

Kerbel, R.S., Implications of immunological heterogeneity of tumors. Nature, 1979, 280, 358-360.

Leith, J.T., Dexter, D.L., DeWyngaret, J.K., et al., differential responses to X-irradiation of subpopulations of two heterogeneous human carcinomas in vitro. Cancer Res., 1982, 42, 2556-2561.

Mann, B.D., et al., Clinical correlations with drug sensitivities in the clonogenic assay. Arch. Surg., 1982, 117, 33-36.

Nowell, P.C., The clonal evolution of tumor cell populations. Science, 1976, 194, 23-28.

Pimm, M.V. and Baldwin, R.W., Antigenic differences between primary methylcholanthrene-induced rat sarcomas and post surgical recurrences. Int. J. Cancer, 1977, 20, 37-43.

Schabel, F.M., Concepts for systemic treatment of micrometastases. Cancer, 1974, 35, 15-24.

Shari, J., and Wolf, W., A model for prediction of chemotherapy response to 5-Fluorouracil based on the differential distribution 5-(18F) Fluorouracil in sensitive vs. resistant lymphocytic leukemia in mice. Cancer Res., 1977, 37, 2306-2308.

Siracky, J., An approach to the problem of heterogeneity of human tumor-cell populations. Br. J. Cancer, 1979, 39, 540-577.

Vindeln, L.L., Hansen, H.H., Christensen, I.J., et al., Clonal heterogeneity of small-cell anaplastic carcinoma of the lung demonstrated by flow-cytometric DNA analyses. Cancer Res., 1980, 40, 4295-4300.

Von Hoff, D.D., et al., Prospective clinical trial of a human tumor cloning system. Cancer Res., 1983, 43, 1926-1931.

Drug Sensitivity of Primary Versus Metastasis

DANIEL D. VON HOFF, MD.

Associate Professor
Department of Medicine

GARY M. CLARK, PH.D.

Associate Professor
Department of Medicine
University of Texas Health Science Center at San Antonio
San Antonio, Texas

INTRODUCTION

The human tumor cloning system introduced by Hamburger and Salmon[6, 7, 14] is an *in vitro* technique which allows direct culturing of patients' tumors. The cloning system has been utilized to attempt to predict for response or lack of response of an individual patient's

Acknowledgment:
Supported by Grant CH162B from the American Cancer Society. We thank Clair Waggoner for her assistance in the preparation of this manuscript.

HUMAN TUMOR CLONING
ISBN 0-8089-1671-8

tumor to a particular anticancer agent. Retrospective as well as prospective clinical trials of the system for that purpose have indicated the system can correctly predict for response of the patient's tumor 64 to 98% of the time and can correctly predict for lack of response of the patient's tumor 84 to 98% of the time.[12, 13, 17, 18] One major question which has arisen with the assay is that if a drug sensitivity profile is obtained with the cloning assay on the primary tumor, will that sensitivity profile be similar to the sensitivity profile of the metastasis or metastases? The purpose of the present study was to determine if drug sensitivity profiles were the same or different in a primary tumor and one of its metastases.

MATERIALS AND METHODS

Patients

A total of 46 patients were entered in this study. To be eligible for the study patients had to be undergoing a diagnostic or therapeutic procedure for their malignancy. In addition, they had to have simultaneous biopsies or removals of both their primary tumor and a metastasis. Overall, 8 of the 46 had had prior chemotherapy.

For a point of reference, in addition to these 46 patients, an additional 25 patients had their tumors (8 fluids, 5 solids) sampled, a single cell suspension prepared and split and processed separately. This group provided a benchmark for assessment of variability in colony formation and drug sensitivity testing which is due to variations in culture technique.

Collection of Tumor Cells

After obtaining informed consent in accordance with federal and institutional guidelines, malignant effusions, ascites, bone marrow-containing tumor cells, and solid tumor specimens were collected from 46 patients undergoing procedures done as part of a diagnostic work-up or as part of treatment for their disease. No major surgical procedures were performed solely to obtain specimens for drug sensitivity testing. Solid tumors or lymph nodes were minced into 2 mm to 5 mm fragments in the operating room and were immediately placed in McCoy's Medium 5A plus 10% heat-inactivated newborn calf serum plus 1% penicillin and streptomycin (all from Grand Island Biological Co., Grand Island, N.Y.) under aseptic conditions. Within 4 hours, these solid tumors were mechanically dissociated with scissors, forced through a No. 50 stainless steel mesh, through 20-, 22-, and 25-gauge needles, and through a nylon mesh; and then washed with Hanks' balanced salt solution as described previously.[12, 13, 17, 18] Ascitic, pleural, pericardial fluids, and bone marrows were obtained by standard techniques. The fluid or marrow was placed in sterile containers containing 100 units of preservative-free heparin (Grand Island Biological Co.) per ml of malignant fluid or marrow. After centrifugation at 150 x g for 10 minutes, the cells were harvested and washed twice in Hanks' balanced salt solution plus 10% heat-inactivated fetal calf serum. The viability of cell suspensions was determined in a hemocytometer with trypan blue. Viability of cells derived from solid tumors ranged from 0 to 100% (median, 26%), while viability of cells from effusions and bone marrows ranged from 0 to 100% (median, 90%).

Drug Sensitivity Studies

Stock solutions of i.v. formulation of both standard and investigational agents were prepared in sterile buffered 0.9% NaCl solution, water, or dimethyl sulfoxide, and stored at -70°C in aliquots sufficient for individual assays. Subsequent dilutions for incubation were made in 0.9% NaCl solution or water. The tumor cells were exposed to standard anticancer agents at the following final concentrations (μg/ml): actinomycin D, 0.01; bis(chloroethyl)nitrourea, 0.10; cis-platinum, 0.20; doxorubicin, 0.04; 5-fluorouracil, 6.00; methotrexate, 0.30, melphalan, 0.30; mitomycin C, 0.10; vinblastine, 0.05; and vincristine, 0.01. The final concentration of investigational drugs (μg/ml) included: bisantrene, 0.50; etiopside, 3.0; and mitoxantrone, 0.05. This single concentration of each drug corresponded to approximately 1/10th of the peak plasma concentrations for each drug in humans.[1]

Tumor cell suspensions were transferred to tubes and adjusted to a final concentration of 1.5×10^6 cells/ml in the presence of the appropriate drug dilution or control medium. In an attempt to perform more drug studies per tumor specimen, only the single concentration of drug listed above was usually used.

Cells were incubated with and without the drug for 1 hour at 37°C in Hanks' balanced salt solution plus 10% heat-inactivated fetal calf serum. They were then centrifuged at 150 x g for 10 minutes, washed twice with the Hanks' balanced salt solution, and cultured as detailed below.

Assay for Tumor Colony Forming Units (TCFU's)

The culture system used in this study has been extensively described elsewhere[1, 2, 6, 7, 14, 18] In brief, cells to be tested were suspended in 0.3% agar in enriched Connaught Medical Research Laboratories Medium 1066 (Grand Island Biological Co.) supplemented with 15% horse serum, penicillin (100 units/ml), streptomycin (2 mg/ml), glutamine (2 mM), $CaCl_2$ (4 mM), and insulin (3 units/ml). Prior to plating, asparagine (0.6mg/ml), DEAE-dextran (0.5 mg/ml; Pharmacia Fine Chemicals, Inc., Piscataway, N.J.), and freshly prepared 2-mercaptoethanol (final concentration, 50 mM) were added to the cells. One ml of the resultant mixture was pipetted onto 1 ml feeder layers in 35 mm plastic Petri dishes (Falcon Plastics). The final concentration of cells in each culture was 5×10^5 cells in 1 ml of agar medium. The feeder layers used in this study consisted of McCoy's Medium 5A plus 15% heat-inactivated fetal calf serum and a variety of nutrients described by Pike and Robinson.[11] Immediately before use, 10 ml of 3% tryptic soy broth (Grand Island Biological Co.), 0.6 ml of asparagine, and 0.3 ml of DEAE-dextran were added to 40 ml of the enriched underlayer medium. Agar (final concentration, 0.5%) was added to the enriched medium, and underlayers were poured into 35 mm Petri dishes. After preparation of both bottom and top layers, the plates were examined under an inverted microscope and on a feature analysis system (FAS II; Bausch and Lomb, Inc.) to ensure the presence of a good single-cell suspension. To be considered evaluable for subsequent growth and drug sensitivity information, the plates had to have excellent single-cell suspensions which were defined as ≤ 10 features (diameter, ≥ 40 μm) on each of 3 control plates. The plates were then incubated at 37°C in a 7% CO_2 humidified atmosphere.

Colonies (>50 cells) usually appeared by day 14 of culture, and the number of colonies on control and drug-treated plates was determined by counting the colonies on an inverted stage microscope at x 30 magnification or by using the feature analysis system. At least 20 tumor colonies/control plate were required for a drug experiment to be considered evaluable for measurement of drug effect.

Data Analysis and Statistical Considerations

The results of the _in vitro_ cloning assay were expressed as percentage of survival of TCFU's for a particular drug relative to its control. This quantity was calculated as the ratio between the mean number of colonies surviving on triplicate plates (treated with a single clinically achievable drug concentration) and the mean number of colonies growing on triplicate control plates. Correlation coefficients between the pairs were performed utilizing the Spearman rank order method.

RESULTS

A total of 29 patients of the 46 (63%) attempted had both their primary tumor and a metastasis successfully cultured simultaneously ($\geq$20 colonies in control plates). In addition, 13 of the 25 patients' tumors (52%), which were collected and divided to obtain a benchmark for assessment of variability in the assay, formed $\geq$20 colonies in control plates of both members of the pair. Table 1 details the num-

TABLE 1
Number of Patients in Each Category by Tumor Type

	Category	
Tumor Type	Specimen Processed Twice	Primary Versus Metastasis
Ovary	5	12
Breast	3	7
Genitourinary (bladder, kidney adrenal, prostate)	1	5
Gastrointestinal (colon, pancreas, liver)	1	1
Lung	1	1
Miscellaneous (endometrium, melanoma, unknown origin, basal cell)	2	3
Totals	13	29

ber of patients in each category by tumor type. Ovarian and breast cancers were the most common tumor types represented. The number of paired drug tests performed on the same specimen processed twice was 21 while the number of paired drug tests performed on the primaries and their metastases was 87.

Table 2 summarizes the correlation between the number of colonies noted in the paired samples with the same specimen processed twice and primary versus metastasis. As noted in that table, there was excellent agreement in colony growth in paired samples when the specimen is split (r = 0.99). The correlation coefficient is much lower for the primary versus the metastasis (0.37). This was true whether the type of metastasis was a solid or an effusion (ascites or pleural effusion).

The correlation in drug sensitivity results between the same specimen processed twice and between primary and metastasis were examined next. Table 3 details the correlation coefficients for percent colonies surviving the various categories. As a reference point, for the same specimen processed twice, the correlation coefficient was 0.83 (i.e., there was very good agreement in the pairs). For the primary versus metastatic pairs, the coefficient was 0.03 (i.e., there was very poor agreement within the pairs). Thus, there was clearly distinct heterogeneity in drug sensitivity results between primary tumors and their metastases.

TABLE 2
Summary of Colony Formation Correlations in Paired Samples

Category	No. of Specimens	Corr. Coeff. for No. of Colonies
Same specimen-processed twice	13	0.99
Primary-metastasis	29	0.37
(metastasis a solid)	16	0.35
(metastasis an effusion)	13	0.38

TABLE 3
Summary of Correlations for Percent Colonies Surviving in Paired Samples

Category	No. of Patients	No. of Paired Drug/Tests	Corr. Coeff. for % Survival
Same specimen processed twice	13	21	0.8300
Primary-metastasis	29	87	0.0301

The category of additional interest for a more in depth study was the primary versus metastasis group. The first question was if the primary tumors were more sensitive overall than their metastatic counterparts. The median percent survival of tumor colony forming units for the primaries was 90% (range: 31 to 208) while the median percent survival of tumor colony forming unit for the metastases was 94% (range: 30 to 708). These numbers do indicate that for large numbers of drug tests (87 overall), the primaries were overall more sensitive than the metastases (p = 0.03 paired T test). However, the ranges for each group were quite large so the total significance of this finding is unknown.

Of additional interest was to determine if the category of antineoplastics made a difference in the correlation between sensitivity (or resistance) of the primary tumor and its metastasis. As noted in Table 4, the best correlation in percent colonies surviving for primary versus metastasis was for the intercalating agents and the alkylating agents. Thus, if sensitivity (or resistance) to intercalating agents or alkylating agents is noted in the culture of the primary tumor, it is likely that sensitivity or resistance will also be the same for the metastatic deposit. This was not the case for the vinca alkaloids and the antimetabolites where there are major discordancies in drug sensitivity (or resistance) results for primary tumors versus their metastases.

TABLE 4

Correlation of Percent Survival for Primary Versus Metastases by Category of Antineoplastic

Category of Antineoplastics	n	Correlation Coefficient for % Survival in Primary vs. Metastasis
Intercalating agents	21	0.66
Alkylating agents	26	0.52
Spindle poisons	7	0.21
Antimetabolites	21	-0.22

DISCUSSION

This study was designed to compare drug sensitivity results obtained from a primary tumor with drug sensitivity results obtained from a metastasis in the same patient. For purposes of a benchmark for comparison, for an additional number of specimens the same tumor was divided and processed twice. Although 71 patients' tumors were cultured (46 primary metastasis plus 25 specimens processed twice, only 42 (59%) formed enough colonies in both members of the pairs to be evaluable for drug sensitivity studies. Thus, the result of this study represents selected groups of tumors which formed colonies in the assay.

This study demonstrated that when the same specimen is processed twice, there is excellent agreement in the number of colonies that are formed with each processing (r = 0.99). Therefore, the assay is quite reproducible in terms of colony growth and hence culturing techniques to produce that growth. When the primary versus metastasis pairs are examined, it is clear that the number of colonies formed per number of cells plated (cloning efficiencies) are quite different (r = 0.37). This difference was also demonstrated by Schlag and Schreml in their primary-metastasis study.[15] This difference in cloning efficiencies has also been shown for primary tumor versus ascites by Epstein and Marcus[4] and for different metastatic sites by Meyskens et al.[10] Since the tumor must be made into a single cell suspension for the drug sensitivity testing, there may be different populations of cells (tumor and non tumor) in the primary versus metastatic specimen. These different populations could affect cell growth, as noted by Buick and colleagues.[3]

When correlations in drug sensitivity results (percent survival) between primaries and their metastases were examined, the correlation coefficient was only 0.03. This compared to a correlation coefficient of 0.83 for drug sensitivity results when the same specimen was processed twice. Based on these findings, it is clear that the drug sensitivity profiles are different for primaries and their metastases. Schlag and Schreml have also reported on correlation for drug sensitivities between a patient's primary tumor and its metastasis.[15] In 10 paired samples, in which drug sensitivity testing was performed, they noted no satisfactory correlation in sensitivity or resistance between primary tumor and metastases. Therefore, for primary versus metastasis, our results are in agreement with those of Schlag and Schreml.

The precise reasons for the heterogeneity between sensitivity results for primary versus metastasis are unknown. However, since tumor heterogeneity has been adequately documented for animal tumor and human tumor cell lines, the above results come as no surprise.[2, 8, 9, 16] This heterogeneity is somewhat discouraging since drug sensitivity results of the primary tumor can not be used to guide treatment of metastatic deposits. Therefore, it was imperative to look for something the primary and the metastatsis might have in common. With further analysis we have found that there is a good correlation in sensitivity results for primary tumors and their metastases with the alkylating or the intercalating agents. However, the correlations between sensititity results for primary tumors and their metastases for the antimetabolites and spindle poisons are very poor.

These results may be of importance in some settings. For example, if drug sensitivity testing is performed on a primary tumor, it would be more likely that the in vitro sensitivity results of the primary would more closely predict for the sensitivity result of the metastasis for both the alkylating and intercalating agents. This would not be the case for the antimetabolites and the spindle poisons. This information might be most useful in the adjuvant situation when the primary tumor is available for drug sensitivity testing and yet we want an antineoplastic agent with activity against both the primary and the microscopic metastases. In this situation, results obtained for alkylating and intercalating agents would be more indicative of the sensitivity of the metastasis(es). This concept that, indeed, the tumor and its metastasis must be more homogenous in terms of sensitivity (or resistance) to certain classes of antineoplastics at least gives some way to approach the tumor with some common factors rather than on the basis of total heterogeneity. In fact, this common sensitivity profile for primary and metastasis may be one of the reasons adjuvant programs succeed. Speculation on the effectiveness of breast cancer adjuvant regimens (which contain alkylators or intercalators) versus the lack of effectiveness of adjuvant colorectal cancer regimens (which contain the antimetabolite 5FU) could be made. Design of adjuvant regimen with these results in mind should be considered.

The drug sensitivity results obtained for the various classes of antineoplastics for the primary versus the metastasis are also of interest in light of results of animal tumor work performed at the Southern Research Institute. That group has documented that mutation to resistance in mouse tumor cell populations depended on the class of chemotherapeutic agent. Mutation to resistance was highest to the mitotic inhibitors, intermediate to the antimetabolites, and lowest to the highly reactive agents (alkylating - intercalating agents).[5] This order of mutation to resistance is similar to the ranking of correlation of drug sensitivities for primaries versus their metastases. If the animal model is applicable to the human situation, perhaps the mutation in the metastases is slower for the alkylating and intercalating agents than it is for the antimetabolites and the spindle poisons. Only serial studies in human tumors can address that speculation. However, if that is indeed the case, therapeutic strategies could be built around that observation.

Overall then, this study has documented that there is indeed considerable heterogeneity in drug sensitivity results when primary versus metastasis are examined. This heterogeneity is less marked for the primary versus metastasis for the alkylating and intercalating agents than for the antimetabolites and spindle poisons. The significance and usefulness of this finding are unknown.

REFERENCES

1. Alberts DS, Chen HSB: Tabular summary of pharmacokinetic parameters relevant to in vitro drug assay, in Salmon S (ed): Cloning of Human Tumor Stem Cells. New York, Liss, 1980, pp 351-359.
2. Barranco SC, Haenelt BR, Gee EL: Differential sensitivities of five rat hepatoma cell lines to anticancer drugs. Cancer Res 38: 656-660, 1978.
3. Buick RS, Fry SE, Salmon SE: Effect of host cell interactions on clonogenic carcinoma cells in human malignant effusions. Br J Cancer 41: 695-704, 1980.
4. Epstein LB, Marcus SG: Review of experience with interferon and drug sensitivity testing of ovarian carcinoma in semisolid agar

culture. Cancer Chemother Pharmacol 6: 273-277, 1981.

5. Griswold DP, Schabel FM Jr, Corbett TH, Dykes DJ: Concepts for controlling drug resistant tumor cells, in Fidler J, White RT (eds): Design of Models for Testing Cancer Therapeutic Agents. New York, Van Nostrand, Reinhold, 1982, pp 215-224.
6. Hamburger AW, Salmon SE: Primary bioassay of human myeloma stem cells. J Clin Invest 60: 846-854, 1977.
7. Hamburger AW, Salmon SE: Primary bioassay of human tumor stem cells. Science (Wash DC) 197: 461-463, 1977.
8. Hart JR, Fidler IJ: The implications of tumor heterogeneity for studies on the biology and therapy of cancer metastasis. Biochem Biophys Acta 651: 37-50, 1981.
9. Hepper GH, Dexter DL, De Nucci T, et al.: Heterogeneity in drug sensitivity assay tumor cell subpopulations of a single mammary tumor. Cancer Res 38: 3750-3763.
10. Meyskens FL, Soehnlen BJ, Saxe DF, et al.: In vitro clonal assay for human metastatic melanoma cells. Stem Cells 1: 61-72, 1981.
11. Pike B, Robinson W: Human bone marrow colony growth in vitro. J Cell Physiol 76: 77-81, 1970.
12. Salmon SE, Alberts DS, Durie BGM, et al.: Clinical correlations of drug sensitivity in the human tumor stem cell assay. Res Result Cancer Res 74: 300-305, 1980.
13. Salmon SF, Alberts DS, Meyskens FL Jr, et al.: Clinical correlations of in vitro drug sensitivity, in Salmon SF (ed): Cloning of Human Tumor Stem Cells: New York, Liss, 1980, pp 223-245.
14. Salmon SE Hamburger AW, Soehnlen B, et al.: Quantitation of differential sensitivity of human tumor stem cells to anticancer drugs. N Engl J Med 298: 1321-1327, 1978.
15. Schlag P, Schreml W: Heterogeneity in growth and drug sensitivity of primary tumors and metastases in the human tumor colony-forming assay. Cancer Res 42: 4086-4089, 1982.
16. Spremulli EN, Dexter DL: Human tumor cell heterogeneity and

metastasis. J Clin Oncol 1: 496-509, 1983.

17. Von Hoff DD, Casper J, Bradley E, et al.: Association between human tumor colony-forming assay results and response of an individual patient's tumor to chemotherapy. Am J Med 70: 1027-1032, 1981.

18. Von Hoff DD, Clark GM, Stogdell BJ, et al.: Prospective clinical trial of a human tumor cloning system. Cancer Res 43: 1926-1931, 1983.

IN VITRO CULTURING OF TUMOR CELLS ON SOFT AGAR

SVEND AA. ENGELHOLM, MD
MOGENS SPANG-THOMSEN, MD
NILS BRÜNNER, MD
INGE NØHR, RESEARCH ASSOCIATE

UNIVERSITY INSTITUTE OF PATHOLOGICAL ANATOMY
UNIVERSITY OF COPENHAGEN
COPENHAGEN, DENMARK

HENRIK ROED, MD
LARS VINDELØV, MD

THE FINSEN INSTITUTE
COPENHAGEN, DENMARK

Introduction

There has been great interest in recent years in the establishment of stromal cell-free cultures of human cancer in vitro (1, 5, 6). Our work in this field involves the investigation of the biological characteristics of tumor cells and the study of the effect of cytotoxic drugs in vitro (3).

This report describes a method for establishing stromal cell-free cultures in which disaggregated tumor cells are grown on top of a layer of hardened agar. A human heterotransplanted small cell carcinoma of the lung grown in nude mice was employed to illustrate the technique. Flow cytometric DNA analysis was used to check the representativeness of the cells grown in vitro.

The results showed that the tumor cells form colonies on agar with a plating efficiency as high as that obtained in agar. The culture method produced a selective growth of tumor cells and flow cytometric DNA analysis indicated that the cells were representative of the parent tumor. Furthermore, the number of colonies on agar showed a linear relationship to the number of tumor cells plated.

Material and methods

The tumor tissue was obtained from a heterotransplanted

HUMAN TUMOR CLONING
ISBN 0-8089-1671-8

human small cell carcinoma of the lung, subclassified as the intermediate type. The tumor was disaggregated to a single cell suspension by a combined mechanical and enzymatic method using long-term trypsinization at 4°C (4).
The disaggregated cells were suspended in 1.0 ml Eagles's minimal essential medium (MEM), containing 20% foetal calf serum, MEM amino acids, MEM vitamins, L-glutamine, glucose 10% (0.5 ml/100 ml), gentamycin (2 µg/ml), and mycostatin (10 U/ml). The cell suspension was plated in 35 mm plastic Petri dishes on top of a layer 1.0 ml of hardned 0.25% agarose (Difco) containing 26% of the culture medium described above supplemented with 13% fetal calf serum, 2.5% sheep red blood cells, and mercaptoethanol (5×10^{-5} M). In addition, a culture assay for tumor colony forming cells was performed by the double-layer agar technique (7). In each Petri dish, 10^3-10^5 cells were plated. After incubation for 10-20 days at 37°C in a 5% CO_2 humidified atmosphere of air, the dishes were examined for tumor colonies using a dissecting microscope. Colonies were defined as aggregates of at least 50 cells.
The colony forming efficiency was calculated as CFE = no. of colonies/no. of plated tumor cells x 100. The number of tumor cells was derived from the number af cells plated corrected for the fraction of diploid stromal cells determined by flow cytometric DNA analysis (4).
Samples for flow cytometric DNA analysis from the solid tumor were obtained by fine-needle aspiration (12). Cells grown in vitro were harvested by decanting the medium into a test tube. After centrifugation (500 g/5 min), the cells were suspended in citrate buffer, frozen in liquid N_2, and stored at -80°C until analysis (9). Before analysis, the samples were stained with propidium iodide (10). Two internal standards were used to calculate the DNA index (11). The percentage of cells in the cell cycle phases and the fraction of diploid stromal cells were determined by statistical analysis of the DNA distribution (2).
Tumor specimens for histology were processed by conventional histological techniques, and were stained with hematoxylin and eosin.
Cell cultures for transmission electron microscopy were fixed by Karnovsky's fixative. The cells were enrolled in 2% noble agar at 40°C followed by fixation in 1% osmium tetroxyd. Dehydration was performed in ethanol (99%) and 3.3 epoxy-propane, and finally the samples were embedded in Vestopal W.

The chromosome analysis of cells grown in vitro was carried out with conventional Giemsa, Giemsa trypsin, and C-banding staining. The tumorgenicity of the cells grown in vitro was tested by inoculating of 10^6-10^7 cells into the flanks of NC/KH (Kommunehospitalet) nude mice.

Results

The CFE of this human small cell carcinoma of the lung was 3.0 ± 0.4 after the plating of 10^4 tumor cells on agar. This CFE is comparable to the value obtained by the double layer agar technique (1.6 ± 0.5). After the plating of 10^3-10^4 tumor cells on agar, a linear relationship was found between the number of cells plated and the number of colonies (fig. 1).

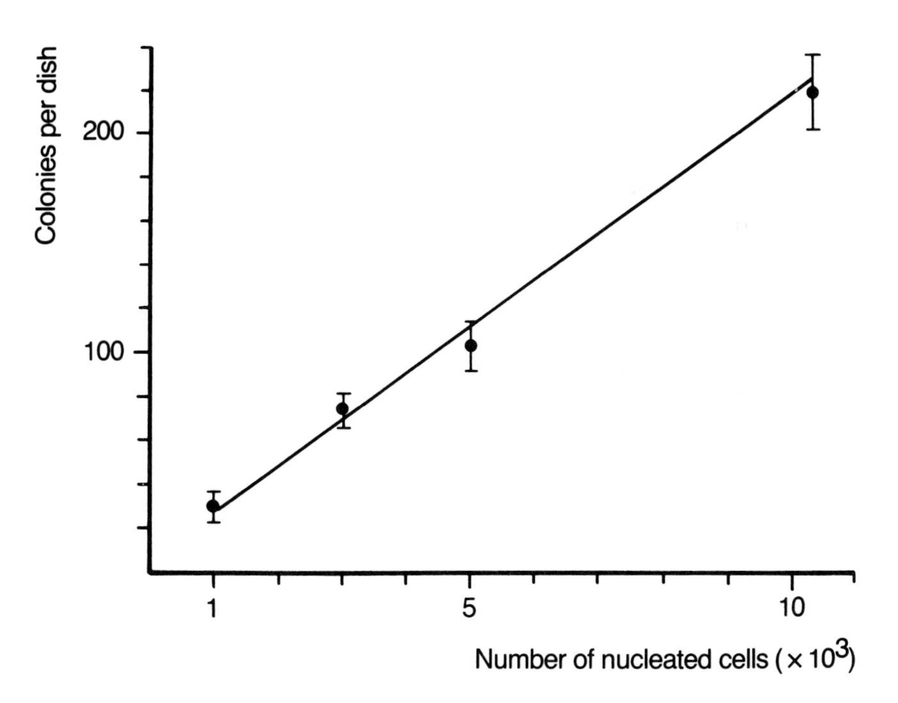

Figure 1 Number of colonies as a function of tumor cells plated. Each point represents the mean of three counted dishes $\pm$ SE. The coefficient of correlation = 0.98.

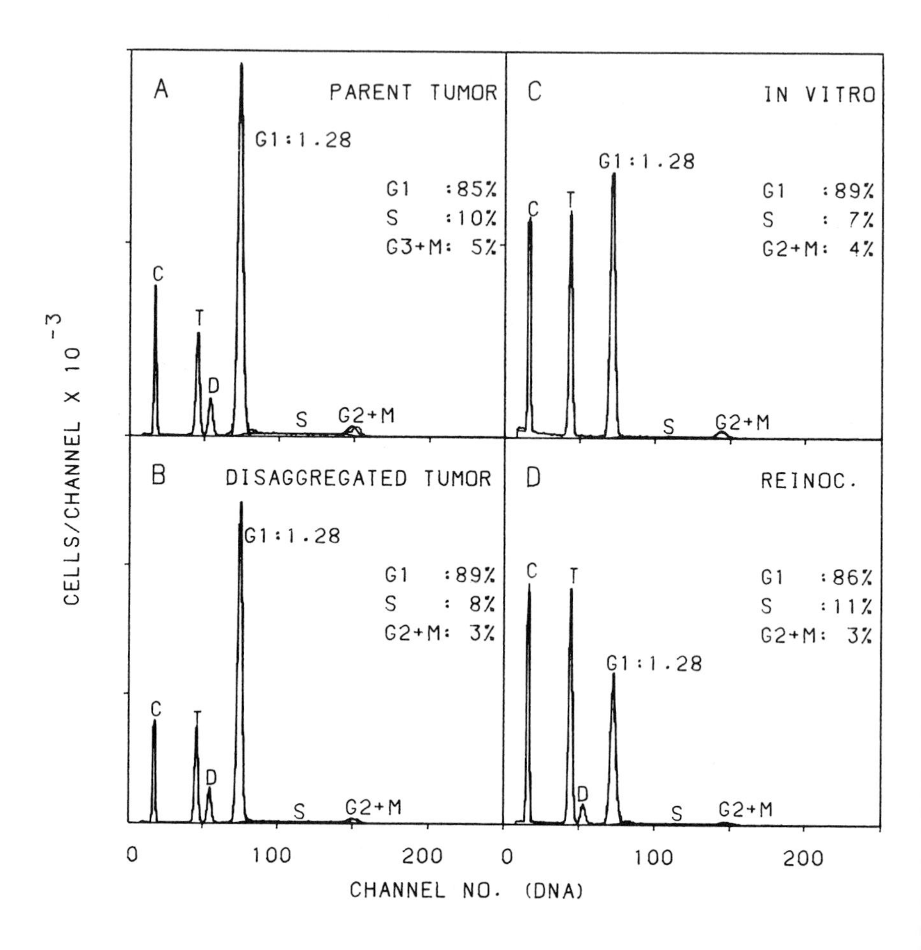

Figure 2 Flow cytometric DNA distribution of a human small cell carcinoma of the lung transplanted into nude mice. The parts of the histograms produced by G_1, S_1 and G_2+ M cells are indicated in the figure. The peaks marked D represents diploid mouse stromal cells, and the peaks C and T are internal standards used to calculate the DNA index. A. The DNA distribution of the parent tumor. Tumor cells were obtained by fine-needle aspiration. B. The DNA distribution of the tumor cells after disaggregation. C. The DNA distribution of the tumor cells after growth on agar in 25 days. The peak D of the stromal cells had disappeared. D. The DNA distribution of the tumor produced by reinoculation of the cells grown in vitro.

The DNA histogram of the tumor used is shown in figure 2 A. As seen in figure 2 B, the disaggregated cells were representative of the parent tumor in DNA index, cell cycle distribution, and fraction of stromal cells.
The culture method was selective for the growth of tumor cells, as demonstrated in figure 2 C. Twenty-five days after plating, the peak representing the stromal cells had disappeared.
Electron microscopy of the cells cultured in vitro demonstrated dense core bodies, and chromosome analysis showed that the tumor cells were of human origin.
Reinoculation of the cells grown in vitro produced tumors in nude mice. Flow cytometric DNA analysis showed a DNA distribution identical to that of the original tumor (fig. 1 D), and histological examination showed that the tumor had retained the appearance of the intermediate type of small cell carcinoma of the lung.

Discussion

This study showed that the in vitro technique described was selective for the growth of tumor cells, and that a linear relationship between the number of tumor cells plated and the number of colonies was found in the range investigated. The CFE obtained was comparable to that of the double layer agar technique.
The cells produced tumors in nude mice after growth on agar, representative of the parent tumor.
The method is applicable for the establishment of continuous stroma-free cell lines in vitro without the use of serum-free medium (1, 5) or other complicated separation techniques (5) because the technique permits the harvesting of viable tumor cells. Tumor cells can be grown in approximately 30 days without adding new medium. Herafter the tumor can be established as a long-term culture without the use of a feeder layer or contiditioned medium (6).
The in vitro method described has some advantages over the double layer technique. The agar dishes can be prepared several days prior to the experiments and kept in a refrigerator. The critical procedure of suspending the tumor cells in agar at 37°C is avoided. The method permits the harvesting of cells for cloning, for continuous growth, etc., by decanting the medium without the use of enzyme
The results presented, obtained on a single heterotransplanted tumor has been confirmed in several other experi-

ments with tumor specimens obtained direct from patients with lung cancer, renal cancer and testicular cancer and on tissue obtained from several heterotransplanted small cell carcinomas of the lung, a malignant lymphoma, a colonic carcinoma, and a malignant melanoma.

The in vitro technique described for the growth of tumor cells on agar permits the harvesting of viable tumor cells after incubation with cytotoxic drugs. The procedure combined with the use of flow cytometric DNA analysis makes it possible to check the identity of the in vitro grown cells and to reveal possible heterogeneity if subpopulations have different DNA index (12). Heterogeneity in sensitivity to chemotherapy is one of the most serious problems in predictive tests. Perhaps more sophisticated predictive tests can be developed by combining of different drug tests based on the growth of tumor cells on agar.

Summary

An in vitro technique for selective growth of tumor cells is described. The method is based on plating of tumor tissue in culture medium supplemented with fetal bovine serum, on top of hardened agar medium containing sheep red blood cells. Under these conditions single tumor cells formed colonies with a plating efficiency as high as with the double agar technique. The procedure enables the harvest of viable cells without the use of enzymes. Flow cytometric DNA analysis demonstrated that the method yielded selective growth of tumor cells. After 3-4 weeks of growth the stromal cells disappeared. Thus, this method is applicable for establishing stroma-free cell lines from human solid tumors. In addition, it is well suited for the detection of chemosensitivity by the clonogenic assay and by measuring the drug-induced cell cycle perturbations using flow cytometric DNA analysis.

This work was supported by grants from the Danish Cancer Society and the Foundation of Købmand Knud Øster-Jørgensen og hustru Maria Øster-Jørgensen for Medical Research.

REFERENCES

1. Carney DN, Bunn PA, Gazdar AF, Pagan JA, Minna JD: Selective growth in serum-free hormone supplemented medium of tumor cells obtained by biopsy from patients with small cell carcinoma of the lung. Proc Natl Acad Sci 78: 5: 3185-3189, 1981.

2. Christensen IJ, Hartmann NR, Keiding N et al.: Statistical analysis of DNA distribution from cell populations with partial synchrony, in Lutz (ed): Pulse Cytometry, Third International Symposium. Ghent, European Press, p 71.
3. Engelholm SA, Spang-Thomsen M, Vindeløv LL: A short-term in vitro test for tumour sensitivity to adriamycin based on flow cytometric DNA analysis. Br J Cancer 47: 497-502, 1983.
4. Engelholm SA, Spang-Thomsen M, Brünner N, Nøhr I, Vindeløv LL: Disaggregation of human solid tumours by combined mechanical and enzymatic methods. Submitted for publication.
5. Fogh J: Human tumor cells in vitro. New York and London, Plenum Press, 1975.
6. Gazdar AF, Carney DN, Russell EK et al.: Establishment of continuous, clonable cultures of small-cell carcinoma of the lung which have amine precursor uptake and decarboxylation cell properties. Cancer Res 40: 3502-3507, 1980.
7. Hamburger AW, Salmon SE: Primary bioassay of human tumor stemm cells. Science 197: 461, 1977.
8. Salmon SE: Cloning of human tumor stem cells. New York, Alan Liss, 1980.
9. Vindeløv LL, Christensen IJ, Keiding N, Spang-Thomsen M, Nissen NI: Long-term storage of samples for flow cytometric DNA analysis. Cytometry 3: 5: 317-322, 1983.
10. Vindeløv LL, Christensen IJ, Nissen NI: A detergent-trypsin method for the preparation of nuclei for flow cytometric DNA analysis. Cytometry 3: 5: 323-327, 1983.
11. Vindeløv LL, Christensen IJ, Nissen NI: Standardization of high-resolution flow cytometric DNA analysis by simultaneous use of chicken and throut red blood cells as internal reference standards. Cytometry 3: 5: 328-331, 1983.
12. Vindeløv LL, Hansen HH, Gersel A et al.: Treatment of small cell carcinoma of the lung monitored by sequential flow cytometric DNA analysis. Cancer Res 42: 2499, 1982.

DRUG APPLICATION TO THE SURFACE OF SOFT-AGAROSE CELL CULTURES

MICHAEL C. ALLEY, PH.D.
Assistant Professor
Department of Pharmacology

MICHAEL M. LIEBER, M.D.
Associate Professor
Department of Urology

Mayo Medical School and Mayo Clinic
Rochester, Minnesota 55905

INTRODUCTION

Evaluation of human tumor cell chemosensitivity using soft-agar colony formation assays generally have utilized one of two standard methods of drug application: 1) "One-hour exposure" followed by washing and resuspension of cells prior to inoculation in culture or 2) "continuous exposure" by incorporation of drug into soft-agar cell suspensions prior to inoculation (Alberts et al, 1980; Soehnlen et al, 1980). The present study was designed to assess possible advantages of an alternate method of drug exposure: that is, drug application to the surfaces of soft-agarose cultures following inoculation. In contrast to other techniques, surface application was viewed as a means to eliminate mechanical manipulation of cells in the presence of drug and to permit all cultures to be set up from a single, "bulk" cell suspension with minimal handling.

MATERIALS AND METHODS

Detailed descriptions of tumor acquisition, digestion, soft-agarose culture, and chemosensitivity testing by the drug incorporation technique have been reported previously (Agrez et al, 1982a,b; Alley et al, 1982; Alley and Lieber, 1984). The same procedures were utilized in the present study with the exception of the following modifications permitted by the culture surface drug application technique: After digestion and washing, cells from each specimen were suspended in "bulk" at a density of 5×10^5 cells/ml; 1 ml

HUMAN TUMOR CLONING
ISBN 0-8089-1671-8

aliquots of cell suspension containing 0.3% Seaplaque agarose were then applied to the base layer of each culture dish with a constant-volume step-syringe (Eppendorf Repeator 4780), the barrel of which was fitted with a large-bore fibrometer tip (Becton Dickinson and Co.). Following agarose gel formation, aliquots (100 μl) of each drug solution (20x final concentration) were applied to surfaces of 3 culture dishes and aliquots of each drug vehicle (water, 0.9% saline, 10% ethanol) were applied to 6 culture dishes. Note that culture dishes set up by the "drug incorporation" and "surface drug application" methodologies contained identical formulation: Base layers (1 ml) contained McCoy's medium supplemented with 9.8% fetal bovine serum, 140 μg sodium pyruvate, 28 μg L-serine, 1.3 μmol L-glutamine, 65 units penicillin, 65 μg streptomycin, 65 μg asparagine, 246 μg DEAE dextran, 0.5% tryptic soy broth, and 0.5% agar; cell layers (1 ml) contained 5×10^5 cells in CMRL 1066 medium supplemented with 10.5% fetal bovine serum, 1.4 units insulin, 0.21 μmol vitamin C, 140 units penicillin, 140 μg streptomycin, 14 μmol L-glutamine, 68 μg asparagine, 35 μmol 2-mercaptoethanol, and 0.3% agarose plus 100 μl drug solution.

For a given specimen culture, chemosensitivity was assessed shortly following exhibition of significant growth in "proliferation control" culture dishes (Alley and Lieber, 1984). Cultures were stained with a metabolizable dye, INT (Alley et al, 1982), and colonies were counted by a computerized image analyzer, the Omnicon Feature Analysis System, Model II (Bausch and Lomb, Inc., Rochester, NY). The evaluable region of each culture dish (35 contiguous fields [each 4.44×3.22 mm^2] equivalent to 500 mm^2 area) was assessed on the basis of a standard colony count program (Kressner et al, 1980). Selective scoring of viable cell groups was achieved in the gray-manual mode with the aid of a scintered glass filter placed between the light source and culture dish. The maximum optical density detection level (lower threshold value) was set to 456; the minimum optical density detection level (upper threshold value) was adjusted to exclude features of non-stained cell groups and debris from analysis (levels ranged from 520-640 depending upon specimen culture opacity). The mean colony count (≥ 60 μ diameter) and standard error of the mean for each group of cultures (6 dishes/control group and 3 dishes/drug-treated group) were computed and tabulated by the analyzer. Colony formation in drug-treated cultures was expressed relative to that in vehicle-treated cultures as percent of control growth.

RESULTS

Preliminary Evaluation of Culture Surface Drug Application

A preliminary assessment of the drug overlay technique was performed with soft-agarose cultures of a human rhabdomyosarcoma continuous cell line (A204). As shown in Table 1, surface application of most clinically useful agents resulted in greater than 70% inhibition of colony formation; in fact, at therapeutically relevant concentrations most agents inhibited colony formation more than 90%. It was not unexpected that cyclophosphamide and procarbazine lacked activity in this *in vitro* system since each of these agents requires metabolic bioactivation (e.g., Lieber et al, 1982; Alley et al, 1984). While melphalan at 0.05 μg/ml was inactive (data not shown), subsequent evaluations revealed that significant inhibition of colony formation required the presence of 1.0 μg/ml or higher concentrations whether melphalan was introduced by culture incorporation or culture surface application. Thus, soft-agarose culture matrix appeared to provide no significant barrier to activity of standard chemotherapeutic agents.

In a subsequent experiment, drug activity following culture incorporation was compared with that following culture surface application for 7 agents in 9 primary human tumor cell cultures (2 colon, 4 kidney, and 3 ovary). Linear regression analysis of paired data is depicted in Figure 1. All entries except 3 (circled) fall within 95% confidence limits of the line, $Y = 0.914\ X + 22.1$, where X represents the percent survival resulting from the drug overlay technique and Y represents the percent survival measured by the drug incorporation technique ($r = 0.780$, $n = 42$, $p < 0.001$). A slope factor of 0.914 ($\pm$ 0.234, 95% CI), coupled with a Y intercept of 22.1 ($\pm$ 13.0, 95% CI) suggests that the culture surface application technique provides a somewhat more sensitive index of drug effect than the culture incorporation technique.

Use of Culture Surface Drug Application in the Chemosensitivity Testing of Primary Tumor Cell Cultures

Culture surface drug application was employed in the assessment of 145 consecutive evaluable human solid tumor specimens. Significant proliferation was observed in 73 specimen cultures, 55 of which were sensitive to one or more chemotherapeutic agents. As shown in Table II, use of surface application in this series of tumor cultures resulted

TABLE I
Sensitivity of Human Rhabdomyosarcoma Cells (A-204) to Chemotherapeutic Agents Applied to Soft-Agarose Culture Surfaces

Agent	Culture Concentration[1] (μg/ml)	Colony Formation[2] (% of Control Growth)
Vehicle Controls	--	100 ± 5.4
Actinomycin D	0.010	0.8 ± 0.2
Bisantrene (ADAH)	0.50	7.1 ± 1.5
Doxorubicin	0.60	2.3 ± 0.3
L-Alanosine	50	3.5 ± 1.4
Acridinyl Anisidide (m-AMSA)	1.0	1.3 ± 0.2
Cytosine Arabinoside (ARA-C)	0.20	1.3 ± 0.5
Diaziquone (AZQ)	1.0	9.9 ± 1.2
Carmustine (BCNU)	2.0	12.5 ± 1.1
Bleomycin	2.0	2.6 ± 1.1
Cyclophosphamide	70	79.0 ± 9.9
Dibromodulcitol (DBD)	5.0	10.4 ± 1.4
Galactitol (DAG)	2.0	7.3 ± 1.0
5-Fluorouracil	10	0.8 ± 0.2
Mitoguazone (MGBG)	50	1.0 ± 0.2
Hydroxyurea (HUR)	60	5.1 ± 1.4
Methotrexate	1.0	0.8 ± 0.2
Mitomycin C	0.040	1.0 ± 0.1
N-Phosphonacetyl-L-aspartic acid (PALA)	200	1.7 ± 0.8
Cisplatin (CDDP)	1.5	14.1 ± 2.2
Procarbazine	5.0	78.5 ± 4.5
Triazinate (TZT)	40	1.3 ± 0.4
Vinblastine	0.050	3.3 ± 0.2
Teniposide (VM-26)	10	1.3 ± 0.2
Etoposide (VP-16)	10	0.4 ± 0.2
Sodium Azide	600	0.7 ± 0.3
Mercuric Chloride	100	0.05 ± 0.05

[1] Culture concentration of each chemotherapeutic agent was selected to approximate the mean plasma concentration present in patients one hour following administration of a maximum tolerated dose.

[2] Tabulated data are the mean ± 1 SEM for each group (n=6 vehicle control cultures; n=3 drug-treated cultures).

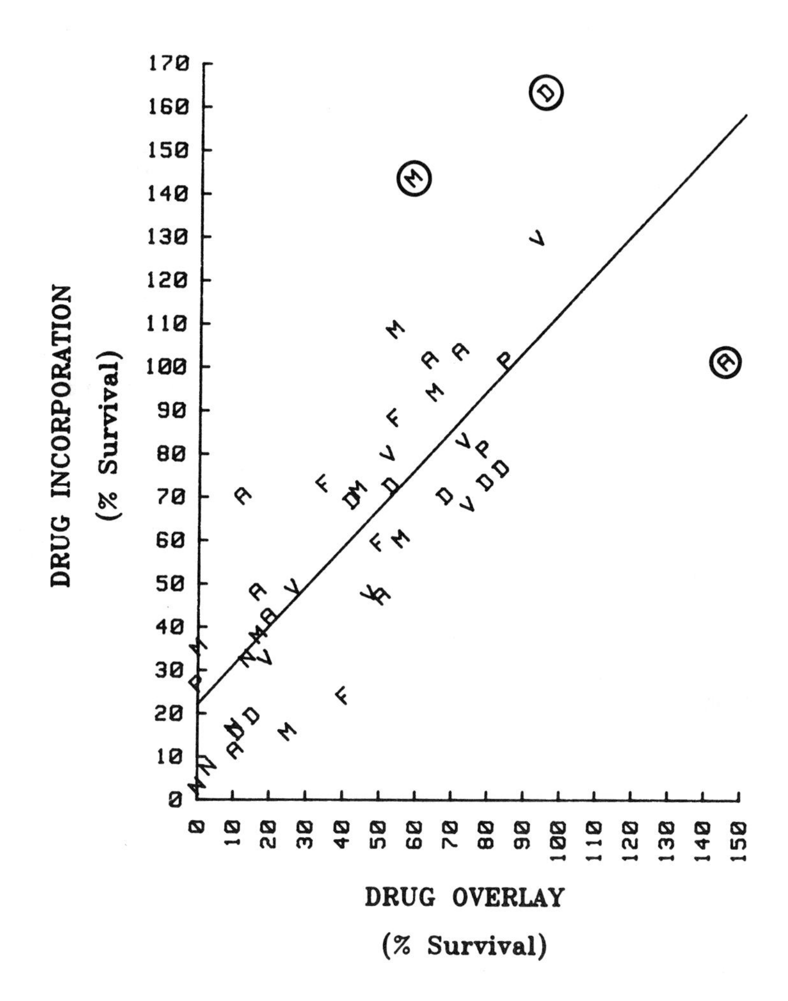

Figure 1. A Comparison of Drug Activities in Primary Human Tumor Cell Culture Following Two Methods of Drug Application. The graph depicts paired mean percent survival data gathered from nine individual specimen cultures (see text). N = sodium azide (600 µg/ml); F = 5-fluorouracil (10 µg/ml); M = mitomycin C (0.04 µg/ml); A = actinomycin D (0.01 µg/ml); D = doxorubicin (0.6 µg/ml); V = vinblastine (0.05 µg/ml); P = cisplatin (1.5 µg/ml).

in higher frequencies of proliferation and chemosensitivity than were observed for the drug incorporation method in an earlier series of specimen cultures. While similar frequencies of drug sensitivity were observed for the two methodologies when data are expressed relative to the number of assays (A), higher frequencies were observed when data is normalized with respect to the number of cultures (C). These findings coupled with those depicted in Figure 1 suggest that *in vitro* tumor cell growth may be enhanced by minimizing mechanical manipulations of cells prior to culture inoculation.

Assessment of Tumor Cell Colony Formation Following Culture Surface Application of Multiple Drug Concentrations

In a subsequent series of 117 primary tumor cell cultures as well as secondary cultures derived from 4 xenograft-passaged human renal carcinomas, each drug was applied at three concentrations. As noted in previous testing of single concentrations, tumor cell cultures exhibited a wide range of sensitivities to a given agent. For example, colony formation by one renal carcinoma (Figure 2A).

TABLE II
Evaluability and Chemosensitivity of Primary Human Tumor Cultures Following Two Methods of Drug Application[1,2]

Criteria	Drug Incorporation	Drug Overlay
Consecutive evaluable specimen cultures (C):	195	145
Cultures exhibiting significant growth and successfully assayed (A):	82 (42% of C)	76 (52% of C)
Drug sensitive:	55 (67% of A; 28% cf C)	55 (72% of A; 38% of C)

[1] Table entries indicate the number and normalized frequencies of specimen cultures meeting each criterion

[2] Sensitivity refers to > 70% inhibition of colony formation by one or more chemotherapeutic agents present at clinically relevant concentrations.

was sensitive to all agents at therapeutically relevant concentrations (1x) except mitoguazone. By contrast, colony formation by another renal carcinoma (Figure 2B) was resistant to all agents at the same respective 1x concentrations following identical applications. At higher concentrations (10x and 100x) actinomycin D, mitoguazone and mitomycin C were "active", whereas L-alanosine, etoposide and vinblastine were inactive. Similar *in vitro* drug sensitivity profiles were observed in subsequent cultures; and, in fact, *in vitro* colony formation by the former specimen was markedly inhibited (<15% of control growth) by lesser concentrations (0.1x) of actinomycin D, mitomycin C, and vinblastine. Coefficients of variation for the colony count of primary, secondary as well as cell line cultures inoculated from a single, "bulk" cell suspension were generally small: less than 20% for colony counts exceeding 80/500 mm^2 (e.g., see Figure 2).

DISCUSSION

A standardized laboratory assay capable of identifying effective chemotherapeutic agents for individual tumor specimens would be a useful adjunct to the clinical management of cancer patients. While the "human tumor stem cell assay" was designed specifically for this purpose (Salmon et al, 1978), certain technical features of the original assay complicate its performance. For example, the conventional protocol requires that prior to setting up bilayer cultures, aliquots of cell suspension be transferred to separate tubes, each containing a different drug, the same drug at different concentrations, as well as respective drug vehicles (Alberts et al, 1980; Soehnlen et al, 1980). While few tubes may be required for small tumor specimens, larger specimens require many tubes. Not only is tube handling cumbersome, but it is our empirical judgement that excessive mechanical manipulation of cells suspended in culture medium at elevated temperatures may lead to diminished *in vitro* cell proliferation. Moreover, addition of a small volume of stock cell suspension (at high density) to each tube followed by thorough mixing and application of three, 1 ml aliquots to individual plates using a different pipet for each tube reduces the accuracy and precision of delivering the same number of cells to each dish of a given specimen culture.

As an alternate approach to culture and chemosensitivity testing in the present study, the potential utility of preparing all cultures from a single, "bulk" cell suspension

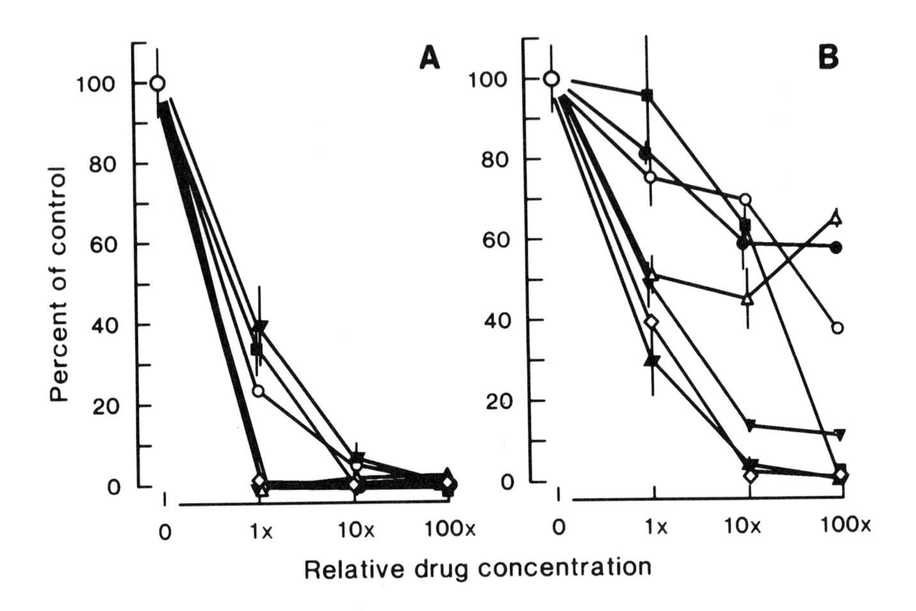

Figure 2. Colony Formation by Two Human Renal Carcinomas (Xenograft-passaged) Following Drug Application to Soft-agarose Culture Surfaces. Colony counts in drug-treated cultures (n=3) are expressed relative to that in vehicle control cultures (○, n=6) as percent. The mean ± 1 SEM for each group is depicted. The following agents were evaluated (1x concentration, µg/ml): ▼ actinomycin D (0.010); ● L-alanosine (5.0); ○ etoposide (1.0); ▰ mitoguazone (50); ◇ mitomycin C (0.04); △ vinblastine (0.050); and ■ mercuric chloride (1.0).

followed by culture surface drug application was assessed. Such a methodology was observed previously to be suitable in human tumor cell line cultures (Alley et al, 1982). In the present study surface application of a large battery of cli-aically useful drugs brought about excellent inhibition of tumor cell proliferation in cell line as well as primary and secondary cultures. The increased frequency of detecting chemosensitive cell cultures following surface drug application appeared to result from improved tumor cell growth (evaluability rate) afforded by fewer mechanical manipulations prior to culture. Thus, the method of drug application appears to be a subtle, but important factor which influences the ease of performance and evaluability of human tumor cell cultures of this type.

SUMMARY

Previous methods of evaluating human tumor cell chemosensitivity using soft-agar colony formation assay have required that cell suspensions be aliquoted into multiple tubes, one for each drug concentration and each drug vehicle prior to culture inoculation. In the present study, the utility of an alternate method of drug exposure was investigated: Application of drug and/or drug vehicle to culture surfaces following cell inoculation. Culture surface drug application was observed 1) to provide a means to avoid mechanical manipulation of cells in the presence of drug and 2) to reduce the number of steps for chemosensitivity testing required by other methodologies. In addition, the preparation of all culture dishes from a single, "bulk" cell suspension for each specimen culture appeared 1) to improve the accuracy and precision of culture inoculation and 2) to facilitate growth in soft-agar colony formation assays.

ACKNOWLEDGEMENTS

The authors wish to acknowledge preparation of human solid tumor specimens by the pathology staff of St. Marys Hospital and Rochester Methodist Hospital, culture and chemosensitivity testing by Mary Adams, Linda Foster, Barbara Furlow, Sue Gossman, Sharon Guy, Dane Mathieson, Cindy Uhl, and Carol White, and manuscript preparation by Shelly Nicklay.

REFERENCES

Alberts, D.S., Chen, H.-S., and Salmon, S.E. In vitro drug assay: Pharmacologic considerations. In S.E. Salmon (ed.), _Cloning of Human Tumor Stem Cells_, _Prog. Clin. Biol. Res._, _Vol. 48_, New York: Alan R. Liss, 1980, 197-207.

Agrez, M.V., Kovach, J.S., Beart, R.W. Jr., Rubin, J., Moertel, C.G., and Lieber, M.M. Human colorectal carcinoma: Patterns of sensitivity to chemotherapeutic agents in the human tumor stem cell assay. J. Surg. Oncol., 1982a, 20:187-191.

Agrez, M.V., Kovach, J.S., and Lieber, M.M. Cell aggregates in the soft agar "human tumor stem-cell assay". Brit. J. Cancer, 1982b, 46:880-887.

Alley, M.C., Uhl, C.B., and Lieber, M.M. Improved detection of drug cytotoxicity in the soft-agar colony formation assay through use of a metabolizable tetrazolium salt. Life Sci., 1982, 31:3071-3078.

Alley, M.C. and Lieber, M.M. Improved optical density of colony enlargement and drug cytotoxicity in primary soft agar cultures of human solid tumour cells. Brit. J. Cancer, February, 1984, in press.

Alley, M.C., Powis, G., Appel, P.L., et al. Activation and inactivation of cancer chemotherapeutic agents by rat hepatocytes cocultured with human tumor cell lines. Cancer Res., February, 1984, in press.

Kressner, B.E., Morton, R.R.A, Martens, A.E., et al. Use of an image analysis system to count colonies in stem cell assays of human tumors. In S.E. Salmon (ed.), Cloning of Human Tumor Stem Cells, Prog. Clin. Biol. Res., Vol. 48, New York: Alan R. Liss, 1980, 179-193.

Lieber, M.M., Ames, M.M., Powis, G., and Kovach, J.S. Drug sensitivity testing in vitro with a liver microsome "activated" soft agar human stem cell colony assay. In I.J. Fidler and R.J. White (eds.), Design of Models for Testing Cancer Chemotherapeutic Agents, New York: Van Norstrand Reinhold, 1982, 12-18.

Salmon, S.E., Hamburger, A.W., Soehnlen, B., et al. Quantitaion of differential sensitivity of human tumor stem cells to anticancer agents. New Engl. J. Med., 1978, 298:1321-1327.

Soehnlen, B., Young, L., and Liu, R. Standard laboratory procedures for in vitro assay of human tumor stem cells. In S.E. Salmon (ed.), Cloning of Human Tumor Stem Cells, Prog. Clin. Biol. Res., Vol. 48, New York: Alan R. Liss, 1980, 331-338.

HUMAN TUMOR CLONING CULTURE SUPPLEMENTS

G. Spitzer[2], E. Singletary, B. Tomasovic, G. Umbach[3], Y. Hu, N. Merchant, V. Hug, J. Ajani

Department of Hematology (G.S., B.T., Y.H), Surgery (E.S.), Gynecology (G.U.), Medical Oncology (N.M., V.H., J.A.), The University of Texas, M. D. Anderson Hospital and Tumor Institute, Houston, Texas.

Introduction

The ability to grow human cells in soft agar culture requires serum as a source of growth factors in addition to an appropriate nutrient media. Transformed cells generally are less dependent on serum than nontransformed cells.[3, 23] The development of a fresh human tumor culture system with high cloning efficiency in a defined medium would allow reliable drug sensitivity testing and the evaluation of biologic growth modifiers on tumor cell proliferation.

Sato and coworkers[1,22,25] have advocated serum replacement in high density culture by hormones and growth factors such as insulin, transferrin, and somatomedins. The approach by Ham's group[9] has been a stepwise reduction of serum in the clonal growth assay by adjustment of low molecular weight components of the nutrient medium. Recently Hamburger[12] reported that

[1]This investigation was supported in part by NIH Grants CA 28153, CA 14528 and Grant 174208 Allotment for the Solid Tumor Cloning Laboratory.

[2]Recipient of a scholarship from the Leukemia Society of America.

[3]Recipient of a postdoctoral fellowship research grant from the Max Kade Foundation, New York.

HUMAN TUMOR CLONING
ISBN 0-8089-1671-8

the amount of serum needed for clonogenic growth of fresh human tumor cells in soft agar could be diminished by the use of culture supplements but not eliminated. She reported a pronounced increase in cloning efficiency in 18 of 21 specimens after the addition of insulin, transferrin, and selenium and in 15 of 17 specimens after the addition of hydrocortisone, insulin, transferrin, estradiol and selenium. This same hormone supplementation supports the growth of fresh human lung cancer cells in serum-free monolayer culture.[3] Growth of established cell lines equivalent to serum substituted cultures can be achieved in predominantly serum-free medium with the following additions in different combinations: selenium, putrescine, bovine serum albumin, transferrin, insulin, triiodothyronine, progesterone, testosterone, follicle-stimulating hormone, thyrotropin-releasing hormone and prolactin.[2,8,15,17,18,20]

Mitogenic polypeptides important for proliferation of normal epithelial cells and fibroblast growth have been investigated.[7] Increased tumor colony growth with epidermal growth factor (EGF)[10,19] or platelet-derived growth factor (PDGF)[4,16,20,24] was demonstrated. Other workers have not consistently reproduced these findings.

Other investigators have reported colony growth stimulation with tumor cell conditioned media (CM). Gazdar et al[6] documented the ability to establish a continuous clonable small cell carcinoma of the lung (SCCL) cell lines for periods of 12 to 36 months from 6 of 11 tumor specimens by conditioned media of other SCCL cultures in contrast to only 2 of 21 tumor specimens supplemented with fetal bovine serum. After 1 to 3 months, the cell lines could be maintained in culture without the conditioned media. This conditioned media improved approximately by two-fold the cloning assay efficiency and colony size of 4 of 10 fresh and cryopreserved small cell carcinoma clinical specimens but did not affect the cloning assay efficiency of the established SCCL cell lines. A high cloning assay efficiency of 70% with fresh biopsy-derived breast cancers in agar was achieved by Hug et al[14] with a modified medium containing hydrocortisone, EGF and pooled conditioned media from 3 human breast carcinoma lines.

Many of the above hormones and growth factors are available in serum but may be at too diluted concentrations or inhibited by other constituents present in the serum. Different batches of conventional serum may also vary in qualitative and quantitative amounts of these substances. To determine if this was a significant limitation of serum culture media, we investigated the addition of these described growth factors to promote cloning efficiency of fresh human tumor cells in serum-rich culture medium. Because of the improvement in plating efficiency and the induction of anchorage independent cell growth with tumor cell products,[5,18,23] we also examined the growth effect of conditioned media derived from human tumor cell lines.

Methods

Malignant tumor specimens with histology confirmed by a pathologist were collected from patients undergoing diagnostic or therapeutic procedures. The majority of these tumor samples were ovary (34%), lung (14%), gastrointestinal (12%) and endometrium (6%). A modification of the Hamburger and Salmon[11] bilayer soft agar colony-forming assay for fresh human tumor cells was used. Briefly, tumor cells from ascitic fluid samples were harvested by centrifugation. Solid tumor specimens were mechanically dissociated with a scapel to 1 mm size. The cells were enzymatically treated with Type III collagenase (Worthington Biochemical Corporation, Freehold, New Jersey) and deoxyribonuclease (Sigma Chemical Corporation, St. Louis, Missouri) at a final strength of 0.75% and 0.005% for 14 to 18 hours at 37 C under continous agitation to form single cell suspension. The cells were then washed in Hank's Balanced Salt Solution. A one ml lower layer of 0.5% agar in culture medium F12 (Ham) (K.C. Biological, Lenexa, Kansas) supplemented with 10% fetal bovine serum (FBS) and the experimental additions were poured onto a 35 mm plastic Petri dish and allowed to solidify at room temperature. The tumor cells were suspended in 0.3% agar and alpha Minimal Essential Medium (K.C. Biological, Lenexa, Kansas) supplemented with 15% FBS and pipetted onto the underlayer. The final concentration of cells in each upper layer was 5×10^5. After solidification of agar, the Petri dishes were examined under an inverted microscope to subtract any tumor cell aggregates larger than 100 µm from the final post-incubation colony count. The cultures were incubated for two to three weeks at 37 C in a humidified atmosphere of 5% CO_2 in 12% O_2. Tumor cell colonies as defined by round cell groups with a minimum diameter of 100 µm were then scored using an Olympus inverted stereoscopic microscope at 40X magnification. The following culture supplements (concentrations and type of solvent in parentheses) were supplied by Collaborative Research, Waltham, Massachussetts: triiodothyronine (100 pm, 0.2 N NaOH), follicle-stimulating hormone (0.5 µg/ml, H_2O), progesterone (1.0 nm, 95% ethanol), testosterone (1.0 nm, 95% ethanol), parathyroid hormone (0.5 µg/ml, 0.1 N acetic acid), thyrotropin-releasing hormone (1.0 nm, 0.1 N acetic acid), fibroblast growth factor (50 ng/ml, H_2O), putrescine (0.1 mM, H_2O) and selenium (30 nm, H_2O). The following culture media additives were from Sigma (St. Louis, Missouri): bovine serum albumin

(5 mg/ml), sheep prolactin (= sheep luteotropic hormone, 10 ng/ml, Sigma hormone solvent). We also investigated the combination of two culture supplements: a mixture of bovine insulin (2.5 μg/ml, dissolved in H_2O, supplied by Collaborative Research) and transferrin (5 μg/ml, dissolved in Hank's Balanced Salt Solution, supplied by Sigma). The above factors are final concentrations based on the total volume of the agar-medium matrix in the Petri dish. All assays were carried out in triplicate.

Results

In the first series of 11 specimens (Table 1), 9 controls showed no growth (less than 5 colonies) and 2 controls had unsatisfactory growth (13 and 14 colonies). Nine of the test cultures with insulin and transferrin had less than 20 colonies. However, two supplemented specimens had a mark increase in colonies as compared to controls (0 to 131 colonies, 14 to 102). Therefore, we modified our control plates with this culture supplementation in our subsequent series of experiments.

TABLE 1
Culture Supplementation With Insulin + Transferrin:

No. of specimens	11
No. ≥ 20 colonies	2
No. < 20 colonies	9
No. ≥ 50 % increase	2
0 → 131	
14 → 102	

In the second experiment using 8 specimens (Table 2), the addition of hydrocortisone increased the number of colonies by 50% or more in only one sample (0 to 264 colonies). The other 7 supplemented samples grew less than 20 colonies.

In the third trial of 16 to 20 (mean 19) specimens, the colonies in the control dishes ranged from 0 to 284 (mean 40). The following factors were tested individually: bovine serum albumin (BSA), putrescine, selenium, triiodothyronine, follicle-stimulating hormone, progesterone, testosterone, parathyroid hormone, thyrotropin-releasing hormone (TRH), prolactin, platelet-derived growth factor and fibroblast growth factor. The only substances that produced a greater than 50% increase in colony formation in 2 samples or more were BSA, progesterone, and TRH. The addition of BSA in 19 specimens (Table 3) stimulated formation of 20 or more colonies in 7 specimens with 2 specimens significantly enhanced (4 to 56 colonies, 1 to 17). Progesterone increased the number of colonies to 20 or more in 7 of 20 specimens (Table 4). A 50% increase or more was seen in 4 cases (0 to 32 colonies, 58 to 88, 4 to 14, and 1 to 7). Culture supplementation with TRH in 18 specimens (Table 5) promoted colony formation of 20 or more in 6 cases with 2 specimens showing a 50% increase or more (4 to 30 colonies, 1 to 13).

TABLE 2
Culture Supplementation With Hydrocortisone:

(Control = 12.5% FCS + Insulin + Transferrin)	
No. of specimens	8
No. ≥ 20 colonies	1
No. < 20 colonies	7
No. ≥ 50 % increase	1
	0 → 264

TABLE 3

Culture Supplementation With Bovine Serum Albumin:

No. of specimens	19
No. ≥ 20 colonies	7
No. < 20 colonies	12
No. ≥ 50 % increase	2
	4 → 30
	1 → 13

TABLE 4

Culture Supplementation With Progesterone:

No. of specimens	20
No. ≥ 20 colonies	7
No. < 20 colonies	13
No. ≥ 50 % increase	4
	0 → 32
	58 → 88
	4 → 14
	1 → 7

TABLE 5

Culture Supplementation With Thyrotropin-releasing Hormone:

No. of specimens	18
No. ≥ 20 colonies	6
No. < 20 colonies	12
No. ≥ 50 % increase	2
	4 → 30
	1 → 13

We next examined the colony formation response of fresh human tumor cells to conditioned media from 11 tumor cell lines: breast (1), ovarian (2), colon (3), myeloma (2), vulvar (1), cervix (1) and malignant melanoma (1). Our initial experiment evaluated conditioned media from the breast cell line MDA435. This conditioned media was harvested from exponentially growing cultures originally seeded at 5 X 10^5 cells/ml at day 2, 4, 6 and 8. After centrifugation and filtration, the CM was then readded to the cultures at 10%, 25% and 50% concentrations. Autostimulation of the breast cell line by CM was maximally enhanced with high CM concentrations and delayed harvesting of the CM after 8 days (Figure 1). This conditioned media was then tested at a final concentrations of 25% in an underlayer on agar cultures of fresh human malignant tumors: breast (2), oat cell (2), melanoma (5), lung (3) and colon (2). Single cell suspensions of the tumor specimens were achieved by overnight enzymatic dissociation with collagenase and DNase. Three of the 14 specimens, breast (1) and oat cell (2), had a 50% or more increase in colony formation (Table 6).

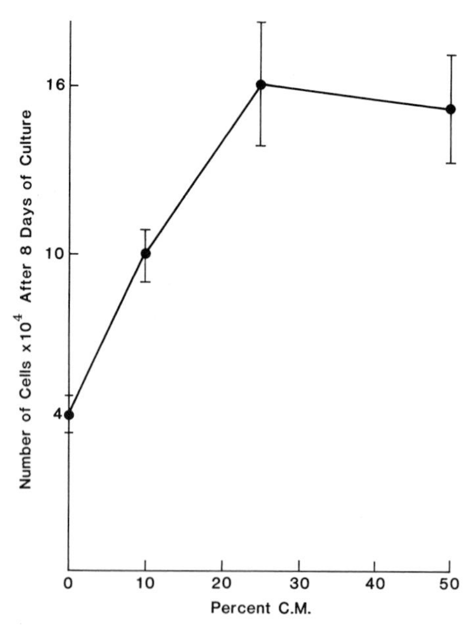

Figure 1. Conditioned media from a breast cell stimulates its own growth.

TABLE 6

Colony Supplementation With Breast Cell Line

No. of specimens	14
No. ≥ 20 colonies	13
No. < 20 colonies	1
No. ≥ 50 % increase	3
Breast	(0 → 58±6)
Oat cell	(61±21 → 91±6)
	(18± 3 → 42±4)

The conditioned media from only one other cell line, ovarian cancer, was successful in growth stimulation. The preparation of the ovarian conditioned media differed from the breast CM in that shorter periods of harvest were found more effective. Culture supplementation with this ovarian conditioned media enhanced colony growth by 50% or more in 6 of 12 fresh human tumor specimens: ovarian (2), lung adenocarcinoma (1), colon (2) and endometrium (1) (Table 7).

TABLE 7

Culture Supplementation With Ovarian Cell Line Conditioned Media:

No. of specimens	12
No. ≥ 20 colonies	4
No. < 20 colonies	8
No. ≥ 50 % increase	6
Ovarian	(57 ± 24 → 275 ± 43)
	(0 → 16 ± 4)
Lung AdenoCa	(19 ± 2 → 63 ± 15)
Colon	(2 ± 2 → 53 ± 21)
	(21 ± 6 → 84 ± 17)
Endometrium	(6 → 14 ± 4)

In 2 specimens, the addition of EGF to this CM revealed no synergism and possible inhibition (Table 8). Comparison of colony formation of crude ovarian CM and its different molecular weight components demonstrated a possible inhibitor with a molecular weight greater than 10,000 daltons (Table 9). This potential antagonistic factor emphasizes the need to include molecular sizing and purification of conditioned media derived from different cell lines to insure an active conditioned media is not overlooked secondary to inhibitors.

TABLE 8

Interactions Between Ovarian Cell Line Conditioned Media (CM) And Epidermal Growth Factor (EGF):

Culture Conditions	Colonies/ 5 X 10^5 cells			
	Mean	SD	Mean	SD
CONTROL	57 ±	24	19 ±	2
EGF	213 ±	27	15 ±	4
CM	272 ±	43	63 ±	15
CM + EGF	154 ±	12	67 ±	14

TABLE 9

Possible Molecular Weight of Inhibitors In Ovarian Cell Line Conditioned Media (CM):

Culture Conditions	Colonies/ 5 X 10^5 cells	
	Mean	SD
Control	11 ±	2
Crude CM	3 ±	2
<1,000 daltons	8 ±	2
1,000-10,000	23 ±	6
>10,000	0 ±	4

Discussion

The only occassional increment in growth realized by us with tissue culture supplementation differs from other studies using hormones or growth factors in several aspects. Most investigators employ established cell lines which have a constant supply of homogenous cells that may have a undergone extensive adaptation and simplification of their growth requirements. Our tumor cells were biopsy-derived from individual patients which would give a high degree of heterogeneity.[13] In contrast to studies using media with lower serum content, our culture media averaged 12.5% FBS. The concentrations of culture supplements we tested were similar but not identical to levels used in other investigations due to the wide range of concentrations that have been reported to be effective eg. transferrin concentrations ranged from 5 μg/ml[2] to 100 μg/ml[9]. Also, the preparations of platelet-derived growth factor and fibroblast growth factor may vary in the individual laboratories. Our stringent criteria in defining colony growth also differs from many other investigators. Colonies were defined as only cell groups with a minimum diameter of 100 μm (approximately 60 cells or more) as opposed to other studies using a definition of 60 μm diameter[19] or 30 cells.[21] Furthermore, cell aggregates of greater than 100 μm before incubation of the cultures were subtracted from the post-incubation colony count to decrease the "clump" error.

In summary, serum-rich culture media may mask many of the diverse cellular growth requirements of the assay system but does not always succeed in supporting an adequate colony formation of fresh human tumor cells. The addition of culture supplements manipulates only one subset of environmental variables that affect cellular multiplication. The analysis of growth requirements of human tumor cells must take into account not only the qualitative presence or absence of a specific growth factor of the culture medium but also the quantitative balance of concentrations of ions, nutrients, hormones and other growth factors. The physiological conditions of the culture assay such as temperature, pH, O_2 tension, osmolarity and extracellular matrix may play a significant role in the optimal growth of different tumor types.

We are skeptical that tissue culture supplementation will

replace the serum requirement for clonal growth of biopsy-derived human tumors. To supply all the growth factors in serum for the numerous tumor types with inherent heterogeneity within the same histology is probably an unrealistic goal. Tumor growth may however be improved by a combination of hormones, nutrients, mitogenic polypeptides and to be discovered tumor growth factors in addition to serum. Since a broad range of tumors are initially responsive to tumor cell conditioned media, we feel this type of supplementation is a more important approach. Our future experiments will evaluate the different growth factors and their molecular size in various cell line CM's. We will also examine the synergistic or antagonistic effects of pooled CM and the other growth factors such as EGF and PDGF.

REFERENCES

1. Barnes D, Sato G: Methods for growth of culture cells in serum-free medium. Biochemistry 102: 255-270, 1980.
2. Bottenstein JE, Sato GH: Growth of a rat neuroblastoma cell line in serum-free supplemented medium. Proc Natl Acad Sci 76: 514-517, 1979.
3. Carney DN, Bunn PA, Gazdar AF, Pagan JA, Minna JD: Selective growth in serum-free hormone-supplemented medium of tumor cells obtained by biopsy from patients with small cell carcinoma of the lung. Proc Natl Acad Sci 78: 3186-3189, 1981.
4. Cowan DH, Graham J: Stimulation of human tumor colony formation by platelet lysates. Stem Cells 1: 288-289, 1982.
5. DeLarco JE, Todaro GJ: Growth factors from murine sarcoma virus-transformed cells. Proc Natl Acad Sci 75: 4001-4005, 1978.
6. Gazdar AF, Carney DN, Russell EK et al: Establishment of continous, clonable cultures of small-cell carcinoma of the lung which have amine precursor uptake and decarboxylation cell properties. Cancer Res 40: 3502-3507, 1980.
7. Halper J, Moses HL: Epithelial tissue-derived growth factor-like polypeptides. Cancer Res 43: 1972-1979,1983.

8. Halper J, Moses HL: Stimulation of soft agar growth of A431 human carcinoma cells by glucocorticoids and diffusable factors. Fed Proc 42: 525, 1983.
9. Ham RG: Importance of the basal nutrient medium in the design of hormonally defined media, in Sato, Pardee, Sirbasku: Growth of Cells in Hormonally Defined Media. Cold Spring Laboratory, 1982, pp 39-60.
10. Hamburger AW, White CP, Brown RW: Effect of epidermal growth factor on proliferation of human tumor cells in soft agar. J Natl Cancer Inst 67: 825-830, 1981.
11. Hamburger AW, Salmon SE: Primary Bioassay of human tumor stem cells. Science 197: 461-463, 1977.
12. Hamburger AW, White CP, Dunn FE, Citron ML, Hummel S: Modulation of human tumor colony growth in soft agar by serum. Int J Cell Cloning 1: 216-229, 1983.
13. Heppner G, Miller BJ: Tumor heterogeneity: biological implications and therapeutic consequences. Cancer Metastasis Review 2: 5-23, 1983.
14. Hug V, Drewinko B, Spitzer G, Blumenschein GD: Improved culture conditions for the in vitro growth of human breast tumors. Proc Amer Assoc Cancer Res 24: 35, 1983.
15. Kaighn EM, Kirk D, Szalay M, Lechner JF: Growth control of prostatic carcinoma cells in serum-free media: Interrelationship of hormone response, cell density, and nutrient media. Proc Natl Acad Sci 78: 5673-5676, 1981.
16. Kohler N, Lipton A: Platelets as a source of fibroblast growth-promoting activity. Exp Cell Res 87: 297-301, 1974.
17. Meyskens FL, Salmon SE: Regulation of human melanoma clonogenic cell expression in soft agar by follicle-stimulating hormone (FSH), nerve growth factor (NGF) and melatonin (MTN). Proc Amer Assoc Cancer Res 21: 199, 1980.
18. Moore MAS, Williams M, Metcalf D: In vitro colony formation by normal and leukemic human hematopoietic cells: Interaction between colony-forming and colony stimulating cells. JNCI 50: 591-602, 1973.
19. Pathak MA, Matrisian LM, Magun BE, Salmon SE: Effect of epidermal growth factor on clonogenic growth of primary human tumor cells. Int J Cancer 30: 745-750, 1982.
20. Ross R: Physiological quiescence in plasma-derived serum: Influence of plateltet-derived growth factor on cell growth in culture. J Cell Physiol 97: 497, 1978.
21. Salmon SE: Background and overview, in Salmon: Cloning of Human Tumor Stem Cells. Alan R. Liss, 1980, pp 3-13.
22. Sato GH, Pardee AB, Sirbasku DA: Growth of Cells in Hormonally Defined Media. Cold Spring Harbor Conference

on Cell Proliferation 9: 1-624, 1982.
23. Sporn MB, Todaro GJ: Autocrine secretion and malignant transformation of cells. N Engl J Med 303: 878-880, 1980.
24. Westermark B, Wasteson A: A platelet factor stimulating human normal glial cells. Exp Cell Res 98: 170-174, 1976.
25. Wu R, Sato G: Replacement of serum in cell culture by hormones: A study of hormonal regulation of cell growth and specific gene expression. J Toxic Environ Health 4: 427-448, 1978.

EXPERIENCE WITH THE HUMAN TUMOR CLONING ASSAY

Jazbieh Moezzi and Martin J. Murphy, Jr.

The Bob Hipple Laboratory for Cancer Research
Wright State University School of Medicine
Dayton, Ohio, USA

ABSTRACT

We correlated the differential cytology of viable malignant tumor cells with their cloning efficiency. Twenty-nine tumor specimens which were cloned in vitro were evaluated (22 primary solid tumors and 7 malignant effusions). Solid tumor cell viability ranged from 15 to 70% (median = 44%) based on cytology and from 12 to 92% (median = 66%) based on trypan blue dye exclusion indicating that more accurate evaluation of viability was achieved by cytological techniques. There was no significant correlation between percentage of cytologically viable cells and the number of tumor colonies formed. This lack of correlation was remarkably similar in both primary and secondary malignant tumor samples. Cell differentiation was evaluated by Papanicolaou-stained, cytocentrifuged cell suspensions. The percentages of viable malignant cells were plotted against the number of colonies grown. Average tumor colony counts ranged from 7 to 99 (median = 40) per plate, and the proportion of viable malignant cells ranged from 23 to 80% from solid primary tumors and from 5 to 51% in effusions. These data suggest that with a modified agar system in which the top layer is liquid, there is a significant correlation between the number of viable malignant cells plated and the formation of malignant clones from solid primary tumors and a variable correlation for malignant effusions regardless of the identity of the background population of cells. This assay is therefore highly clonogenic when one considers only malignant and viable cells identified by cytologic criteria from solid primary tumors. Furthermore, this suggests the uniform distribution of a clonogenic stem cell population at the primary tumor site.

HUMAN TUMOR CLONING
ISBN 0-8089-1671-8

INTRODUCTION

Malignant tumor cells obtained from effusions or from solid tumors reflect a multitude of cellular morphologies, and cell suspensions derived from these clinically obtained specimens provide cell populations with varying cell viability. It would be difficult, therefore, to establish a firm correlation between the overall number of viable cells found in a mixture of different cell morphologies and the number of malignant clones which, after all, consist of a subgroup of tumor stem cells within the melange of tumor cells. One of the basic problems would be the inherent inaccuracies using trypan blue dye exclusion (TBE) as the sole index of cell viability. Several authors have already shown that there is a poor correlation when the TBE method is used and when the reproductive integrity of malignant cells in control cultures is then examined and following exposure to drugs or enzymes (1,2). We have examined the reproductive integrity of malignant cells and their response to chemotherapy drugs as well as to enzymes and compared this with analyses of the true viability of the cell population as adjudged by the morphology of the cells stained with Papanicolaou (3,4) and Wright-Giemsa. Using well-established pathological criteria for the identification of cellular death, we have studied the clonogeneity of the actual and real percentage of malignant viable cells in the specimen, and based on these observations we conclude that there is no significant relationship between measurements obtained by TBE and actual tumor cell viability and clonogeneity. Using more accurate criteria of actual cell death and cell viability, our results indicate that there is in fact a high degree of clonogeneity that may be correlated with the actual number of viable malignant cells.

MATERIALS AND METHODS

Sample Collection

Solid tumor samples were obtained from specimens of malignant tissue taken at the time of surgery and transported in chilled, sterile McCoy's 5A tissue culture medium (GIBCO, Grand Island, NY) and were processed within 12 hours of surgical removal. Fluid effusions were anticoagulated with preservative-free heparin (20 IU/ml) and were all processed within four hours of obtainment.

Tumor Preparation

Upon receipt, solid tumors were examined and freed of fibroadipose and necrotic tissue and mechanically were minced with sterile scalpels into 0.5 to 1 mm^3 pieces, in medium consisting of RPMI 1640 (Flow Laboratories, McLean, VA) with 10% fetal calf serum (GIBCO) and 25 mM HEPES buffer (GIBCO) to which was added 0.8% (W/V) collagenase II (Sigma Chemical Co., St. Louis, MO) and 0.002% DNase I (Sigma) as already has been described by Slocum et al. (5). After incubation in a humidified atmosphere of 5% CO_2 in a 37°C incubator, the digested tissue was washed with fresh medium and passed through 100 mm mesh stainless steel screens. The cell suspension was spun at 300 x g for 5 min., the supernatant removed and the cell button washed with fresh medium and examined for viability. Fluid effusions were also passed through 100 mm mesh stainless steel screens and the cell suspensions centrifuged at 300 x g for 10 min. and then washed with the preparation medium and resuspended for counting and viability determinations.

Cytospin Slide Preparation for Cell Morphology

After the number of viable cells had been estimated by trypan blue dye exclusion, an aliquot of cell suspension was also cytocentrifuged in a Shandon cytocentrifuge and stained with a modification of the original Papanicolaou method and with Wright-Giemsa. Slides prepared in this fashion were screened for tumor and atypical cells and the percentage of malignant cells was estimated by counting 1000 cells/slide. The cell viability was also estimated by enumerating the dead and severely injured cells by using morphologic criteria for nuclear and cytoplasmic cell death (6).

Clonal Assay for Tumor Cells

Our studies use a modification of the method originally described by Hamburger and Salmon (7), excluding mouse spleen cell conditioned medium and $CaCl_2$ and by adding 25 mM HEPES buffer (GIBCO) to the medium. DEAE-dextran was omitted since it had no significant effect on the assay (8). Briefly, one ml aliquots of cell suspensions (5 x 10^5 nucleated cells) in RPMI 1640 (Flow) containing 12% horse serum with no agar were poured over 1 ml of the same medium containing 0.75% agar in 35 mm Petri dishes. All plates were prescreened and examined for cell clumps and clusters and satisfactory plates were then incubated at 37°C in 5% CO_2 which was humidified to saturation. Culture plates were examined every 2-3 days for growth and by 2-3 weeks plates were visually scored under an inverted microscope.

Slide Preparation for Colony Morphology

Colonies were defined as aggregates of greater than or equal to 30 cells in dark dense colonies and equal to or greater than 20 cells in colonies with large clear cells with uniform cellular structure and defined borders. These measured equal to or greater than 70 microns in diameter. Colonies with unusual or pleomorphic appearances were harvested with a finely drawn Pasteur pipet tip and placed on an albumin coated slide and stained with Papanicolaou and Wright-Giemsa so as to exclude macrophage colonies. Using these morphological criteria for identification we avoided the obvious pitfall of scoring the occasional macrophage colony which, of course, does grow in the top liquid layer. The plated cell numbers were adjusted so that two or more colonies did not merge, and all colonies therefore were distinct. This was further proven by mixing two or three different malignant cell populations, all with different cell morphologies, in one culture plate and noting that there was no merging of different cell morphologies within the same colonies. Approximately 95% of all cases studied reveal colonies under the inverted microscope which numbered less than 200/plate and only 1% of all cases ever produced greater than 500 colonies per plate.

RESULTS

In this study, 22 primary solid tumors and 7 malignant effusions have been successfully cloned using a modification of the liquid top layer (9). In eight of these cases, the soft agar techniques of Hamburger and Salmon (7) were performed in parallel. The solid primary tumors reflected 8 breast carcinoma, 4 colon and 4 lung cancers, 3 ovarian cancers, 1 hypernephroma and 1 melanoma. (Table I).

Our study is primarily based on the absolute percentage of viable cells and the absolute number of viable malignant cells as estimated by cell morphology obtained from cell suspensions at the time of plating. These differential counts, prepared on cytocentrifuged and stained preparations enabled the exclusion of cells with demonstrable morphologic manifestations of cellular death including cytoplasmic as well as nuclear lesions (6). Using these criteria for cell death, the accurate evaluation of cell viability as determined by cytological examination was 0 to 36% (median = 16%) lower than cell viability estimates obtained by the TBE method. These differences are however even more pronounced in solid primary tumors. Figure 1 reveals the lack of correlation between trypan blue determination of cell viability and tumor cell clonogeneity in liquid top layers. These results compare favorably with those of Cowan et al. (10).

Table I

Viability and Morphology of Malignant Solid Tumors and Effusions

Solid Tumors	*Tumor Source	TBE Viab (%)	Cytology Viab (%)	Large Viab Cells (%)	Differential of Viable Cells (%) Malignant & Atypical	PMN‡ and Macrophages[C]	Lymphocytes	Fibro-blast	No. of Tumor Colonies/ 5 x 10⁵ Cells†
1068	Ov.	71	50	31	45	19	20	16	14.5 ± 2.5
1064	Br.	69	45	15	65	10	23	2	70.5 ± 16.5
1062	Br.	64	44	20	80	16	4	0	54 ± 10
1061	Br.	82	60	70	78	7	13	2	83 ± 6
1053	Ov.	76	58	23	66	20	4	10	57.5 ± 7.5
1042	Ov.	69	43	19	49	31	9	11	31 ± 1
1041	Co.	48	20	21	27	25	7	21	26.5 ± 3.5
1040	Br.	65	35	17	23	23	27	27	15 ± 3
1035	Lu.	40	39	75	28	51	8	13	28.5 ± 4.5
1034	Br.	60	44	20	70	11	25	4	99.5 ± 5.5
1026	Br.	50	32	8	50	25	24	1	7 ± 1
1019	S.	83	47	40	68	0	28	0	60 ± 1
1013	Co.	73	59	20	25	50	24	0	10.5 ± 1.5
1012	Lu.	48	30	4	28	56	6	10	11 ± 2
1008	Lu.	12	15	7.5	78	12	6	4	74.5 ± 9.5
1007	Br.	51	41	ND	58	17	21	5	50 ± 0
1075	Br.	75	47	17	41	14	42	5	41 ± 2
1076	Lu.	36	35	6	33	34	33	0	20 ± 1
1083	Br.	66	55	25	65	18	14	3	77 ± 5
1091	Co.	92	70	25	29	11	60	0	35 ± 0
1092	Ki.	55	45	16	59	20	20	1	60 ± 1
1095	Co.	81	46	32	49	14	30	6	42 ± 2
Effus.*					Malignant & Atypical	PMN‡ and Macrophages[C]	Lymphocytes	Mes.‡ Cells	
1067	AF/Co.	92	85	8	51	22	9	18	70 ± 8
1023	AF/Ov.	99	95	23	17	55	28	0	70.5 ± 9.5
573	PF/Br.	95	95	10	5	23	72	0	16.5 ± 1.5
751	PF/Br.	99	86	18	45	29	10	16	18 ± 3
1077	PF/U.	99	99	35	20	25	45	10	8 ± 2
987	PF/U.	98	95	ND	5	18	65	12	9.5 ± 2.5
1006	PF/Ov.	100	100	60	36	19	25	20	62.5 ± 6.5

*Ov.: Ovary; Br.: Breast; Co.: Colon; Lu.: Lung; S.: Skin; Ki.: Kidney; PF: Pleural Fluid; AF: Ascitic Fluid; U.: Unknown

ND - Not determined

†Mean ± S.E.

‡PMN: Polymorphonuclear Leukocytes; Mes: Mesothelial Cells

In Figure 2 the percentage of absolute viable cells as determined by cell morphology was plotted against the number of colonies grown. There is an obvious lack of correlation indicating variation in the number of tumor cells in the group of viable cells.

Figure 1

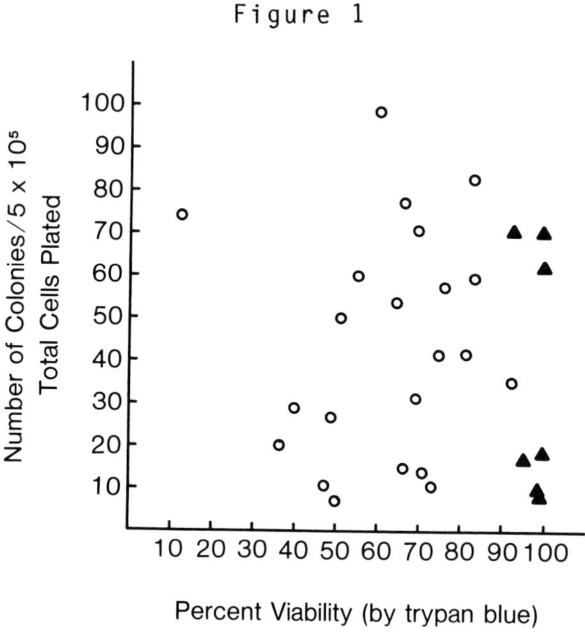

Twenty-nine malignant samples were selected because of their pure and uniform clonal growth. The average number of tumor colonies ranged from 7 to 99 (median = 40) per plate and the proportion of viable malignant cells ranged from 23 to 80% from solid tumors and from 5 to 51% in malignant effusions. Figure 3 shows a significant degree of correlation between the number of viable malignant cells plated and the formation of malignant clones from solid primary tumors and a lack of correlation with clones derived from metastatic effusions. The average plating efficiency was calculated by comparing the number of colonies to the total nucleated cells plated (11) and this ranged from 0.01 to 0.001% (Table II).

Table II

Clonogeneity of Solid Primary Tumors and Malignant Effusions

Sample	Number and Tumor Origin	Cloning Efficiency: Colonies/5 x 10^5 Cells Plated (%)	Absolute Cloning Efficiency: Colonies/Viable Malignant Cells Plated (%)
Solid Tumors (Primary)	8 Breast 4 Colon 4 Lung 3 Ovary 1 Kidney (Hypernephroma) 1 Skin (Melanoma)	0.001 to 0.01	0.002 to 0.02
Malignant effusions (Metastatic)	5 Pleural 2 Peritoneal	0.001 to 0.01	0.008 to 0.08

Figure 2

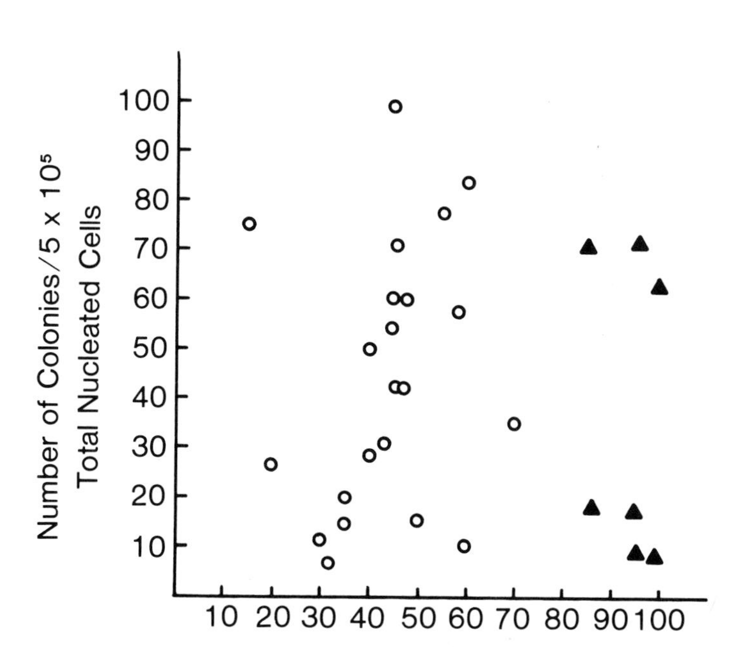

Using the actual number of viable nucleated cells plated, one derives a clear log-log correlation for the number of colonies produced (Figure 4). Furthermore when the number of malignant cells plated are compared to the log of the tumor colonies derived, one expresses another linear correlation (Figure 5).

Figure 3

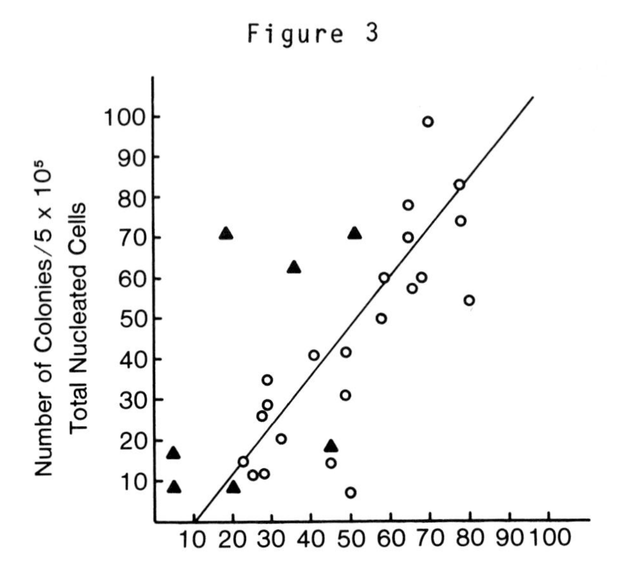

DISCUSSION

Using the top liquid layer in which to grow tumor clones, one may more easily pluck these colonies and derive excellent cytological morphologies. Using this direct method, a linear and proportional distribution has been noted between the number of viable malignant cells and the number of malignant clones derived therefrom.

Although some cellular migration and the formation of cellular aggregates have been described (9,12), the examination of morphology of true colonies and subsequent cytological confirmation of these colonies has removed any false positives from these present data.

Figure 4

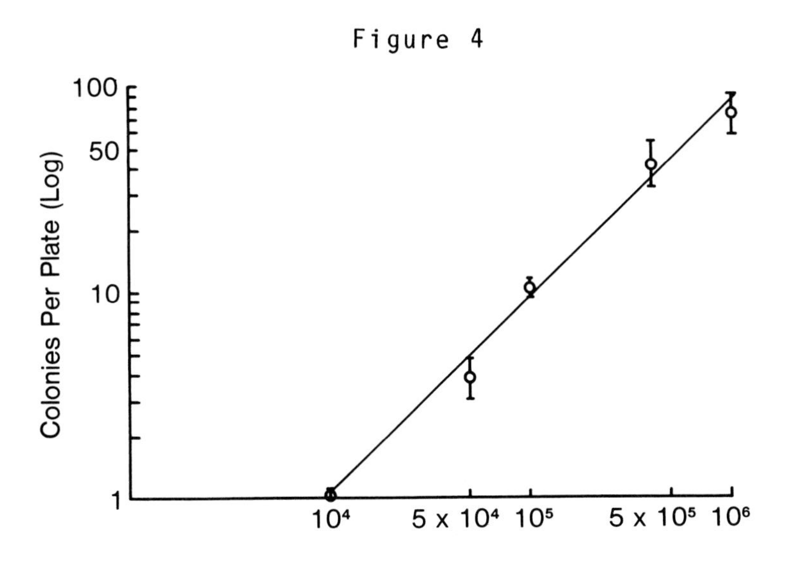

Parallel to the cytological studies we estimated the number of large, viable cells with a hemocytometer but these figures were not reliable since they also counted macrophages and normal epithelial cells (12) which were often observed with several kinds of cancer. In this study we closely scrutinized the usefulness and accuracy of the trypan blue dye exclusion technique and compared it with the estimates of viability obtained utilizing morphological evidence of cellular death (6). Whilst the latter method could hardly be used for routine analysis due to the excessively long time required for cytological screening, nonetheless it was found to be more accurate than TBE. We have shown therefore that there is a lack of correlation between the number of viable cells as determined by trypan blue or by simple cell morphology and actual clonogeneity. This lack of correlation is ascribable to variation in the absolute number of tumor cells within the group of overall viable cells.

One of the most significant observations of this present study is the correlation between the number of viable malignant cells plated and the clonogeneity of the tumor stem cells cloned derived

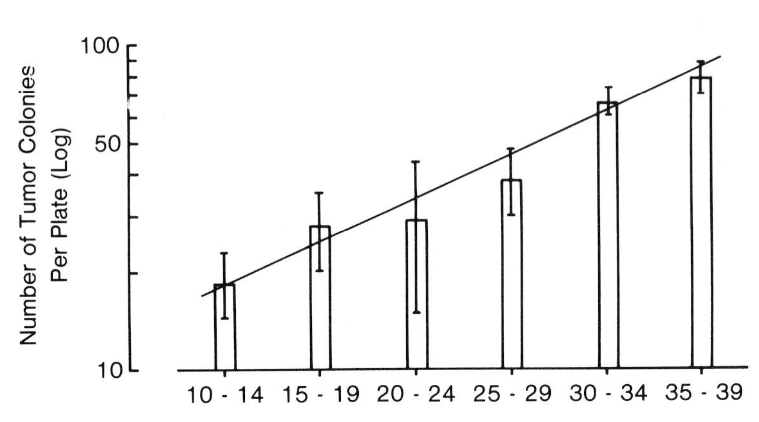

therefrom when one examines solid tumors. An interesting corollary is that there is apparently a lack of correlation between the clonogeneity seen from cells obtained from malignant effusions. One must therefore not only consider cell viability, one must also consider the absolute number of viable malignant cells and the proportion that they represent amongst a non-malignant cellular background. These current studies indicate that the most significant feature is the absolute number of viable malignant cells which are plated.

Routine Papanicolaou-stained cytocentrifuged cell preparations provided us with a very accurate means of ruling out pseudocolonies, cellular aggregates and preexisting colonies which otherwise might confuse and obscure accurate colony counts. These studies also showed only variable correlation between cells from malignant effusions and clonogeneity (Table II). This lack of correlation in malignant effusions might represent the actual physiological difference that tumor cells reflect when found free-floating in visceral cavity effusions. The possibility exists that by the time the effusion is obtained the growth-stimulating properties known to exist in effusions may have already stimulated tumor stem cell growth. This could provide for a very irregular growth pattern such as has been encountered (13).

It should also be recalled that the cellular population in fluid effusions is quite different from that seen in solid tumors with the former consisting mainly of mesothelial cells, macrophages and leukocytes whereas the latter consists of fibroblasts, fibrocytes, endothelial cells, normal epithelial cells, smooth muscle cells, macrophages, leukocytes and fat cells. These differences in the cellular populations may play an important facultative role. Also the number of tumor cells found in effusions varies according to the advancement of visceral involvement.

In conclusion, the assay as employed has been useful in studying cell morphology and the growth characteristics of malignant cells. Also it has been very useful in confirming the clonogeneity of viable malignant cells when derived from solid primary tumors. This might suggest a uniform distribution of self-renewing stem cells within the primary tumors.

ACKNOWLEDGEMENTS

The technical excellence of N. Acharya, D. Dewey and J. Kwak is gratefully appreciated. These studies were supported by the Bob Hipple Memorial Committee for Cancer Research.

REFERENCES

1. Warters, R.L., and Hofer, K.G. The in vivo reproductive potential of density separated cells. Exp. Cell. Res., 87. 143-151; 1974.

2. Weisenthal, L.M.; Dill, P.L.; Kurnick, N.B.; and Lippman, M.E. Comparison of dye-exclusion assays with a clonogenic assay in the determination of drug-induced cytotoxicity. Cancer Res., 43, 258-264; 1983.

3. Papanicolaou, G.N. Atlas of Exfoliative Cytology. Cambridge, Mass, Harvard University Press; 1954.

4. Luna, L.G. Special Techniques in: Manual of Histologic Staining Methods of the Armed Forces Institute of Pathology, pp. 70-71; New York; McGraw Hill Book Co.; 1968.

5. Slocum, H.K.; Pavelic, Z.P.; Kanter, P.M.; Nowak, N.J.; and Rustum, Y.M. The soft agar clonogenicity and characterization of cells obtained from human solid tumors by mechanical and enzymatic means. Cancer Chem. 6, 219-225; 1981.

6. Koss, L. G. Diagnostic Cytology and its Histopathologic Basis, pp. 57-60; Philadelphia; J.B. Lippincott Co.; 1979.

7. Hamburger, A.W. and Salmon, S.E. Primary bioassay of human tumor stem cells. Science, 197, 461-463; 1977.

8. Hug, V.; Spitzer, G.; Drewinko, B.; and Blumenschein, G.R. Effect of Diethylaminoethyl-dextran on colony formation of human tumor cells in semisolid suspension cultures. Cancer Res., 43, 210-213; 1983.

9. Friedman, H.M. and Glaubiger, D.L. Assessment of in vitro sensitivity of human tumor cells using [^{3}H] thymidine incorporation in a modified human tumor stem cell assay. Cancer Res., 42, 4683-4689; 1982.

10. Cowan, J.D.; Van Hoff, D.D.; Neuenfeldt, B.; Mills, G.; and Clark, G. Predictive value of trypan blue exclusion cell viability measurements on colony formation in human tumor cloning assay. (In Press).

11. Meyskens, F.L., Jr.; Soehnlen, B.J.; Saxe, D.F.; Casey, W.J.; and Salmon, S.E. In vitro clonal assay for human metastatic melanoma cells. Stem Cells, 1, 61-72; 1981.

12. Thomson, P. and Meyskens, F.L., Jr. Method for measurement of self renewal capacity of clonogenic cells from biopsies of metastatic human malignant melanoma. Cancer Res., 42, 4606-4613; 1982.

13. Buick, R.N.; Fry, S.E.; and Salmon, S.E. Effect of host cell interactions on clonogenic carcinoma cells in human effusions. Br. J. Cancer, 41, 695-704; 1980.

Interlaboratory Reproducibility Of The Human Tumor Colony Forming Assay (HTCA): The Southeastern Cancer Study Group (SECSG) Experience

DAN W. LUEDKE, M.D.*

FRANCIS J. CAREY, M.D.
ST. LOUIS VAMC AND ST. LOUIS UNIVERSITY
ST. LOUIS, MISSOURI

AWTAR KRISHAN, PhD.
UNIVERSITY OF MIAMI MEDICAL SCHOOL
MIAMI, FLORIDA

DARWIN CHEE, PhD.
ONCOLOGY LABORATORIES, INC.
WARWICK, RHODE ISLAND

BARBARA CHANG, M.D.
AUGUSTA VAMC
AUGUSTA, GEORGIA

ROBERT FRANCO, M.D.
UNIVERSITY OF CINCINNATI
CINCINNATI, OHIO

HARVEY B. NIELL, M.D.
UNIVERSITY OF TENNESSEE
MEMPHIS, TENNESSEE

MICHAEL E. JOHNS, M.D.
UNIVERSITY OF VIRGINIA
CHARLOTTESVILLE, VIRGINIA

DANIEL E. KENADY, M.D.
UNIVERSITY OF KENTUCKY
LEXINGTON, KENTUCKY

KARIM ZIRVI, PhD.
NEW JERSEY COLLEGE OF MEDICINE
NEWARK, NEW JERSEY

HUMAN TUMOR CLONING
ISBN 0-8089-1671-8

INTRODUCTION

In July 1981, an ad hoc committee was formed by the Southeastern Cancer Study Group (SECSG) to investigate group involvement with the Human Tumor Colony Forming Assay of Hamburger and Salmon (HTCFA).[1] An organizational meeting and workshop were held in October 1981, with investigators attending from ten HTCFA laboratories within the group. It was immediately apparent that numerous fundamental differences existed among the various laboratories in the performance of the assay. These included enzymatic versus mechanical dissociation of the tumor specimen, contents of the culture media, duration of drug incubation, criteria for tumor growth, and definition of a tumor colony. Because of these and other differences in performance of the assay, the workshop participants agreed that a reproducibility study was necessary before clinically relevant procotols could be generated. Following lengthy discussion, assay methods were adopted that would be strictly followed and provisions were made for a common source of critical materials.

MATERIALS AND METHODS

Bacto Agar (Difco, Detroit, Michigan), RPMI media with fetal calf serum (GIBCO Laboratories, Grand Island, New York), trypsin (GIBCO Laboratories), and DNase (Sigma Chemical, St. Louis, Missouri) were provided from a single source (the University of Miami). Each laboratory purchased 40 mesh Cellectors (E-C Apparatus, St. Petersburg, Florida), and Nitex 40 micron mesh (Tetko, Incorporated, Elmsford, New York). Drugs for testing were prepared in a single laboratory (St. Louis Veterans Administration Medical Center) in unit dose, frozen at -70°C and shipped in dry ice to each participating laboraotry; new batches were prepared every three months. Drugs tested were (Bleo) bleomycin (Bleoxane, Bristol Laboratories, Syracuse, New York), (Mito) mitomycin (Mutamycin, Bristol), (CDDP) cisplatin (Platinol, Bristol), (Mtx) methotrexate (Mexate, Bristol), (Melph) melphelan (Burroughs Wellcome Co., Research Triangle Park, NC), and (HU) hydroxyurea (Sigma Chemical). All drug

*For the SECSG, Birmingham, Alabama, (Supported in part by NCI Grants #RIO-CA-16214 and UIO-CA-19657).
We wish to express our appreciation to the competent and patient technologists of the member laboratories, to Mr. Bruce Harrison (Pharmacist), to Mrs. Jean Schlueter (Data Manager), and to Ms. Angelia Estes (Secretary).

concentrations were such that 0.1 c.c. diluted in 4ml of media produced a concentration equal to 1/10 of the peak plasma level achievable in humans.[2]

A two layer culture technique similar to that described by Hamburger and Salmon[1] was utilized to incubate cells. Prior to each trial, a bottom layer was prepared for each culture plate which consisted of 1ml of RPMI media containing 10% fetal calf serum (FCS) and 0.5% agar placed in a 35mm Costar petri dish (K.C. Biological, Lenexa, Kansas). This was refrigerated in a moist atmosphere until use.

The SOTO tumor, derived from a squamous cell carcinoma of the human cervix, was grown in one laboratory (University of Miami) in athymic mice. Discreet nodules were harvested under sterile conditions, sliced into 1mm cubes, and pooled. Equal portions were randomly selected and placed in ten sterile plastic tubes containing RPMI media, which were then sealed and packed in separate mailing boxes containing wet ice. One tumor sample was retained in the Miami laboratory, while the other nine were shipped by overnight mail to the other participating laboratories.

The following day all ten laboratories processed and plated the tumor by the adopted assay method. Tumor pieces were removed from the RPMI media and transferred to a sterile petri dish. Following addition of 1ml of Hanks balanced salt solution (HBSS Gibco) and 10% FCS the tumor fragments were finely minced using scissors or crossed scalpels. The resulting suspension was then transferred to a sterile glass ehrlenmeyer flask with stirring bar. After the addition of 30ml of a solution containing 0.25% trypsin and 0.002% DNAse, the suspension was incubated at 37°C for 45 minutes, with continuous gentle stirring. Following incubation, any remaining large clumps were removed by filtration through a 40 mesh screen. The suspension was then transferred to a 50 ml conical tube and enzymatic digestion halted by filling the tube with HBSS and 10% FCS. Following a 15 minute centrifugation at 80g at room temperature, the supernate was removed, the cells resuspended in the HBSS with 10% FCS and the centrifugation step was repeated. Following removal of the supernate, the cells were resuspended in 20 ml of RPMI media with 10% FCS, held at 37°C. The cell suspension was then filtered through a 40 micron Nitex filter. Trypan blue (Gibco Laboratories) viability was determined within 5 minutes at 1 part cell suspension to 3 parts 0.4% trypan blue under 400X magnification. Cells were then diluted to $5X10^{5}$ per ml in plating media and kept at 37°C. From this suspension 4cc aliquots were removed and placed in each of ten 17X100mm plastic snap cap tubes. Each drug was warmed to room temperature and 0.1cc dispensed to its respective tube, with 0.1ml of HBSS dispensed to the control tubes. Melted agar was kept at 45°C and 0.45ml was added to each tube, mixed, and 1ml of the mixture plated on each of 4 petri dishes (prepared with bottom layer). The process was repeated for all drugs and controls. Three petri dishes from each drug and control were incubated in a humidified atmosphere of 5% CO_2 at 37°C for two weeks. One petri dish from each variable was preserved with a 1ml mixture of 50% glycerol (Sigma Chemical), 5% DMSO (Sigma Chemical), and 45% H_2O at 45°C; these plates were the 0-time controls.

At the end of the two week incubation period, colony counts were visually determined for all plates (0-time control, control, and drug treated).

A colony was defined as a cluster of cells greater than 50 micron in diameter. Colony counts were expressed as the difference between the total colony count on a plate and the corresponding 0-time control colony count. The mean colony count of the control plates had to be at least 30 in order for drug sensitivity to be evaluated.

RESULTS

The results of five separate tumor trials are evaluable for reproducibility. Three additional trials are not evaluable due to failure of all specimens to grow in vitro, media contamination, or tumor contamination respectively. In addition all laboratories were not successful in testing each drug in all trials. In all trials, the tumor was defined as being sensitive to a drug if the mean number of colonies formed on a drug treated sample was less than 30% of the control. The tumor was defined as being resistent to a drug if the mean number of colonies formed on a drug treated sample was 30% or more of the control.

Sensitivity data were analyzed for the level of agreement achieved among the laboratories. Each drug treated tumor specimen was labeled as sensitive or resistent. Within each trial agreement among laboratories was sought for each of the drugs tested. The total number of times agreement was achieved was then determined for all of the drugs tested in the five evaluable trials. This number was divided by the total number of possible agreements (i.e. total number of drug tests in all five of the trials) which gave a "correlation coefficient". Table 1 illustrates the level of agreement on tumor sensitivity achieved for each of the drugs and for all of the drugs successfully tested. A high level of agreement was noted among all drugs tested, with melphelan, a relatively unstable drug, having the least agreement. As shown in Table 2, this agreement was consistent among participating laboratories; no single laboratory deviated dramatically from the others. If the majority of laboratories found the tumor resistent to a given drug within a trial, a high level of agreement was found among laboratories, with a correlation coefficient of 0.92. Less agreement was evident if the tumor was found to be sensitive by a majority of laboratories, with the correlation coeffient being 0.79.

Despite attaining agreement on sensitivity, discrepancies were present. Cloning efficiency markedly varied among laboratories and trials (Table 3). Tumor viability, as determined by the trypan blue viability test, also varied (Table 4). Furthermore, as seen in Table 5, Soto Tumor showed resistence to the concentration of drugs tested in the majority of trials. However, this was inconsistent, as a different pattern of sensitivity was observed in trials 3 and 4, from that seen in trials 1,2, and 5. This was not due to differences in drug lots. Also apparent in Table 5 is variability in the number of times a given drug was actually tested. This was due, in part, to contamination problems and inadequate cell yield to test all drugs.

TABLE 1

Inter-Laboratory Agreement
On Sensitivity Of Soto Tumor

Mito	Mtx	CDDP	Bleo	Melph	H.U.
*25/28**	23/24	23/27	25/29	19/24	24/25

Possible Agreements = 155

Actual Agreements = 137

Correlation Coefficient = .88

*Number of times agreement on drug sensitivity was achieved.

**Number of times drug successfully tested in all trials.

TABLE 2

Individual Agreement With The Majority
On Sensitivity Of Soto Tumor

Drug	Lab #1	Lab #2	All Others
Mito	*6/6**	6/6	13/16
Mtx	6/6	6/6	11/12
CDDP	6/6	4/6	13/15
Bleo	5/6	5/6	13/15
Melph	4/5	4/4	11/15
H.U.	5/5	3/4	16/16
Total	32/34	28/32	77/89
Correlation Coefficient	= 0.94	0.88	0.87

*Number of times a laboratory agreed with the majority.

**Number of times drug successfully tested in that laboratory.

TABLE 3

SECSG HTCFA
REPRODUCIBILITY STUDY
CLONING EFFICIENCY

Trial No.	Mean Cloning Efficiency	Range *
1	0.099%	0.009 - 0.14%
2	0.06%	0.01 - 0.2%
3	0.013%	0.006 - 0.03%
4	0.029%	0.005 - 0.08%

*This does not include values from those laboratories having no growth.
Data available from only 4 trials.

TABLE 4

SECSG HTCFA
REPRODUCIBILITY STUDY
TRYPAN BLUE VIABILITY

Trial No.	Mean Viability	Range
1	80%	69-89%
2	79%	57-100%
3	45%	30-60%
4	53%	26-100%

TABLE 5

SOTO
DRUG SENSITIVITY

Trial	Mito	Mtx	CDDP	Bleo	Melph	H.U.
1	*1/5**	0/4	0/4	0/4	2/4	0/4
2	0/5	1/2	0/5	0/5	0/3	0/3
3	4/5	0/5	3/5	4/5	4/5	0/5
4	6/7	0/7	2/7	3/7	5/6	1/7
5	0/6	0/6	0/6	0/6	1/6	0/6

*Number of times tumor was sensitive.

**Total number of times drug was successfully tested.

Discussion

When principal investigators from the SECSG member laboratories initially met to formulate joint studies utilizing the HTCFA, reproducibility of results appeared to be a relatively simple task. This illusion disppeared very quickly. Considerable time, energy, and individual laboratory expense (as well as some group funds) have been expended to date, with a number of lessons learned. In so far as the HTCFA may be of value and the applicability of the assay to the cooperative group process relevant, discussion of results of the SECSG effort should focus on those lessons learned, with possible application to future studies by this and any other consortium of laboratories.

Reproducibility of results proceeds from common methods and technique. As mentioned in the introduction, wide variation in the actual performance of HTCFA existed among member laboratories prior to the organizational meeting and workshop. The HTCFA is tedious, and time consuming, with a number of steps which require interpretation by the technologist; these factors may also impede reproducibility.

In order to ensure procedures and techniques were alike among the laboratories, a second workshop, sponsored by Oncology Laboratories, Inc. was held. Each technologist performed the HTCFA, using the SOTO tumor with investigators and technologists comparing technique. In spite of previous attempts at comparability, numerous differences were found. These differences in procedures and techniques may or may not contribute to differences in assay results. At the close of this workshop, all participants agreed that performing the assay in an interactive fashion was critical to any subsequent reproducibility study conducted with laboratories geographically dispersed. Further trials are necessary to confirm this observation.

Selection of a relevant model is also important for demonstrating reproducibility of the HTCFA. Since procedural differences existed in all phases of the assay, participating investigators agreed to utilize a tumor system which necessitated performance of the entire assay. The SOTO tumor, grown in the athymic mouse fulfilled this requirement. However, significant problems exist with this tumor including variability in cloning efficiency and drug sensitivity among the various trials. These problems greatly inhibit statistical analysis of data collected from different trials and have prevented the establishment of standards which each laboratory must meet. Future trials are designed to alleviate some of these problems by having each laboratory process three specimens per trial and to test fewer drugs but at three concentrations each. Two of the drugs which will be omitted are melphelan, due to instability, and methotrexate, which appears to be inactive in this system. These changes will allow sufficient data acquisition to allow statistical analysis of each trial and eliminate the need to pool data from different trials. It will also provide information on consistency of results within each of the participating laboratories.

Data generated from the SECSG experience is similar to that of the EORTC, which conducted HTCFA reproducibility studies utilizing the human adenocarcinoma cell line WiDr.[3] The EORTC had inconsistencies in cloning efficiency and drug sensitivity which were not dissimilar to what was observed in this study. EORTC also used WiDr grown in monolayer, which fails to test the full HTCFA method, since disaggregation of solid tumor was not necessary.

The experience of the SECSG, supported by that of the EORTC, suggests that results of drug sensitivity testing in the HTCFA are reproducible from one laboratory to another if the definition of sensitivity is as described above. With many differences in technique resolved, with implementation of procedure revisions, and with the experience gained with the assay, SECSG investigators are cautiously optimistic that further reproducibility of results in the HTCFA can be achieved.

REFERENCES

1. Hamburger A, Salmon S, Kim M, et al: Direct cloning of human ovarian carcinoma cells in agar. Cancer Res 38: 3438-3444, 1978.

2. Alberts D. Unpublished data.

3. Rozencweig M: Quality control study of the human tumor clonogenic assay. Proc AACR (Abs) 24: 313, 1983.

Quality Control of a Multicenter Human Tumor Cloning System: The Southwest Oncology Group Experience

GARY M. CLARK, Ph.D.

Research Associate Professor
Department of Medicine

DANIEL D. VON HOFF, M.D.

Associate Professor
Department of Medicine
University of Texas Health Science Center
San Antonio, Texas 78284

INTRODUCTION

The human tumor cloning assay is an in-vitro soft agar technique, developed by Hamburger and Salmon, which purportedly allows the growth of clonogenic cells from human tumor specimens.[2,3] This assay offers a new strategy for screening chemotherapeutic agents for in-vitro activity against human tumors. Three new anticancer drugs, mitoxantrone, bisantrene, and 4' deoxydoxorubicin, have been brought to clinical trials partially on the basis of activity demonstrated in this in-vitro system.[6,8,9]

See Appendix for a list of participating institutions and investigators

Supported in part by grant 3 U10 CA 32102

HUMAN TUMOR CLONING
ISBN 0-8089-1671-8

Members of the Southwest Oncology Group considered that the cloning assay might aid in its selection and evaluation of new agents entering clinical trials. The majority of group institutions were interested in participating in such an effort with independent laboratories in each institution. But before incorporating the cloning assay for screening new agents in a number of laboratories, it was necessary to demonstrate that the system is reproducible across screening laboratories with a sufficient level of inter-laboratory quality control. This report summarizes the results of the Southwest Oncology Group's quality control experiments involving 18 cloning laboratories.

MATERIALS AND METHODS

Study Design

A standard human cell line was prepared at a reference laboratory. This cell line together with four concentrations of four standard antitumor drugs were mailed in a frozen state (1 day delivery) to each of 18 cloning laboratories including the reference laboratory itself. Cells were plated and incubated with the various drugs according to the protocol described below. After the plates were counted in the participating laboratories, they were shipped back to the reference laboratory where they were recounted manually by a technician.

Culture Material

All assays were performed using the HE_p-2 cell line obtained from the American Type Culture Collection. This cell line was established from epidermoid carcinoma tissue from the larynx of a 56-year-old Caucasian male.[4]

Drugs

The 4 antitumor agents used for these experiments were: adriamycin (.01, .1, 1, 10 ug/ml), cis-platinum (.02, .2, 2, 20 ug/ml), 5-fluorouracil (.6, 6, 60, 600 ug/ml), and vinblastine (.05, .5, 5, 50 ug/ml). The final concentrations were prepared in the central reference laboratory.

Cloning Assay

Cells were cultured in soft agar in triplicate as described by Hamburger and Salmon.[2,3] Single cell suspensions were suspended in

0.3% agar enriched with CMRL-1066 (Grand Island Biological Co., Grand Island, NY) supplemented with 15% heat inactivated horse serum, penicillin (100 units/ml), streptomycin (2 mg/ml), glutamine (2 mM), calcium chloride (4 mM) and insulin (3 units/ml). Prior to plating, asparagine (0.6 mg/ml), DEAE-dextran (0.5 mg/ml) (Pharmacia Fine Chemicals, Inc., Piscataway, NY) and freshly prepared 2-mercaptoethanol (50 mM) were added to the cells. One ml of the resultant mixture was pipetted onto 1 ml feeder layers in 35mm plastic petri dishes. The final concentration of cells in each culture was 1.0×10^4 cells in 1 ml of agar medium.

The underlayers consisted of McCoy's Medium 5A plus 15% heat inactivated fetal calf serum and a variety of nutrients as described by Pike and Robinson.[5] Immediately before use, 10 ml of 3% tryptic soy broth (Grand Island Biological Co.), 0.6 ml of asparagine, and 0.3 ml of DEAE-dextran were added to 40 ml of the enriched McCoy's 5A medium. Agar (0.5% final concentration) was added to the enriched medium and underlayers were poured in 35 mm petri dishes.

After preparation of both bottom and top layers, the plates were examined under an inverted stage microscope to assure the presence of a good single cell suspension. The plates were then incubated at 37°C in a 7% CO_2 humidified atmosphere.

Drug Sensitivity Studies

Prior to plating in agar, cell suspensions were transferred to tubes with a final concentration of 6.0×10^4 cells/ml in the presence of the appropriate drug with the drug tested at 4 concentrations in a series of replicate tubes. Cells were incubated with the drug at 37°C for 60 minutes in McCoy's 5A medium. The cells were then centrifuged at 150 xg for 10 minutes. After the post incubation centrifugation, the human tumor cells were washed twice with McCoy's 5A medium and prepared for culture as described above.

After 12 to 14 days in culture the number of tumor colony forming units (TCFUs defined as > 50 cells) on the triplicate control plates and the triplicate drug plates were counted. TCFUs on the triplicate plates were averaged to obtain one data point for the drug treated plates and another for the control, and dose-response curves were constructed. In-vitro sensitivity was defined in terms of the ratio between the number of TCFUs surviving on the drug-treated plates and those on control plates.

Statistical Analysis

Rather than rely on a single methodology, 4 quality control criteria were used to evaluate the participating laboratories. Each criterion placed emphasis on a different aspect of quality control.

Coefficient of Variation for Control Growth

We have previously shown that the standard deviation of control growth is highly correlated with the average number of control TCFUs.[1] The coefficient of variation (CV) is a measure of variability that is most appropriate when the standard deviation increases proportionally with the mean. It is defined as the ratio between the standard deviation of the number of TCFUs on the triplicate control plates divided by the average control growth. The National Cancer Institute currently uses the CV as a quality control criterion in its program to screen new agents with the human tumor cloning assay. The CV for control growth is required to be less than 50 in order for an assay on a fresh human tumor to be considered evaluable.[6] Since this quality control study used a cell line which theoretically ought to be more homogeneous and easier to clone than fresh human tumors, we required that the CV be less than 30 for a laboratory to be considered acceptable.

Sensitivity Disagreements

The remaining 3 quality control criteria are all based on the concept of percent survival of TCFUs in the presence of an antitumor agent. The cell line was defined to be sensitive to a particular concentration of a given drug if the percent survival was less than or equal to 30% and resistant if greater than 30%.

When conducting this type of quality control study, it is necessary to have a "gold standard" for comparing the participating laboratories. The true sensitivity pattern of the cell line must be determined at the time of the experiment. We chose to use the results from the two laboratories most experienced in the primary culture of human tumors in soft agar, namely the University of Arizona Health Sciences Center and the University of Texas Health Science Center at San Antonio. The sensitivity pattern of the cell line based on the results from these two laboratories is shown in Table 1.

TABLE 1
Drug Sensitivities Based on Experienced Laboratories

Adriamycin		Cis-platinum		5-FU		Vinblastine	
0.01	R	0.02	R	0.6	R	0.05	R
0.1	R	0.2	R	6.0	R	0.5	R
1.0	S	2.0	S	60	R	5.0	S
10	S	20	S	600	R	50	S

R = resistant (survival < 30%)
S = sensitive (survival > 30%)

The cell line was resistant to the two lowest concentrations of all 4 agents. It was sensitive to the 2 highest concentrations for 3 of the 4 drugs. Although it was resistant to all concentrations of 5-FU, there was a dose-response relationship across the 4 concentrations. The lack of sensitivity against 5-FU might be explained by the results of Yang and Drewinko[10] who found that 5-FU had a dramatic decrease in efficacy after 2 weeks regardless of storage conditions. Using the results in Table 1 as the "gold standard", the number of disagreements for each laboratory was determined. More than 25% disagreement was considered unacceptable.

IC70 Index

As a third criterion, the inhibitory concentration 70 (IC70) for a particular drug was defined as the lowest concentration that produced at least 70% kill, or equivalently, the lowest concentration at which the cell line was sensitive to that agent. Thus, the IC70s were: 1 ug/ml for adriamycin; 2 ug/ml for cis-platinum; > 600 ug/ml for 5-FU; 5 ug/ml for vinblastine (Table 1). An IC70 Index was constructed for each laboratory. For each drug the index had a value of 0 if the concentration was correctly identified; 1 if the laboratory disagreed by one concentration; and 4 if the laboratory disagreed by 2 or more concentrations. These values were summed for the 4 drugs. An IC70 Index greater than 4 was considered unacceptable. This allowed a laboratory to miss the IC70 by one concentration for each of the 4 drugs, or to miss the IC70 by 2 concentrations for just one of the drugs.

Survival Agreement Index

The 2 previous criteria evaluated the laboratories on their ability to declare the cell line sensitive or resistant, but did not utilize the actual percent survivals. Survival of 35% and 95% were equally correct as long as the "gold standard" was resistance. The final quality control criterion was based on the actual percent survivals and closely resembles a standard deviation. Within a laboratory, the percent survival for each of the 16 drug concentrations was compared with the "gold standard" from the 2 experienced laboratories. The difference was squared, summed over the 16 drug concentrations, and divided by 16. The Survival Agreement Index (SAI) was defined as the square root of this quantity. Thus, small values of the SAI indicate close agreement with the "gold standard", while large values indicate disagreement. Values greater than 30 were considered unacceptable.

RESULTS

Eighteen laboratories participated in this study. The control growth was quite variable among the laboratories ranging from 26

TCFUs to over 4,000. The study protocol was quite explicit in most areas, but did allow flexibility with regard to the day the cells were harvested. Therefore, some of this variability could possibly be related to the day after passage at which the cells were harvested in the individual laboratories. Comparisons between the reported counts and the manual counts at the reference laboratory were remarkably good with a correlation coefficient of .99, thereby virtually eliminating counting errors as a source of variability.

The 18 laboratories have been randomly coded and designated A through R. Table 2 presents the results of applying the 4 quality control criteria to the data from these 18 laboratories. A total of 10 laboratories were considered unacceptable on one or more of the 4 quality control criteria. Only 4 laboratories had CVs greater than 30 for control growth. There was no relationship between large CVs and the number of control TCFUs. One laboratory did not report the individual plate counts so neither the standard deviation nor the coefficient of variation could be determined. Thus, 13 of the 18 laboratories were rated acceptable on this quality control criterion.

TABLE 2
Results of Quality Control Experiment

Lab.	CV for Control Growth	Sensitivity Disagreements	IC70 Index	Sensitivity Agreement Index
A*	35*	4/16	4	34*
B*	5	1/16	1	50*
C*	30*	7/16*	10*	38*
D*	37*	4/8 *	8*	45*
E*	31*	7/16*	7*	33*
F	11	0/16	4	19
G*	4	5/14*	9*	50*
H	3	4/16	3	28
I	16	0/16	0	12
J*	19	6/16*	9*	31*
K	8	4/16	4	27
L*	6	6/16*	10*	48*
M*	27	2/16	1	34*
N	14	1/16	0	12
O	10	1/16	1	22
P	4	4/16	3	23
Q	NA	1/16	1	22
R*	13	3/16	5*	27

* = Unacceptable rating
NA = Individual plate counts not available.

Six laboratories had more than 25% sensitivity disagreements and were considered unacceptable based on this criterion. Three of these laboratories were among those with unacceptably high CVs. Seven laboratories had an IC70 Index greater than 4. There was a high correlation between the number of sensitivity disagreements and the IC70 Index, although the IC70 Index did identify one additional unsatisfactory laboratory. The Sensitivity Agreement Index was the most severe quality control criterion. Nine laboratories had a Sensitivity Agreement Index greater than 30 and were rated unsatisfactory. Two of these 9 had satisfactory ratings on the other 3 criteria.

The 10 laboratories that were rated unacceptable on one or more of the 4 quality control criteria were then invited to perform the experiment again under identical conditions using the same protocol. The results of the second attempts are summarized in Table 3. Three laboratories had unacceptably high CVs on control growth; one laboratory had an unacceptable number of sensitivity disagreements and did not satisfy the IC70 Index criterion; and one laboratory had an unacceptable Sensitivity Agreement Index. Four of the 10 laboratories were acceptable on all criteria on the second attempt. Thus, after 2 attempts, a total of 12 of 18 laboratories satisfied all 4 quality control criteria.

TABLE 3

Results of Quality Control Experiment--Second Attempt

Lab.	CV for Control Growth	Sensitivity Disagreements	IC70 Index	Sensitivity Agreement Index
A*	32*	3/16	3	22
B	16	0/16	0	13
C*	9	1/16	1	37*
D	8	4/16	3	28
E+	8	1/10	4	19
G*	4	5/16*	6*	28
J*	41*	2/16	2	24
L	11	1/16	1	22
M*	58*	2/16	2	24
R	24	1/16	1	14

* = Unacceptable rating
+ = Did not complete all tests

DISCUSSION

This quality control experiment was just the first phase of a quality control program within the Southwest Oncology Group. The first task was to develop reasonable quality control criteria. Since the control growth is used to compute the percent survival for each of the drug sensitivity evaluations, it is most important that the estimate of control growth be very reliable. A large CV for control growth indicates that something is wrong with that experiment. The single cell suspension may not have been prepared adequately, the number of cells plated in the triplicate petri dishes may not have been the same, or there may have been an error in counting the number of TCFUs. Whatever the problem, the result is an imprecise estimate of control growth, and therefore, an imprecise estimate of percent survival. Experiments with large CVs should probably be discarded even in routine, day to day assays.

Establishing quality control criteria for the results of drug sensitivity testing requires that a "gold standard" be determined. Two options were considered for the present study. First, the median percent survival was computed across all 18 participating laboratories for each concentration of each drug. This had the disadvantage of including the results from laboratories that would eventually be rated unsatisfactory. The next alternative was to identify laboratories that were experienced in the primary culture of human tumors in soft agar. The University of Arizona Health Sciences Center and the University of Texas Health Science Center at San Antonio satisfied this criterion and the results from these two laboratories were averaged to establish the "gold standard" for this study. Even though these two "gold standards" were quite comparable with a correlation coefficient of .84, neither technique was entirely satisfactory. In the future, the "gold standard" should be established in a single reference laboratory by performing the experiment in parallel 3 or 4 times and averaging the results.

The 3 quality control criteria that were used to evaluate drug sensitivity testing placed emphasis on different aspects of the assay. Since the results for a particular agent are often reported simply as resistant or sensitive, the first criterion addressed the ability of the laboratories to distinguish between resistance and sensitivity. However, when screening a new agent with the assay, it may be more important to examine dose-response curves and identify which concentrations of that agent produce sensitivity. The IC70 Index addressed this issue, but in fact, added only marginal information in addition to that provided by the number of sensitivity disagreements. The Sensitivity Agreement Index demands much more precision of the laboratories, but is an accurate gauge only if the "gold standard" has been defined accurately. In the present study, the Sensitivity Agreement Index identified 9 out of 10 laboratories that were unacceptable on one or more of the quality control criteria.

There were a number of unexpected complications during this study. The one-day mailing and delivery of culture materials and drugs was not always accomplished as planned. Delivery problems resulting in thawed drugs and nonviable cells necessitated the remailing of drugs to 5 institutions. Cells were remailed to 8 institutions a total of 13 times. Such problems should be anticipated when designing future multi-institutional experiments requiring standardized protocols and materials. The apparent resistance of the cell line to 5-FU was puzzling until the results of Yang and Drewinko[10] were presented. Their finding that both 5-FU and melphalan showed a 50% decrease in efficacy after 2 weeks regardless of storage conditions must be taken into account in future studies. A pilot study utilizing the proposed protocol on the same culture material before and after shipment might be well advised before launching into a full-scale experiment.

In future studies, both positive and negative controls should be included among the drugs to be tested. This was accomplished fairly well in the present experiment because the concentrations selected did span the entire dose-response curve for 3 of the drugs. Both the drugs and the drug concentrations should be blinded in future studies. Although there was no indication that knowing the drug concentrations introduced bias into the present study, it will be especially important to blind the positive and negative controls. It would also be useful to include blinded, duplicate drugs at the same concentration. This would allow reproducibility to be assessed within a laboratory as well as between laboratories. It is also important to note that cell lines are not quite the same as fresh human tumors. Admittedly the logistics of performing this type of quality control experiment with human tumors are quite formidable, but perhaps effusions or other fluid specimens could be incorporated into future experimental designs.

Even though this study was just the first phase of a quality control program, the results do indicate that the human tumor cloning assay can be performed at several institutions and yield reproducible results. This could have a profound impact on the National Cancer Institute's program to screen new potential anticancer agents. Each year over 10,000 new compounds are available for screening. In order for a compound to be evaluated in a clinical trial, it must pass through several steps. First, it must show activity in a pre-screen and then demonstrate activity against a panel of human tumors. At this stage, it must be successfully formulated and complete a thorough toxicology evaluation. Currently, only 3 to 4% of the compounds that are screened against the human tumor panel, are evaluated in clinical trials. In order to bring approximately 20 new agents into the clinic each year, 500 to 600 compounds must be evaluated and have activity in a pre-screen and be tested against the human tumor panel. Multi-center, cooperative groups that have the capability to perform the human tumor cloning assay in a quality controlled manner could help identify these 500 to 600 compounds. If each of the 12 laboratories that satisfied the quality control criteria in this study could screen 100 compounds each year, and if the activity rate for these compounds is just 10%, then the Southwest Oncology Group alone could screen approximately one-fourth of the required number of drugs. It might also be anticipated that agents that are active in the human tumor cloning assay might have a better likelihood of activity against the human tumor panel since an evaluation with this assay will include a variety of human tumors. In this case, the cooperative groups could be an even greater asset to the screening effort.

APPENDIX

The following investigators participated in this study: Cancer Center of Hawaii, Noboru Oishi; Cleveland Clinic Foundation, G. Thomas Budd; Henry Ford Hospital, Richard C. Klugo, Robert M. O'Bryan, Eugene J. Tilchen; Louisiana State University Medical Center, Dominic Fann, Lee Roy Morgan; Mt. Sinai Medical Center, Tom Anderson; Ohio State University, Antionette Beynen, George E. Milo; Seattle Tumor Institute of Swedish Hospital Medical Center, Gary E. Goodman; Scott and White Clinic, Juan Caraveo, William McCombs; Tulane University, Carl M. Sutherland; University of Arizona Health Sciences Center, David S. Alberts, Sydney E. Salmon; University of Kansas Medical Center, Bruce F. Kimler, Chan H. Park; University of Michigan Hospital, Ron B. Natale; University of Mississippi Medical Center, John P. Kapp, Ralph B. Vance; University of Oregon Health Science Center, Bruce W. Dana; University of South Dakota School of Medicine, David Elson; University of Southwest Louisiana, John M. Rainey; University of Texas Health Science Center at Galveston, Victor Gupta; University of Texas Health Science Center at San Antonio, Hannah Sullivan; Wayne State University Medical Center, Charles D. Haas, Greg Kyle.

REFERENCES

1. Clark GM, Von Hoff DD: Statistical considerations for in vitro/in vivo correlations using a cloning system, in Hill DT (ed): Human Tumor Drug Sensitivity Testing In Vitro - Techniques and Clinical Applications. New York, Academic Press, 1983, pp. 225-233
2. Hamburger AW, Salmon SE: Primary bioassay of human tumor stem cells. Science 197:461-462, 1977

3. Hamburger AW, Salmon SE: Primary bioassay of human myeloma stem cells. J Clin Invest 60:846-854, 1977
4. Moore AE, Sabachewsky L. Toolan HW: Culture characteristics of four permanent lines of human cancer cells. Cancer Res 15:598-602, 1955
5. Pike B, Robinson W: Human bone marrow colony growth in agar gel. J Cell Physiol 75:77-81, 1970
6. Salmon SE, Liu RM, Casazza AM: Evaluation of new anthracycline analogs with the human tumor stem cell assay. Cancer Chemother Pharmacol 6:103-110, 1981
7. Shoemaker RH, Wolpert-DeFillipes MK, Makuch RW, Venditti JM: Use of the human tumor clonogenic assay for new drug screening. Proc AACR 24:311, 1983
8. Von Hoff DD, Coltman CA Jr, Forseth B: Activity of mitoxantrone in a human tumor cloning system. Cancer Res 41:1853-1855, 1981
9. Von Hoff DD, Coltman CA Jr, Forseth B: Activity of 9-10 anthracene-dicarboxaldehyde bis((4,5-dihydro-1 H-imidazol-2-yl) hydrazone) dihydrochloride (CL216,942) in a human tumor cloning system. Cancer Chemother Pharmacol 6:141-144, 1981
10. Yang LY, Drewinko B: The stability of the lethal efficacy of antitumor drugs. Proc AACR 24:315, 1983

III. STUDIES OF SPECIFIC TUMOR TYPES

CONTINUOUS *IN VITRO* GROWTH OF T-CELL ACUTE LYMPHOBLASTIC LEUKEMIA OF CHILDHOOD (T-ALL) IN AN HYPOXIC ENVIRONMENT

Stephen D. Smith, M.D.,Roger A.Warnke, M.D., Margaret Shatsky, Ph.D., Michael P. Link, M.D., Pamela S. Cohen, M.D., and Bertil E. Glader, Ph.D., M.D.

Departments of Pediatrics and Pathology
Stanford University School of Medicine
Stanford, California, and
Division of Hematology/Oncology
Children's Hospital at Stanford
Palo Alto, California.

INTRODUCTION

Normal peripheral blood T-lymphocytes proliferate *in vitro* in response to mitogens (or specific antigens). Once activated, long-term growth of these cells is possible if the cultures are supplemented every 3-4 days with T-cell growth factor (TCGF, Interleukin-2, IL2) (13). Malignant cells from patients with T-cell leukemia differ from normal peripheral blood T-lymphocytes in that they respond directly to IL-2. (11,14) Such proliferation is however, dependant on the presence of IL-2 and growth of these malignant T-cells has been maintained for 3-4 months. The cells grown by this method are phenotypically relatively mature T-cells for they are TdT negative and E-rosette positive (11,14).

Supported by PHS Grant Nos. CA34710 and CA34233 awarded by the National Cancer Institute, DHHS and the American Cancer Society Grant CH182A. Dr. Smith is a scholar of the Leukemia Society of America.

HUMAN TUMOR CLONING
ISBN 0-8089-1671-8

In the past, a few T-cell lines have been established from children with T-cell acute lymphoblastic leukemia (T-ALL) (1,2,7,8,10). These cell lines proliferate without the presence of IL-2 and long-term growth has been established. Recently, we reported a method of colony formation and subsequent cell line establishment from children with non-Hodgkin's lymphoma. (15,16) In the present report we describe the results of culturing samples from 7 children with T-cell ALL. Each sample formed colonies on agar and 3 cell lines were established.

MATERIALS AND METHODS

Source of Malignant T-Cells.
The data described are derived from 7 patients, age 6 to 18 years, with T-cell acute lymphoblastic leukemia. The protocol procedures were approved by the Medical Committee for the Use of Human Subjects in Research and informed consent was obtained from the parents or patients.

The clinical and pathologic data on the patients are presented in Table I. Each patient had a mediastinal mass and the malignant cells rosetted with sheep RBCs. (3) Within four hours of collection, malignant lymphoid cells were separated by Ficoll-Hypaque density sedimentation and washed twice with McCoy 5A media containing 15% fetal calf serum, penicillin (50 U/ml) and streptomyin (50 μg/ml) (complete media).

TABLE 1

DEMOGRAPHIC DATA OF 7 CHILDREN WITH T-CELL ACUTE LYMPHOBLASTIC LEUKEMIA

PATIENT	AGE (YEARS)/SEX	CLINICAL STATUS	MATERIAL ANALYZED
1	16,M	Relapse	Bone marrow
2	18,F	Relapse	Bone marrow
3	12,M	Relapse	Peripheral blood
4	10,M	At diagnosis	Bone marrow
5	10,M	At diagnosis	Pleural effusion
6	6,M	At diagnosis	Peripheral blood
7	13,F	At diagnosis	Bone marrow

Establishment and Maintenance of Cell Lines

The basic methodology for culturing malignant T-cells was a modification of the methods previously reported (15,16). Briefly, the separated malignant cells were warmed to 37°C in complete media and pipetted repeatedly to breakup cell aggregates. A single cell suspension was mixed with agar (0.3%) and plated onto 35 mm petri dishes containing a previously prepared feeder layer. Two types of feeders were utilized. One contained a combination of complete media with 1 X 10^6 peripheral blood leukocytes, 10% human plasma and 0.5% agar. The second feeder type contained a combination of complete media with 10% normal human serum and 0.5% agar (serum feeders). The petri dishes were incubated at 37°C in an Heraeus incubator gassed with 5% O_2, 6% CO_2, and 89% N_2. Neither mitogens, thiols, antifungal agents nor interleukins were added to this assay. Colonies were plucked after 18 days of growth, suspended in complete media (without agar) and transferred to fresh feeders. Cells were observed every 3-5 days and were transferred to fresh feeders when the media appeared acidic or when the cell number approached 5 x 10^5 cells/cc.

Studies to Characterize Cell Surface Antigens

Monoclonal antibodies to Leu-1 (Pan-T), Leu-2a (T cytotoxic/suppessor), Leu-3a (T helper), Leu-4 (Pan-T), Leu-5 (E-rosette receptor), Leu-6 (thymocyte), Leu-9 (Pan-T), CALLA and HLA-DR were generously provided by Becton-Dickinson Corporation (Mountain View, CA). OKT_9 and OKT_{10} were provided by Ortho Pharmaceuticals (Raritan, NJ) and TO15 (Pan-B) was supplied by David Mason (Oxford University). Surface immunoglobulin was detected by the $F(AB')_2$ fragments of FITC-conjugated goat anti-human immunoglobulin (TAGO, Burlingame, CA).

Cell surface and cytoplasmic antigens on the individual colonies were determined by immunohistochemical staining (18). Discrete colonies were aspirated from the agar with a finely drawn pipette. Colonies were cytocentrifuged onto glass slides and fixed in acetone for 5 minutes. The slides were washed with modified PBS (3 minutes) and then sequentially incubated with 1) a specific monoclonal antibody (30 minutes), 2) biotin conjugated horse anti-mouse IgG (30 minutes) (Vector Labs, Burlingame, CA), 3) avidin-horseradish peroxidase (30 minutes) (Vector Labs, Burlingame, CA), 4) Diaminobenzidine (DAB) (5 minutes), and 5) copper sulfate (5 minutes). The slides were washed with modified PBS between each incubate. Methylene blue was used as a counterstain. Cells were scored as positive which demonstrated dark staining on their cell membrane or cytoplasm.

Once cell lines were established, cell surface antigens were identified by the binding of monoclonal antibody as detected by indirect immunofluorescence. Separated lymphoblasts (10^6) were placed in plastic tubes and incubated with 1 μg of purified monoclonal antibody at 4^0 C. After 20 minutes incubation, cells were washed twice in cold phosphate buffered saline (PBS) with 0.02% sodium azide. Next the washed cells were incubated (4^0C, 20 minutes) with fluorescein isothiocyanate-conjugated goat anti-mouse immunoglobulin (Tago, Burlingame, CA), washed twice, fixed in 1% formaldehyde in PBS and analyzed for fluorescent staining with a fluorescence activated cell sorter (FACS IV, Becton-Dickinson Electronics Laboratory, Mountain View, CA). The intensity of fluorescence of 10,000 cells in each sample was determined. The results were expressed as percent positive cells compared to background fluorescence when non-reactive myeloma protein was utilized in place of specific antibody.

Enzyme Analysis

After centrifugation on a Ficoll-Hypaque density gradient, the interface mononuclear cells were washed three times and suspended at a concentration of 0.5 - 1.5 x 10^7 cells/ml in 150 M K-phosphate buffer (pH 7.1). The cells were lysed by alternate freezing and thawing and membranes were removed by centrifugation. Protein determinations were made on the enzyme extracts by the method of Lowry (6). Adenosine deaminase (ADA) and nucleoside phophorylase (NP) activities were determined according to established methods (4,5,17). The resulting enzyme activity was expressed in units defined as Δ OD over 1 minute divided by the protein concentration.

Assays for terminal deoxynucleotidyl transferase (TdT) activity were performed using reagents available from Bethesda Research Laboratories, Bethesda, MD. The procedure was performed according to the manufacturer's directions.

RESULTS

Initially, T-ALL cells were cultured _in vitro_ on feeders containing 1 X 10^6 leukocytes in 0.5% agar. On such feeders, both normal myeloid colonies (CFU-C) and T-ALL colonies readily grew and colonies were identified on the basis of morphology. (15) Subsequently, immunohistochemical staining showed that a few of these colonies contained predominately B-cells. It was observed that a few Mu+, Kappa+, Lambda-, HLA-Dr+ colonies sporadically grew on such feeders.

Because of the lack of specificity of this assay methodology, subsequent T-ALL samples were plated on feeders which

TABLE 2

CLONING EFFICIENCY (C.E.) OF CLINICAL SAMPLES AND CORRESPONDING CELL LINES.

	Initial C.E.	Cell Line C.E.
Sample #1	0.005%	---
K-T-1	---	20%
Sample #2	0.004%	---
SUP-T2	---	40%
Sample #3	0.002%	---
SUP-T3	---	55%

contained filtered human serum and no leukocytes. Such cultures have had a low cloning efficiency (approximately 0.005%) but the cells grown have been exclusively T-cells. On a morphologic basis (all cultures tested) colonies grown on serum feeders have been comprised mostly of immature lymphoblasts. No CFU-C nor fibroblast colonies have been identified. With immunohistochemical staining with monoclonal antibodies (4 cultures tested), colony cells have been Leu 1+, Leu 5+, Leu 9+, HLA-DR- and TO15-. In addition, morphologic and immunohistochemical staining has shown that some colonies possessed a subpopulation of normal monocytes and mature T-lymphocytes. The significance of these cells is not known and experiments evaluating the influence of normal cells on T-ALL cloning efficiency are now in progress.

The initial cloning efficiency (C.E.) of T-ALL cells was very low (Table 2). After 18 days of initial growth, individual colonies were transferred in complete media to fresh feeder and continued growth occurred in 3 samples. The transferred colonies divided slowly in liquid and cultures usually had to be maintained for 3-4 months before sufficient cells (5×10^7) were available for complete analysis. When cells were replated back in agar (on serum feeders) 1×10^5 cells, a very high C.E. occurred (Table 2). The secondary C.E. of the cell lines was often 10,000 times greater than the initial C.E. of the patient sample. Such results indicate that an inhibitor to cell growth may be present in the patient's sample or that most of the original tumor cells lacked the capacity for colony growth.

TABLE 3

REACTIVITY* OF MONOCLONAL ANTIBODIES WITH MALIGNANT CELLS

	K-T-1	SUP-T2	SUP-T3
Leu 1	+	+	+
Leu 2a	-	+/-**	+
Leu 3a	-	-	+
Leu 4	-	-	-
Leu 5	+	+	+
Leu 6	-	-	+
Leu 9	+	+	+
OKT_9	-	+	-
OKT_{10}	-	-	+
CALLA	-	-	-
HLA-Dr	-	-	-
SIg	-	-	-

* Reactivity = >20% positive cells
** Initial analysis positive but subsequent analyses negative X 2.

Results of the immunophenotyping studies of the cell lines are shown in Table 3. Each cell line was positive for multiple T-cell markers and each rosetted with sheep RBCs. Each cell line was positive for pan-T lymphocyte markers (Leu 1,5,9) and negative for Calla, HLA-Dr and SIg. None expressed Leu 4 -- a marker of mature T-cells. The K-T-1 cell line was negative for all other markers tested and was classified as an early thymocyte cell line. The SUP-T2 reacted with Leu 2 initially but repeat analyses were negative twice. Such results could represent cell antigen changes in culture or a technical problem during immunophenotyping studies. SUP-T-2 expressed OKT_9 and appears to be an early thymocyte cell line. SUP-T3 expressed Leu 2a, Leu 3a and Leu 6, all markers of the mid-thymocyte stage of T-cell differentiation. The immunophenotype of these cells clearly classified them as immature thymocytes and not as mature T-cells.

The results of the enzyme analysis of the cell lines are shown in Table 4. These data show very high ADA activity for each cell line tested. In our laboratory, ADA levels of normal peripheral blood mononuclear cells (after Ficoll-Hypaque gradient separation) is 5.4 Eu/mg protein. ADA activity is known to be high in immature T-cell malignancies and the values obtained on these T-cell lines are the highest

TABLE 4

ADENOSINE DEAMINASE (ADA) AND PURINE NUCLEOSIDE PHOSPHORYLASE (NP) ACTIVITY IN T-CELL LINES.

	ADA (Eu/protein)	NP (Eu/protein)
K-T-1	149	0.20
SUP-T-2	69	0.14
SUP-T-3	166	0.19

obtained in our laboratory. Nucleoside phosphorylase (NP) activity was low in each cell line. Normal peripheral blood mononuclear cells have a value of 0.36 Eu/mg protein and values below 0.20 Eu/mg protein are characteristically associated with T-cell leukemia. Repeated enzyme analysis showed the same general results and the values expressed in Table 4 are the average of two separate analyses performed at least one month apart. No spontaneous pattern of change in ADA or NP activity was noted with continuous growth of the cell lines *in vitro*. TdT activity was also present in each cell line with greater than 90% of the cells demonstrating nuclear fluorescence.

DISCUSSION

The methodology used in this report combined the use of colony formation on agar and continuous growth of cell lines in liquid. The initial cloning efficiency was low but when T-ALL cells were plated on serum feeders, selective growth of malignant T-cell colonies resulted.

Colonies readily grew after 18 days of culture in a hypoxic environment. Passage of single colonies in attempts to establish a cell line is time consuming and difficult work. Screening of initial colony growth by immunohistochemical staining has significantly improved our assay. Currently a panel of monoclonal antibodies (Leu 1, Leu 6, Leu 9, HLA-Dr and TO15) is used to screen for antigens present on the initial colony cells. Colony cells can then be classified as B-lymphocytes (Leu 1-, Leu 6-, Leu 9-, HLA-Dr+, TO15+) or mature T-lymphocytes (Leu 1+, Leu 6-, Leu 9+, HLA-Dr-, TO15-). The specific immunophenotype of patient's bone marrow cells can be compared to the colony reactivity with specific monoclonal antibodies. When the initial SUP-T3 colonies were shown to be Leu 1+, Leu 6+, Leu 9+, HLA-Dr-, T015-, significant time was then devoted to passage of these colonies and establishment of an immortal cell line resulted.

Once a cell line was established, the cloning efficiency of that line was determined by replating the cells in agar with a serum feeder. A high cloning efficiency occurred, often 10,000 fold greater than the initial cloning efficiency. The reasons for the initial low cloning efficiency of the patient samples are under investigation. Since the cell line cloning efficiency is high, the nutritional needs of the cells in culture are probably being met. However, the basic nutrients needed for initial colony formation may be different from that of cell lines. Clearly, patient's samples contain immunocompetent cells (which are not present in the cell lines) which may block initial T-ALL colony formation. The presence of such cells may be the reason why only colonies from patient's at relapse have been successfully established as cell lines (Table 1) (15, 16).

Proof that the colonies grown represent malignant T-ALL colonies is based on morphology (all samples tested), immunohistochemical staining (4 of 7 samples tested) and analysis of established cell lines (3 cell lines established). Since the cell lines provide an unlimited number of cells for analysis, they provide the best means of identification of the character of the colony cells. The three T-ALL lines have the immunophenotype of immature T-cells. In the classification of T-cell differentiation proposed by Schlossman (12), K-T-1 and SUP-T2 have cell surface antigen present on early thymocytes while SUP-T3 has antigens present on mid-thymocytes.

Enzyme analysis showed that the cells lines were TdT positive and possessed high ADA and low NP activity. This pattern of enzyme activity is similar to that of immature thymocytes and has been observed in other T-lymphocyte cell lines established from children with T-ALL.

The cells plated _in vitro_ grew in an hypoxic environment without the presence of conditioned media, mitogens, or interleukins. McCoy 5A media with serum (human and fetal calf) were the only nutritional supplements added. The procedure used and the cells grown are different from a recent report from Ruscetti and Gallo (13). In their report, T-cell malignancies were cultured _in vitro_ in the continued presence of partially purified T-cell growth factor (interleukin 2, IL-2). Human IL-2 is an immunoenhancing protein which is produced by helper T-cells following mitogen stimulation. In the presence of IL-2, 8 of 11 T-ALL samples proliferated and these cells were absolutely dependent on a fresh supply of IL-2 administered every 3-4 days (11,13,14). The cells were not grown as colonies and growth continued for only four months _in vitro_. The cells grown were TdT- and E-rosette+ typing them as mature T-cells (14). The cultured cells reported in the present study are TdT+ and express the surface immunophenotype of immature T-cells (Table 3). Growth of these cells was independent of any added specific growth factor and all the lines have been growing for longer than four months. These T-cell characteristics are similar to the IL-2 independent T-ALL cell lines established by other investigators (1,2,7,8,11).

REFERENCES

1. Adams RA, Pthier L, Flowers A et al: The question of stem-lines in human acute leukemia. Exp Cell Res 62:5-10, 1972.
2. Foley GE, Lazarus H, Farber S, et al: Continuous culture of human lymphoblasts from peripheral blood of a child with acute leukemia. Cancer 18:522-529, 1965.
3. Froland SS: Binding of sheep erythrocytes to human lymphocytes. A probable marker of T-lymphocytes. Scand J Immunol 1:269-274, 1972.
4. Hopkinson PJ: Further data on the adenosine deaminase (ADA) polymorphism and a report of a new phenotype. Ann Human Genet 32:361-365, 1969.
5. Kalckar HM. Differential spectrophotometry of purine compounds by means of specific enzymes. J Biol Chem 167:427-443, 1947.
6. Lowry AL, Randall RJ: Protein measurement with the folin phenol reagent. J Biol Chem 167:193:263, 1951.
7. Minowada J, Moore GE: T-lymphocyte cell lines derived from patients with acute lymphoblastic leukemia. In: _Comparative Leukemia Research 1973. Leukemogenesis_ (eds: Y. Ito and RM Dutcher). Tokyo. University of Tokyo Press; Basel:Karger, 1975. pp 251-261.
8. Minowada J, Ohnuma T, Moore GE: Rosette-forming human lymphoid cell lines. I. Establishment and evidence for origin of thymus-derived lymphocytes. J Nat Cancer Inst 49:891-895, 1972.
9. Miyoshi I, Hiraki S, Tsubota T et al: Human B-cell, T-cell and null cell leukemia cell lines devised from acute lymphoblastic leukemias. Nature 267:843-844, 1977.
10. Morikawa S, Tatsumi E, Baba M et al: Two E-rosette forming lymphoid cell lines. Int J Cancer 21:166-170, 1978.
11. Poiesz BJ, Ruscetti FW, Mier JW, Woods AM, Gallo RC: T-cell lines established from human T-lymphocytic neoplasias by direct response to T-cell growth factor. Proc Natl Acad Sci 77:6815-6819, 1980.

12. Reinherz E, Kung P, Goldstein G, Levey R and Schlossman S: Discrete stage of human intrathymic differentiation: Analysis of normal thymocytes and leukemic lymphoblasts of T-cell lineage. Proc Natl Acad Sci 77:1588-1592, 1980.
13. Ruscetti FW, Gallo RC. Human T-cell lymphocyte growth factor: Regulation of growth and function of T-lymphocytes. Blood 57:379-394, 1981.
14. Ruscetti FW, Poiesz BJ, Tarella C, Gallo RC: T-cell growth factor and the establishment of cell lines from human T-cell neoplasias. In: M.A.S. Moore (ed) Progress in Cancer Research and Therapy. Vol. 23. Maturation Factors and Cancer. Raven Press, N.Y. pp 153-166.
15. Smith SD, Rosen D: Establishment and characterization of a human null-cell lymphoblastic lymphoma cell line (K-LL-3). Int J Cancer 23:494-503, 1979.
16. Smith SD, Wood GW, Fried P and Lowman JT: In vitro growth of lymphoma colonies from children with non-Hodgkin's lymphoma. Cancer 48:2612-2623, 1981.
17. Smyth JF, Poplack DG, Holiman BJ, Leventhal BG, Yarbro G. Correlation of adenosine deaminase activity with cell surface markers in acute lymphoblastic leukemia. J Clin Invest 62:710-712-1978.
18. Wood G, Warnke R: Suppression of endogenous avidin-binding activity in tissues and its relevance to biotin-avidin detection systems. J Histochem Cytochem 29:1196, 1981.

Factors Influencing The Clonogenicity Of Human Lung Cancer Cells

PETER R. TWENTYMAN AND GERALD A. WALLS
Medical Research Council
Clinical Oncology and Radiotherapeutics Unit,
Hills Road, Cambridge, England.

INTRODUCTION

We are interested in studying tumor cell heterogeneity in human lung cancer. In order to carry out the studies it has been our aim to isolate clonal subpopulations of cells from lung cancer material either in established culture or directly from clinical specimens. As there is undoubtedly a large degree of selection involved in any clonal isolation procedure (especially where the cloning efficiency is low), we believe that an optimisation of cloning conditions will result in the widest range of clone types being available for further study. In this paper we report the results of some of our early studies on the influence of various factors upon the clonogenicity of human lung cancer cells.

MATERIALS AND METHODS

The two basic cloning methods which have been used in these studies are those described by Courtenay and Mills (1978) and by Carney et al (1980). These will be referred to for convenience as the 'Courtenay' assay and the 'Carney' assay.

1. The 'Courtenay' assay

The medium used in this assay was modified Hams F12 supplemented with 15% fetal calf serum and with penicillin and streptomycin (all supplied by Gibco Biocult Ltd.). Red cells were prepared by carrying out a cardiac puncture on August rats using preservative-free heparin, separating out the red cells by centrifugation, rinsing three times with phosphate-buffered saline and resuspending to the original blood volume in medium. The red cell suspension was then heated to 44°C for 1 hour and stored at 4°C for up to one month. An 8x dilution in medium was carried out immediately before use of the red cells. A 6% solution of Agar Noble (Difco) was prepared and sterilized by boiling for 15 minutes. This was then diluted x 10 in prewarmed medium (44°C) to give a final concentration of 0.6 mg/ml and the solution kept at 44°C until required. Suspensions of the test cells in medium were prepared (as below) at 2.5 x the required final concentration and kept at 37°C.

HUMAN TUMOR CLONING
ISBN 0-8089-1671-8

Immediately before setting up the cloning tubes, 2.0ml of the test cell suspension was added to 0.5 ml of red cell suspension followed by 2.5 ml of 0.6% agar solution. Aliquots of 1ml of this suspension were then placed into each of 3 or 4 Falcon 2051 sterile plastic tubes. These tubes were stood in crushed ice until the agar set. They were then each gassed for 6 seconds using a plugged Pasteur pipette attached to a gas line from a cylinder containing 90% nitrogen, 5% oxygen and 5% carbon dioxide and the top of each tube 'snapped' closed. The tubes were placed in racks in plastic cake boxes which were then gassed with the same mixture for 10 minutes before being sealed and incubated at 37°C. After 7 and 14 days of incubation, 1ml of medium was added to the agar plug in each tube and the tubes and boxes regassed. In those tubes where incubation continued for 4 weeks (ie. in experiments with material other than established cell lines), 1ml of medium was removed at 21 days and 1ml of fresh medium was added. At the end of the incubation period, the agar plug was tipped out from each tube into the inverted lid of a 5cm plastic petri dish. The base of the dish was then pushed down onto the plug so that the agar spread in a thin layer. Colonies containing >50 cells were counted under an inverted microscope. In some experiments, heavily-irradiated (100 Gray of 250kV x-rays) cells were included in the live cell suspensions. The gas mixture was also varied in some experiments and tubes were gassed with either 95% air plus 5% CO_2 or 95% nitrogen plus 5% CO_2 as alternatives to the usual mixture. The effect of omitting the red cells was also studied.

2. The 'Carney' assay

The medium used in this assay was RPM1 1640 supplemented with 20% heat-inactivated fetal calf serum and with penicillin and streptomycin (all supplied by Gibco Biocult Ltd.) A 5% solution of LGT Agarose ("Seaplaque" - Marine Colloids Inc.) in water was prepared and sterilized by boiling for 15 minutes. This was then diluted x 10 in prewarmed medium (44°C) to give a final concentration of 0.5%, and 2ml aliquots of this agarose were pipetted into 35mm plastic petri dishes (Falcon) and placed in the refrigerator to set. A 6% solution of agarose was also prepared, diluted x 10 with medium to give a final concentration of 0.6% and kept at 44°C. Suspensions of the test cells in medium were prepared (as below) at 2 x the required final concentration and kept at 37°C. Immediately before plating, 2.5ml of the test suspension was mixed with 2.5ml of 0.6% agarose solution and aliquots of 1ml were then pipetted into each of 3 or 4 of the 35 mm petri dishes containing the agarose 'bottom layer' and newly removed from the refrigerator. The plates were placed into plastic sandwich boxes which were then gassed for 10 minutes with a mixture of 95% Air and 5% CO_2 amd incubated at 37°C. Plates were examined and colonies (>50 cells) counted under the inverted microscope at 7-8, 14-15 and 21-23 days after plating (N.B. The agarose used in our studies is slightly different for that used by Carney et al (1980) and we are currently carrying out a comparison of the 2 types in this assay).

3. Preparation of cell suspensions

The established cell lines used in this study were NCI-H69, a small cell lung cancer line kindly supplied by Dr. D. Carney; POC, a small cell lung cancer line kindly supplied by Dr. M. Ellison; MOR a lung adenocarcinoma line kindly supplied by Dr. M. Ellison, and COR-L23 a large cell anaplastic lung cancer cell line derived in our own laboratory. In addition, we used samples of small cell lung cancer material (COR-L24(N), COR-L32 and COR-L42) maintained in vitro in our laboratory from clinical samples but not yet established as cell lines. Single cell suspensions were prepared from all of these by treatment for 15 minutes with 0.4% trypsin and 0.02% versene in P.B.S. and subsequently either pipetting or pulling the suspension in and out of a 10ml syringe fitted with a 25 gauge needle. If any clumps were seen, the suspension was passed one or more times through tightly packed cotton gauze in a sterile glass funnel and re-examined. The pleural effusion (COR-L51) from a patient with small cell lung cancer was cleared of red cells by centrifugation on Ficoll, passed through cotton gauze and pipetted. The clinical sample of normal bone marrow was prepared from a large hip-replacement specimen, agitated for 2 hours in medium containing 1mg/ml of neutral protease (Sigma Type IX) and cleared of red cells by centrifugation on Ficoll. Cloning of these last 2 specimens was always carried out on aliquots stored in liquid nitrogen and thawed immediately before assay.

Xenografts of NCI-H69 and COR-L23 were grown in nude mice. Tumours were disaggregated to a single cell suspension by mincing with scissors, agitation in medium containing 1mg/ml neutral protease for 2 hours and filtration through gauze. Numbers of live cell plated varied between 200 (for established cell lines), 10^5 (for non-established in vitro cultures) and 10^6 (for normal bone marrow).

RESULTS

The data for experiments comparing the two cloning assays are shown in Tables 1-3. In each table, two 'head to head' experiments are shown, ie. the left hand column for the 'Courtenay' assay is the same experiment as the left hand column for the 'Carney' assay. Plating efficiencies are always expressed as absolute fractions and NOT as percentages.

TABLE 1
CLONING COMPARISON - PLATING EFFICIENCY OF ESTABLISHED CELL LINES

	Courtenay		Carney	
NCI-H69 Small Cell	0.46	0.39	0.19	0.009
POC Small Cell	0.71	0.61	0.30	0.04
L23 Large Cell Anaplastic	0.19	0.21	0.08	0.05
MOR Adenocarcinoma	0.07	0.11	0.05	0.003

All values are fractional (not %) plating efficiencies

TABLE 2
CLONING COMPARISON - PLATING EFFICIENCY OF RECENT 'IN VITRO' SPECIMENS OF SMALL CELL LUNG CANCER

	Courtenay		Carney	
COR - L32	0.27	0.091	0.0003	0.00024
COR - L24 (N)	0.007	0.006	0.00002	0.0
COR - L42	0.003	0.015	0.00005	0.0

All values are fractional (not %) plating efficiencies

TABLE 3
CLONING COMPARISON - PLATING EFFICIENCY DIRECT FROM CLINIC

	Courtenay		Carney	
COR - L51 pleural effusion	0.00013	0.00016	0.0	0.0
NBM 1 normal bone marrow	0.00005	0.00003	0.0	0.0

All values are fractional (not %) plating efficiencies

It may be seen that, in all cases, a higher plating efficiency was produced in the 'Courtenay' assay. In general, also the colonies were larger and contained more cells. For the 'Carney' assay, colonies were usually larger and more numerous at day 14 than at day 7, but often a decline in numbers and a general deterioration in quality of colonies was seen by day 21. The effect of changing gas mixture and the presence or absence of rat red cells in the 'Courtenay' assay is shown in Table 4.

TABLE 4

EFFECT OF RED CELLS AND GAS MIXTURE IN THE COURTENAY ASSAY

SPECIMEN	RBC	% OXYGEN IN GAS		
		0	5	20
NCI-H69 small cell line	+	0.43	0.45	0.28
	-	0.09	0.02	0.00
COR-L23 large cell line	+	0.15	0.13	0.12
	-	0.11	0.08	0.07
MOR adenocarcinoma line	+	0.17	0.16	0.10
	-	0.15	0.15	0.04
NCI-H69 Xenograft	+	0.40	0.44	0.26
	-	0.18	0.07	0.05
COR-L23 Xenograft	+	0.20	0.18	0.10
	-	0.04	0.01	0.00
COR-L51 small cell pleural effusion direct from clinic	+	5.0×10^{-5}	6.5×10^{-5}	0.8×10^{-5}
	-	0.0	0.0	0.0

All values are fractional (not %) plating efficiencies.

It may be seen that for the NCI-H69 small cell line, the presence of red cells is a very important factor. In addition, the low oxygen tensions give much better plating efficiencies than that obtained in 20% O_2. The effect of red cells appears less important for the two non-small cell lines, but for each of them, the combination of no red cells and high oxygen is a particularly bad combination. The results for NCI-H69 xenograft cells are similar to those for the in vivo line. For COR-L23, however, there is a much greater dependance upon red cells for the xenograft. The clinical small cell specimen also shows a large dependence upon red cells and a lower cloning efficiency at 20% oxygen than at lower tensions.

The influence of heavily irradiated feeder cells upon the plating efficiency of NCI-H69 in the presence or absence of red cells is shown in Table 5.

TABLE 5

EFFECT OF HR CELLS ON NCI-H69 CLONING IN THE COURTENAY ASSAY

No of HR cells	0	10^3	10^4	10^5	10^6
+RBC	0.86	0.84	0.82	0.87	0.09
-RBC	0.11	0.13	0.22	0.74	0.02

All values are fractional (not %) plating efficiencies

It may be seen that, in the presence of red cells, there is no effect of adding HR cells until a deleterious effect is seen at 10^6 HR cells due, presumably, to medium depletion. In the absence of red cells, a small positive effect is seen with 10^4 HR cells, and 10^5 HR cells raises the plating efficiency almost to the level seen with RBC. A rapid fall is again seen with 10^6 HR cells.

DISCUSSION

In these studies the cloning efficiency was higher in the assay of Courtenay and Mills (1978) than in that of Carney et al (1980). Although we may have introduced some small modifications of the two methods in our laboratory, it does not seem likely that these would result in major changes in the results. The plating efficiency of line NCI-H69 is, indeed, much higher in both the assays used here than the plating efficiency originally quoted by Gazdar et al (1980) for this line. This is probably due to the higher passage number at which our studies were carrried out, but also provides good evidence for the potentially high plating efficiency obtainable with either method. It should, however, be noted that the "Courtenay" assay can, apparently, produce colonies from normal bone marrow. This is, perhaps, not surprising as the use of rat red cells was first advocated by Bradley et al (1971) for improving the clonogenicity of normal mouse marrow. In these studies, however, the use of exogenous colony stimulating factor was required in addition to red cells for colony formation to occur. We are currently studying a larger number of normal bone marrows to find out the range of plating efficiencies obtained in the 'Courtenay' assay. In the specimen reported here, however, although many diffuse 'colonies' were seen, and not counted, the colonies which were counted had a great morphological similarity to colonies produced by small cell lung cancer specimens. This may prove a problem when using the 'Courtenay' assay to isolate colonies of small cell lung cancer from involved marrow specimens. We believe that the generally higher plating efficiency which we have seen for all specimens in the 'Courtenay' assay is likely to result from the 'replenishable' nature of the assay (ie. weekly medium addition) and from the 'feeder' effect of the rat red cells.

Our findings regarding the effect of red cells and gas mixture are in good agreement with those reported by Tveit et al (1981) for human melanoma xenografts. They found that the presence of red cells was the most important factor but also that 5% oxygen produced better plating efficiencies than 20% oxygen. It is interesting that, in our studies, we find gassing with oxygen-free nitrogen to be at least as good as 5% oxygen. The residual oxygen present after this gassing is unlikely to be greater than 1% and we are presently determining more specifically the lower limit for colony formation.

The use of heavily irradiated feeder layers in clonogenic assays is common although there have been few studies in specific human tumour cell clonogenic assays to determine optimal numbers. When the number of live cells plated is greater than 10^4 (as in our studies using material other than established lines) it may be unecessary. The experiment reported in this paper indicates that even for a line such as NCI-H69 with a very high plating efficiency, the use of 10^4 HR cells is inadequate for optimal cloning. The required number for lines of low plating efficiency may be much higher. The use of red cells clearly, however, obviates the requirement for HR cells, at least for line NCI-H69.

ACKNOWLEDGEMENTS

We thank Professor Norman Bleehen and Dr. Hugo Baillie-Johnson for providing clinical specimens of lung cancer material. We are also grateful to Dr. Adi Gazdar of the N.C.I.-Navy Medical Oncology Branch for cytological evaluation of cultured material.

REFERENCES

Bradley, T.R., Telfer, P.A. and Fry, P. The effect of erythrocytes on mouse bone marrow colony development in vitro. Blood, 1971, 38, 353-359.

Carney, D.N., Gazdar, A.F. and Minna, J.D. Positive correlation between histological tumor involvement and generation of tumor cell colonies in agarose in specimens taken directly from patients with small-cell carcinoma of the lung. Cancer Res., 1980, 40, 1820-1823.

Courtenay, V.D. and Mills, J. In vitro colony assay for human tumours grown in immune-supressed mice and treated in vivo with cytotoxic agents. Br. J. Cancer, 1978, 37, 261-268.

Gazdar, A.F., Carney, D.N., Russell, E.K. et al. Establishment of continuous, clonable cultures of small-cell carcinoma of the lung which have amine precursor uptake and decarboxylation cell properties. Cancer Res., 1980, 40, 3502-3507.

Tveit, K.M., Endresen, L., Rugstad, H.E., Fodstad, O. and Pihl, A. Comparison of two soft-agar methods for assaying chemosensitivity of human tumours in vitro: malignant melanomas. Br. J. Cancer, 1981, 44, 539-544.

The Human Tumor Cloning Assay in Maxillofacial Surgery

HANS-ROBERT METELMANN[a], JÜRGEN BIER[a], CLAUDIA METELMANN[b] and DANIEL D. VON HOFF[c]

[a]Department of Maxillofacial Surgery, Klinikum Steglitz & [b]Institute of Medical Microbiology, Free University of Berlin, Berlin, F.R.Germany
[c]University of Texas Health Science Center at San Antonio, Medicine/Oncology, San Antonio, TX, USA

In treating tumors of the oral and facial region, it is often difficult to decide whether to establish the indication for surgical treatment or for chemotherapy. Operative removal of the tumor still offers the best chances of success for the majority of patients in our field. However, there are many patients who cannot be expected to submit to mutilation through a radical, curatively intended tumor resection. Most patients are not able to endure the necessary removal of the tongue with all the accompanying functional losses. Of particular interest are patients who would be endangered by the general medical risks of a long-term surgical intervention.

To dispense with operative therapy and initiate chemotherapy in such situations requires a serious decision, in which a drug predictive test can provide decisive aid. The human tumor cloning assay fulfills basic prerequisites for such a predictive test (1-5) and to develop a reliable test system is the main area of application of the cloning assay in our field.

As first and main prerequisite, the assay as described comprehensively by Hamburger and Salmon in 1977 (6, 7) renders it possible to successfully clone malignant tumors from all areas of the face and the oral cavity in vitro. In accordance, Mattox and Von Hoff have previously reported similar results (8, 9). In our laboratory 64 tumors, 74% of all specimens obtained, were cultured with more than 5 typical tumor cell colonies per plate and an average plating efficiency of 0.039%. 33 of the patients' tumors (38%) had sufficient growth (more than 30 colonies) to perform drug sensitivity studies.

The best cloning results were achieved for endophytic carcinomas, hidroadenocarcinomas and cervical metastases in general.

HUMAN TUMOR CLONING
ISBN 0-8089-1671-8

There is some confirmatory evidence that colonies successfully grown in soft agar are composed of the patient's malignant cells (8, 10, 11). Electron microscopy of cells in colonies from squamous cell carcinomas demonstrated characteristic tonofilaments. Inoculation studies with athymic nude mice revealed the proof that colonies of a cultured carcinoma of a tongue were malignant. Occasionally we succeeded in detecting epithelial pearls in vitro.

Since failures of cloning were often connected with contamination of the plates by microorganisms, a second prerequisite is the preparatory decontamination of tumor samples. Primary tumors in the oral cavity, particularly those growing in an ulcerative manner, generally show considerable colonisation with blastomyces and pyogenic germs. We have carried out preliminary experiments with an in vivo decontamination procedure by short-term implantation in normally immunocompetent mice.

The technical procedure is simple. Solid tissue samples with a maximal diameter of 4 mm can best be placed in the retroperitoneal space of anesthetized CF 1 mice. During the implantation period, the number of contaminating microorganisms is reduced by 98% within 72-96 hours. Nevertheless, the malignancy of the tumor is preserved, as substantiated by its ability to grow in athymic nude mice. No serious histological alterations are noticeable in the 3 day period. Cell vitality decreases only slightly, presumably because the necrotic cells are cleared away by the host. When the tumor samples were removed after 3 day implantation, 72% were free of microorganisms, 5% remained contaminated and 23% were lost through death of the mouse or histolysis. The successfully decontaminated tumor material could be further processed in the typical manner in the tumor cloning assay. Of 7 squamous cell carcinomas cleansed by the implantation method, 3 tumors showed clonal growth and were suitable for drug testing.

As a third prerequisite, we were able to demonstrate the selectivity of the cloning assay for malignant tumor cells in our special field. In cloning attempts with 46 chronically proliferative inflammatory processes, including cystic sacs, dental granulomas, fistulae and hyperplasias, and 20 benign tumors, including hemangioma, fibroma, osteoma, lipoma and fibroadenoma, we did not find in any of the cases more than 5 colonies per plate.

A further decisive prerequisite: it is possible to test all cytostatic drugs that are of importance in maxillofacial surgery. Cis-platinum, methotrexate, bleomycin, Fluorouracil, vincristine and the newly developed peplomycin permit screening without difficulty. Cyclophosphamide and DTIC whose anti-tumor activity depends upon microsomal activation, can be evaluated in vitro (12, 13) after 1 hour incubation with 100 μg/ml of a commercially available S9 liver preparation from rats (Litton Bionetics, Kensington, MD). Another possibility to apply these drugs is to use 4 HPCyclophosphamide as the effective metabolite of CTX that is available ready-made (Asta-Werke, Brackwede, Germany) or to activate DTIC in a simple procedure by 1 hour exposure to white light (14, 15).

The fifth prerequisite is the pars-pro-toto meaning of the drug test results; i.e., the maxillofacial squamous cell carcinomas in their various extensions and metastases usually show the same cytostatic sensitivity pattern. In an initial experiment, tumor samples from 3 different areas of a primary maxillary carcinoma and 2 contralateral cervical lymph node metastases of the same tumor were subjected to separate cloning and drug testing in parallel. Not surprisingly, we found different plating efficiency. But all tumor areas showed the same order of drug sensitivity. Since one part of the tumor represents the whole tumor from this point of view, we have called it pars-pro-toto behavior. We verified this observation in 8 patients and found a poor correspondence in only 1 case.

The precise place the cloning assay will have in predicting whether a patient with a maxillofacial cancer will or will not respond to an anticancer drug has yet to be defined. Certainly today the treatment for patients with cancer in our field is to utilize drug combinations. The cloning assay has multiple problems in duplicating pharmacology of single agents, so the testing of combinations of drugs is frought with even more problems. Realizing this problem exists, an in vivo in vitro correlation study has been instituted to explore the reliability and suitability of the assay as a drug predictive test.

All patients had tumors of the oral cavity, exclusively pathologically documented squamous cell carcinomas. Though, in principle, surgery was indicated, they were subjected to chemotherapy because of an internistic, anesthetic or psychiatric contraindication. Chemotherapy was carried out with a combination regimen including cis-platinum, methotrexate and bleomycin. In vitro drug tests were set up prior to chemotherapy employing at least 2 samples from different areas of the tumors in separate assays. Single drug concentrations in vitro corresponded to 1/10 of the peak plasma concentrations that is clinically achieved in man (16). A more than 70% decrease in tumor-colony-forming units on drug treated plates compared with control plates was set as a definition of drug efficiency in vitro. The drug efficiency in vivo was assessed after a 3 month clinical course.

At this initial stage of the study, tumors from 10 patients had sufficient growth (> 30 colonies) on control plates and enough cells to perform drug sensitivity tests. For 1 patient in vitro drug prediction was incomprehensible, because 3 different assays performed in parallel with tumor material from different areas of the carcinoma yielded 3 different predictions of drug sensitivity. In 4 patients where treatment was ineffective the test had correctly predicted the failure. In 4 patients with partial or clinically complete remission, the test indicated effectiveness for at least one of the drugs in use.

The last drug test concerned an 86-year-old man in poor general condition with a carcinoma in the anterior floor of the mouth and the advanced stage of the disease. Because of an internistic contraindication the patient was subjected to chemotherapy. Two individual drug tests were performed, and the results consistently showed all drugs to be ineffective. However, after completion of the 3rd therapy cycle, we found to our surprise an almost complete remission of the cancer. It has been possible to maintain this condition up to the present time, and now, 14 months after commencement of therapy, the general condition of the patient is satisfactory.

This false-negative test result casts some uncertainty upon the cloning assay as a predictive test. On the whole, the human tumor cloning assay fulfills many basic prerequisites for clinical application in maxillofacial surgery. It remains the most important task of our further research, however, to improve the reliability of cytostatic testing in view of the decisions contingent upon these test results.

REFERENCES

1. *Salmon SE, Soehnlen BJ, Durie BGM et al.: Clinical correlations of drug sensitivity in tumor stem cell assay (Abstract). Proc Am Assoc Cancer Res 20:340, 1979*
2. *Salmon SE, Alberts DS, Durie BGM et al.: Clinical correlations of drug sensitivity in the tumor stem cell assay, in Mathe G (ed): Recent Results in Cancer Research. New York, Springer Verlag, 1980, pp 300-305*
3. *VonHoff DD, Casper J, Bradley E et al.: Association between human tumor colony forming assay results and response of an individual patient's tumor to chemotherapy. Am J Med 70: 1027-1032, 1981*

4. Salmon SE, Von Hoff DD: In vitro evaluation of anticancer drugs with the human tumor stem cell assay. Semin Oncol 8: 377-385, 1981
5. VonHoff DD, Mattox DE, Cortes E et al.: The human tumor stem cell assay system: A potential aid for intraarterial drug trials in head and neck cancer. Rev Sud Oncol 3: 15-19, 1979
6. Hamburger A, Salmon SE: Primary bioassay of human myeloma stem cells. J Clin Invest 60: 846-854, 1977
7. Hamburger A, Salmon SE: Primary bioassay of human tumor stem cells. Science 197: 461-463, 1977
8. Mattox DE, VonHoff DD: Culture of human head and neck cancer stem cells using soft agar. Arch Otolaryngol 106: 672-675, 1980
9. Mattox DE, VonHoff DD: In vitro stem cell assay in head and neck squamous carcinoma. Amer J Surg 140: 527-531, 1980
10. Harris GJ, Zeagler J, Hodach A et al.: Ultrastructural analysis of colonies growing in a human tumor cloning system. Cancer 50: 722-726, 1982
11. VonHoff DD, Johnson GE: Secretion of tumor marker in the human tumor stem cell system (Abstract). Proc Am Assoc Cancer Res 20: 51, 1979
12. Kovach JS, Ames MW, Powis G et al.: Use of a liver microsome system in testing drug sensitivity of tumor cells in soft agar (Abstract). Pro Am Assoc Cancer Res 21: 257, 1980
13. Metelmann H-R, VonHoff DD: Application of a microsomal drug activation system in a human tumor cloning assay. Invest New Drugs 1: 27-32, 1983
14. Loo TL, Houshoulder GE, Gerulath AH et al.: Mechanism of action and pharmacology studies with DTIC (NSC 45388). Cancer Treat Rep 60: 149-153, 1976
15. Metelmann H-R, VonHoff DD: In vitro activation of Dacarbazine (DTIC) for a human tumor cloning system. Int J Cell Clon 1: 24-32, 1983
16. VonHoff DD, Forseth B, Metelmann H-R et al.: Direct cloning of human malignant melanoma in soft agar culture. Cancer 50: 696-701, 1982

PARAMETERS FOR PREDICTING GROWTH OF SQUAMOUS CELL CARCINOMA OF THE HEAD AND NECK IN THE HUMAN STEM CELL ASSAY (HTSCA)

JULIE A. KISH, M.D.

Assistant Professor of Medicine

JOHN CRISSMAN, M.D.

Professor of Pathology

CHARLES D. HAAS, M.D.

Associate Professor Medicine

JOHN ENSLEY, M.D.

Assistant Professor of Medicine

GLEN CUMMINGS, PH.D.

Assistant Professor of Medicine

ARTHUR WEAVER, M.D.

Professor of Surgery

JOHN JACOBS, M.D.

Assistant Professor of Surgery

MUHYI AL-SARRAF, M.D.

Professor of Medicine
Wayne State University
Detroit, Michigan

HUMAN TUMOR CLONING
ISBN 0-8089-1671-8

ABSTRACT

Establishment of sufficient growth of squamous cell carcinoma of the head and neck *in vitro* for drug testing is fraught with difficulty. Specimens from 49 patients were processed for human tumor stem cell assay (HTSCA), 42 specimens were from resections and 7 from biopsies. Forty-seven of these were evaluated histologically for growth predictive parameters including cytoplasmic and nuclear differentiation, mitoses, inflammatory response, vascular invasion and pattern of invasion. Ten of 49 (20%) specimens produced sufficient colonies for drug testing (≥30 colonies/plate); 10 (20%) had marginal growth (6-29 colonies/plate); 21 (42%) had no growth (0-5 colonies/plate) and 6 (14%) were contaminated. Only 2 (5%) specimens had insufficient cells to plate. Nine of ten samples that were adequate for drug testing were from specimens with little or no morphologic evidence of cytoplasmic keratinization. This was significant at p=0.01. All of the specimens that grew adequate colonies for drug testing had 1-5 mitoses per hpf (p=0.03). Forty percent of tumors that had evidence of vascular invasion produced sufficient colonies for drug evaluation (p=0.04). No significant correlation could be made between nuclear differentiation, inflammatory response, or pattern of invasion with adequate growth in the HTSCA.

INTRODUCTION

The human tumor stem cell assay (HTSCA) was developed to individualize a patient's chemotherapeutic regimen for their neoplasm.[3,8] The proportion of all human adult tumors that grow with a plating efficiency adequate for assessment of drug activity (20-30 colonies/500,000 cells plated) is less than 50% in most series.[6,9,10,11] Squamous cell carcinomas of the head and neck are particularly difficult to culture in this system; the percentage of adequate growth of these tumors varies from 0-25%, compared to squamous cell carcinoma of the lung which averages sufficient colony growth in approximately 50% of the cases.[14] In view of this poor growth rate, 49 head and neck specimens submitted for HTSCA were carefully evaluated histologically to identify parameters which predicted for growth.

MATERIALS AND METHODS

Forty-nine specimens were obtained fresh from the operating room and transported to the laboratory in enriched Hank's solution. Specimens were minced and teased in transport media to allow release of single cells. Collagenase was used in a small number of cases. The cell suspensions were then strained through 60 micron screens and aspirated through #21 and #25 gauge needles to produce single cell suspension. These procedures were performed under sterile conditions in a laminar flow hood. Cell counts were performed by hemocytometer, and trypan blue dye exclusion was done to assess viability. Cell suspensions were then manipulated to a concentration of $3x10^6$cells/ml in enriched Hank's solution. Aliquots containing $1.5x10^6$ cells/ml were added to 1.0 ml of enriched Hank's solution then placed in tubes without drugs (controls) and into tubes containing various chemotherapeutic agents. These were incubated at 37°C for 1 hour then washed x 2 with enriched Hank's basic salt solution. The cells were then plated in triplicate in a two layer soft agar system at 0.5 x 10^6cells/plate as described by Hamburger and Salmon with the deletion of mercaptoethanol and asparagine. The plates were then incubated at 37°C in humidified 7.5% CO_2 in room air.

Control plate colonies were manually counted weekly with an inverted phase microscope. Clumps and aggregates were evaluated by comparison of colony forming plates with a "fixed cold control" plate. This was a randomly selected formalin fixed refrigerated control plate. Subtraction of clumps was carefully performed. Colonies were defined by light microscopic morphology and were required to consist of at least 30 cells. Random Papanicolaou stains were performed on colonies to ensure tumor presence. Adequate growth in our laboratory was defined as at least 30 colonies per control plate. Control plates with 6 to 29 colonies were defined as marginal growth and drug sensitivities could not be performed. No growth was defined as 0-5 colonies per control plate. Contamination was determined after plating.

The original histologic specimens were evaluated for the histologic parameters cited in Table 1 by a pathologist who was unaware of the performance in the HTSCA. The tumor cell population was examined for: 1) cytoplasmic kera-

TABLE 1

Histologic Grading Score
Squamous Cell Carcinoma

Cytoplasmic Keratinization	Well Formed Pearls 50%	Attempts at Pearl Formation 20-50%	Poor 5-20%	None
Nuclear Differentiation	Few Enlarged Nuclei 75% Mature	Moderate Number Enlarged Nuclei 75-50% Mature	Numerous En- larged Nuclei 25-50% Mature	Anaplastic, Little Evidence of Mature Nuclei
Mitoses #	0-1	1-3	3-5	5
Inflammatory Response	Marked Con- tinuous Rim	Moderate Patchy	Slight, Few Small Patches	None
Vascular Invasion	Not Identified	-	-	Present
Pattern of Invasion	Pushing Borders	Solid Cords	Thin, Irr. Cords	Single Cells

TABLE 2

Head and Neck Cancer
Stem Cell Growth in Relation to Site

Site	No.	a	b	c	d	e	Overall Growth*	Adequate for Drug Testing**
Oral Cavity	6	4	2	-	-	-	0/6 (0%)	0/6 (0%)
Tongue	8	3	1	-	2	2	4/7 (57%)	2/8 (25%)
Tonsil	5	3	-	1	-	1	1/5 (20%)	1/5 (20%)
Glottic	19	8	1	1	5	4	9/19 (47%)	4/19 (21%)
Supraglottic	5	1	1	-	2	1	3/5 (60%)	1/5 (20%)
Others	6	2	1	-	1	2	3/6 (50%)	2/6 (33%)

One tonsil specimen and one glottic specimen had insufficient cells for plating.

a - no growth
b - contaminated
c - insufficient
d - marginal
e - adequate
* marginal (d) + adequate (e)
** >30 colonies/plate

TABLE 3

Head And Neck Cancer
HTSCA
Specimens Adequate For Drug Testing

Sample Number	Colony Count	Colony Efficiency
1	32	0.006
2	35	0.007
3	43	0.0086
4	54	0.01
5	61	0.012
6	62	0.0124
7	66	0.0132
8	68	0.013
9	77	0.015
10	141	0.028

tinization, 2) nuclear differentiation and 3) number of mitoses. The tumor-host interface was evaluated by the 1) degree of inflammatory response, 2) presence of vascular invasion and 3) the tumor growth pattern.

RESULTS

Forty-nine specimens were submitted for HTSCA from two Wayne State University affiliated hospitals (Harper-Grace Hospital, Veterans Administration Medical Center, Allen Park). Forty-two were from surgical resections and 7 from biopsies. The primary sites varied, with larynx the most common primary site (Table 2). When the tissue was harvested from a major resection, lymph node metastases were preferentially used for HTSCA. No specimens were grossly contaminated prior to plating, although some contamination in mucosal or ulcerated specimens was expected.

Ten of forty-nine (20%) had sufficient cells for drug testing ($\geq$30 colonies/control plate), and an additional ten specimens had marginal growth (6-29 colonies/control plate). This produced an overall growth rate of 41%. Twenty-one of forty-nine specimens (42%) had no growth (0-5 colonies/plate). Only two specimens of forty-nine (4%) had insufficient cells to plate and six of forty-nine (12%) were contaminated as judged by bacterial or fungal growth. The cloning efficiencies overall ranged from 0.0018 to 0.028. The median cloning efficiency for samples with 30 colonies or greater was 0.012. The specimens with marginal growth all had cloning efficiencies less than 0.005. The colony counts for samples adequate for drug testing ranged from 32-141 (Table 3). Those samples with marginal growth had colony counts from 8-27. Many of the well differentiated tumors produced significant amounts of debris that might have been misinterpreted as colonies if the formalin fixed cold controls had not been available for comparison.

Of the ten specimens that grew, 8 were from resections and 2 from biopsies. Of the samples with insufficient cells for plating, one was from a biopsy and one was from a resection. Contamination occurred in 6 biopsy specimens from the upper areo-digestive tract - lip, buccal mucosa, and larynx.

Histologic evaluation was performed on 47/49 samples.

Two samples submitted only had focal areas of tumor on the slide and could not be fully evaluated. The lack of cytoplasmic differentiation was identified by the paucity of intracellular and extracellular keratin formation. Its absence was statistically significant (p=0.01) in those specimens with greater than 30 colonies/plate as measured by the chi square test. Three successfully cultured specimens had no attempts at cytoplasmic keratinization; six had poor cytoplasmic keratinization (Table 4). Forty-seven percent (9/19) of evaluable samples with little or no keratin formation grew, whereas 3% (1/28) of the evaluable samples with moderate to prominent keratin production grew. Six of ten specimens adequate for drug testing had 3-5 mitoses per hpf while four of ten had 1-3 mitoses per hpf. This achieved statistical significance by chi square analysis (p=0.03) (Table 5). Nuclear differentiation did not correlate with colony formation (Table 6). The inflammatory response in the tumors that grew consisted predominately of a few small patches of lymphocytes; however, this was not statiscally significant (Table 7). The presence of vascular invasion was significant (p=0.04) with six of 15 (40%) samples demonstrating adequate growth (Table 8). The manner of tumor invasion of tissue (pushing borders, solid cords, thin cords, or single cells) was evenly distributed among those that grew vs. those that did not grow (Table 9).

DISCUSSION

Achieving a rate of 20% of samples adequate for drug testing is comparable to that achieved by others.[4,7,14] It is not surprising that the more poorly differentiated tumors demonstrated a higher cloning efficiency in this assay than the well differentiated tumors since proliferative capacity is a function of the stem cell and reflects a lack of differentiation. Selby and Buick have clearly delineated the reciprocal relationship that exists between the proliferative potential of a human tumor population and its degree of cellular differentiation.[11] The work of Meyer, _et al_ with breast cancer differentiation and estrogen receptor status supports this premise.[9] Mattox noted that as tumor differentiation of squamous head and neck cancer went from poor to well differentiated, the percent of

TABLE 4

Head And Neck Cancer

Stem Cell Growth In Relation To Cytoplasmic Keratinization

Degree of Cytoplasmic Keratinization	No.	a	b	c	d	e	Overall Growth*	Adequate for Drug Testing**
High	16	7	2	-	7	-	7/16 (46%)	0/16 (0%)
Moderate	12	6	2	-	3	+1	4/12 (33%)	1/12 (8%)
Low	11	4	+1	-	-	6	6/11 (50%)	6/11 (50%)
None	8	4	-	1	-	3	3/8 (37%)	3/8 (37%)
						p=0.01		

a - no growth
b - contaminated
c - insufficient
d - marginal
e - adequate
* marginal (d) + adequate (e)
** >30 colonies/plate

TABLE 5

Head And Neck Cancer

Stem Cell Growth In Relation To Histological Grading

Mitoses *	No.	a	b	c	d	e	Overall Growth*	Adequate for Drug Testing**
0-1	13	7	2	-	4	-	4/13 (31%)	0/13 (0%)
1-3	15	7	-	-	4	4	8/17 (47%)	4/17 (24%)
3-5	14	5	2	-	1	6	7/15 (47%)	6/15 (40%)
>5	5	2	1	1	1	- p=0.03	1/6 (17%)	0/6 (0%)

a - no growth
b - contaminated
c - insufficient
d - marginal
e - adequate
* marginal (d) + adequate (e)
** > 30 colonies/plate

TABLE 6

Head And Neck Cancer

Stem Cell Growth In Relation To Histological Grading

Nuclear Differentiation	No.	a	b	c	d	e	Overall Growth*	Adequate for Drug Testing**
Few Enlarged Nuclei 75% Mature	4	1	1	-	1	1	2/4 (50%)	1/4 (25%)
Mod. Enlarged Nuclei 75-50% Mature	12	6	-	-	4	2	6/12 (50%)	2/12 (17%)
Numerous Enlarged Nuclei	22	11	4	-	4	3	7/22 (32%)	3/22 (14%)
Anaplastic	9	3	-	1	1	4	5/9 (56%)	4/9 (44%)

a - no growth
b - contaminated
c - insufficient
d - marginal
e - adequate
* marginal (d) + adequate (e)
** >30 colonies/plate

TABLE 7

Head And Neck Cancer
Stem Cell Growth In Relation To Histological Grading

Inflammatory Response	No.	a	b	c	d	e	Overall Growth*	Adequate for Drug Testing**
Marked Continuous Rim	6	4	-	-	2	-	2/6(33%)	0/6(0%)
Moderate Patchy	9	2	2	-	4	1	5/9(56%)	1/9(11%)
Slight Few Small Patches	24	11	3	1	1	8	9/24(38%)	8/24(33%)
None	8	4	-	-	3	1	4/8(50%)	1/8(13%)

a - no growth
b - contaminated
c - insufficient
d - marginal
e - adequate
* marginal (d) + adequate (e)
** > 30 colonies/plate

TABLE 8

Head And Neck Cancer
Stem Cell Growth In Relation To Histological Grading

Vascular Invasion	No.	a	b	c	d	e	Overall Growth*	Adequate for Drug Testing**
Not Present	32	14	5	-	9	4	13/32 (41%)	4/32 (13%)
Present	15	7	-	1	1	6	7/15 (47%)	6/15 (40%)
						p=0.04		

a - no growth
b - contaminated
c - insufficient
d - marginal
e - adequate
* marginal (d) + adequate (e)
** > 30 colonies/plate

TABLE 9

Head And Neck Cancer

Stem Cell Growth In Relation To Histological Grading

Pattern of Invasion	No.	a	b	c	d	e	Overall Growth*	Adequate for Drug Testing**
Pushing Borders	7	3	-	-	3	1	4/7 (57%)	1/7 (14%)
Solid Cords	15	8	3	-	1	3	4/15 (27%)	3/15 (20%)
Thin, Irreg. Cords	16	8	2	-	3	3	6/16 (38%)	3/16 (19%)
Single Cells	9	2	-	1	3	3	6/9 (67%)	3/9 (33%)

a - no growth
b - contaminated
c - insufficient
d - marginal
e - adequate
* marginal (d) + adequate (e)
** > 30 colonies/plate

overall growth in the HTSCA went from 71% to 40%.[8] The histologic labels of well, moderate and poorly differentiated squamous tumors are based on a conglomerate of parameters and subjective review. The Jacobson grading scale has been used to quantify histologic grading for head and neck tumors as the Gleason system does for prostate cancer. Johns, using the Jacobson grading, found no statistical correlation between cloning efficiency and overall score on the modified Jacobson grading.[5]

Our study attempted to dissect the overall histological grade utilizing the components of the modified Jacobson scale and to identify the specific parameters that might correlate with growth in the assay. The meticulous examination of tumor histology led to a statistically significant correlation of adequate growth in the HTSCA and the amount of cytoplasmic keratinization, number of mitoses and presence of vascular invasion. Three of ten samples adequate for drug testing had no keratin; 6/10 had very small amounts (5-20%). Overall nine of nineteen (47%) samples with little to no keratin grew, whereas only 1/28 samples (3%) with moderate to prominent keratinization grew. The number of mitoses seen in the samples that were adequate for testing ranged from 1-5/hpf (p=0.03). This reflects the rapid turnover seen in this poorly differentiated tumor population. There were only five samples with greater than 5 mitoses per hpf. Forty percent of specimens with 3-5 mitoses per hpf were adequate for testing. Vascular invasion is considered to be a manifestation of aggressive tumor behavior and its ability to metastasize. Of those tumors adequate for drug testing in the HTSCA, 40% (6/15) had vascular invasion compared to 13% (4/32) that did not (p=0.04). Therefore we can conclude that cytoplasmic keratinization is the most important histological parameter predicting for growth in the HTSCA and tumors with little or no keratin have a higher probability of growing in the assay than not. In addition, tumors with rapidly proliferating cell populations as evidenced by frequent mitoses and vascular invasion also have an increased frequency of growth in the HTSCA.

In vivo, however, well differentiated tumors of the head and neck with a slower cell turnover metastasize and kill the host. Therefore, stem cells must be present and these tumors should grow in the assay. The question arises whether our inability to clone well differentiated head and neck tumors in the HTSCA reflects an actual decreased number of stem cells, an altered form of the stem cell, or an iatrogenic influence produced by single cell suspension preparation. Analysis of tumor populations by flow cytometry and thymidine labelling should provide a better definition of the differences between well differentiated and poorly differentiated tumors. Increasing the yield of single cell suspensions with improved disaggregation techniques should lead to an increase in the number of stem cells harvested. The physical environment for growth *in vitro* may not adequately reflect the *in vivo* milieu of the well differentiated tumors. Further evaluation of the growth environment is essential and currently ongoing.

CONCLUSION

The HTSCA has tremendous potential for the treatment of head and neck cancer. Poorly differentiated squamous carcinomas of the head and neck from resections with little to no keratinization, frequent mitotic figures and vascular invasion have higher frequency of growth in the HTSCA. The current techniques for single cell suspension preparation and the soft agar system of Salmon and Hamburger do not allow for sufficient growth of well differentiated head and neck tumors. Our role is to further delineate the characteristics of the varied head and neck tumor populations and to identify the most appropriate disaggregation techniques and the most suitable growth environment for all squamous carcinomas of the head and neck.

REFERENCES

1. Cobleigh, M., Gallagher, P., McGuire, W., et al.: Culture of Squamous Cell Carcinoma of the Head and Neck (H/N) in the Human Tumor Clongenic Assay (HTSCA). Proc. AACR 24:313, 1983.
2. Crissman, J.D., Gluckman, J., Whiteley, J., et al.: Squamous Carcinoma of the Floor of the Mouth. Head and Neck Surgery, 3:2-7, 1980.
3. Hamburger, A. and Salmon, S.: Primary Bioassay of Human Tumor Stem Cells. Science, 197:461-463, 1977.
4. Johns, M.E.: The Clonal Assay of Head and Neck Tumor Cells: Results and Clinical Correlation. Laryngoscope, 92:(28)1-26, 1982.
5. Johns, M.E. and Mills, S.E.: Cloning Efficiency: A Possible Prognostic Indicator in Squamous Cell Carcinoma of the Head and Neck. Cancer 52:1401-1404, 1983.
6. Johnson, P. and Kassof, A.H.: The Role of the Human Tumor Stem Cell Assay in Medical Oncology. Archives of Int. Medicine, 143:111-114, 1983.
7. Mattox, D.E. and Von Hoff, D.D.: In Vitro Stem Cell Assay in Head and Neck Squamous Carcinoma. Am. J. Surgery, 140:527-540, 1980.
8. Mattox, D.E. and Von Hoff, D.D.: Culture of Human Head and Neck Cancer Stem Cells Using Soft Agar. Archives Otolaryngology, 106:672-674, 1980.
9. Meyer, J., Rao, B., Stevens, S. and White, W.: Low Incidence of Receptor Status in Breast Carcinomas with Rapid Rates of Cellular Replication. Cancer, 40:2290-2298, 1977.
10. Salmon, S.E., Hamburger, A.W., Soehnlen, B., et al.: Quantitation of Differentiation Sensitivity of Human Tumor Stem Cells to Anti-Cancer Drugs. NEJM, 298: 1321-1327, 1978.
11. Selby, P., Buick, R. and Tannock, I.: A Critical Appraisal of the Human Tumor Stem Cell Assay. NEJM, 308(3):129-133, 1983.
12. Von Hoff, D., Casper, J., Bradley, E., et al.: Association Between Human Tumor Colony Forming Assay Results and Response of an Individual Patient's Tumor to Chemotherapy. Am. J. Medicine, 70:1027-1032, 1981.

13. Von Hoff, D.D.: Sent This Patient's Tumor for Culture and Sensitivity. Editorial NEJM, 308(2):154-155, 1983.
14. Von Hoff, D.D., Clark, G.M., Stagdill, B.J., et al.: Prospective Clinical Trial of a Human Tumor Cloning System. Cancer Research, 43(4):1926-1931, 1983.

Novel *In Vivo* and *In Vitro* Models for the Study of Human Ovarian Cancer

THOMAS C. HAMILTON
ROBERT C. YOUNG
ROBERT F. OZOLS

Medicine Branch
Division of Cancer Treatment
National Cancer Institute
9000 Rockville Pike
Bethesda, MD 20205

I. MODEL SYSTEMS OF OVARIAN CANCER

Many experimental model systems have been developed to study various aspects of ovarian cancer including (1) etiology and biology with particular reference to the role of hormones in the development and progression of the disease, (2) selection of new anticancer drugs and individualization of chemotherapy, (3) mechanisms of drug resistance and its pharmacologic modulation, and (4) the potential role of new therapeutic modalities such as monoclonal antibodies and stimulators of the immune system. Table 1 lists several of these model systems along with their applications and experimental shortcomings.

The investigation of drug resistance mechanisms in human ovarian cancer and the pharmacological reversal of resistance to increase the effectiveness of the commonly used anticancer drugs: melphalan, Adriamycin, and cisplatin has been facilitated by the development of new experimental model systems of human ovarian cancer. We have developed *in vitro* and *in vivo* models of the human disease consisting of human ovarian cancer cell lines[2,6,9,11,12,16] and an intraperitoneal xenograft model[12] of human ovarian cancer which parallels the clinical course of the disease in patients. This transplantable tumor produces host death from respiratory compromise and/or bowel obstruction due to massive ascites and intraabdominal carcinomatosis.

HUMAN TUMOR CLONING
ISBN 0-8089-1671-8

TABLE 1: OVARIAN CANCER MODELS

I. In Vitro Models

A. Human tumor stem cell assay (HTSCA)[7,15,18]
uses: individualization of therapy; screening new drugs; establishing dose-response relationships.
limitations: technical problems limit its general applicability to individualizing therapy; small colony size precludes most biochemical studies.

B. Human ovarian cancer cell lines[1,2,6,9,11,12,16]
uses: study of biology; resistance mechanisms; pharmacological manipulation of resistance; screening new drugs.
limitations: potential for in vitro selection procedures to pertubate drug sensitivity/resistance phenotype; cannot distinguish between tumor directed cytotoxicity and general normal tissue toxicity.

C. Animal tumor cell lines[14,17]
uses: mechanisms of drug resistance; pharmacologic studies.
limitations: derivation often different than that of common epithelial tumors of the human ovary; animal origin.

D. Normal progenitor models[8] (rat ovarian surface germinal epithelium)
uses: study the stepwise progression toward malignancy; significance of hormones on growth and function.
limitations: of animal origin.

II. In Vivo Models

A. Subcutaneous xenografts in immunodeficient or immunodeprived hosts[4,19]
uses: screening new drugs, characterizing tumor cell biochemistry, storage of tumor cells for use in HTSCA.
limitations: may not be clinically representative of the human disease.

B. Xenografts at immunoprivileged sites[3] uses and limitations: as II. A.

TABLE 1 (Con't)

C. Intraperitoneal model[12]
uses: study biology and metastasis, drug resistance and its modulation, influence of hormones, screening new drugs.
limitations: various selection procedures for initiation could have altered character of the cells.

D. Animal models[13,14,17]
uses: hormonal relationships, carcinogenesis, pharmacology.
limitations: animal origin, incompletely studied.

II. NEW OVARIAN CANCER CELL LINES

These new cell lines may be divided into three categories (1) those derived from untreated patients (2) those from patients refractory to combination chemotherapy, and (3) those with resistance induced or selected *in vitro* by stepwise incubation of a sensitive cell line with individual drugs of importance for the treatment of ovarian cancer (see Table 2). These cell lines have been characterized for their drug sensitivity/resistance profiles and their cross-resistance to other drugs and irradiation[2,6,9,11,12,16]. In addition, the OVCAR cell lines have been evaluated for steroid hormone receptors, and the receptors in NIH:OVCAR-3 studied in some detail. This cell line and the subcutaneous tumors it forms in athymic mice contain androgen and estrogen receptors based on specificity (androgenic and antiandrogenic compounds or estrogenic and antiestrogenic compounds, respectively), sedimentation profiles in low-salt (7-9S binding) and high salt (4-5S) sucrose density gradients, dissociation constants (androgen 250pM and estrogen 9.6pM), binding site concentrations (30 fmoles/mg cytosol protein), and for the estrogen receptor, the ability to interact with a monoclonal antibody generated against purified estrogen receptor from human breast cancer cells[10,11].

The cell lines (Table 2) show a varying propensity to clone in agarose and form tumors in nude athymic mice with only NIH: OVCAR-2, a cell line from an extensively treated patient (cyclophosphamide, cisplatin, hexamethylmelamine, and irradiation), lacking these features. NIH:OVCAR-2, however, does have an abnormal karyotype, appears malignant cytologically, and is of particular interest due to its relative insensitivity to Adriamycin even though the patient was not treated with Adriamycin[9].

TABLE 2: OVARIAN CARCINOMA CELL LINES

From Untreated Patients

A1847[a]	A2780[a]	NIH:OVCAR-5

From Refractory Patients

Cell Line	Drug Exposure In Vivo
NIH:OVCAR-2	Cyclophosphamide, Cisplatin, Hexamethylmelamine, Irradiation
NIH:OVCAR-3	Cyclophosphamide, Cisplatin, Adriamycin
NIH:OVCAR-4	Cyclophosphamide, Cisplatin, Adriamycin

Resistance Induced In Vitro

Parent Cell Line	Drug Resistance			
	Melphalan	Thiotepa	Adriamycin	Cisplatin
A1847	+(1847ME)	+(1847TH)	+(1847^{AD})	+(1847^{CP})
A2780	+(2780ME)	+(2780TH)	+(2780^{AD})	+(2780^{CP})

[a] Kindly provided by Dr. Stuart Aaronson, NCI

The cell lines established from untreated patients are relatively more sensitive to standard chemotherapeutic agents than the cell lines from refractory patients. In addition, it is noteworthy that the cell lines with resistance induced *in vitro* spanned the range of sensitivity/resistance observed *in those* cell lines from refractory patients, thus demonstrating the relevance of the former for the study of drug resistance. This system of cell lines of varying levels of complexity has some important advantages for the study of drug resistance. A cell line with moderate resistance induced *in vitro* to a single agent is a convenient system in which *to initially* probe for resistance mechanism(s). The significance of such mechanism(s) of resistance can then be compared to that which develops in the more complex *in vivo* mileau following simultaneous exposure to multiple *drugs by* the use of cell lines from heavily treated refractory patients. This approach has led to the discovery and determination of the significance of transport phenomena as a mechanism of Adriamycin resistance in some human ovarian cancer cell lines and ways this resistance may be pharmacologically completely or partially reversed[16]. Similar approaches are in use to study alkylating agent and cisplatin resistance and ways to pharmacologically increase the effectiveness of these drugs[6].

The use of cell lines for the above studies has circumvented some problems inherent in working with fresh tumor cells in a clonogenic assay. Cell lines permit more technical manipulations and allow for expansion and confirmation of results. In addition, the generally higher cloning efficiency of cell lines also favors the more precise and confident determination of dose response relationships than when working with fresh specimens where only 30-50 colonies are routinely observed.

While the advantages of cell lines for the investigation of drug resistance and its pharmacological manipulation are clear, this system also has some potential limitations which may affect correlation with *in vivo* results. The primary potential shortcoming of drug sensitivity studies in cell lines from individual patients is lack of knowledge as to the influence of selection processes operative in adaptation to growth *in vitro* upon the drug sensitivity/resistance phenotype. The human tumor stem cell assay as applied to fresh tumors partially deals with this problem but as stated above, suffers other inherent problems. In an alternative approach, one group has used nude athymic mice for storage of human tumors which may be removed at will from animals and used fresh as in the HTSCA[19]. In this manner the influence of *in vitro* selection processes are decreased and the investigator may repeat experiments and thereby expand and confirm results obtained on an individual patient's tumor. Of significance these workers also have shown that if tumor bearing animals are treated with a drug, the pattern of response in the subcutaneous tumor is similar to the results obtained on that tumor in the HTSCA[19].

III. TRANSPLANTABLE INTRAPERITONEAL NUDE MOUSE MODEL OF HUMAN OVARIAN CANCER

An important limitation of any *in vitro* (HTSCA or cell line) system is that cell killing either by a drug or by use of a standard chemotherapeutic drug with a pharmacologic agent which increases the anticancer drug's effectiveness gives no indication of whether cytotoxicity is tumor cell specific or the agent or combination is a general poison and thus likely to produce increased toxicity in patients. Thus, *in vivo* models in which differential cytotoxic effects may be evaluated are needed to compliment observations made in cell lines. Subcutaneous human ovarian cancer xenografts in immunodeficient or immuno-deprived hosts[4,19] and xenografts at immunoprivileged sites[3] have been used to evaluate anticancer effects of various agents. These systems generally rely on change in tumor size for an indication of response and bear little if any resemblance to clinical ovarian cancer. We have addressed this problem by developing an intraperitoneal model of human ovarian cancer[12]. The approach utilized was to sequentially select the NIH:OVCAR-3 cell line for in vivo growth (formation of subcutaneous tumors in nude athymic mice), and the capacity of tumor cells from these xenografts to exhibit substrate independent growth (clone in agarose). The agarose colonies (harvested in mass) were grown in monolayer and selected a second time for in vivo growth as above. The xenografts were dissociated with collagenase and tumor cells (separated from host cells on Ficoll gradients) injected intraperitoneally into female nude athymic mice. In approximately 40 days, animals developed distended abdomens. Examination of the peritoneal cavity at this stage revealed tumor foci on all peritoneal surfaces, the diaphragm, and the viscera. Bulky disease and ascites were present as were microscopic pulmonary metastases. Animals were found ultimately to die from complications of massive intraperitoneal disease[12]. This model should prove useful for the study of drug resistance and its manipulation, evaluation of new drugs, and novel treatment approaches including monoclonal antibodies[1], bacterial toxins[5], Mullerian inhibition factor, and natural killer like cells.

IV. CONCLUSIONS

We have developed a complex array of human ovarian cancer cell lines with which to study drug resistance and its pharmacologic manipulation. In addition, a unique intraperitoneal model of human ovarian cancer has been established which may help determine the relative effects of new therapeutic modalities upon tumor cells versus normal tissues. Initial studies in these in vitro/in vivo systems have demonstrated the potential for pharmacological manipulation of both Adriamycin and alkylating agent resistance.

REFERENCES

1. Bast RC, Freeney M, Lazarus H, et al: Reactivity of a monoclonal antibody with human ovarian carcinoma. J Clin Invest 68: 1331-1337, 1981.

2. Behrens BC, Louie KG, Hamilton TC, et al: Resistance and cross-resistance of human ovarian cancer cell lines to Adriamycin, melphalan and irradiation. Proc Amer Assoc Cancer Res 25: (in press) 1984.
3. Bodgen AE, Cobb WR, Costanza ME, et al: Drug sensitivity profiles of primary tumors and their metastases: 6 day subrenal capsul assay. Proc Amer Soc Clin Oncol 1: 12, 1982.
4. Davy M, Mossiage J, Johannessen JV: Heterologous growth of human ovarian cancer. Acta Obstet Gynecol Scand 56: 55-59, 1977.
5. FitzGerald DJP, Trowbridge IS, Pastan I, et al: Enchancement of toxicity of antitransferrin receptor antibody-Pseudomonas exotoxin conjugates by adenovirus. Proc Natl Acad Sci 80: 4134-4138, 1983.
6. Green JA, Vistica DT, Young RC, et al: Melphalan resistance in human ovarian cancer: potentiation of melphalan cytotoxicity by nutritional and pharmacologic depletion of intracellular glutathione levels. Proc Amer Assoc Cancer Res 25: (in press) 1984.
7. Hamburger AW, Salmon SE, Kim MB, et al: Direct cloning of human ovarian cancer cells in agar. Cancer Res 38: 3438-3444, 1978.
8. Hamilton TC, Davies P, Griffiths K: Estrogen receptor-like binding in the surface germinal epithelium of the rat ovary. J Endocr 95: 377-385, 1982.
9. Hamilton TC, Foster BJ, Grotzinger KR, et al: Developmet of drug sensitive and resistant human ovarian cancer cell lines. Proc Amer Assoc Cancer Res 24: 313, 1983.
10. Hamilton TC, McKoy WM, Grotzinger KR, et al: A model system for the investigation of androgen and estrogen action in human ovarian carcinoma. Endocrinol 112 Suppl: 184, 1983.
11. Hamilton TC, Young RC, McKoy WM, et al: Characterization of a human ovarian carcinoma cell line (NIH:OVCAR-3) with androgen and estrogen receptors. Cancer Res 43: 5379-5389, 1983.
12. Hamilton TC, Young RC, Rogan AM, et al: A unique intraperitoneal model of human ovarian cancer. Proc Amer Assoc Cancer Res 25: (in press) 1984.
13. Jabara AG: Induction of canine ovarian tumors by diethystilbestrol and progesterone. Aust J Exp Biol Med Sci 40: 139-152, 1962.
14. Marchant J, Cotchin E: Animal tumors of the female reproductive tract: spontaneous and experimental. New York, Springer-Verlag, 1977, pp 1-25.

15. Ozols, RF, Willson JKV, Grotzinger KR, et al: Cloning of human ovarian cancer cells in soft agar from malignant effusions and peritoneal washings. Cancer Res 40: 2743-2747, 1980.
16. Ozols RF, Rogan AM, Hamilton TC, et al: Verapamil plus Adriamycin in refractory ovarian cancer: design of a clinical trial on basis of reversal of Adriamycin resistance in human ovarian cancer cell lines. Proc Amer Assoc Cancer Res 25: (in press) 1984.
17. Sekiya S, Endoh N, Kikuchi Y: In vivo and in vitro studies of experimental ovarian adenocarcinoma in rats. Cancer Res 39: 1108-1112, 1979.
18. Selby P, Buick RN, Jannock I: A critical appraisal of the "human tumor stem cell assay." N Engl J Med 308: 129-134, 1983.
19. Taetle R, Howell SB, Giuliani FC, et al: Comparison of the activity of doxirubicin analogues using colony-forming assays and human xenografts. Cancer 50: 1455-1461, 1982.

The Clonogenic Characterization Of Prostate Cancer From Bone Marrow Aspirates

Frederick R. Ahmann, M.D.

Staff Physician
Tucson VA Medical Center
and
Assistant Professor of Medicine
University of Arizona School of Medicine
Tucson, Arizona 85724

While significant therapeutic and diagnostic advances have been achieved in many human malignancies over the past 20 years, there has been no significant improvement in the prognosis or understanding of prostatic carcinoma. This year approximately 25,000 men will be diagnosed as having incurable prostatic carcinoma (1) and few promising therapeutic developments are on the horizon to aid them. It is hoped that new laboratory research techniques will provide information to allow us to improve the clinical approach to this common malignancy.

Traditional laboratory techniques cannot be readily applied to investigations of prostatic carcinoma. There are no satisfactory animal models (2,3) and information obtained from such models in other tumors has inconsistently correlated with the same diseases in higher animals (3). There are only three accepted human prostatic cancer cell lines (4), and success with implanting human prostate cancer into athymic nude mice has been limited (5).

A productive research technique being increasingly utilized over the past two decades has been clonogenic assays. The application of this method to the study of human hematopoietic cells has led to major insight into the understanding of normal hematopoietic cell physiology and of a variety of hematopoietic disorders including chronic myelogenous leukemia, acute myelogenous leukemia, refractory anemias, aplastic anemia, myelofibrosis, polycythemia vera, and

HUMAN TUMOR CLONING
ISBN 0-8089-1671-8

pre-leukemia (6-9). These culturing methods in semi-solid media have now been successfully applied to human solid tumors (10). Experiments primarily with soft agar clonogenic culture in solid tumors have defined cytogenetic abnormalities, new tumor markers cell interactions, and a large number of pharmacologic tumor-host growth modifiers including clinically used anti-cancer compounds (11-18).

Because prostatic carcinoma has the common clinical characteristic of diffusely metastasizing to bone and because bone marrow aspirates are safe, relatively non-morbid and yield cells which are technically easy to work with, we chose to perform bone marrow (BM) aspirates on patients with metastatic prostatic carcinoma to bone and attempt to characterize prostatic carcinoma in clonogenic culture from these specimens. We report here our experience in 27 consecutive BM aspirates.

MATERIALS AND METHODS

Patient Selection: Patients were identified as potential candidates for bone marrow aspirate from the medical oncology patient and out-patient populations at the Tucson VA Medical Center and at the University of Arizona Cancer Center. All patients had histologically proven prostatic carcinoma and had evidence of diffuse (> 15 sites) bony metastases on radio-nucleotide bone scanning. Patients were also selected for BM aspirate at times in their clinical course when their disease was not in a remission.

Biopsy Procedure, Site Selection, and Pathologic Studies: Bone marrow aspirates were performed with either an Illinois or Rosenthal 18 guage needle via standard technique (19). An initial 1/2cc of marrow aspirate was immediately prepared for smear and clot section examination after Wright staining by the Department of Pathology. An additional marrow aspirate 5-10cc was then drawn into a 10cc syringe containing 1/2cc of preservative free sodium heparin. All aspirates were performed from either the left or right posterior superior iliac crest or from the sternum in the 4th intercostal space. Recent (performed within five weeks of the time of BM aspirate) radionucleotide bone scans were available in all patients.

Cell Separation, Culture Technique, and Colony Scoring: The cell separation method with Ficoll-Hypaque and clonogenic culture technique are described in detail elsewhere (18,20). In brief, the bone marrow "buffy" coat is collected and colony formation scoring. Three specimens had bacterial contamination and the culture plates were discarded. The first two specimens plated were not felt to have been accurately scored because of inexperience, at the time, of identifying tumor colonies (CFU- GMs) and hence these results were regarded as inevaluable. Of the remaining 22 evaluable platings, three plates displayed no tumor colony formation (13%). An additional five patient specimens (23%) displayed only tumor cluster formation. Tumor colony formation was scored in 14 of the 22 evaluable platings (64%). A mean tumor colony count of 26 was found in these 14 specimens with the range of colony counts being 1 to 181 per plate.

Morphologic characterization studies were performed on the tumor colonies by the methods described in ten cases. The staining characterization of the cells in these colonies was as follows: PAS positive, peroxidase negative, and non-specific esterase negative. Only a rare colony was acid phosphatase positive. The prostatic specific antigen stain was positive from the cells of these platings. Under inverted microscopy in four patients' platings, larger cells were observed within the tumor colonies. Staining of these colonies revealed that these larger cells were esterase positive.

Cytogenetic analyses have been successfully performed on the colonies of three patients' cultures. All three revealed hypodiploid numbers of chromosomes. A more detailed

TABLE 1

Clonogenic Results from 27 Bone Marrow Aspirates in Patients with Metastatic Prostatic Carcinoma

1. Mean cell yield.		4.8×10^7
2. Inevaluable for colony count.		5 (18%)
a) Contamination	3 (60%)	
b) Early work	2 (40%)	
3. Evaluable for colony count.		22 (82%)
a) No growth	3 (13%)	
b) Cluster formation only	5 (23%)	
c) Colony formation	14 (64%)	
i. Mean colony count	26	
ii. Range	1 to 181	

resuspended in media with 10% fetal calf serum. This suspension is then layered over a Ficoll- Hypaque solution (Pharmacia) for 40 minutes. The cells at the interface are harvested, washed twice with PBS, and are counted and evaluated for viability via trypan blue dye exclusion. 10^6 nucleated bone marrow cells are then plated in 35mm Petri dishes in 1 ml of 0.3% agar layered over a 1 ml of 0.5% underlayer containing an admixture of fetal bovine serum and growth factors. The plates are cultured in an incubator at 37° C in a humidified 7.5% CO_2 environment. Plates are examined serially with an inverted-phase contrast microscope for evidence of cluster and/or colony (> 30 cells) formation.

Morphologic Markers, Cytogenetic Analysis, Electron Microscopy and Self Renewal Studies: Soft agar culture material can be prepared for staining via several methods including the "dry slide" method (21), direct staining in the culture plate (22), and "plucking" of the individual colonies (23) and fixation directly on a glass slide or by processing in paraffin. Stains utilized included the following: peroxidase, periodic acid Schiff, acid phosphatase with and without tartrate, Wright-Giemsa, non-specific esterase and prostatic specific antigen. All of the above are standard techniques (24). The prostatic specific antigen is a recently described and commercially available peroxidase anti-peroxidase method (25). Electron microscopy and cytogenetic studies were carried out in collaborating laboratories at the University of Arizona School of Medicine. The self-renewing experimental technique is also described in detail elsewhere (23). Individual colonies are "plucked" from the semi-solid agar culture overlayer, resuspended as single cells via mechanical disaggregation through small needles, and placed back into soft agar clonogenic culture in microtiter wells.

RESULTS

Between January 1983 and November 1983, 27 bone marrow aspirates were performed on patients with metastatic prostatic carcinoma who fit the criteria noted above and the results are displayed in Table 1. The mean yield of nucleated cells obtained from the aspirates after the separation procedure described above was 4.8×10^7 cells with a range of 9.0×10^6 to 2×10^8. There were no complications or significant morbidity associated with any of the BM aspirations. Five patient specimens (18%) were inevaluable for characterization was performed in one patient. This profile revealed 44 chromosomes with a deletion of the long arm of chromosome 12 and 18 as well as a single double minute. An ultrastructural study of the cells "plucked" from one colony revealed cells with microvilli, prominent nucleoli, a high nuclear cytoplasmic ratio, and prominent cytoplasmic vacuolization. A successful self renewing experiment was accomplished in one of three attempts. In this case, a third generation of colonies resulted when the colonies from the first replating were again disaggregated into single cells and replated for a second time in a semi-solid media.

Unexpectedly significant granulocyte-macrophage colony formation (CFU-GM) was a significant finding in these experiments. As displayed in Table 2, the plating of 10^6 nucleated bone marrow cells using the method described resulted in significant CFU-GM counts as scored on day 7. There was an association between tumor colony formation and CFU-GM formation.

DISCUSSION

The application of clonogenic culture techniques to fresh human prostatic cancer cells would hold significant potential of increasing insight into this common malignancy. Bone marrow aspirates are an attractive tumor sampling procedure for patients with prostatic carcinoma with the commonly seen circumstances of diffuse bony metastases. Our results support the potential utility of this methodology. Tumor colony formation was seen in 64% of aspirates placed in culture with colony formation occurring in some patients who had either no tumor cells identified on routine pathologic screening of the aspirates and clot sections or who had radionucleotide bone scans indicating no disease activity at the aspiration sites. The primary limitation with this method is an

TABLE 2
Tumor Colony & CFU-GM Correlations in 12 Samples

	# of Samples	CFU-GM Mean	CFU-GM Range
≥ 3 Tumor Colonies (Range 3-181)	6	163	65-226
< 3 Tumor Colonies (Range 0-1)	6	47	0-131

inability to quantify the number of tumor cells plated from an individual aspirate. Hence, plating efficiencies could neither be calculated or predicted and many culture plates did not achieve a significant enough number of tumor colonies to have permitted experiments exploring potential biologic modifiers of colony formation. Additional cell separation techniques and methods to more accurately enumerate the number of tumor cells in the original bone marrow aspirates will need to be utilized before such experiments can be routinely successfully performed.

Nevertheless, several interesting observations resulted from this work. The tumor colonies which formed were consistent with malignancy orginating from the prostate based on their staining characterization (PAS positive, peroxidase negative, non-specific esterase negative, and occasional prostatic specific antigen positivity), cytogenetic analyses, ultrastructural studies and self-renewing potential. Observations were recorded consistent with tumor colonies being stimulated in some cases by macrophages. Additionally, other cellular elements were identified in the bone marrow (CFU-GM) which appeared to correlate with tumor colony formation. The relationships of these normal bone marrow cellular populations to malignant cell growth need to be further explored as they potentially relate to prostatic cancer's inherent characteristic of metastasizing diffusely to bone.

This work is also supportive of the ability of clonogenic culture materials to be utilized successfully in cytogenetic analyses. There has been little chromosome characterization available to this time on patients with prostatic carcinoma and clearly this method will aid in additional work in this area. The tumor colonies which form also offer the opportunity to pursue the characterization of specific clonogenic populations of prostatic carcinoma with immunologic and other markers and may help identify new immunologic and enzymatic markers specific for prostatic cancer. Further work in this area would also be enhanced by improved separation methods.

REFERENCES

1. Silverberg E: Cancer Statistics, 1981. CA-A Journal for Clinicians. 31:13-28, 1981.
2. Alison MR and Wright NA: Growth kinetics. In: Prostate Cancer. Recent results in cancer research.

Ed. Duncan. Springer-Verlan, Berlin, 1981. page 29-43.

3. Connors TA and Phillips BJ: Screening for anti-cancer agents. The relative merits of in-vitro and in-vivo techniques. Biochem. Parm. 24:2217-2224, 1975.
4. Webber MM: In-vitro models for prostatic cancer. In: Models for Prostate Cancer. Eds: Murphy. Alan R. Liss, N.Y., 1980. page 133-147.
5. Gittes RF. The nude mouse-its use as tumor bearing Model of the Prostate. In: Models for prostate cancer. Ed: Murphy, N.Y., Alan R. Liss, Inc. 1979, page 31-37.
6. Metcalf D: Hemopoietic colonies. Recent results in cancer research. Springer-Verlag, Berlin, 1977. page 1-127.
7. Metcalf D: Detection and analysis of human granulocyte monocytic precursors using semi-solid cultures. Clinics in Haematology. 16:263-285, 1979.
8. Moor MAS, Spitzer G, Williams N, Metcalf D, and Buckley J: Agar culture studies in 127 cases of untreated acute leukemia: The prognostic value of reclassification of leukemia according to in-vitro growth characteristics. Blood 44:1-18, 1974.
9 Greenberg PL, Nichols W, Schrier SL: Granulopoiesis in acute myeloid leukemia and pre-leukemia. NEJM 284:1125-1132, 1974.
10. Salmon SE: Background and Overview: In: Cloning of Human Tumor Stem Cells. Ed: Salmon. Alan R. Liss, N.Y., 1980, page 1-13.
11. Hamburger AW, Salmon SE, Kion MB, Trent JM, Soehnlen BJ, Alberts DS and Schmidt HJ: Direct cloning of human ovarian carcinoma cells in agar. Cancer Res 38:3438-3444, 1978.
12 Buick RN, Stanisic TH, Fry SE, Salmon SE, Trent JM and Krasovich P: Development of an agar-methylcellulose clonogenic assay for cells in transitional cell carcinoma of the human bladder. Cancer Res. 39:5051-5056, 1979.
13 Meyskens FL, Soehnlen BJ, Saxe DF, Casey WF, and Salmon SE: In-vitro clonal assay for human metastatic melanoma cells. Stem Cells 1:61-72, 1981.
14 Gazdar AF, Carney DN, and Minna JD: In-vitro study of the biology of small cell carcinoma of the lung. Yale J. of Biol. and Med. 54:187-193, 1981.
15 Cowan DH and Graham J: Alteration of human tumor colony formation by platelet derived growth factors. Submitted, Cancer Research, 1983.

16. Buick RN, Fry SE and Salmon SE: Effect of host-cell interaction on clonogenic carcinoma cells in human malignant effusions. Brit. J. Cancer 41:695-704, 1980.
17. Meyskens FL and Salmon SE: Modulation of clonogenic human melanoma cells by FSH, melatonin and nerve growth factor. Brit. J. Cancer. 43:1211-115, 1981.
18. Salmon SE, Hamburger AW, Soehnlen BJ, Durie BGM, Alberts DS and Moon TC: Quantitation of differential sensitivities of human tumor stem cells to anti-cancer drugs. NEJM 298:1321-1327, 1978.
19. Williams WJ: Examination of the bone marrow. In: Hematology. Eds: Williams, Beutler, Erslew and Randles. McGraw-Hill. N.Y., 1977, pages 25-32.
20. Loos J and Roos D: Ficoll-Isopaque gradients for the determination of density distributions of human blood lymphocytes and other reticulo-endothelial cells. Expl. Cell Res. 86:333-341, 1974.
21. Salmon SE, and Buick RN: Preparation of permanent slides of intact soft agar colony cultures of hematopoietic and tumor stem cells. Cancer Res. 39:1133-1136, 1979.
22. Salmon SE: Morphologic studies of tumor colonies. In: Cloning of human tumor stem cells. Ed: Salmon. Alan R. Liss, N.Y., 1980. page 135-151.
23. Meyskens FL, Thomson SP, and Moon TE: Self-renewal of melanoma colonies in agar: Similar colony formation from replating primary colonies of different sizes. JCI (submitted), 1983.
24. Sheehan DC and Hrapcha BB: Theory and practice of histotechnology. Mosby, St. Louis, 1980, page 1-481.
25. Nadji J. Tabei SZ, Castro A, Chu TM, Murphy GP, Wang MC and Morales AR: Prostatic specific antigen. An immunohistologic marker for prostatic neoplasms. Cancer 48:1229-1232, 1981.
26. Trent JM: Protocols of procedures and techniques in chromosome analyjsis of tumor stem cell cultures in soft agar. In: Cloning of human tumor stem cells. Ed: Salmon. Alan R. Liss, N.Y., 1980, page 135-151.

Outcome of "Non-Malignant" Specimens Submitted for Soft-Agar Clonogenic Assay

VERNON K. SONDAK
Division of Surgical Oncology

ARMANDO E. GIULIANO
Associate Professor of Surgery
Division of Surgical Oncology

DAVID H. KERN
Associate Research Oncologist
Division of Surgical Oncology and Surgical Service

Division of Surgical Oncology, John Wayne Clinic
Jonsson Comprehensive Cancer Center
UCLA School of Medicine
Los Angeles, California 90024
and
Surgical Service
Veterans Administration Medical Center
Sepulveda, California 91343

INTRODUCTION

The soft-agar clonogenic assay, developed by Hamburger and Salmon,[5] has received widespread attention as an in vitro method for determining tumor-cell sensitivity to cytotoxic drugs. High correlations between in vitro and in vivo responses were obtained by

Supported by Grants CA09010, CM07420, CA29605, and CA12582, awarded by the National Cancer Institute, DHHS; and by VA Medical Research Services.

HUMAN TUMOR CLONING
ISBN 0-8089-1671-8

a number of investigators.[8,10,18] Dogma states that non-neoplastic cells cannot proliferate in soft agar under assay conditions;[21] thus growth in soft agar was correlated with malignancy or malignant potential.[11,17] Numerous studies have confirmed the malignant origin of colonies formed in the clonogenic assay.[2,4,6,7,12,13,19,20] Although several investigators have described nonmalignant cells that grew and formed colonies in soft agar,[1,12,13,15] there have been no reports describing the fate of tissues submitted for clonogenic assay that were found later not to be malignant pathologically.

Since 1979, the UCLA Human Tumor Cloning Laboratory has received specimens from over 2200 patients with known or suspected malignant tumors. A portion of each specimen is examined by a pathologist and a copy of the surgical pathology report is always obtained from the referring institution. A total of 38 assays were performed on tissues later determined not to be malignant. This report will examine the outcome of these assays and discuss the implications for routine chemosensitivity testing.

MATERIALS AND METHODS

Clonogenic assays were performed on freshly excised surgical specimens as previously described.[8] Single-cell suspensions were prepared by enzymatic digestion, and the nucleated cell yield was determined with a hemocytometer. Trypan blue dye exclusion was used to estimate the percentage of viable cells. If fewer than 1×10^6 viable cells were recovered, the assay was terminated due to "insufficient viable cells." Otherwise, viable cells were plated in soft agar at a concentration of 4×10^5 viable cells per well. Stock solutions of cytotoxic agents known to be active in vitro were added to some wells for chemosensitivity testing. Control wells were plated without cytotoxic drugs and inspected at regular intervals for colony formation. The assay was reported as unsuccessful if no growth took place and as insufficient growth if less than 30 colonies per control well were observed. If 30 or more colonies per control well were present, an automated colony counting system incorporating an inverted microscope was used to calculate percentage inhibition of growth for each cytotoxic drug tested. Greater than 50% inhibition of colony formation compared to control was considered in vitro sensitivity.

Some of the specimens were tested in a rapid assay using tritiated thymidine uptake as the endpoint for measuring growth in soft agar.[16] More than 300 counts per minute (CPM) above system background in control wells was considered as successful growth. In vitro sensitivity for this assay was defined as 80% or greater inhibition of thymidine uptake compared to control wells. Growth in soft agar is a requisite for a successful assay in both systems.

RESULTS

Growth in Soft Agar and In Vitro Sensitivities

After a review of over 1800 specimens submitted for chemosensitivity testing, 38 (1.7%) were found not to be malignant on histopathologic examination. Twelve of the 38 (31%) fulfilled the criteria for growth in soft agar (formation of 30 or more colonies or incorporation of greater than 300 CPM of thymidine above background). Twenty-six of the specimens were tested in the standard soft-agar clonogenic assay, and 10 (38%) produced growth of 30 or more colonies per control well. Two additional assays showed growth of less than 30 colonies per control well (27 and 28 colonies per control well), and sensitivity data were obtained for one drug in the latter instance (see Tables 1 and 2).

One or more cytotoxic agents caused 50% or greater inhibition of colony formation (in vitro sensitivity) in three of the eleven assays that yielded drug sensitivity data: a hibernoma (lipoma of brown fat) was sensitive to Adriamycin and cis-platinum; a histologically uninvolved axillary lymph node from a patient with known melanoma was sensitive to BOLD (bleomycin, vincristine, BCNU, DTIC); and a recurrent desmoid tumor responded in vitro to Adriamycin, cis-platinum, and methotrexate. The remaining seven assays revealed no cytotoxic agents with in vitro activity.

Of twelve specimens processed with radioactive thymidine incorporation, two incorporated sufficient label to fulfill the criteria for growth in soft agar (Table 3). Both specimens were pathologically uninvolved lymph nodes sampled from patients with known lymph node involvement with melanoma elsewhere. The in vitro sensitivity patterns were similar to those observed for proven cases of melanoma metastatic to lymph nodes.

Nucleated Cell Yields and Cytologic Correlation

Viable nucleated cells were recovered from 34 of 38 nonmalignant specimens. Twenty-eight of these yielded at least 1×10^6 viable cells, the minimum required for assay performance. Cell viability ranged from 5 to 100%, but in most instances was in excess of 70% by trypan blue dye exclusion. Laboratory technicians are instructed to count only large nucleated cells consistent with malignant cells, but in retrospect there is no way to determine the precise nature of the cells counted.

TABLE 1

"Nonmalignant" Specimens That Did Not Form Colonies in the Soft-Agar Clonogenic Assay

Operative Dx	Histologic Dx	Specimen Weight (grams)	Total Nucleated Cell Yield	Viability	Assay Result
Malignant paraganglioma, retroperitoneum	Benign paraganglioma	1.0	2×10^6	5%	no growth
Possible carcinoma, lung	Necrotizing coccidiomycosis granuloma	4.0	1.2×10^7	44%	contaminated
Probable sarcoma, shoulder	Chronic inflammation, giant cell reaction	8.3	8×10^5	100%	insufficient viable cells
Conjunctivitis, eye	Conjunctivitis	0.25	2.4×10^6	100%	no growth
Anterior mediastial mass	Thymoma	4.7	9.8×10^7	88%	no growth
Malignant pleural effusion	Pleural effusion, no malignant cells	300 cc	4.2×10^5	100%	insufficient viable cells
Metastatic melanoma, subcutaneous	Lipoma	16.3	0		insufficient viable cells

Breast cancer	Fat necrosis	1.0	0		insufficient viable cells
Malignant pleural effusion	Pleural effusion, no malignant cells	1200 cc	1.2×10^5	92%	insufficient viable cells
Malignant ascites	Ascites, no malignant cells	1000 cc	1.4×10^7	71%	no growth
Malignant ascites	Ascites, no malignant cells	800 cc	8×10^5	50%	insufficient viable cells
Malignant pleural effusion	Pleural effusion, no malignant cells	not recorded	0		insufficient viable cells
Breast cancer	Fat necrosis	3.1	5×10^5	40%	insufficient viable cells
Metastatic breast cancer, subcutaneous	Connective tissue, no tumor seen	0.5	0		insufficient viable cells

TABLE 2
"Nonmalignant" Specimens that Formed Colonies in the Soft-Agar Clonogenic Assay

Operative Dx	Histologic Dx	Specimen Weight (grams)	Total Nucleated Cell Yield	Viability	Assay Result
Anaplastic lung cancer	Hamartoma	1.2	1.7×10^7	42%	CONTROL: 73 colonies/well
Probable sarcoma, neck	Hibernoma	3.0	2×10^6	100%	CONTROL: 345 colonies/well
Malignant thymoma	Thymic cysts with epithelial atypia	0.6	1.6×10^6	75%	CONTROL: 34 colonies/well
Metastatic breast cancer, femur	Bone currettings, no tumor seen	2.8	4.8×10^6	54%	CONTROL: 28 colonies/well
Metastatic melanoma, axilla	Lymph node, no tumor seen	1.4	3×10^6	100%	CONTROL: 72 colonies/well
Desmoid tumor, back	Desmoid (aggressive fibromatosis)	10.3	5.3×10^7	81%	CONTROL: 61 colonies/well
Leiomyosarcoma, uterus	Huge uterine leiomyoma	16.4	5×10^6	100%	CONTROL: 136 colonies/well

Leiomyosarcoma, uterus	Huge uterine leiomyoma	17.5	2×10^6	100%	CONTROL: 52 colonies/well
Possible sarcoma, abdomen	Inflammation, no evidence of malignancy	4.4	3.5×10^6	86%	CONTROL: 67 colonies/well
Metastatic seminoma, lymph nodes	Lymph node, marked necrosis, no tumor seen	3.0	6×10^7	12%	CONTROL: 50 colonies/well
Cystosarcoma phylloides, breast	Benign cystosarcoma phylloides	7.5	1.1×10^8	72%	CONTROL: 27 colonies/well (insufficient growth for drug testing)
Metastatic melanoma, subcutaneous	Lipoma	2.6	1.6×10^6	80%	CONTROL: 62 colonies/well

TABLE 3
Outcome of "Nonmalignant" Specimens Submitted for Soft-Agar Thymidine-Labelling Assay

Operative Dx	Histologic Dx	Specimen Weight (grams)	Total Nucleated Cell Yield	Viability	Assay Result
Metastatic melanoma	Lymph node, no tumor seen	0.9	2.7×10^7	85%	CONTROL: 5442 cpm/well
Acute myelogenous leukemia	Necrotizing granuloma, lung	2.1	1.4×10^7	77%	No growth; CONTROL: 59 cpm/well
Metastatic testicular cancer, mediastinum	Neurofibroma	1.8	1×10^5	100%	Insufficient viable cells
Possible ovarian cancer	Torsion, ovarian cyst	2.9	2.9×10^6	100%	Contaminated
Breast cancer	Breast, no tumor seen	4.6	7×10^5	100%	No growth; CONTROL: 161 cpm/well
Metastatic melanoma, subcutaneous	No tumor seen	0.3	5.8×10^7	86%	No growth; CONTROL: 114 cpm/well

Metastatic melanoma, neck	Lymph node, no tumor seen	2.7	1.4×10^8	85%	CONTROL: 630 cpm/well
Adenocarcinoma of rectum	Villous adenoma, no atypia	2.6	4.2×10^7	91%	No growth; CONTROL: 40 cpm/well
Ovarian thecoma	Benign ovarian cystadenothecoma	6.0	4.1×10^6	82%	CONTROL: 47 cpm/well
Ovarian cancer	Necrotic debris, no viable epithelial cells	1.7	4.2×10^6	52%	CONTROL: 93 cpm/well
Malignant pleural effusion	Pleural effusion, no malignant cells	1100 cc	6×10^6	83%	CONTROL: 40 cpm/well
Possible sarcoma	Neurofibroma	6.0	3.6×10^6	94%	CONTROL: 98 cpm/well

cpm = counts per minute

A smear was prepared from a drop of the single-cell suspension and stained with Wright-Giemsa stain before four specimens were plated in soft agar. The four specimens had been interpreted as a necrotizing granuloma, a neurofibroma, an uninvolved subcutaneous tissue nodule and a cervical lymph node. The smears were reviewed by an experienced cytopathologist without knowledge of the assay outcome. In one instance (Table 3, Case 7), multinucleated atypical cells consistent with melanoma were seen. This specimen exhibited uptake of thymidine in soft agar. A second smear (Table 3, Case 6) contained scattered clumps of atypical cells suspicious for, but not diagnostic of, melanoma, and the remaining two smears contained no cytologically malignant cells. None of these latter three exhibited growth in soft agar. Preparation of smears from the cell suspension has now become routine for all specimens received in our laboratory.

DISCUSSION

The malignant origin of colonies formed in soft agar from single-cell suspensions of human tumors has been demonstrated by light and electron microscopy, karyotype studies, and measurement of tumor markers.[2,4,6,7,12,13,19,20] Initially it was thought that growth in soft agar was a property of malignant cells[21] or of cells with malignant potential.[11,17] Recently, several laboratories have documented the growth in soft agar of a variety of nonmalignant specimens, such as fibrocystic breast disease,[12] prostatic hyperplasia,[13] parathyroid adenomas and hyperplasia,[1] and benign pediatric tumors.[15] No instances of normal tissue component growth have been reported under standard assay conditions, when either colony formation[9,21] or thymidine uptake[3,16] has been used to measure growth.

In the course of 4 years of chemosensitivity testing at UCLA, over 2200 specimens of presumably malignant tumors have been received. Thirty-eight, or 1.7%, were subsequently classified as nonmalignant when representative samples were examined histologically. Thirteen of these specimens exhibited sufficient growth in soft agar to allow for chemosensitivity testing. The diversity of tissue types encountered in this study precludes a single explanation for this phenomenon, but several possibilities suggest themselves.

Sampling error, or the submission of a non-representative portion of the tumor to the pathologist, could result in a diagnosis of "no tumor seen," even though the assay specimen actually contained malignant cells. In two cases in which a permanent cytology record was available in the form of Wright-Giemsa stained smears, atypical or frankly malignant cells were seen. If more such smears had been prepared, other cases might have been found. "Sampling error" of bone-marrow specimens from patients with neuroblastoma has been documented.[19]

The tendency of many surgeons to send the majority of a small tumor specimen for clonogenic assay, rather than to the pathologist, can contribute to this problem. If the specimen is a small lymph node or subcutaneous nodule, the pathologist may not receive any tumor-bearing material. Likewise, with malignant effusions, only 20-50 milliliters may be sent for cell block analysis, while several hundred milliliters are sent for assay. Effusions with low concentrations of malignant cells may be mistaken for benign by the cytopathologist working with a limited sample. Further support for sampling error is the fact that specimens taken from patients with known malignant tumors displayed sensitivity patterns similar to those for that tumor (e.g., melanoma sensitive to BOLD) (see Table 4). To minimize sampling errors, surgeons must submit truly representative portions of the tumor for pathologic confirmation in every case. Furthermore, preparation of stained cytology smears that are retained in a permanent file permits retrospective confirmation of the malignant nature of the tumor specimen. These smears have become the routine in our laboratory.

Sampling error cannot account for all the instances in which benign tissues grew successfully in our series and in other reports from the literature. Some tumors that appear benign histologically may possess some or all of the clinical characteristics of malignancy, including the ability to grow in soft agar. The huge uterine leiomyomas, cystosarcoma phylloides, and aggressive fibromatosis specimens that exhibited growth may fall into this category. The sensitivity patterns of these lesions paralleled those of soft-tissue sarcomas, lending credence to this theory. Growth in soft agar may serve as an indicator of malignancy in these borderline lesions.[17]

Frank malignancy is clearly not an absolute requirement for growth in soft agar. An intrinsic loss of control over cellular proliferation — without actual malignant transformation — may be sufficient to permit sustained growth in agar. The pulmonary hamartoma and the hibernoma may represent lesions in this category, which is perhaps the most interesting in its biologic implications.

To our knowledge, normal cells have not grown in either the clonogenic assay[9,21] or the thymidine labelling assay.[3,15] It is possible that fibroblasts, lymphocytes, or normal stromal or parenchymal elements may account for the colony formation in some assays. Clumping artifact — the aggregation of non-proliferating cells — may occur with some benign tissues and lead to a false impression of growth in the clonogenic assay. Serial counting of colonies beginning within 24 hours of plating may minimize this artifact. Assays demonstrating large numbers of colonies within 24 hours of plating (too early for significant tumor

TABLE 4
Drug Sensitivities for "Nonmalignant" Specimens
Demonstrating In Vitro Sensitivity
(50% or Greater Inhibition of Colony Formation
or 80% or Greater Inhibition of Thymidine Uptake)
to One or More Cytotoxic Agents

Histologic Dx	Sensitive Drugs ($\geq$ 50% inh)	Resistant Drugs
Hibernoma	Adriamycin (63%) Cis-Platinum (50%)	Methotrexate
Lymph node, no tumor seen (known melanoma)	BOLD (83%)	Bleomycin DTIC
Aggressive fibromatosis (recurrent)	Adriamycin (68%) Cis-Platinum (74%) Methotrexate (50%)	Vinblastine, Cis-Platinum, Methotrexate, CYVADTIC
Lymph node, no tumor seen (known melanoma)*	BOLD (98%) Bleomycin (95%)	DTIC

*Thymidine-labeling assay

Abbreviations: BOLD = combination Bleomycin, Vincristine, BCNU, DTIC; CYVADTIC = combination L-PAM, Vincristine, Adriamycin, DTIC.

growth) are considered unsatisfactory. The thymidine labelling assay, because it measures thymidine uptake by dividing cells, is not susceptible to clumping artifact.

CONCLUSIONS

Of over 2200 specimens processed for in vitro chemosensitivity testing, 1.7% were subsequently classified as nonmalignant. Almost one-third of these grew in soft agar. This finding underscores the fact that not enough is known about the determinants and prerequisites for growth in soft agar. It also illustrates the rarity of this phenomenon in routine tumor chemosensitivity testing. When combined with the wealth of reports documenting the malignant origin of colonies that form from known malignant specimens, it is unlikely that non-neoplastic cells account for significant growth in standard clonogenic assays.

To maximize the utility of the clonogenic assay in clinical medicine, every specimen submitted for testing should also have a representative portion sent for a pathologist's examination by the operating surgeon. For very small tumors, this requirement may decrease the cell yield available for assay, but, as we have illustrated, it is necessary. A cytologic record of each assay should be kept in a permanent file. If pathologic results in conflict with the initial diagnosis are subsequently discovered, comparison with the cytology may be helpful. Further investigation is warranted into the determinants of growth in soft agar by both benign and malignant cells.

REFERENCES

1. Bradley EC, Reichert CM, Brennan MF, Von Hoff DD: Direct cloning of human parathyroid hyperplasia cells in soft-agar culture. Cancer Res 40:3694-3969, 1980.
2. Buick RN, Stanisic TH, Fry SE, Salmon SE, Trent JM, Krasovich P: Development of an agar-methyl cellulose clonogenic assay for cells in transitional cell carcinoma of the human bladder. Cancer Res 39:5051-5056, 1979.
3. Friedman HM, Glanbiger DL: Assessment of in vitro drug sensitivity of human tumor cells using [^{3}H] Thymidine incorporation in a modified human tumor stem cell assay. Cancer Res 42:4683-4689, 1982.
4. Hamburger AW, Salmon SE: Primary bioassay of human myeloma stem cells. J Clin Invest 60:846-854, 1977.
5. Hamburger AW, Salmon SE: Primary bioassay of human tumor cells. Science 197:461-463, 1977.

6. Hamburger AW, Salmon SE, Kim MB, Trent JM, Soehnlen BJ, Alberts DS, Schmidt HT: Direct cloning of human ovarian carcinoma cells in agar. Cancer Res 38:3438-3443, 1978.
7. Harris GJ, Zeagler J, Hodach A, Casper J, Harb J, Von Hoff DD: Ultrastructural analysis of colonies growing in a human tumor cloning system. Cancer 50:722-726, 1982.
8. Kern DH, Bertelsen CA, Mann BD, Campbell MA, Morton DL, Cochran AJ: Clinical application of the clonogenic assay. Ann Clin Lab Sci 13:10-15, 1983.
9. Pavelic ZP, Slocum H, Rustum YM, Creaven PJ, Karakousis C, Takita H, Mittelman A: Growth of human solid tumor cell colonies in soft agar. Proc Am Asssoc Cancer Res Am Soc Clin Oncol 20:329, 1980.
10. Salmon SE, Hamburger AW, Soehnlen BJ, Durie BGM, Alberts DS, Moon TE: Quantitation of differential sensitivity of human tumor stem cells to anticancer drugs. N Engl J Med 298:1321-1327, 1979.
11. San RHC, Laspia MF, Soiefer AT, Maslawky CJ, Rice JM, Williams GM: A survey of growth in soft agar and cell surface properties as markers for transformation in adult rat liver epithelial-like cell cultures. Cancer Res 39:1026-1034, 1979.
12. Sandbach J, Von Hoff DD, Clark G, Cruz AB, O'Brien M: Direct cloning of human breast cancer in soft agar culture. Cancer 50:1315-1321, 1982.
13. Sarosdy MF, Lamm DL, Radwin HM, Von Hoff DD: Clonogenic assay and in vitro chemosensitivity testing of human urologic malignancies. Cancer 50:1332-1338, 1982.
14. Seemayer TA, Knack J, Wang N-S, Ahmed MN: On the ultrastructure of hibernoma. Cancer 36:1785-1793, 1975.
15. Shoemaker RH, Ingel HJ, McLachlan SS, Hartfiel JL: Assay of drug sensitivity of pediatric solid tumors cloned in semi-solid medium. Proc Am Assoc Cancer Res Am Soc Clin Oncol 21:252, 1981.
16. Tanigawa N, Kern DH, Hikasa Y, Morton DL: Rapid assay for evaluating the chemosensitivity of human tumors in soft agar culture. Cancer Res 42:2159-2164, 1982.
17. Tucker RW, Sanford KK, Handleman SL, Jones GM: Colony morphology and growth in agarose as tests for spontaneous neoplastic transformation in vitro. Cancer Res 37:1571-1579, 1977.
18. Von Hoff DD, Casper J, Bradley E, Sandach JL, Jones D, Makuch R: Association between human tumor colony-forming assay and response of an individual patient's tumor to chemotherapy. Am J Med 70:1027-1032, 1981.

19. Von Hoff DD, Casper J, Bradley E, Trent JM, Hodach A, Reichert C, Makuch R, Altman A: Direct cloning of human neuroblastoma cells in soft agar culture. Cancer Res 40:3591-3597, 1980.
20. Von Hoff DD, Forseth B, Metelmann H-R, Harris G, Rowan S, Coltman CA: Direct cloning of human malignant melanoma in soft agar culture. Cancer 50:696-701, 1982.
21. Yuhas JM, Li AP, Martinez AO, Ladman AJ: A simplified method for production and growth of multicellular tumor spheroids. Cancer Res 37:3639-3643, 1977.

IV. PHARMACOLOGY

Recent Results of New Drug Screening Trials With a Human Tumor Colony Forming Assay

ROBERT H. SHOEMAKER, MARY K. WOLPERT-DEFILIPPES
NANNETTE R. MELNICK, JOHN M. VENDITTI, RICHARD M. SIMON
National Cancer Institute
Bethesda, Maryland 20205

DAVID H. KERN
University of California at Los Angeles
Los Angeles, California 90024

MICHAEL M. LIEBER
Mayo Foundation
Rochester, Minnesota 55901

WILLIAM T. MILLER
VSE Corporation
Alexandria, Virginia 22303

SYDNEY E. SALMON
University of Arizona
Tucson, Arizona 85724

DANIEL D. VON HOFF
The Cancer Therapy and Research Foundation of South Texas
San Antonio, Texas 78229

INTRODUCTION

The human tumor colony forming assay has been extensively evaluated as a tool for predicting the clinical response of individual patients' tumors to specific anticancer drugs. Studies conducted in a number of independent laboratories have indicated substantial predictive value of the test particularly with respect to predicting clinical resistance [2,4,11,14]. The assay has also been used in pre-clinical evaluations of antitumor drugs discovered in the National Cancer Institute (NCI) *in vivo* screening program[12], for analog comparisons[5,7] and as a means of generating data for guiding compounds into Phase II clinical trials in particular tumor types [8,13]. In this paper we will describe recent results of an NCI program to apply the human tumor colony forming assay (HTCFA) to *de novo* screening of unknown compounds. This project was initiated in the fall of 1980 and

HUMAN TUMOR CLONING
ISBN 0-8089-1671-8

has involved four centers working in a cooperative contract-based effort directed by the NCI. Results of studies aimed at evaluation of the feasibility of using the assay for screening, examination of the validity of the assay, and evaluation of the potential of the assay for identifying novel antitumor drug leads have been described previously[10]. Briefly, the assay was deemed to be feasible for screening with a limited number of tumor types on a moderate scale. Tumor types found to be most suitable for use in screening operations included breast, colorectal, kidney, lung, melanoma, and ovary. The validity of the assay for use in drug screening was supported by the demonstrated ability to detect the majority of established agents and by evidence of both intra- and inter-laboratory reproducibility. Identified limitations of the assay included a lack of sensitivity to compounds which require metabolic activation, such as cyclophosphamide, and insensitivity to certain antimetabolites, apparently because of interference by medium constituents. Initial screening results supported the potential of the assay for identifying antitumor drug leads not detected by the conventional *in vivo* screening system. In this report we will describe further studies designed to evaluate the drug screening potential of the assay.

FLOW OF DRUGS THROUGH DCT SCREENS, 1984

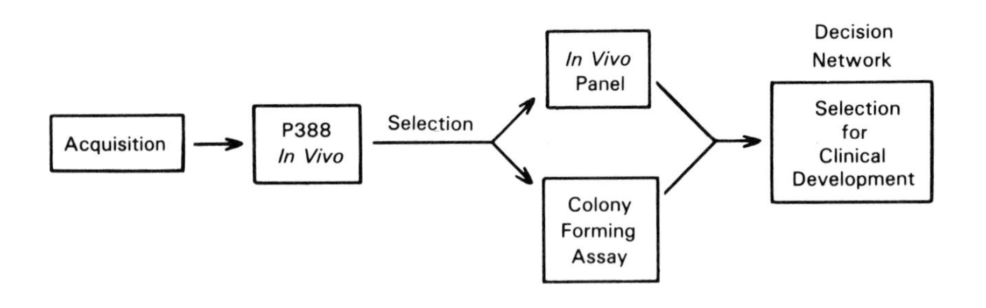

Figure 1. Schematic illustration of the current new drug screening program utilized by the Division of Cancer Treatment, NCI.

Figure 1 illustrates the flow of compounds through the current NCI screening system. Under the current screening program approximately 10,000 new compounds are acquired on an annual basis and subjected to testing in the in vivo P388 leukemia pre-screen. Conventionally, compounds active in the P388 pre-screen are advanced for further testing and evaluation in the in vivo tumor panel. Compounds lacking activity are not ordinarily considered for further development. Compounds showing significant activity in the in vivo tumor panel are presented to the NCI Decision Network Committee for consideration for development to clinical trials. As shown in Figure 1, the human tumor colony forming assay has been viewed as offering an alternative route for bringing compounds into clinical trials in man. In order to define the ability of the human tumor colony forming assay to detect compounds which are missed by the current in vivo screening program, drug screening trials have focused on testing of compounds which were negative in the in vivo P388 leukemia pre-screen.

MATERIALS AND METHODS

Drug sensitivity studies were performed essentially as described by Salmon et al.[6], except that drug exposure was by the "continuous" method. Cell suspensions were prepared from fresh tumor specimens by mechanical or enzymatic methods or a combination of both, as appropriate to the tumor specimen. Five hundred-thousand nucleated cells were combined with the test compound or appropriate vehicle control and plated in the upper layer of the two layer system. For each experiment a total of 12 control plates were utilized. Six plates were used to estimate control growth. These were incubated without perturbation for 7-21 days in the same manner as the drug treated plates. One plate was stained with tetrazolium chloride[1] 24 hours after plating and utilized to establish the initial "colony" count. Three plates were treated with 100 ug/ml mercuric chloride (positive control). The remaining two plates were monitored during the incubation period to determine the optimal time for counting colonies. Triplicate plates were employed for testing of unknown compounds.

Colony counting was performed using a Bausch and Lomb Omnicon FAS-II image analysis system[3] after in situ staining with tetrazolium chloride. A minority of the experiments reported here (approximately 10%) were performed prior to our implementation of the vital staining procedure. Objects presenting a roughly circular profile with a diameter of 60 micrometers or greater were scored as colonies. Experimental and patient demographic data were collected on a disk integral to the image analysis system and transmitted to

NCI via telecommunications linkage. To be considered evaluable, each experiment had to meet four quality control criteria. The initial plate count had to be less than or equal to 47, a net increase of at least 70 colonies over the course of the incubation period was required, the coefficient of variation of the control group had to be less than or equal to 50%, and the positive control compound had to be active. The rationale and requirement for the use of these quality control measures has been described previously [9,10]. Only results from evaluable experiments are presented here.

The two-stage screening procedure used in these studies is illustrated in Figure 2. All compounds are first tested at a a concentration of 10 ug/ml in fifteen different tumor specimens (Stage I). The requirement for initial testing in fifteen specimens was established to obtain a high probability of detecting compounds with a true in vitro response rate of 20% or greater. Compounds showing activity (reduction of colony formation to 30% or less of the control value) in two or more tumors are advanced for additional testing (Stage II). This additional testing consists of five dose-response experiments in each of the six tumor types selected for use in screening (breast, colorectal, kidney, lung, melanoma, and ovary).

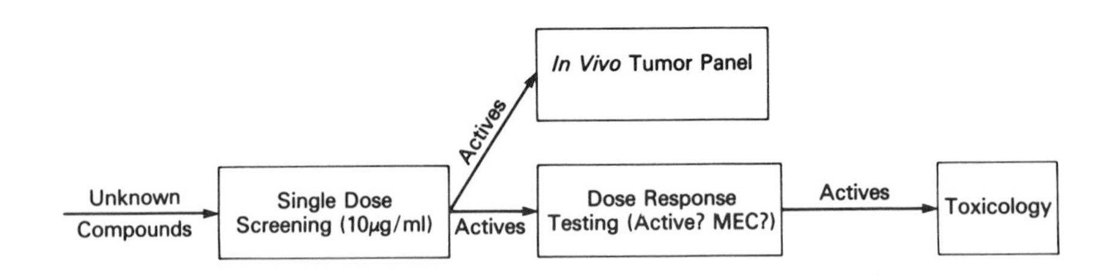

Figure 2. Schematic illustration of the two stage screening procedure used in new drug screening with the HTCFA.

The purposes of Stage II testing are: to confirm the activity observed in Stage I; to estimate the minimum effective concentration; and to define the spectrum of in vitro antitumor activity. Scheduling of testing in both stages is assisted by an interactive computer program which coordinates and maximizes the efficiency of the multi-laboratory effort.

RESULTS

A total of 103 compounds which were negative in the *in vivo* P388 mouse leukemia pre-screen have been tested in the human tumor colony forming assay. While the majority of these compounds were also negative in the HTCFA, compounds showing substantial activity were also observed. In Stage I testing 78/103 (76%) compounds were negative. Twenty-five compounds (24%) were active, with *in vitro* response rates ranging from 13%-100%. These 25 compounds were advanced for further evaluation in Stage II. At this point in time, compounds have reached varying stages of completion. None is yet complete in Stage II testing. Ten of the 25 compounds which entered Stage II testing have overall *in vitro* response rates less than 20% when data from Stage II is pooled with Stage I testing using 10 ug/ml as a reference dose i.e., in Stage II testing tumors are considered to "respond" if colony formation is reduced to 30% or less of the control value at a concentration of 10 ug/ml or less. Data for the 14 compounds which have retained an *in vitro* response rate of 20% or greater are summarized in Table 1.

TABLE 1
Screening Data for Selected Compounds Identified as Active in the HTCFA Which Were Negative in the *In Vivo* Murine P388 Leukemia Pre-Screen

Compound	Response Rate (%)	IC_{70} Range (ug/ml)
A	28/28 (100)	0.1-8
338698	26/28 (93)	0.3->10
B	22/27 (82)	0.1->10
C	22/30 (73)	0.1->10
343549	14/26 (54)	4->10
D	7/18 (39)	<10->10
E	9/28 (32)	<10->10
343633	10/34 (29)	3->10
340307	11/41 (27)	0.1->10
343529	9/35 (26)	3->10
343385	7/28 (25)	5->10
339675	7/28 (25)	<10->10
343378	6/25 (24)	4->10
F	4/18 (22)	<10->10

Dose-response data are summarized in Table 1 in terms of the observed range of IC_{70} values, i.e., the concentration of test compound observed to reduce colony formation to 30% of the control value in a particular tumor. For several compounds IC_{70} ranges from <10->10 ug/ml are listed, indicating that dose-response testing has not been performed and in Stage I testing both sensitive (responding at 10 ug/ml) and resistant (not responding at 10 ug/ml) tumors were observed. Compounds designated by alphabetic characters in Table 1 were obtained by NCI under commercial discreet agreements and cannot be further identified. While additional Stage II testing will be required to make definitive statements regarding selectivity of agents for particular tumor types, two basic patterns of activity are emerging. Compounds are observed with activity distributed uniformly across tumor types (broad spectrum activity) and a few compounds are observed for which there is preliminary evidence of selective activity. These patterns are illustrated in Table 2.

TABLE 2
Patterns of Antitumor Activity Observed in the HTCFA

Response Rates by Tumor Type

Compound	BRE	COL	KID	LUN	MEL	OVA	OVERALL	(%)
338698	3/3	3/3	2/2	3/4	5/5	10/11	26/28	93
343549	2/3	2/5	-	3/4	4/7	3/7	14/26	54
343513	0/2	1/4	0/2	3/3	1/7	0/10	5/28	18
H	0/6	0/5	0/1	3/6	2/10	0/11	5/39	13

In order to generate data to help explain the basis for the HTCFA activity of these compounds which had previously been found inactive *in vivo*, a colony forming assay was devised using the same cell type used for *in vivo* testing (P388) and an *in vitro* assay protocol as similar as possible to that used in the HTCFA. All of the compounds found to have activity in the HTCFA (Stage I) as well as 41 HTCFA negative compounds (selected on the basis of sample availability) were tested in this P388 colony forming assay (P388 CFA). Table 3 presents a comparison of activity in the two assays.

TABLE 3

Comparison of *In Vitro* Screening Results for Compounds Which Were Negative in the *In Vivo* P388 Pre-Screen

		Human Tumor Colony Forming Assay		
		+	-	
P388 Colony Forming Assay	+	9	3	12
	-	16	38	54
		25	41	66

TABLE 4

In Vivo Activity of Compounds Active in the HTCFA

	Experimental Model*									
Compound	PS	B1	LE	CD	C8	M5	LL	C2	LK	MB
338600	-		-	-	-			-	-	-
338698	-									
H	-	-	-	-	-	-			-	-
E	-		-		-					-
339675	-	-	-	-				-	-	
C	-	-	-	-	-	-	-	-	-	-
B	-		-	-	-	-	-		-	-
340307	-		-	-	-	-		-	-	-
I	-									
341964	-									
343513	-	-	+	-	-	-	-	-	-	-
343522	-	-	-	-	-	-	-	-	-	-
343556	-									

* Experimental models are abbreviated as follows: PS=P388 Leukemia; B1=B16 Melanoma; LE=L1210 Leukemia; CD=Murine Mammary Tumor; C8=Murine Colon Tumor; M5=Murine Sarcoma; LL=Lewis Lung Tumor (Murine); C2=Human Colon Tumor Xenograft (Subrenal Capsule Assay); LK=Human Lung Tumor Xenograft (Subrenal Capsule Assay); MB=Human Mammary Tumor Xenograft (Subrenal Capsule Assay).

In vivo testing in the NCI Tumor Panel was performed on all of the HTCFA active compounds which were available in adequate quantity. To date, the gram quantities required for *in vivo* testing have been acquired for only nine compounds and testing of these remains incomplete. The available results are summarized in Table 4.

DISCUSSION

The results presented here support and extend our earlier data[10] indicating that the HTCFA is effective in identifying antitumor drug leads which would be missed by the conventional NCI *in vivo* screening system. In our two stage screening procedure 25/103 compounds were found to be active in Stage I and 14 of these have retained *in vitro* response rates of 20% or greater following additional testing in Stage II. Six of these 14 compounds are of special interest because of their lack of activity in the P388 CFA. This finding indicates that these compounds were inactive in the *in vivo* P388 model because of a lack of sensitivity of the P388 cells rather than because of a lack of *in vivo* bioavailability. Studies are currently in progress to gain insight into the reasons for the lack of *in vivo* activity of the P388 CFA active compounds. To evaluate the possible role of *in vivo* metabolic inactivation in this phenomenon, compounds will be tested in a colony forming assay incorporating a metabolic activating/inactivating system. Should activity be lost under these conditions, this will support the notion that the lack of *in vivo* activity is the result of metabolic inactivation. Other possibilities exist, such as rapid excretion, which are beyond the scope of *in vitro* studies.

To date, only one of the HTCFA active compounds has shown activity *in vivo*. This compound (NSC 343513) has a close structural analog which was identified in the NCI preclinical testing data base. The analog (NSC 169875) is a commercially available tranquilizer marketed under the brand name Triperidol. These compounds share patterns of activity in the *in vivo* Tumor Panel in that both have reproducible activity against L1210 but lack activity in the P388 model. We plan to evaluate Triperidol and several other structural analogs in the HTCFA before making a decision regarding further development of NSC 343513.

Although the new drug screening trials reported here have only included 103 compounds, several interesting antitumor drug leads have been identified. The HTCFA active compounds which lack activity in the P388 CFA would seem to be of special interest. These compounds have clearly different activities in human tumors versus the murine P388 cells and may represent a type of compound not previously identified in the NCI screening effort. Compounds showing evidence for selective activity in particular tumor types also deserve special interest. Although statistically significant evidence for selectivity has been obtained for several compounds, including NSC 343513 illustrated in Table 2, we feel that additional testing is required before making definitive statements regarding the selective potential of these compounds. Future drug screening efforts will continue to focus on evaluation of P388 prescreen negative compounds and on development of the data required to prioritize the available leads for further development.

ACKNOWLEDGEMENTS

We would like to acknowledge the valuable contributions of Drs. Michael Alley, Gary Clark, John Kovach, Thomas Moon, Donald Morton, Jerry Phillips, Mr. Robert Brennan, Ms. Mary Ann Campbell and Ms. Rosa Liu to the design of this project and to the conduct of the work. In addition we would like to acknowledge the dedicated work of the technical personnel at each of the participating institutions. This work was supported by contracts N01-CM0-7419, N01-CM0-7420, N01-CM0-7327, N01-CM1-7497, and N01-CM0-7251 from the National Cancer Institute.

REFERENCES

1. Alley, M. C., Uhl, C. B., and Lieber, M. M. Improved detection of drug cytotoxicity in the soft-agar colony formation assay through the use of a metabolizable tetrazolium salt. Life Sci. 31:3071-3078, 1982.
2. Bertelsen, C. A., Sondak, V. K., Mann, B. D., Korn, E. L, and Kern, D. H. Chemosensitivity testing of human solid tumors - A review of 1582 assays with 258 clinical correlations. Cancer (in press).
3. Kressner, B. E., Morton, R. R. A., Martens, A. E., et al. Use of an automated image analysis system to count colonies in stem cell assays of human tumors. In Cloning of Human Tumors, Salmon, S. E. (Ed.) Alan R. Liss, Inc. New York, 1980, p.p. 179-193.
4. Moon, T. E., Salmon, S. E., White, C. S., et al. Quantitative association between the in vitro human tumor stem cell assay and clinical response to therapy. Cancer Chemother. Pharmacol. 6:211-218, 1981.
5. Rossof, A. H., Johnson, P. A., Kimmell, B. D., et al. In vitro phase II evaluation of cisplatin and three analogs against various human carcinomas. Proc. Am. Soc. Clin. Oncol. 2:C-147, 1983.
6. Salmon, S. E., Hamburger, A. W., Soehnlen, B., et al. New Engl. J. Med. 298:1321-1327, 1978.
7. Salmon, S. E., Liu, R. H., and Casazza, A. M. Evaluation of new anthracycline analogs with the human tumor stem cell assay. Cancer Chemother. Pharmacol. 6:103-110, 1981.
8. Salmon, S. E., Meyskens, F. L., Alberts, D. S., et al. New drugs in ovarian cancer and malignant melanoma: In vitro phase II screening with the human tumor stem cell assay. Cancer Treat. Rep. 65:1-12, 1981.
9. Shoemaker, R. H., Wolpert-DeFilippes, M. K., and Venditti, J. M. Potentials and drawbacks of the human tumor stem cell assay. Behring Inst. Mitteilungen (in press).
10. Shoemaker, R. H., Wolpert-DeFilippes, M. K., Kern, D. H., et al. Application of a human tumor colony forming assay to new drug screening (submitted for publication).
11. Tveit, K. M., Fodstad, O., Lotsberg, J. et al. Colony growth and chemosensitivity in vitro of human melanoma biopsies. Relationship to clinical parameters. Int. J. Cancer. 29:533-538, 1982.

12. Von Hoff, D. D., Coltman, C. A., and Forseth, B. Activity of mitoxantrone in a human tumor cloning system. Cancer Res. 41:1853-1855, 1981.
13. Von Hoff, D. D., Coltman, C. A., and Forseth, B. Activity of 9-10 anthracenedicarboxaldehyde bis [(4,5-dihydro-1 H-imidazol-2-yl) hydrazone] dihydrochloride (CL 216,942) in a human tumor cloning system. Leads for Phase II trials in man. Cancer Chemother. Pharmacol. 6:141-144, 1981.
14. Von Hoff, D. D., Clark, G. M., Stogdill, B. J., et al. Prospective clinical trial of a human tumor cloning system. Cancer Res. 43:1926-1931, 1983.

STUDIES OF CLONOGENIC HUMAN TUMOR CELLS BY THE COURTENAY SOFT AGAR METHOD

KJELL M. TVEIT

RESEARCH ASSOCIATE, M.D., PH.D.
NORSK HYDRO'S INSTITUTE FOR CANCER RESEARCH

LIV ENDRESEN

RESEARCH FELLOW, M.S.
DEPARTMENT OF CLINICAL PHARMACOLOGY
THE NATIONAL HOSPITAL
OSLO 1, NORWAY

ALEXANDER PIHL

PROFESSOR OF CANCER BIOLOGY, M.D., PH.D.
NORSK HYDRO'S INSTITUTE FOR CANCER RESEARCH
THE NORWEGIAN RADIUM HOSPITAL
OSLO 3, NORWAY

It is generally believed that only a fraction of the cells in a human tumor (termed stem cells) is capable of indefinite multiplication and that these cells are responsible for the growth and dissemination of the tumors. The tumor cells forming colonies in semi-solid media, i.e. the clonogenic cells, are believed to represent the tumor stem cells.

The method most commonly used for the study of clonogenic human tumor cells *in vitro*, is that described by Hamburger and Salmon (1977). In 1978, Courtenay and Mills (1978) presented a different cultivation method which has been used as a basis for a chemosensitivity assay for human tumor cells

HUMAN TUMOR CLONING
ISBN 0-8089-1671-8

(Tveit et al., 1980; 1982). The Courtenay method which has been employed in our laboratory since 1978, seems to offer several advantages, but has so far only been used in a few laboratories. It may therefore be warranted to discuss this method and the data obtained with it in some detail.

METHODOLOGY

The distinguishing features of the Courtenay and Mills (C-M) culture procedure are that it employs a hypoxic (3-5% O_2) atmosphere, addition of rat erytrocytes a replenishable liquid top medium, and tubes, instead of Petri dishes. The procedure used in our laboratory is somewhat modified compared to the original one (Courtenay & Mills 1978), and is performed as described below.

Culture method. To each of a series of culture tubes placed in a rack, is added 0.2 ml of a suspension of washed rat red blood cells (RBC), heated to 44°C for 1 h and diluted 1:8 in complete medium (Hams F12, supplemented with 15% foetal calf serum and antibiotics). Then, 0.2 ml of the suspension of properly diluted tumor cells, treated or untreated, is added to each tube. Finally, 0.6 ml of a 0.5% agar (Bacto) in complete medium is added at a temperature of 44°C. The rack with all tubes is immediately shaken and put on ice water to permit the agar to solidify . After 5 minutes, caps are put on the tubes in the open position, and the tubes are placed in an incubator controlling the exact concentration of O_2 (5%), CO_2 (5%), and N_2 (90%). On the next day, the tubes are sealed by pressing down the caps. After 5-7 days, 1 ml of medium is added to each tube, and, after an additional week, another 1 ml is added if it is necessary to keep the cultures for more than 2 weeks.

Preparation of cells and drug treatment. Tumor cells from human tumor specimens are disaggregated by either a mechanical procedure alone (by the use of a Stomacher) or by a combination of a mechanical and an enzymatic procedure (0.1% collagenase, 1000 U/ml hyaluronidase, 0.02% DNase). In all cases clumps of cells are removed by filtering the cell suspension through a 30 or 45 µm nylon mesh. After centrifugation, complete medium is added. In drug sensitivity studies, 5×10^5 cells per ml are treated with drugs at different (at least 4) concentrations and incubated for 1 h at 37° C in an atmosphere of 5% O_2, 5% CO_2 and 90% N_2 under constant shaking. The cells are washed twice in phosphate buffered saline and are suspended in medium to give a final cell concentration of 10^5 viable cells/ml. This gives 2×10^4 viable cells in each culture.

Colony counting. After cultivation for 2 or 3 weeks, the liquid top medium is removed and the clump of agar is squeezed between the top and the bottom of a contact Petri dish. Colonies (more than 30 cells) are counted in a stereo microscope. Plating efficiency (PE) is calculated as the number of colonies in percentage of the number of viable cells plated. In the chemosensitivity experiments, more than 30 colonies are required in control cultures to evaluate colony inhibition (% of control).

IDENTITY OF THE CLONOGENIC CELLS

A basic question is whether the colonies observed in the cultures are derived from malignant cells, and, if so, whether they represent the tumor stem cells. This aspect we have studied primarily in melanomas.

Morphological studies, employing light microscopy, as well as electron microscopy, have shown that the cells composing the colonies formed in the C-M method, have similar characteristics as the tumor cells of the original tumors (Tveit *et al.*, 1982). In melanomas, structures such as premelanosomes, melanosomes and melanine granules have been found within individual cells, proving their origin.

Chromosome analyses carried out on colonies cultivated further *in vitro* showed the presence of aneuploid cell populations (Tveit & Pihl, 1981). Preliminary studies of the DNA content of colonies and of the original cell suspensions, have shown that primarily the aneuploid cells in the tumor form colonies, whereas the diploid cells usually are lost during cultivation (Pettersen *et al.*, unpublished).

The presence in the colonies of malignant cells with unlimited growth potential has been shown by establishment of continuous *in vitro* cell lines, and by the fact that inoculation of the cells into athymic nude mice resulted in tumors that could be serially heterotransplanted (Tveit *et al.*, 1981b; Tveit & Pihl, 1981).

COLONY-FORMING ABILITY

A main problem in soft agar assays is that only a limited number of human tumors will form colonies and that the PEs in general are low. This limits the usefulness of such methods and raises the question whether or not the colonies formed are representative of the tumor cells.

By June -83 we have cultivated in soft agar more than 600 human tumor specimens of different origins. The fraction of tumors growing in the C-M culture method is shown in TABLE 1. Altogether, 64% of the tumors examined formed more than 10 colonies. This is rather similar to what has been found in other large materials employing different soft agar methods (Kern *et al.*, 1982; Von Hoff *et al.*, 1983). Melanomas gave rise to colonies more

TABLE 1.
COLONY FORMATION OF HUMAN MALIGNANT TUMORS IN THE COURTENAY SOFT AGAR METHOD

TUMOR	NO. OF TUMORS FORMING > 10 COL.	%
Malignant melanoma	242/340	71
Breast carcinoma	73/151	48
Ovarian carcinoma	39/58	67
solid	27/41	66
ascites	12/17	71
Bladder carcinoma	13/24	54
Soft tissue sarcoma	14/21	67
Glioma	5/7	71
Total	386/601	64

frequently and of larger size than any other tumor type. Less than 50% of breast carcinomas grew in the system. In the whole material about 50% of the tumors gave sufficient growth to permit chemosensitivity studies (> 30 colonies in controls). In melanomas this percentage is > 60, whereas in breast carcinomas it is only about 40. The PEs were in most cases in the range 0.1 - 10%, which is higher than what has been found in other methods (Kern *et al.*, 1982; Von Hoff *et al.*, 1983).

It should be realized that the present data on colony formation in general are based on 2×10^4 viable cells seeded, which is only about 1/10 of the cells used in other methods. This implies that tumors with PEs < 0.05 % are recorded as "not growing" in our system. Moreover, by using such small cell numbers, less cells are needed for cultivation and more tests (more drugs and more concentrations per drug) can be performed with this method on the tumor samples available.

When both the C-M method and the Hamburger and Salmon (H-S) method were applied on the same tumor material (patients' tumors, xenografts and cell lines), in most cases examined the PEs were higher in the C-M method (Tveit *et al.*, 1981a; Tveit, 1983; Tveit *et al.*, 1983). The difference was pronounced in melanomas, but small in other tumors.

The presence of RBC and low oxygen concentration are in most cases responsible for the differences in colony growth seen in the two methods (Tveit *et al.*, 1981a). The RBC seem to be more important than the oxygen concentration, but the factors combined give the highest PEs. Recently, we have also found that some of the factors used in the standard H-S procedure (ascorbic acid, insulin) may in fact inhibit colony formation of human tumor cells (Endresen *et al.*, 1983; Tveit *et al.*, 1983). By omitting some of the "enrichments" more colonies were formed.

For the purpose of chemosensitivity measurements it is essential that the PEs are independent of the number of cells seeded. This has been found to be the case in the C-M method which give a linear relationship between the number of cells plated and the number of colonies formed, except at high cell concentrations where starvation inhibits colony formation (Tveit *et al.*, 1981a). In contrast, in many cases when the H-S method was used on the same tumor cell samples, the relationship was found to be non-linear at low cell concentrations ($<10^4$ cells). In the H-S system there was an apparent need for a conditioning factor, permitting colony formation at low cell densities. We have also found that RBC are extremely useful as such a conditioning factor. In fact, addition of RBC always ensures a linear relationship, and gives more reproducible data for colony formation (Tveit et al., 1983).

The present study shows that the fraction of tumor cells capable of forming colonies *in vitro*, is not a defined cell population containing a certain number of cells, as it may be strongly influenced by the microenvironment. When tumor cells are cultivated *in vitro*, subpopulations of cells are selected on the basis of the microenvironments. If a small population of clonogenic cells selected by one method is representative of a larger population of clonogenic cells selected by another method, the properties assessed by different methods will be similar. However, it is also conceivable that the few clonogenic cells examined may represent a small subpopulation which is not representative of the larger population of cells, and in this event the results will not be the same with different methods. Even in a tumor which gives similar PEs in different methods, the population of clonogenic cells studied might be dissimilar. *A priori* it is likely that the system which permits growth of a larger population of the clonogenic tumor cells, will also give the most reliable sensitivity data.

CHEMOSENSITIVITY

In chemosensitivity measurements, as in other biological assays, a meaningful result is only obtained in a dose range where the parameter measured varies with the dose, i.e. under conditions where there is a clear dose-response relationship. It is therefore necessary to secure, in each case, a dose-effect curve. Therefore, in chemosensitivity experiments we routinely use at least 4 concentrations of the drugs, covering a considerable dose range.

If high enough drug concentrations are used, a complete inhibition of colony formation should be obtained, as is found in radiosensitivity experiments (Selby et al., 1983). It has become increasingly clear that many of the plateau-shaped curves previously published are due to the presence of cell clumps in the cell suspension seeded. In our experiments we have tried to eliminate the presence of cell clumps by 3 measures: 1. Filtering the cell suspension through a nylon mesh. 2. Examination of cultures on day zero (cultures without RBC which would complicate observations). 3. The use of a positive control where the cells are treated with a compound at a concentration that will completely inhibit colony formation, but will not eliminate preexisting clumps. Since we started our in vitro studies in 1978 we have used for this purpose the toxic protein abrin at concentrations of 1-10 μg/ml.

When the cells are treated with different cancerostatic agents, and the cells are cultivated as described above, a differential sensitivity is found for different tumors of the same histological type. Figure 1 shows dose response curves for

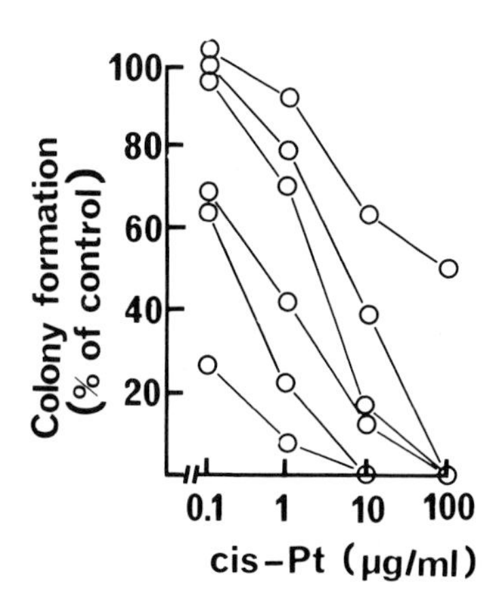

Figure 1. Dose-response curves for six patients' melanomas exposed in vitro to cis-Platinum.

6 patients' melanomas exposed in vitro to cis-platinum. Similar slopes and range of curves are seen for most other drugs and tumor types.

Recently we have found that not only the colony forming ability, but also the sensitivity to cytotoxic agents is dependent on the micro-environment (the culture conditions). Both in our previous work on melanoma xenografts (Tveit et al., 1981a) and in our recent investigation on a range of different types of tumor cells and drugs (Endresen et al., 1983; Tveit et al., 1983), we have found that when drug-treated cells are cultivated concurrently in the two different methods, the apparent sensitivity is in general lower in the C-M than in the H-S method (Figure 2). Preliminary data show that several of the components used in the methods may influence the apparent chemosensitivity (Endresen et al., 1983; Tveit et al., 1983).

The finding that a different chemosensitivity pattern can be found when different culture conditions are used, has important implications. Many workers have modified the original H-S procedure in attempts to increase the PEs. The

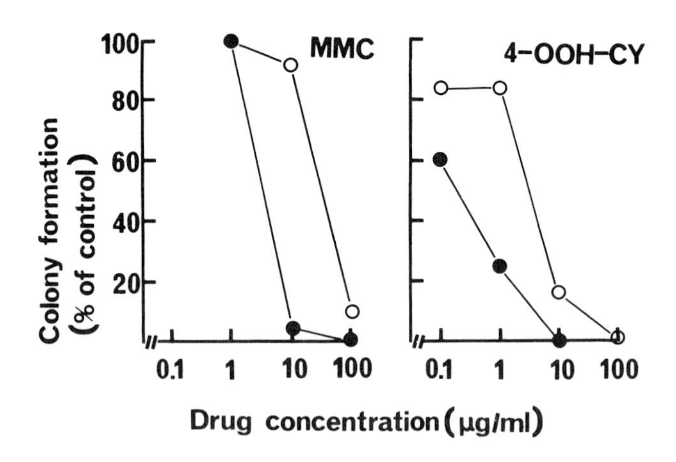

Figure 2. Dose-response curves for a neuroblastoma cell line exposed in vitro to Mitomycin-C (MMC) and 4-hydroperoxy-cyclophosphamide (4-OOH-CY), and cultivated in the C-M method (open circles) and in the H-S method (closed circles).

question whether the modifications introduced have also altered the apparent chemosensitivity, has not been adequately examined.

It follows that results obtained with different modifications may not be directly comparable. Certainly this is true for the H-S and C-M methods which differ in many important respects. The chemosensitivity data obtained with these two methods can not be directly compared.

CRITERIA FOR EVALUATION OF SENSITIVITY

A shortcoming of the *in vitro* chemosensitivity assays is that it is impossible to mimic *in vitro* the drug exposure *in situ*. In a tumor the concentrations and the exposure time at the cellular level after administration of different drugs, are not known. The approach used by us and in most work on colony-forming assays, involving a one h exposure to drugs, as well as the procedure using continuous exposure to drugs during the cultivation period, represent highly artificial conditions for drug exposure.

Most workers use a cut-off point *in vitro* of 1/10 of the plasma concentration x time product, as a basis for evaluation of *in vitro* sensitivity, and classify the tumors as sensitive if colony formation is inhibited by more than 50-70% at this concentration (Meyskens *et al.*, 1981; Von Hoff *et al.*, 1981). However, most probably, dissimilar relationships between the plasma concentrations and the concentrations at the cellular level exist for different drugs with different mechanisms of action, different pharmacology and different ability to penetrate into the tumor cells. The procedure currently used for evaluation of sensitivity may therefore not be optimal for all drugs and tumors.

Our finding of different apparent chemosensitivity when two different methods were used, shows that the criteria for *in vitro* sensitivity/resistance used by workers employing the H-S method, can not be assumed to be valid when the C-M culture method is performed. In both methods a careful and extensive calibration of the *in vitro* data has to be performed for each drug and each tumor type.

CORRELATIONS TO IN VIVO RESPONSES

The use of in vitro assays as predictive tests is based on the assumption that the patients' response to antineoplastic agents is largely determined by the inherent chemosensitivity of the tumor cells, and not by host factors (pharmacokinetics, drug penetration). Furthermore, the use of clonogenic assays presupposes that the colony-forming cells reflect the chemosensitivity of the tumor stem cells. These basic assumptions can be tested on animal tumors, human tumor xenografts, and patients' tumors. Since all chemosensitivity studies in vitro are performed under highly artificial conditions, it is necessary to perform a calibration of the in vitro data by means of in vivo responses.

In our study using the C-M method for cultivation of human melanomas, we made a first line calibration of the chemosensitivity by means of human melanoma xenografts, grown in athymic nude mice, where the in vivo response to a series of different cancerostatic agens can be assessed quantitatively and reproducibly (Tveit et al., 1980). In that study, and also in that of Bateman et al. (1980), a close relationship was found between the in vivo response and the in vitro sensitivity of the disaggregated cells. The establishment of this relationship forms the basis for the clinical application of the method.

On the basis of the in vitro/in vivo correlations in xenografts, we have quantitated in patients' melanomas the in vitro sensitivity in terms of expected growth delay (EGD; Tveit et al., 1982). In previously untreated patients we performed 76 comparisons between the quantitatively measured in vitro sensitivity and the in vivo response (Tveit et al., 1983). The data are shown in TABLE 2. It is seen that 10 patients with partial response (PR) or complete response (CR) all had values higher than 2.0 and 7 of 9 patients with no change (NC) or mixed response (MR) had values in the same range. On the other hand, 53 of 57 patients with progressive disease (PD) had sensitivities lower than 2.0.

The data indicate that patients with EGD< 2.0 have little chance of having an objective clinical

TABLE 2.
RELATIONSHIP BETWEEN IN VITRO SENSITIVITY AND CLINICAL RESPONSE IN MELANOMA PATIENTS

IN VITRO SENSITIVITY (EGD)	CLINICAL RESPONSE		
	PD	NC,MR	PR,CR
0-0.9	13	0	0
1.0-1.9	40	2	0
2.0-2.4	4	5	3
2.5-2.9	0	2	6
3.0-3.5	0	0	1

response, whereas patients with EGD $\geq$ 2.0 have a high probability of having a clinical response. The calibrated assay, as used here, is able to select patients with tumors that are resistant and who should not have standard chemotherapy. In many cases it is also capable of selecting an effective drug.

The problem in melanomas is that all available drugs are rather ineffective. In melanomas, as well as in other resistant tumor types, the most realistic goal of the in vitro chemosensitivity testing is to reduce the number of ineffective and toxic chemotherapy courses by detecting in advance the chemoresistant tumors. In more sensitive tumor types (e.g. breast and ovarian carcinomas), the goal of improving the response rate and extend the survival time is more important.

CONCLUSIONS

The Courtenay and Mills soft agar method is simple to perform. It includes few "enrichments" compared to other methods used for cultivation of human tumor cells. The rat erythrocytes and the low oxygen concentration employed have been shown to stimulate colony formation of tumor cells from human solid tumors. The culture method give more optimal culture conditions for human melanomas than other methods examined. The method may also give more colonies in other tumor types, but the difference is not pronounced. In many tumor types colony formation is still inadequate, and the method should certainly be improved to permit more tumors to grow. A calibrated chemosensitivity assay based on the Courtenay and Mills culture method predicts with high accuracy the in vivo resistance and sensitivity of human melanomas.

Experiments with two different soft agar methods on the same tumor samples show that the population of clonogenic cells in a tumor may be strongly influenced by the microenvironment. Thus, the clonogenic ability, revealed by the plating efficiency, depends on the culture conditions applied. Moreover, different culture methods give different apparent sensitivity to cancerostatic agents. This has important implications for the use of clonogenic assays to predict clinical chemosensitivity. At present it is not possible to conclude which culture conditions and which methods that most properly reflect the properties of the in vivo stem cells that are responsible for tumor progression and dissemination. This has to be established in future studies.

REFERENCES

Bateman, A.E., Selby, P.J., Steel, G.G. & Towse, G.D.W. In vitro chemosensitivity tests on xenografted human melanomas. Br. J. Cancer, 1980, 41, 189-198.

Courtenay, V.D. & Mills, J. An in vitro colony assay for human tumours grown in immune-suppressed mice and treated in vivo with cytotoxic agents. Br. J. Cancer, 1978, 37, 261-268.

Dendy, P.P. & Hill, B.T. Human Tumour Drug Sensitivity Testing In Vitro. Techniques and Clinical Applications. London/New York, Academic Press, 1983.

Endresen, L., Tveit, K.M., Rugstad, H.E. & Pihl, A. Comparative studies of in vitro chemosensitivity of human tumor cell lines by two soft agar methods. Proc. Predictive Drug Testing on Human Tumor Cells, Zurich, 1983.

Hamburger, A.W. & Salmon, S.E. Primary bioassay of human tumor stem cells. Science, 1977, 197, 461-463.

Kern, D.H., Campbell, M.A., Cochran, A.J., Burk, M.W. & Morton, D.L. Cloning of human solid tumors in soft agar. Int. J. Cancer, 1982, 30, 725-729.

Meyskens, F.L., Moon, T.E., Dana, B. et al. Quantitation of drug sensitivity by human metastatic melanoma colony-forming units. Br. J. Cancer, 1981, 44, 787-797.

Selby, P.J., Buick, R.N. & Tannock, I. A critical appraisal of the "human tumor stem cell assay". N. Engl. J. Med., 1983, 308, 129-134.

Tveit, K.M. Evaluation of the Courtenay assay for drug sensitivity prediction in vivo. In P.P. Dendy & B.T. Hill (Eds.), Human Tumour Drug Sensitivity Testing In Vitro. Techniques and Clinical Applications. London/New York, Academic Press, 1983, 305-316.

Tveit, K.M.,Endresen, L, Gundersen, Vaage, S., Davy, M., Rugstad, H.E. & Pihl, A. Clonogenic assay methods in the study of chemosensitivity of human tumors. Proc. 13th Int. Congress of Chemotherapy, part 224, 9-12, 1983.

Tveit, K.M., Endresen, L, Rugstad, H.E., Fodstad, Ø. & Pihl, A. Comparison of two soft agar methods for assaying chemosensitivity of human tumours in vitro: Malignant melanomas. Br. J. Cancer, 1981a, 44, 539-544.

Tveit, K.M., Fodstad, Ø., Lotsberg, J., Vaage, S. & Pihl, A. Colony growth and chemosensitivity in vitro of human melanoma biopsies. Relationship to clinical parameters. Int. J. Cancer, 1982, 28, 533-538.

Tveit, K.M., Fodstad, Ø., Olsnes, S. & Pihl, A. In vitro sensitivity of human melanoma xenografts to cytotoxic drugs. Correlation to in vivo chemosensitivity. Int. J. Cancer, 1980, 26, 717-722.

Tveit, K.M., Fodstad, Ø, & Pihl. The usefulness of human tumor cell lines in the study of chemosensitivity. A study of malignant melanomas. Int. J. Cancer, 1981b, 28, 403-408.

Tveit, K.M. & Pihl, A. Do cell lines in vitro reflect the properties of the tumours of origin? A study of lines derived from melanoma xenografts. Br. J. Cancer, 1981, 44, 775-786.

Von Hoff, D.D., Casper, J., Bradley, E., Sandbach, J., Jones, D. & Makuch, R. Association between human tumor colony-forming assay results and response of an individual patients' tumor to chemotherapy. Am. J. Med., 1981, 70, 1027-1032.

Von Hoff, D.D., Clark, G.M., Stogdill, B.J. *et al.* Prospective clinical trial of a human tumor cloning system. Cancer Res., 1983, 43, 1926-1931.

DOES THE RELATIVE SURVIVAL OF TUMOR COLONY FORMING UNITS ADEQUATELY REFLECT DRUG EFFECT ON PROLIFERATIVE CAPACITY?

THOMAS E. MOON
Research Professor of Medicine
Assistant Director
Cancer Center

STEPHEN R. RODNEY
STEPHEN P. THOMSON
RUTH S. SEROKMAN
Research Associates
Cancer Center

FRANK L. MEYSKENS, JR.
Associate Professor
Department of Internal Medicine

University of Arizona Cancer Center
Tucson, Arizona 85724

INTRODUCTION

The human tumor clonogenic assay (HTCA) measures the ability of tumor cells to form multi-cellular growth units (tumor colony forming units or TCFU) in semi-solid agar media (1-3). The effect of drug exposure on the proliferation of TCFU is quantitated by use of the Relative Survival for specified drug concentrations (4-6). Relative Survival is commonly defined as the ratio of the number of TCFU after drug exposure divided by the number of control TCFU not exposed to drug and varies between 0.0 and 1.0 (0% and 100%). A Relative Survival less than 30% at selected drug concentrations represents a substantial reduction in the number of colonies and is associated with an increased probability of clinical response following the patient's treatment with the drug (7-10). The existance of such an association is accompanied by false positive and false negative rates. We questioned whether the definition of the Relative Survival might be modified to improve its value to indicate changes in TCFU proliferation and its clinical prognostic value.

The four definitions of Relative Survival for a specified drug concentration that we considered include:

HUMAN TUMOR CLONING
ISBN 0-8089-1671-8

1. $\frac{\text{\# TCFU after drug exposure}}{\text{\# control TCFU}}$

   for TCFU of diameter $\geq$ 60 μm:

2. $\frac{\text{\# cells in TCFU after drug exposure}}{\text{\# cells in control TCFU}}$

   for TCFU of diameter $\geq$ 60 μm;

3. $\frac{\text{\# TCFU after drug exposure}}{\text{\# control TCFU}}$

   only for TCFU of diameter much greater than 60μm;

4. $\frac{\text{\# TCFU after drug exposure}}{\text{\# control TCFU}}$

   only for TCFU whose cells have doubled a large number of times.

PATIENTS AND METHODS

Surgical biopsies were handled aceptically by the clinical pathologist and a viable portion of tumor tissue was promptly transferred to the cloning laboratory at the University of Arizona Cancer Center. Tumor types studied included metastatic melanoma, ovarian and lung cancer. Techniques for tumor dissagregation, drug incubation and culture in the HTCA have been detailed elsewhere (2). All drug treated and control cells were processed simultaneously. After applying the upper agar layer, plates were examined to assure that they were composed of single cell suspensions. Cultures were examined thoroughly with inverted microscopy and were generally ready for TCFU counting on day 10-14 after plating. The Bausch & Lomb Omnicon Model "FAS II" was utilized as the automated TCFU counter in these experiments (2,11). TCFU at least 60 μm in diameter were sized and counted to yield a frequency distribution by diameter separated into one of six non overlapping size bins. The range in the number of control TCFU counted was 336, 385 and 682 for the 3 melanoma patients, 105 and 399 for the 2 ovarian cancer patients, and 162 and 721 for the 2 lung cancer patients.

The number of cells in a TCFU of specific size was calculated using the equation:

$$\text{No. Cells per TCFU} = (2.40)\ \frac{(\text{TCFU Diameter})^{2.378}}{(\text{Cell Diameter})^{2.804}}.$$

The derivation of this quantitative relationship has been previously reported (12). The frequency distribution of the number of tumor cells in TCFU of the selected sizes was calculated using the above equation. Also, TCFU were classified by the number of cell doublings and their frequency distribution obtained.

RESULTS AND DISCUSSION

The frequency distribution of TCFU by their diameter is shown in Figure 1 for a melanoma patient. A total of 385 control TCFU $\geq$ 60 μm were counted for the patient. The frequency distribution of 153 TCFU that grew following 1 hour exposure with 0.1 mg/ml velban is shown. The corresponding Relative Survival using Definition 1 was 39.7%. The frequency distribution of 41 TCFU following 0.02 mg/ml continuous exposure to vinzolidine is also shown. The corresponding Relative Survival was 10.6% using Definition 1.

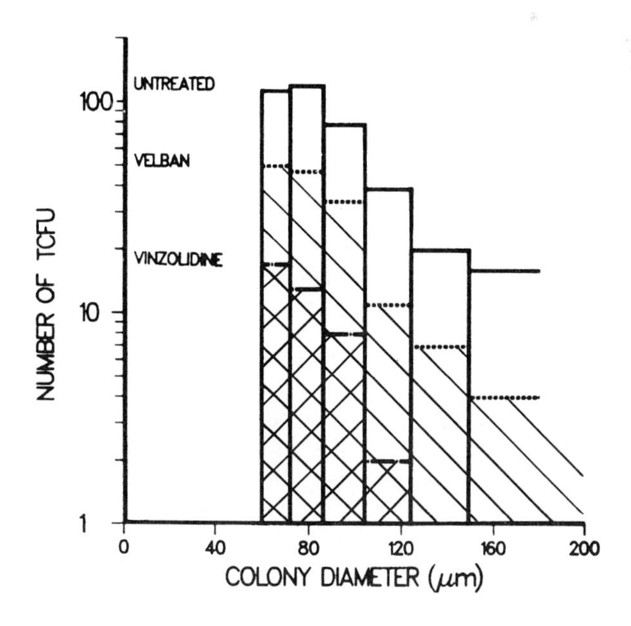

Figure 1 - Frequency Distribution of TCFU.

The frequency distribution of control TCFU shows an inverse relationship between TCFU number and diameter. A similar shape in TCFU distribution is seen following drug exposure. However, the number of TCFU is decreased for all diameters following drug exposure. It will be noted that exposure to vinzolidine resulted in only 1 TCFU in the 124-148 μm diameter range and 0 TCFU of diameter 149 μm.

Table 1 shows the range in the percentage frequency distribution of untreated control and drug exposed (treated) TCFU for all patient's biopsies. The frequency distribution by TCFU diameter was similar for all patients evaluated. Sixty to 86% of all control TCFU had a diameter between 60 to 85 μm, 14 to 30% of all TCFU had a diameter between 86 to 123 μm and 0 to 9% of TCFU had a diameter of at least 124 μm. The range of frequency distribution for drug exposed TCFU with a Relative Survival (RS) of 50 to 70% was similar to the frequency distribution of control TCFU. However, drug exposed TCFU that showed a Relative Survival less than 30% had a statistically different range of frequency when compared to control TCFU ($p < .05$). The TCFU with reduced proliferation (less than 30% Relative Survival) had a significant increase in the percentage of TCFU with diameter 60-85 μm and was 73% to 96% versus 60-86% for the controls. Conversely, drug exposed TCFU with larger diameter had a significant decrease in percentage of TCFU

Table 1

FREQUENCY DISTRIBUTION OF TCFU

Colony Diameter (μm)	RELATIVE FREQUENCY		
	Untreated 100% RS (%)	Treated 50-70% RS (%)	Treated 1-30% RS (%)
60-85	60-86	66-91	73-96
86-123	14-30	9-29	4-24
124	0-9	0-9	0-2

and was 4 to 24% versus 14-30% for TCFU with diameter 86-123 μm and was 0-2% versus 0-9% for TCFU with diameter 124 μ m.

Table 2 illustrates the comparison between Relative Survival as defined by Definition 1 (for which TCFU are counted) verses Definition 2 (for which the total number of cells within TCFU are counted). Table 2 represents the data for 1 patient with ovarian cancer but similar results were observed for all other patients. The 2 columns with the heading Number TCFU reflect the frequency distribution of total number of TCFU and were included in the data shown in Table 1. Using Definition 1 we observed a 15% Relative Survival following 1 hour exposure of 5 mg/ml of bisantrene. The two columns with the heading Number Cells reflects the frequency distribution for control and treated TCFU classified by the total number of cells within TCFU. Using Definition 2, a 12% Relative Survival was observed following the same drug concentration exposure. This data reflects a general pattern that the Relative Survival has defined by Definition 2 yields a similar but slightly smaller value than calulating the Relative Survival using Definition 1. Thus, the value of Relative Survival doesnot appear to be improved by use of Definition 2. Table 2 also illustrates the decrease in the relative frequency of larger TCFU. This decrease is observed both for control as well as treated TCFU. Comparing treated with untreated TCFU one also

Table 2

FREQUENCY DISTRIBUTION OF TCFU AND CELLS IN TCFU

Relative Frequency

	Number TCFU		Number Cells	
TCFU Diameter (μm)	Untreated (%)	Treated (%)	Untreated (%)	Treated (%)
60-85	67	79	44	61
86-123	27	19	39	33
≥ 124	6	2	17	6
		15%RS		12%RS

observes an increase in the frequency of smaller TCFU and a decrease in the frequency of larger TCFU following drug exposure.

Calculation of the number of cells in TCFU permits the classification by the number of cell doublings in TCFU. Figure 2 illustrates the relationship between TCFU diameter, cell diameter, and number of cell doublings in TCFU. Panel C in Figure 2 relates to a tumor specimen with a 22.5 μm mean cell diameter while Panels D and E relate to specimens with tumor cell diameters of 17.95 m and 13.60μm diameter. Comparing Panels C and A shows that three cell doublings will result in a colony with a 65 m diameter while 7 doublings yield a 209 m diameter colony. Panels D and E show that as mean cell diameter decreases, the number of cells and thus the number of cell doublings decreases for TCFU of similar diameters. Comparison of Panels C, D, and E with Panel A shows that a colony of 60 diameter may represent TCFU that have doubled 3, 4, or 5 times. Use of a 60 μm diameter to define a colony will yield TCFU with very different frequency distribution of cell doublings.

The frequency distribution of TCFU by number of TCFU doublings is shown in Figure 3 for the same melanoma patient previously shown in Figure 1. An inverse relationship is observed between number control TCFU (top solid lines) and number of cell doublings. While over 100 TCFU composed of cells that doubled twice were observed only 16 TCFU whose cells doubled 6 or more times were observed. TCFU exposed to vinzolidine yielded a decreased number of TCFU but a similar inverse relationship. No TCFU whose cells doubled 6 or more times were observed after exposure to vinzolidine in this patient's assay. Drug exposure resulted in a smaller proportion Survival (3%) for TCFU containing cells with 5 or more doublings than for those that doubled 2 or 3 times (15% Survival).

The bottom Panel of Figure 4 shows the proportion Survival of melanoma TCFU exposed to vinzolidine by number of cell doublings. The overall Relative Survival for all drug exposed TCFU $\geq$ 60μm was 10.6%. In contrast, the proportion Survival for TCFU whose cells doubled only twice was 17% and continually decreased as number of doublings increased. TCFU that doubled 5 times had a 3% Survival and TCFU that doubled 6 or more times only had a 1% Survival.

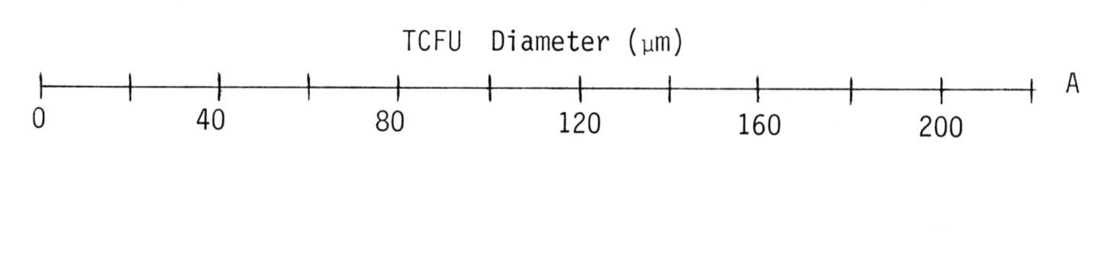

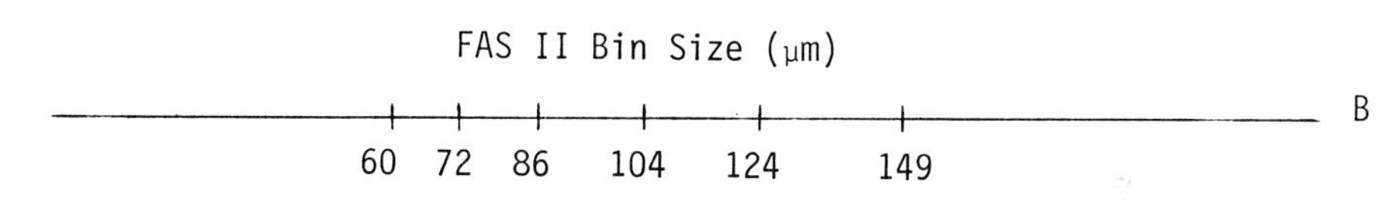

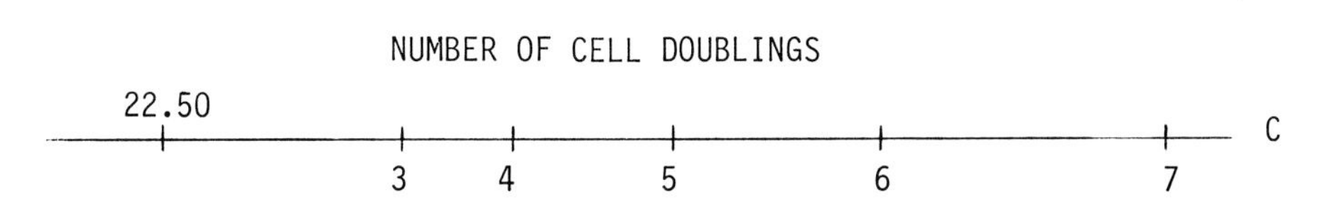

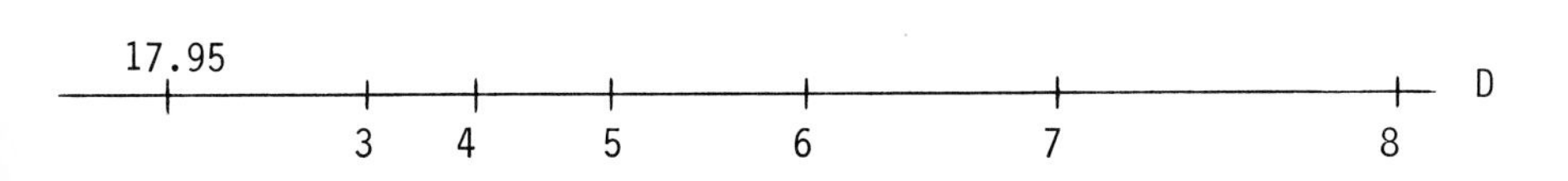

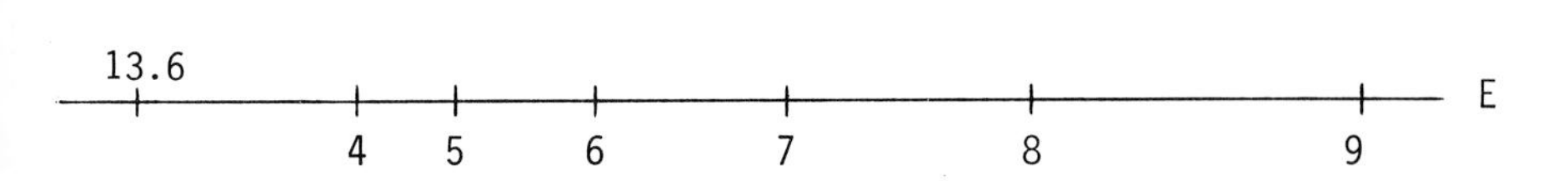

Figure 2 - TCFU Classification by Diameter and Number of Cell Doublings.

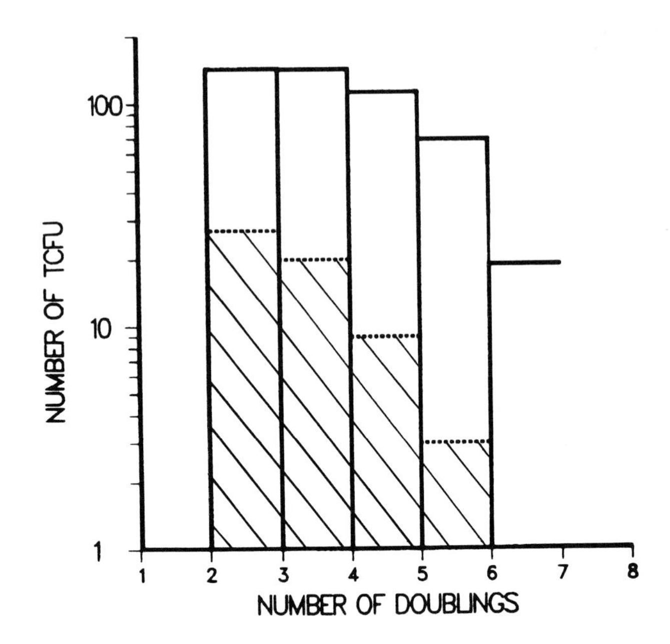

Figure 3 - Frequency Distribution of TCFU.

The gradient of TCFU drug sensitivity shown, in Figure 4, was consistently observed in all patient's assays. The top Panel for Figure 4 shows the proportion Survival by number of cell doublings from the patient with ovarian cancer shown in Table 2. While the Relative Survival for all TCFU $\geq 60\,\mu m$ was 15%, a gradient of drug sensitivity similar to the bottom panel of Figure 4 was observed. TCFU whose cells doubled the fewest (4) times had a proportion Survival of 20% and for cells that doubled many (7) times, the proportion Survival was 6%. However, all TCFU seen in the top panel of Figure 4 had cells that doubled 4 or more times while only 29% of the TCFU seen in the bottom panel had cells that doubled 4 or more times.

Use of the Omnicon FAS II counter provides a reliable method to calculate the frequency distribution of TCFU by their diameter (11). Use of the derived quantitative relationship between number of cells within a TCFU with TCFU diameter and mean diameter of cells further permitted the calculation of the frequency distribution of TCFU by the corresponding number of cell doublings. Both methods of summarizing TCFU frequency indicated decreased cell divisions following drug exposure.

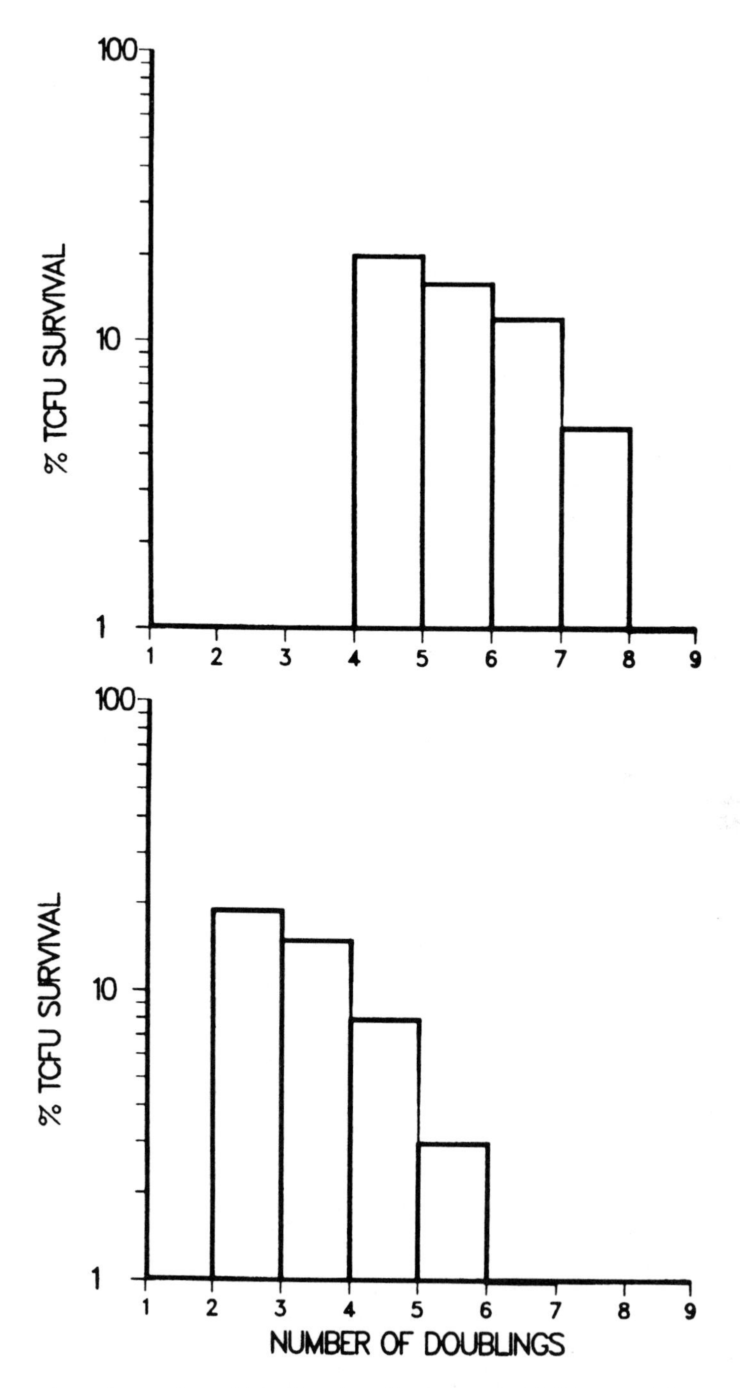

Figure 4 - Drug Sensitivity by Number of TCFU Doublings

This decrease in the number of cell divisions was observed in all size TCFU and was greatest for larger TCFU. There appears to be three hypotheses that could, in part, explain for the observed differences in drug sensitivity by TCFU size. These include:

1. Differential sensitivity by TCFU size;
2. An increased cell doubling time;
3. An increased period of delay prior to initiation of growth.

The data available in this report does not permit a selection of one or more of the hypotheses. In fact, additional experiements will be required to identify if any of these hypotheses describe the actual mechanism accounting for the observed differences in drug sensitivity by colony size.

The information presented in this report provided indirect evidence that there may not be a clonal hierarchy for tumor cells. The observed lack of cytotoxic therapy to cure cancer patients suggests that stem cells are less sensitive to such therapy than other less proliferative tumor cells. However, a greater reduction in the proportion of large TCFU relative to smaller ones was consistently observed for all patients regardless of the type of cancer. This inverse relationship between TCFU size and drug effect suggests the lack of a clonal hierarchy for tumor cells.

The commonly used definition (see Definition 1) of Relative Survival does not completely reflect the frequency distribution by TCFU size and the observed differences in drug sensitivity by TCFU size. An improved definition of Relative Survival may be to include only TCFU that represent 5 or more cell doublings following drug exposure. However, the HTCA methods must be modified to permit greater numbers of large TCFU (13). Unfortunately, the information available in this report is not only modest in the number of patients considered, but also information on clinical response relative to drugs actually tested in vitro was not available for these patients. Only through the collection of information on additional patients and the correlation between the results of their in vitro clonogenic assay and corresponding clinical therapeutic outcome will information become available to objectively identify a better definition for Relative Survival.

ACKNOWLEDGEMENTS

These studies were supported in part by grants CA-23074, CA-17094, CA-21839 and CA-27502 from the National Cancer Institute. The authors thank Mrs. Susan Davidson for assistance in manuscript preparation.

REFERENCES

1. Hamburger AW and Salmon, SE: Primary bioassay of human tumor stem cells. Sciences 197:461-463, 1977.

2. Salmon, SE: Cloning of Human Tumor Stem Cells (ed.), New York, Alan R. Liss, Inc., 1980.

3. Meyskens, FL JR, Soehnlen BJ, Saxe SF, Casey, WJ and Salmon SE: In vitro clonal assay for human metastatic malignant melanoma. Stem Cells 1: 61-72, 1981.

4. Moon TE, Salmon SE, White CS, Chen H-SG, Meyskens FL, Durie BGM, and Alberts DS: Quantitative association between the in vitro human tumor stem cell assay and clinical response to cancer chemotherapy. Cancer Chemother. and Pharmacol. 6:211-218, 1981.

5. Alberts DS, Salmon SE, Chen HSG, Moon TE, Young L., Surwit EA: Parmacologic studies of anticancer drugs using the human tumor stem cell assay. Cancer Chemother. and Pharmacol. 6:253-264, 1981.

6. Salmon SE. Applications of thehuman tumor stem cell assay to new drug evaluation and screening. In: Cloning of Human Tumor Stem Cells, New York, Alan R. Liss, Inc., 1980, pp. 291-312.

7. Salmon SE, Hamburger AW, Soehnlen B, Durie BGM, Alberts DS, Moon TE: Quantitation of differential sensitivity of human tumor stem cells to anticancer drugs. New Engl. J. Med. 298:1321-1327, 1978.

8. Salmon SE, Alberts DS, Meyskens FL, Durie BGM, Jones SE, Soehnlen B, Young L, Chem HSG, Moon TE: Clinical correlations of in vitro drug sensitivity. In: Cloning of Human Tumor Stem Cells, New York, Alan R. Liss, Inc., 1980, pp. 223-245.

9. Von Hoff DD, Clark GM, Stogdill, BJ, Sarosdy MF, O'Brien MT, Casper JT, Mattox DE, Page CP, Cruz AB, Sanbach JF: Prospective clinical trial of a human tumor cloning system. Cancer Res. 43:1926-1931, 1983.

10. Meyskens FL, Loescher L, Moon TE, Salmon SE: A prospective trial of single agent chemotherapy for metastatic malignant melanoma directed by in vitro colony survival in a clonogenic assay. Proc. Amer. Assoc. for Cancer Res., San Diego, 1983, 24:143, Abstract No. 567.

11. Salmon SE, Young L, Lebowitz M, Thomson S, Einsphar J, Tong T and Moon TE: Evaluation of an Image Analysis System For Counting Human Tumor Colonies. Proc. Fourth Conf. on Human Tumor Cloning, Jan. 8-10, 1984, Tucson, Arizona.

12. Meyskens FL, Thomson SP and Moon TE: Quantitation of a number of cells within tumor colonies in semi-solid media and their growth as oblate spheroids. Cancer Research 1984, 44:271-277.

13. Thomson SP, Meysken FL Jr, Hickie RA and Sipes NJ: Colony Size, Linearity of Formation and Drug Survival Curves Can Depend on the Number of Cells Plated in the Clonogenic Assay. Proc. Fourth Conf. on Human Tumor Cloning, Jan. 8-10, 1984, Tucson, Arizona.

Pharmacologic Modulation of Adriamycin Resistance in Ovarian Cancer

ALFRED M. ROGAN
THOMAS C. HAMILTON
ROBERT C. YOUNG
ROBERT F. OZOLS

Medicine Branch
Division of Cancer Treatment
National Cancer Institute
9000 Rockville Pike
Bethesda, MD 20205

INTRODUCTION

Anthracylines are among the most extensively used families of chemotherapeutic agents with clinical effectiveness against a wide spectrum of human malignancies. The utility of anthracyclines in the treatment of solid tumors such as ovarian cancer is frequently limited due to the presence of endogenous resistance to the drug or to the rapid development of acquired resistance which results in only a transient response to therapy. Thus, the response rate to Adriamycin is 30% in ovarian cancer patients without prior treatment and the median duration of responses is only three months[15]. An understanding of the mechanism involved in resistance to anthracylines may facilitate the search for pharmacologic means to alter this resistance and thereby improve the clinical utility of these agents. We have used the human tumor stem cell assay (HTSCA)[9], a murine model of ovarian cancer[8] and human ovarian cancer cell lines[2,5,11] to study dose response relationships and the pharmacologic modulation of Adriamycin resistance in ovarian cancer. The results of these studies have formed the rationale, in part, for clinical trials of intraperitoneal therapy and of combination therapy with verapamil plus Adriamycin in refractory ovarian cancer patients.

RESISTANCE TO ANTHRACYCLINES

Anthracyclines may produce cell death by a number of mechanisms which include free radical damage, intercalation into DNA, and binding to membranes[15]. However, the development of resistance to anthracyclines has not been shown to be

HUMAN TUMOR CLONING
ISBN 0-8089-1671-8

the result of interference with these mechanisms of cytotoxicity. On the other hand, experimental studies *in vitro* have shown that anthracycline resistance in animal tumor cells is associated with changes in drug transport. It has been demonstrated that Adriamycin resistant variants of P388 murine leukemia cells accumulate less drug than their sensitive counterparts and that this difference results from an apparent more active efflux of Adriamycin in the resistant cell lines[6,7].

The observation that resistance to Adriamycin in P388 can be reversed by calcium channel blockers and calmodulin inhibitors and that reversal of resistance is accompanied by increased accumulation of Adriamycin in the resistant cells further supports the causal relationship between transport changes and resistance in these animal tumor cells *in vitro*[13,14]. Furthermore, *in vivo* experiments have shown that animals inoculated with daunomycin-resistant Ehrlich ascites cells and subsequently treated with daunomycin and verapamil survived longer than control animals treated with daunomycin alone[12]. The ability of verapamil, a calcium channel blocker commonly used to treat arrhythmias and angina, to reverse resistance to anthracyclines in this experimental *in* vivo setting suggests that circumvention of anthracycline resistance by verapamil is due to a differential effect of increased anthracycline cytotoxicity in tumor cells compared to normal tissues.

LABORATORY AND CLINICAL STUDIES

Dose Response Relationships. The demonstration of different dose response relationships to Adriamycin in ovarian cancer cells obtained from previously treated and untreated patients and the correlation between clinical results and pharmacologic levels of Adriamycin indicates that resistance to Adriamycin, in some clinical situations, can be overcome by increasing the concentration of Adriamycin such as is achievable by intraperitoneal therapy[10]. We have found that tumor cells from relapsed Adriamycin treated patients were extremely refractory to Adriamycin in the HTSCA and even after exposure to 10 ug/ml, a dose level 10X greater than achievable by intravenous therapy, significant cytotoxicity was not observed. In contrast, specimens from patients with progressive disease after treatment with a non-Adriamycin chemotherapy regimen, while resistant to Adriamycin using standard HTSCA criteria, were sensitive following exposure to concentrations of Adriamycin which while not achievable by intravenous therapy were potentially achievable by intraperitoneal therapy[9]. These observations, in part, formed the rationale for intraperitoneal therapy with Adriamycin in the latter group of patients[10]. A marked pharmacologic advantage (ratio of peak i.p. level to peak plasma level of 474) was produced by intraperitoneal administration of Adriamycin. Concentrations required to cause significant cell killing in the HTSCA were achieved[9] intraperitoneally in this group of patients in whom i.v. Adriamycin has been demonstrated to be an ineffective salvage therapy. Ten patients refractory to non-Adriamycin systemic chemotherapy were treated with i.p. Adriamycin and three responses were observed in patients with minimal residual disease (< 2 cm masses) which is consistent with previous studies with i.p. Adriamycin in murine teratoma[8]. We found that Adriamycin administered i.p. in this system only penetrated into the outermost 4-6 cell layers of intraperitoneal masses[8]. These clinical and laboratory findings suggest that the value of intraperitoneal therapy will be in patients with non-bulky disease and especially in those with minimal disease who have had a good but only partial response to a non-Adriamycin induction chemotherapy regimen.

Adriamycin Resistance in Human Ovarian Cancer Cell Lines. Since the small size of the tumor colonies in the HTSCA precludes most biochemical studies of the mechanisms of drug resistance, we established and developed human ovarian cancer cell lines with varying degrees of resistance to Adriamycin[2-5,11]. Cell lines have been established from untreated patients and from patients refractory to Adriamycin containing and non-Adriamycin containing combination chemotherapy regimens[3,4]. In addition, we have taken ovarian cancer cell lines A1847 and A2780 derived from untreated ovarian cancer patients[1] and by stepwise incubation with Adriamycin selected for or induced resistance toward that individual drug[3]. These cell lines exhibit a wide range of resistance to Adriamycin which varies from 2-5 times in the cell lines from refractory patients to 150X for 2780^{AD} an Adriamycin resistant variant of A2780 produced by the stepwise incubation process. Consistent with clinical observations we have found NIH:OVCAR-2, derived from a patient treated with a non-Adriamycin regimen, to be relatively resistant to Adriamycin in comparison to cell lines from untreated patients.

These cell lines have proven useful for the study of Adriamycin resistance in ovarian cancer. In a moderately (6X) resistant cell line, 1847^{AD} produced by the stepwise incubation procedure *in vitro*, it was found that, as with Adriamycin resistant animal tumor cell lines, Adriamycin resistant human ovarian cancer cells accumulated four times less Adriamycin than the parental sensitive cell line. The mechanism of reduced Adriamycin accumulation was the result of enhanced efflux. More importantly, both the accumulation of and the sensitivity to Adriamycin could be completely restored by verapamil[11]. In 2780^{AD}, (a highly Adriamycin resistant variant) verapamil, however, only partially restored sensitivity to Adriamycin. Likewise in the cell lines from Adriamycin resistant patients verapamil only enhanced Adriamycin cytotoxicity two to three fold in contrast to the four to six fold increase seen in 1847^{AD} and 2780^{AD}. It is apparent from these results that although transport phenomena are a mechanism of Adriamycin resistance in human ovarian cancer, other mechanisms are also operative as manifest by (1) failure of verapamil to completely restore sensitivity to Adriamycin in 2780^{AD}, and (2) the lesser efficacy of verapamil in potentiating Adriamycin cytotoxicity in the cell lines from resistant patients[11]. Of potential clinical significance was the failure to see augmentation of Adriamycin cytotoxicity by verapamil in human fibroblasts derived from a patient whose cancer cell line responded to this combination[11]. Such findings suggest that verapamil may increase the effectiveness of Adriamycin without increased toxicity to normal tissues.

Adriamycin Plus Verapamil in Refractory Ovarian Cancer Patients. These laboratory studies on complete and partial reversal of Adriamycin resistance in human ovarian cancer cells using levels of verapamil (200-1000 ng/ml) achievable in patients, have led to a clinical trial of verapamil plus Adriamycin in refractory ovarian cancer patients[11]. The approach is to treat patients with verapamil for 24 hours and thus achieve the highest possible steady state circulating verapamil level prior to administration of Adriamycin. Patients first receive a 0.15 mg/kg bolus of verapmil then a constant infusion of 0.005 mg/kg/min escalated to 0.015 mg/kg/min if tolerated. After the initial 24 hours, the verapamil is continued in combination with Adriamycin (50 mg/M^2) administered over a subsequent 24 hours, and this treatment phase is followed by an additional 24 hours of verapamil in order to inhibit Adriamycin efflux. These cycles are administered every 28 days. Thus far, four such cycles have been administered and verapamil levels of at least 200 ng/ml were achieved. In addition, the treatment does not appear to potentiate the myelotoxicity or gastrointestinal toxicity of Adriamycin[11], although these are still preliminary observations.

CONCLUSIONS

These studies demonstrate that resistance to Adriamycin can be pharmacologically modulated in human ovarian cancer cells. The experimental studies on dose response relationships to Adriamycin in the HTSCA have led to phase I-II trials of intraperitoneal Adriamycin in selected ovarian cancer patients. Similarly, the observation that verapamil can reverse Adriamycin resistance in some human ovarian cancer cell lines has led to an ongoing trial of verapamil plus Adriamycin in relapsed patients. The availability of human ovarian cancer cell lines with different patterns of sensitivity/resistance to Adriamycin together with the recent development of a novel intraperitoneally transplantable nude mouse model of human ovarian cancer[5] with ascites and intraabdominal carcinomatosis should facilitate the search for other mechanisms of Adriamycin resistance and alternate ways in which it can be pharmacologically reversed.

REFERENCES

1. Eva A, Robbins KC, Andersen PR, et al: Cellular genes analogous to retroviral oncogenes are transcribed in human tumor cells. Nature 295: 116-119, 1982.
2. Green JA, Vistica DT, Young RC, et al: Melphalan resistance in human ovarian cancer: potentiation of melphalan cytotoxicity by nutritional and pharmacologic depletion of intracellular glutathione levels. Proc Amer Assoc Cancer Res 25: (in press) 1984.
3. Hamilton TC, Foster BJ, Grotzinger KR, et al: Development of drug sensitive and resistant human ovarian cancer cell lines. Proc Amer Assoc Cancer Res 24: 313, 1983.
4. Hamilton TC, Young RC, McKoy WM, et al: Characterization of a human ovarian carcinoma cell line (NIH:OVCAR-3) with androgen and estrogen receptors. Cancer Res 43: 5379-5389, 1983.
5. Hamilton TC, Young, RC, Rogan AM, et al: A unique intraperitoneal model of human ovarian cancer. Proc Amer Assoc Cancer Res 25: (in press) 1984.
6. Inaba M, Johnson RK: Uptake and retention of Adriamycin and daunorubicin by sensitive and anthracycline-resistant sublines of P388 leukemia. Biochem Pharm 27: 2123-2130, 1978.
7. Inaba M, Kobayashi H, Sakurai Y, et al: Active efflux of daunomycin and Adriamycin in sensitive and resistant sublines of P388 leukemia. Cancer Res 39: 2200-2203, 1979.

8. Ozols RF, Locker GY, Doroshow JH, et al: Pharmacokinetics of Adriamycin and tissue penetration in murine ovarian cancer. Cancer Res 39: 3209-3214, 1979.
9. Ozols RF, Willson JKV, Weitz MD, et al: Inhibition of human ovarian cancer colony formation by Adriamycin and its major metabolities. Cancer Res 40: 4109-4112, 1980.
10. Ozols RF, Young RC, Speyer, JL, et al: Phase I and pharmacologic studies of Adriamycin administered intraperiteonally to patients with ovarian cancer. Cancer Res 42: 4265-4269, 1982.
11. Ozols RF, Rogan AM, Hamilton TC, et al: Verapamil plus Adriamycin in refractory ovarian cancer: design of a clinical trial on basis of reversal of Adriamycin resistance in human ovarian cancer cell lines. Proc Amer Assoc Cancer Res 25: (in press) 1984.
12. Slater LM, Murray SL, Wetzel MW: Verapamil restoration of daunorubicin responsiveness in daunorubicin resistant Ehrlich ascites carcinoma. J Clin Invest 70: 1131-1134, 1982.
13. Tsuruo T: Reversal of acquired resistance to vinca alkaloids and anthracycline antibiotics. Cancer Treat Reports 67: 889-894, 1983.
14. Tsuruo T, Iida H, Tsukagoshi S, et al: Protection of vincristine and Adriamycin effects in human hemopoietic tumor cell lines by calcium antagonists and calmodulin inhibitors. Cancer Res 43: 2267-2272, 1983.
15. Young RC, Ozols RF, Myers CR: The anthracycline antineoplastic drugs. New Eng J Med 305: 139-153, 1981.

MODULATION OF IN-VITRO ACTIVITY OF ADRIAMYCIN® AND VINBLASTINE BY VERAPAMIL

Gary E. Goodman, M.D., M.S.

Medical Oncology
The Swedish Hospital Tumor Institute
Seattle, WA 98104

INTRODUCTION

De novo and acquired drug resistance to cytotoxic drugs is a major problem in the clinical treatment of human cancers. Many tumors are drug resistant at the time of presentation, and others which are initially sensitive later relapse, displaying a broad spectrum of drug resistance.

The mechanisms of drug resistance have been investigated using drug-resistant cell lines. P388 murine leukemia, Ehrlich ascites carcinoma and other lines have been subcloned in increasing concentrations of Adriamycin® (ADR), vincristine (VCR) and other cytotoxic drugs resulting in relatively resistant cell lines. Tsuruo et al. recently reported that Verapamil, the calcium influx inhibitor, augmented the cellular accumulation of VCR and ADR in both sensitive and VCR/ADR resistant P388 cell lines.[1] These authors found that Verapamil enhanced the cytotoxicity of VCR and ADR approximately four-fold and two-fold respectively against the standard P388 cell line. More pronounced, however, Verapamil enhanced the cytotoxicity of VCR and ADR in P388 drug-resistant cell lines with a 40- and 10-fold respective increase in sensitivity, i.e., it reversed drug resistance.[2,3]

Slater et al. reported somewhat similar results with Ehrlich ascites carcinoma.[4] Verapamil reversed the resistance to ADR in resistant cells but had a minimal effect on the degree of cytotoxicity in sensitive cells.

Verapamil is one of a number of calcium ion antagonists shown to have significant effects on membrane ion flux.[5] It and the calmodulin inhibitors modulate the slow calcium-dependent inward current system and are being used clinically in the treatment of supraventricular tachyarrhythmias[6] and angina pectoris[7]. Pharmacokinetic studies of Verapamil in patients with cardiovascular disease have shown that after oral administration peak plasma concentrations of up to 1 ug/ml can be achieved.[8] Sustained plasma concentrations of 200-500

HUMAN TUMOR CLONING
ISBN 0-8089-1671-8

ng/ml are possible with daily dosing. Norverapamil, an active metabolite, also occurs in similar plasma concentrations. These plasma concentrations are in the range of those found to be active in vitro by Tsuruo et al.

Because of potential clinical usefulness in drug-resistant human cancers, we have initiated an evaluation of Verapamil in combination with cytotoxic drugs using the human tumor cloning assay.

METHODS

The cloning techniques utilized were similar to those described by Hamburger and Salmon.[9] In all cases, epidermal growth factor (100 ng/ml), hydrocortisone (50 ng/ml), Selenium (1.75 ng/ml) and transferrin (5 ug/ml) were added to the underlayer as growth supplements. The upper layer consisted of 50% enriched CRML and 50% mixed effusion fluid (500 ml of cell-free ascites or pleural fluid from 10 random patients was mixed, aliquoted into 500-ml bottles and frozen until used; details to be published separately).

All single-cell suspensions were prepared by mechanical means (minced with scalpels and passed through #27 needles). If an adequate single-cell suspension was not obtained, cell aggregates were treated with 0.8% collagenase-II for two hours and 0.05% DNase-II for an additional 15 minutes. Cells were exposed to either ADR 1.0, 0.1 and 0.01 ug/ml, or VBL 0.5, 0.05 and 0.005 ug/ml, with and without Verapamil 1.0 ug/ml for one hour. Verapamil was not tested alone. After drug exposure, cells were washed and plated in triplicate in the culture media described. Colonies >70 ul or >50 cells were counted on an inverted-phase microscope after three weeks' incubation at 37°C in 7% CO_2. Only samples with growth of >30 colonies per control plate were considered evaluable for this analysis.

Table 1 illustrates the tumor types, previous treatment, clinical response and mean number of colonies per control plate. Ten samples were obtained from malignant effusions and two from solid tumors (#355 and #542). Only three samples were from patients receiving previous treatment, and only one patient had failed previous ADR.

Percent survival was calculated by expressing the average number of colonies in drug plates as a percentage of the mean number of colonies in the control plates.

TABLE 1 - TUMOR CHARACTERISTICS

Sample#	Tumor Type	Previous Rx	Response	Mean Colonies/ control plate
355	Lung	None	--	51
357	Ovary	None	--	87
365	Breast	None	--	53
368	Breast	CMF, ADR	PR	420
460	Breast	None	--	54
478	Breast	Tamoxifen	PR	1641
495	Ovary	None	--	112
530	Melanoma	None	--	310
542	Neurosarcoma	None	--	33
543	Lung	None	--	30
552	Breast	None	--	386
681	Endometrium	Megesterol	ADJ.	123

RESULTS

Ratios between percent survival of drug/Verapamil treated and drug-only treated cells are illustrated in Tables 2 and 3. A ratio of >1 means Verapamil augmented the growth inhibitory effect of the cytotoxic drug as compared to drug alone. A ratio of <1 means Verapamil improved growth as compared to drug alone. Table 2 illustrates the results of treatment with VBL. Most of the ratios are close to unity. Comparisons were made by two-way analysis of variance. There were no significant differences between the VBL and VBL/Verapamil treated cells. $p>0.05$

In the Adriamycin studies as seen in Table 3, again most ratios were near unity. Only sample #495 showed a significant difference between ADR treatment alone and ADR/Verapamil at a p value of <0.05. This sample was from a patient with ovarian cancer without previous treatment.

TABLE 2
MODULATION OF VINBLASTINE ACTIVITY WITH VERAPAMIL

Sample #	Percent Survival: Drug / Drug + Verapamil		
	Vinblastine ug/ml		
	0.5	0.05	0.005
355	51/51	82/78	90/98
357	92/95	122/94	72/50
365	11/6	64/30	--/17
368	54/68	115/107	133/109
460	46/41	24/43	96/35
478	49/49	61/59	59/65
495	100/--	102/--	82/--
530	66/84	87/68	94/73
542	--	--	--
543	62/60	130/120	123/92
552	95/86	80/57	82/68
681	114/121	--	--

TABLE 3
MODULATION OF ADRIAMYCIN ACTIVITY WITH VERAPAMIL

Sample#	Percent Survival: Drug / Drug + Verapamil		
	Adriamycin ug/ml		
	1.0	0.1	0.01
355	0/0	43/53	75/76
357	5/1	144/157	93/115
365	4/3	47/34	60/--
368	0/0	67/76*	99/89
460	30/17	117/179*	7/13
478	63/47	77/68	60/53
495	70/60	96/63*	110/74**
530	--	--	--
542	9/13	34/14	15/14
543	49/83	105/140	140/171
552	13/8	57/51	66/54
681	121/139	--	--

*$p<0.05$ **$p<0.01$

DISCUSSION

In the small number of tumors studied, we found that Verapamil enhanced the growth inhibitory activity of ADR in only one tumor. There was no modulation of VBL effect. However, Tsuruo and Slater have both reported Verapamil's greatest effect was in reversing the acquired resistance in drug-resistant lines. There was a lesser effect in the sensitive parent cell lines. The murine P388/ADR and P388/VCR cell lines acquire resistance by increasing the active cellular efflux of drug. Tsuruo reported that Verapamil inhibited this augmented drug efflux. Hence, Verapamil may only be effective in reversing a specific mechanism of drug resistance, i.e., augmented drug efflux.

Only one of the tumors we studied was obtained from a patient previously sensitive but now resistant to ADR. The other specimens were from patients without prior cytotoxic drug treatment. The tumor cells we studied may be more analogous to the standard Ehrlich ascites carcinoma line where drug sensitivity is unaffected by Verapamil and where mechanisms of drug resistance may be unrelated to drug transport. We are presently testing tumors obtained from patients formerly sensitive but now resistant to ADR or the Vinca alkaloids. This clinical situation is more analogous to that of the P388 and Ehrlich drug-resistant cell lines.

Our studies also used a concurrent one-hour exposure to drug and Verapamil. We are now examining the effect of prolonged exposure to Verapamil since rapid return of augmented drug efflux has been shown after Verapamil is removed from the culture media.[3]

Verapamil may prove to be a useful compound in investigating mechanisms of drug resistance. In our limited studies, it is unclear if Verapamil effects drug sensitivity in human cancers in the same way as in more homogeneous cell lines. Its usefulness in the treatment of heterogenous polyclonal human tumors which probably have multiple mechanisms of drug resistance remains to be established.

ACKNOWLEDGMENT

We thank W. James for her splendid secretarial assistance, and John Crowley, Ph.D., for statistical support.

REFERENCES

1. Tsuruo, T., Iida, H., Tsukagoshi, S., and Sakurai, Y. Increased accumulation of Vincristine and Adriamycin in drug-resistant P388 tumor cells following incubation with calcium antagonists and calmodulin inhibitors. Cancer Res. 42:4730-4733, 1984

2. Tsuruo, T., Iida, H., Tsukagoshi, S., and Sakurai, Y. Overcoming of Vincristine resistance in P388 leukemia in vivo and in vitro through enhanced cytotoxicity of Vincristine and Vinblastine by Verapamil. Cancer Res. 41: 1967-1972, 1981.

3. Tsuruo, T. Reversal of acguired resistance to Vinca alkaloids and anthracycline antibiotics. Cancer Treat. Rep. 67:889-894, 1983.

4. Slater, L.M., Murray, S.., Wetzel, M.W., Wisdom, R.M., and DuVall, E.M. Verapamil restoration of daunorubicin responsiveness in daunorubicin-resistant Ehlrlich ascites carcinoma. J. Clin. Invest. 70:1131-1134, 1982.

5. Kohlhardt, M., Bauer, P., Krause, H. and Fleckenstein, A. Differentiation of the transmembrane Na and Ca channel in mammalian cardiac fibers by the use of specific inhibitors. Pfluegers Arch. Gesamte Physiol. Menescher Tiere 335:309-322, 1972.

6. Waxman, H.L., Myerburg, R.J., Appel, R. and Sung, R.J. Verapamil for control of ventricular rate in paroxysmal supraventricular tachycardia and atrial fibrillation or flutter - A double-blind randomized cross-over study. Ann. Intern. Med. 94:1-6, 1981.

7. Simoons, M.L., Tooms, M., Lubsen, J. and Hugenholtz, P.G. Treatment of stable angina pectoris with Verapamil hydrochloride - A double-blind cross-over study. Eur. Heart J. 1:269-274, 1980.

8. Tartaglione, T.A., Pieper, J.A., Lopez, L.L. and Mehta, J. Pharmacokinetics of Verapamil and non-verapamil during long-term oral therapy. Res. Comm. in Chem. Path. and Pharmacol. 40:15-27, 1983.

9. Hamburger, A.W. and Salmon, S.E. Primary bioassay of human tumor stem cells. Science 197:461-463, 1977.

CONCOMITANT VERAPAMIL TREATMENT INCREASES DOXORUBICIN SENSITIVITY OF HUMAN TUMOR CLONOGENIC CELLS

Bruce W. Dana
Jan Woodruff

Division of Hematology & Medical Oncology,
Oregon Health Sciences University,
Portland, OR 97201

INTRODUCTION

Recent studies suggest that acquired resistance of tumor cells to anthracyclines and vinca alkaloids is reversible by concomitant exposure to calcium transport inhibitors. Tsuruo *et al* reported that the toxicity of vincristine or doxorubicin for P388 leukemia cells resistant to either drug alone is enhanced by simultaneous treatment with verapamil or other calcium antagonists (1,2). The mechanism of this effect may be an inhibition of outward drug transport by resistant tumor cells, leading to higher intracellular drug levels, and enhanced cytotoxicity (3). Slater *et al* (4) have reported similar results using Ehrlich ascites carcinoma cells in BALB/c mice. Animals bearing a daunorubicin-resistant tumor survived a mean of 21.7 days when treated with daunorubicin alone, similar to untreated controls. Verapamil alone had no effect, but combined verapamil-daunorubicin treatment increased mean survival to 44.0 days. The concentration of

HUMAN TUMOR CLONING
ISBN 0-8089-1671-8

daunomycin required to inhibit 50% of DNA or RNA synthesis in drug-resistant cells was reduced from 4-6 ug/ml to 1-2 ug/ml (identical to concentration required in drug-sensitive cells) when verapamil was present. Ramu *et al* (5) confirmed that anthracycline-resistant P388 cells could be rendered sensitive to anthracyclines by calcium antagonists, but suggest that the cause is an alteration in the cell membrane due to changes in lipid metabolism induced by the calcium antagonist.

Since acquired drug resistance is a major problem limiting the effectiveness of anticancer treatment for human cancers, we examined whether the synergistic effect of verapamil and doxorubicin against murine tumor cell lines could be reproduced in primary human cancer cells. Our findings suggest that in some patients either acquired or spontaneous doxorubicin resistance may be reversible.

MATERIALS AND METHODS

Tumor cells: Fresh tumor cells were obtained from seven cancer patients undergoing routine diagnostic or therapeutic procedures. Specimens were promptly transferred from the operating room to the laboratory in transport media (McCoy's 5A media with 5% fetal calf serum). Specimens were minced to 5mm pieces sterilely and stored overnight at 4°C in the same media.

Drugs: Doxorubicin and verapamil were obtained from the hospital pharmacy. Both were diluted to a stock solution of 100 ug/ml with sterile water and stored in aliquots at -70°C.

Clonogenic assay: Minced tumor specimens were processed within 24 hours of procurement. Cell suspensions were prepared by mechanical disaggregation with crossed scalpels in McCoy's 5A media, followed by passage through 23g needles. Clumps were removed by passage through 25 micron nylon mesh, and the cells were washed, counted, and assessed for viability with Trypan blue.

The tumor cells were exposed to drugs for 60', plated in soft agar as described by Salmon *et al* (6) and incubated in a humidified incubator at 37°C in 5% CO_2.

Tumor colonies (aggregates > 20 cells) were enumerated from coded plates by inverted microscopy at 14-21 days.

Statistics: Significance of differences was assessed with the t test.

RESULTS

Seven tumors were successfully cloned. Four histologic types of malignancy were included (Table I); two or seven patients had prior exposure to doxorubicin.

TABLE I

Patient Characteristics

Patient	Tumor Type	Prior Doxorubicin
DD	Melanoma	-
ES	Ovarian	+
MS	Lung adenocarcinoma	-
LA	Lung adenocarcinoma	-
LS	Melanoma	-
SO	Melanoma	-
JR	Lymphoma	+

Five of seven tumor specimens were obtained from metastatic sites (Table II). Viability of the prepared cell suspensions varied from 20-89% and the cloning efficiencies were low (0.00013-0.00492).

TABLE II

Tumor Characteristics

Patient	Type-Source	Viability	Cloning Efficiency ($x10^{-4}$)
DD	Mel-node	28	1.3
ES	Ov-node	20	49.2
MS	Lu-effusion	83	6.0
LA	Lu-primary	27	7.6
LS	Mel-node	89	6.4
SO	Mel-node	78	6.0
JR	Lym-node	82	4.0

All tumors were resistant to doxorubicin(0.1ug/ml) (Table III). Verapamil (1.0ug/ml) did not alter clonogenicity of any tumor compared to controls. Concomitant verapamil-doxorubicin exposure, however, produced significant reduction in clonogenicity in 4/7 tumors, including two with spontaneous doxorubicin resistance and two with acquired resistance. Using 50% colony survival as a criterion of chemosensitivity, three of seven doxorubicin-resistant tumors were rendered sensitive by verapamil treatment.

TABLE III

Effect of concomitant verapamil-doxorubicin

	Clonogenicity (% controls)		
Patient	Doxorubicin (0.1ug/ml)	Verapamil (1.0ug/ml)	Verapamil + Doxorubicin
DD	91	132	31*
ES	99	102	54*
MS	87	100	50*
LA	71	89	115
LS	71	94	128
SO	116	130	67
JR	84	69	31*

*P $\leq$ 0.05

Similar results were obtained when the verapamil concentration was increased to 10ug/ml (Figure I). There was no effect of verapamil alone on tumor clonogenicity and no increased synergism with doxorubicin.

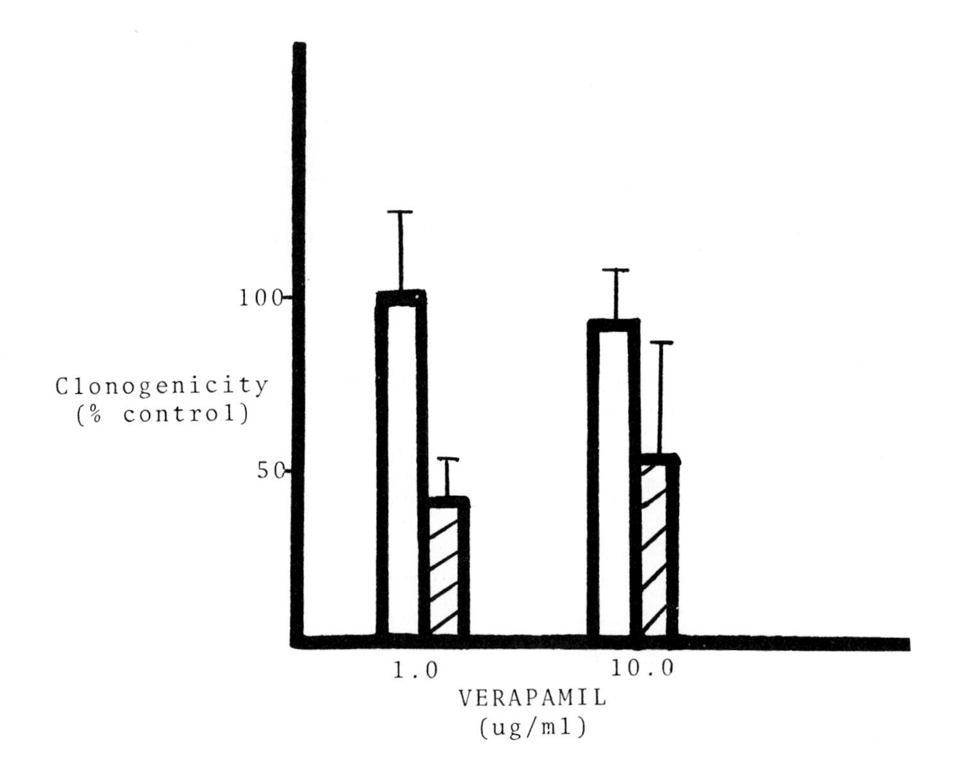

Figure 1 - Primary human tumor specimens were cloned in soft agar after a 60' exposure to verapamil alone (open bars) or verapamil and doxorubicin (0.1ug/ml) *(shaded bars). Brackets indicate mean + S.D.*

DISCUSSION

Our results suggest that calcium antagonist-mediated reversibility of anthracycline resistance in murine tumor cells extends also to some primary human cancer cells. This effect occurs in multiple types of human cancer, and occurs both in cells with spontaneous resistance and in those with resistance acquired during treatment. The serum levels of verapamil required to produce this effect are clinically achievable, since the usual clinical range is 1-3ug/ml (7).

The mechanism of this synergistic interaction is unknown, though Tsuruo *et al* (3) have produced evidence to implicate an inhibition of active anthracycline efflux by the calcium antagonist. P388 leukemia cells resistant to anthracyclines have lower intracellular drug levels than sensitive cells exposed to the same extracellular concentration. Resistant cells are also able to extrude drug more rapidly when the anthracycline is pre-loaded into the cells. Several antagonists of calcium transport and inhibitors of calmodulin were able to inhibit this outward efflux of doxorubicin and to increase intracellular concentrations. These, in turn, correlated with enhancement of doroxurubicin cytoxicity in the resistant cell lines. Whether the calcium-calmodulin membrane complex is directly involved in the active drug efflux, and whether such a mechanism is present in human cancer cells remains to be determined.

Our results suggest that relatively non-toxic adjuctive treatment with calcium antagonists might markedly broaden both the spectrum and the degree of activity of some anticancer agents. If further *in vitro* trials are promising, clinical trials of this concept would seem warranted.

ACKNOWLEDGEMENT

Supported in part by a grant from the Medical Research Foundation of Oregon.

REFERENCES

1. Tsuruo T, Iida H, Tsukagoshi S, et al. Overcoming of vincristine resistance in vivo and in vitro through enhanced cytotoxicity of vincristine and vinblastine by verapamil. Cancer Res. 41:1967-1972 (1981).
2. Tsuruo T, Iida H, Nojiri M, et al. Circumvention of vincristine and Adriamycin resistance in vitro and in vivo by calcium influx blockers. Cancer Res. 43:2905-2910 (1983).
3. Tsuruo T, Iida H, Tsukagoshi S, et al. Increased accumulation of vincristine and Adriamycin in drug-resistant P388 tumor cells following incubation with calcium antagonists and calmodulin inhibitors. Cancer Res. 42:4730-4733 (1982).
4. Slater LM, Murray SL, Wetzel MW et al. Verapamil restoration of daunorubicin responsiveness in daunorubicin-resistant Ehrlich ascites carcinoma. J. Clin. Invest. 70:1131-1134 (1982).
5. Ramu A, Shan T, Glaubiger D. Enhancement of doxorubicin and vinblastine sensitivity in anthracycline-resistant P388 cells. Cancer Treat. Rep. 67:895-899 (1983).
6. Salmon SE, Hamburger AW, Soehnlen B et al. Quantitation of differential sensitivity of human-tumor stem cells to anticancer drugs. N. Eng. J. Med. 298:1321-1327 (1978).
7. Chen C, Gettes LS. Effect of verapamil on rapid Na^+ channel-dependent action potentials of K^+ - depolarized ventricular fibers. J. Pharmacol. Exp. Ther. 209:415-421 (1979).

Potentiation of Melphalan Cytotoxicity in Human Ovarian Cancer Cell Lines

ROBERT F. OZOLS
JOHN A. GREEN
THOMAS C. HAMILTON
DAVID T. VISTICA
ROBERT C. YOUNG

Division of Cancer Treatment
National Cancer Institute
9000 Rockville Pike
Bethesda, MD 20205

I. INTRODUCTION

Melphalan is among the most active agents used in the treatment of ovarian cancer. Approximately 40% of previously untreated patients respond to therapy with melphalan as a single agent. However, acquired drug resistance frequently develops with repeated courses of therapy and limits the median duration of response to 12-17 months. Only 5-15% of all responding patients are alive 5 years after initiation of treatment[10]. The manner in which ovarian cancer cells become resistant to melphalan and other alkylating agents has not been established. Based on studies in experimental tumors it appears that resistance to alkylating agents is multifactorial and a variety of mechanisms have been proposed, including: (1) a decrease in the net accumulation of drug (2) an increase in the capacity to repair alkylating agent damage (3) metabolism of the active form of the drug to a less cytotoxic intermediate and (4) protection of critical cellular sites from the damaging effects of the drug. The latter two mechanisms have been associated with the non protein sulfhydryl content of resistant cells (see review by Vistica[9]) which is frequently two-fold higher than in sensitive cells. This correlation between intracellular glutathione levels and melphalan cytotoxicity has been extensively studied by Vistica and co workers in L1210 leukemia[6-9]. They demonstrated a two fold increase in glutathione content in an L1210 cell line which was 4-5 times more resistant to melphalan than the parent sensitive cell line. These workers also examined the effects of either nutritional

HUMAN TUMOR CLONING
ISBN 0-8089-1671-8

or pharmacologic depletion of glutathione levels upon melphalan cytotoxicity in these cell lines. Incubation of the variant resistant cell lines in a cystine free medium reduced the glutathione content to the same levels as in the sensitive cell line and completely restored sensitivity to melphalan. Nutritional deprivation of glutathione levels in the parent sensitive cell line was also accompanied by an increased cytotoxicity of melphalan.

Similar results were observed _in vitro_ with buthionine sulfoximine, an inhibitor of γ-glutamyl cysteine synthetase, an enzyme involved in the synthesis of glutathione. This synthetic amino acid was developed by Meister and his colleagues[2]. When melphalan resistant L1210 cells were exposed to this drug at a concentration which was only minimally cytotoxic the intracellular content of glutathione was reduced and sensitivity to melphalan restored. These reductions in glutathione content were not associated with an increase in melphalan transport. Melphalan resistant L1210 cells, however, were shown to detoxify melphalan to a less cytotoxic intermediate and this dechlorination was associated with the glutathione content of the cells[7]. The mechanism of the increased dechlorination has not been established.

II. GLUTATHIONE LEVELS IN HUMAN OVARIAN CANCER

The potentiation of melphalan cytotoxicity in human ovarian cancer cell lines by nutritional and pharmacologic depletion of intracellular glutathione levels has recently been reported by us[1]. Cell lines OVCAR 2-4 were established from patients who were clinically resistant to alkylating agents. The detailed characteristics of these cell lines have previously been reported[3,4]. Cell line A1847 was established from an untreated ovarian cancer patient and kindly provided by Dr. Stuart Aaronson, Laboratory of Cellular Molecular Biology, DCCP, National Cancer Institute. This cell line was made resistant to melphalan by increasing the concentration of the drug in the growth medium from 10^{-8} M to 10^{-5} M over a 6 month period. The resistant variant termed 1847ME, was 3.3 fold more resistant to melphalan than the sensitive cell line from which it was derived. In addition, the degree of resistance to melphalan in 1847ME was similar to that observed in the cell lines OVCAR 2-4 established from alkylating agent resistant patients.

Similar to the reported results in melphalan resistant L1210 leukemia and other alkylating agent resistant experimental tumors[9], 1847ME also had a two fold higher intracellular glutathione content than the parent sensitive cell line, A1847. The OVCAR cell lines also had similar elevations in glutathione content compared to alkylating agent sensitive cell lines. The net intracellular accumulation of melphalan was identical in A1847 and in 1847ME.

Incubation of 1847ME in cystine free medium for 24 hours decreased glutathione levels by 88% and completely reversed resistance to melphalan. Nutritional depletion of glutathione levels in the sensitive cell line also potentiated the cytotoxicity of melphalan. Since nutritional deprivation of cystine is not practical in ovarian cancer patients we also examined the effects of buthionine sulfoximine upon melphalan cytotoxicity in the ovarian cancer cell lines. Glutathione levels were reduced by 75% following incubation of 1847ME cells with buthionine sulfoximine and melphalan sensitivity was completely restored. Synergism between melphalan and buthionine sulfoximine was also observed in the OVCAR cell lines at concentrations of buthionine sulfoximine which were minimally cytotoxic. In addition, there was detoxification of melphalan in the 1847ME resistant cell line.

III. CONCLUSIONS

These results demonstrate a correlation between melphalan cytotoxicity and glutathione levels both in drug resistant L1210 leukemia and in human ovarian cancer cell lines. The mechanisms of the apparent protective effect of glutathione upon the cytotoxic effects of melphalan remain to be established. It is possible that the increased glutathione levels facilitate the dechlorination of melphalan to a less cytotoxic intermediate by glutathione linked-S-transferases. It is also possible that the increased levels of glutathione may protect critical intracellular sites from alkylating agent damage. In particular, elevated glutathione levels may interfere with the production of interstrand, intrastrand or DNA-protein crosslinks produced by alkylating agents. Studies are currently in progress to determine what specific role glutathione has upon conferring protection from the cytotoxic effects of melphalan.

Regardless of the mechanisms involved, the demonstration that pharmacologic manipulation of glutathione levels results in increased cytotoxicity is of potential major clinical importance in the treatment of ovarian cancer. The toxicity of buthionine sulfoximine in large animals remains to be determined as does the specificity of the potentiating effects of buthionine sulfoximine upon melphalan cytotoxicity in normal versus malignant cells. Studies are currently in progress in a novel transplantable nude mouse model of human ovarian cancer which produces ascites and intraabdominal carcinomatosis[5] to determine whether survival can be enhanced by combination therapy with melphalan and buthionine sulfoximine. If such a synergistic effect can be established *in vivo* it would provide strong evidence that clinically beneficial enhancement of melphalan cytotoxicity by glutathione depletion is a distinct possibility.

There may be other mechanisms of alkylating agent resistance which are not associated with glutathione levels. Drug transport, however, does not appear to be a major determinant of melphalan cytotoxicity in human ovarian cancer cells as shown by the absence of any difference in melphalan accumulation in the 1847ME cell line compared to the sensitive cell line[1]. The availably of a series of human ovarian cancer cell lines with varying degrees of resistance to melphalan should facilitate the evaluation of alternative mechanisms of alkylating agent resistance and pharmacologic ways in which this resistance can be reversed.

REFERENCES

1. Green JA, Vistica DT, Young RC, et al: Melphalan resistance in human ovarian cancer: potentiation of melphalan cytotoxicity by nutritional and pharmacologic depletion of intracellular glutathione levels. Proc Amer Assoc Cancer Res (in press) 1984.
2. Griffith AW, Meister A: Potent and specific inhibition of glutathione synthesis by buthionine sulfoximine (S-n-butyl homocysteine sulfoximine). J Biol Chem 253: 7558-7560, 1979.
3. Hamilton TC, Foster RJ, Grotzinger KR, et al: Development of drug sensitive and resistant human ovarian cancer cell lines. A model system for investigating new drugs and mechanisms of resistance. Proc Amer Assoc Cancer Res 24: 313, 1983.
4. Hamilton TC, Young RC, McKoy WM, et al: Characterization of a human ovarian cancer cell line (NIH:OVCAR-3) with androgen and estrogen receptors. Cancer Res 43: 5379-5389, 1983.
5. Hamilton TC, Young RC, Rogan AM, et al: A unique intraperitoneal model of human ovarian cancer. Proc Amer Assoc Cancer Res (in press) 1984.
6. Suzukake K, Petro BJ, Vistica DT: Reduction in glutathione content of L-PAM resistant L1210 cells confers drug sensitivity. Biochem Pharmac 31: 121-124, 1982.

7. Suzukake K, Petro BJ, Vistica DT: Dechlorination of L-phenylalanine mustard by sensitive and resistant tumor cells and its relationship to intracellular glutathione content. Biochem Pharmacol 32: 165-167, 1983.
8. Vistica DT, Somfai-Relle S, Suzukake K, Petro BJ: Inhibition of glutathione synthesis by S-n-butyl homocysteine sulfoximine and sensitization of murine tumor cells resistant to L-phenylalanine mustard. J Cell Biochem (Suppl) 6: 375, 1982.
9. Vistica DT: Cellular pharmacokinetics of the phenylylalanine mustards. Pharmac Ther 22: 379-405, 1983.
10. Young RC: Chemotherapy of ovarian cancer. Past and Present. Semin Oncol 2: 267-276, 1975.

ANTICALMODULIN AGENTS AS INHIBITORS OF HUMAN TUMOR CELL CLONOGENICITY

Robert A. Hickie, David J. Klaassen, Gary Z. Carl

Department of Pharmacology and Division of Oncology
College of Medicine, University of Saskatchewan
Saskatoon, Saskatchewan, Canada S7N 0W0

Frank L. Meyskens, Jr., Kristie L. Kreutzfeld,
Steven P. Thomson

Department of Internal Medicine, Division of Hematology and Oncology
University of Arizona Cancer Center
Tucson, Arizona, U.S.A. 85724

INTRODUCTION

Since cancer is characterized by unregulated cell growth, it is of considerable interest to investigate changes in putative endogenous cell growth regulators in order to identify underlying mechanisms which would be feasible targets to exploit in developing new and more selective cancer chemotherapeutic agents. In our recent studies, we have been particularly interested in examining the role that altered calmodulin (CaM) levels may play in neoplastic cell growth (20, 46-48).

CaM is a small Ca^{2+}-binding protein (molecular weight about 17,000), which is considered to be a key intracellular Ca^{2+} receptor mediating Ca^{2+}-regulated enzymes and processes in eukaryotic cells, including those affecting cell replication and proliferation either directly or indirectly (8,10, 20,30,50). For example, CaM can influence the metabolism and cellular levels of chemical substances that are believed to play a role in cell growth, particularly during the prereplicative (G_1) phase of the cell cycle; these include the cyclic nucleotides (19,27,40,50), polyamines (14,29, 40,42) and prostaglandins (2,3,15,33,52). CaM also ensures that the mitogenic intracellular Ca^{2+} spike during the late G_1 phase (just prior to DNA synthesis) is short-lived by promoting the outward transport of Ca^{2+} through the plasma

HUMAN TUMOR CLONING
ISBN 0-8089-1671-8

membrane via activation of Ca^{2+}, Mg^{2+}-ATPase (13,43). Glycogen metabolism and cell motility is also modulated by CaM through the activation of phosphorylase kinase, glycogen synthase kinase, and myosin light chain kinase (10,30). In addition to regulating the phosphorylation of appropriate protein substrates, CaM has been shown to modulate phosphatase activity, most notably histone dephosphorylation (51). The regulation of microtubular function, both in relation to the cytoskeleton and mitotic apparatus, has been shown to involve CaM (7,25,30,44), whose main effect is to promote the disassembly of the tubulin dimers. This action provides a positive signal for DNA synthesis (11) which can be negated by microtubule stabilizers such as $Taxol^{R}$ (7,24). The Ca^{2+}-CaM complex has been found to be necessary in order for DNA synthesis to proceed in hepatocytes (1); these workers showed that replication and proliferation can be induced in quiescent (G_o) hepatocytes by adding exogenous CaM in the presence of Ca^{2+}.

Mounting evidence suggests that the relative autonomy which cancer cells exhibit to endogenous growth regulators (13,40,50) may be associated with an increase, during the G_1 phase of the cell cycle, of a factor (or factors) that promotes the progression of cells through the cycle. In view of the apparent involvement of CaM as an inducer of replication and proliferation in normal cells, we investigated changes in CaM levels in neoplastic cells using the rat hepatoma/liver model (46,48). The results of these studies indicated that the levels of CaM, measured by both enzymatic assay and by radioimmunoassay, were increased significantly (50-200%) in the hepatomas with the size of increase being directly proportional to the growth rate of the tumor. Other workers have also reported substantial increases in CaM in other hepatomas (28), in human breast cancers (39) and in a number of transformed cell lines (9,26,41,45). The latter studies showed that CaM levels are usually elevated two to four-fold in all exponentially growing transformed cells irrespective of the method of transformation; furthermore, these increased CaM levels were found to be due specifically to increased rates of CaM synthesis in transformed cells. In non-transformed cells it should be noted that CaM levels do not increase to the same extent as in transformed cells, even during the exponential growth phase. Changes in CaM levels during the cell cycle have been reported by Chafouleas and co-workers (7,8) using a synchronized Chinese hamster ovary cell line ($CHO-K_1$). These workers found that a sharp increase in the cellular CaM concentration accompanies the G_1/S phase transition, suggesting that CaM is required for progression into and through the S phase; the antiCaM drug W-13 was shown to inhibit this progression. These investigators also reported that CaM is essential for quiescent cells (G_o) to re-enter the cell cycle. Other studies on the same cell line (38) confirmed that CaM is required for progression through the S phase; if the anti-CaM drug W-7 is applied at the S phase, the cell cycle was blocked in late G_2 or early M phases. Taken together, the results of the foregoing studies suggest that CaM is involved in both DNA synthesis and in mitosis.

If excessive levels of CaM contribute to unregulated tumor growth, we postulate that it should be possible to inhibit the replication and proliferation of neoplastic cells by using agents that selectively bind to (and reversibly inactivate) the abnormally high levels of intracellular CaM. To test this hypothesis, we initiated studies to evaluate the growth-inhibitory efficacy of antiCaM agents on human cancer cells using the soft agar clonogenic assay (18,37). In our preliminary studies we examined the effect of a phenothiazine derivative (trifluoperazine) and two naphthalene sulfonamides (W-7 and W-13) on a human breast cancer cell line, MDA-MB-231 (20,47). The results showed that all three drugs effectively inhibited the colony formation of the breast cancer cells in a concentration-dependent manner. Continuous exposure to the drugs was found to be considerably more effective than 1 hour exposure. In view of these promising findings, we have extended our recent investigations to other chemical classes of antiCaM agents and to additional human tumor cell types. This paper discusses the results of this study.

EXPERIMENTAL

Materials and Methods

Tumor Cells

The human tumor cell lines used in this study include: MDA-MB-231 (breast) (4,5); UACC 81.46A (melanoma); RPMI 8226 (myeloma); NCI 417 (lung); WiDr (colon); and, HEY (ovary). The breast, myeloma and colon cell lines were obtained from the American Type Culture Collection (ATCC), Rockville, Md. All three cell lines grew optimally in medium RPMI 1640. The melanoma cell line was sub-cultured from a subcutaneous nodule of metastatic melanoma taken from a female patient at the University of Arizona Cancer Center in Tucson. This cell line was cultured in Ham's F10 medium (low tyrosine). The lung cancer cells used in this study were supplied by Dr. Desmond Carney, NCI, Bethesda, Md. These cells grew readily in RPMI 1640 as floating cell aggregates (6). The ovarian cancer cell line was provided by Dr. Ronald Buick, Ontario Cancer Institute, Toronto, Ontario. The medium used to culture these cells was RPMI 1640 with added glutamine (4 mM).

Drugs

AntiCaM agents representing five chemical types were tested on each of the six cancer cell lines outlined above, these include: trifluoperazine (a phenothiazine) (34-36, 49); W-7 (a naphthalene sulfonamide i.e., N-(6-aminohexyl)-5-chloro-l-n.s.) (21-23,36); haloperidol (a butyrophenone) (34-36); penfluridol (a diphenylbutylpiperidine) (34-36); and calmidazolium (an imidazolinium chloride) (16,17). Stock solutions (1-5 mM) of the drugs were prepared by dissolving the drugs in ethanol. Appropriate aliquots of the stock solutions (or dilutions thereof) were added to the clonogenic assay media such that the ethanol concentration did not exceed 0.02% v/v, a concentration which had no discernible effect on colony formation.

Clonogenic Assay

The culture system used to clone the human tumor cell lines for this study was adapted from the double-layered soft agar system developed by Hamburger and Salmon (18); conditioned medium was also not added (47). For breast myeloma, lung, colon and ovarian cancer cell lines, the basal medium added to the 0.5% soft agar in the bottom ("feeder") layer (1 ml) was RPMI 1640 containing 10% heat-inactivated fetal calf serum (FCS) and 1% penicillin-streptomycin (p/s). With the ovarian tumor cells, the basal medium also contained glutamine (4 mM). For the melanoma, the basal medium added to the agar in feeder layer was Ham's F10 containing 10% FCS and gentamicin (20 μg/ml). Single cell suspensions of the tumor cells were prepared in appropriate concentrations and plated with the upper layer (1 ml) consisting of either Ham's F10 medium (plus 10% FCS and gentamicin) with melanoma or RPMI 1640 (plus 10% FCS and p/s) with the other cell lines in 0.3% soft agar, with or without the antiCaM drugs. The 35 mm culture dishes were incubated at 37° in a humidified atmosphere containing 5% carbon dioxide for 7 to 14 days, depending on the cell line. The assays were carried out in quadruplicate and checked routinely after plating for uniform single cell distribution. In order to determine the colony forming efficiency (sometimes referred to as "plating efficiency") as well as proliferative capacity, both the number of colonies and their size were determined routinely. An automated optical image analyzer (Omnicon FAS-II, Bausch and Lomb) was used to count the colonies and to group them into various sizes, namely: 60; 72; 86; 104; 124; and 149 micron diameters. For drug assay studies, the total number of colonies larger than 60 microns were enumerated.

RESULTS AND DISCUSSION

Effect of Cell Number Plated on the Number and Size of Tumor Colonies Formed

To optimize our experimental conditions for the drug studies, we investigated the relationship between the number of tumor cells plated and the number and size of the colonies formed in soft agar (31). FIGURE 1 illustrates the number of colony forming units per plate for lung cells (NCI 417) plated in concentrations ranging from 1×10^3 to 100×10^3 cells per plate. It is evident from this data that the number of colonies formed increases initially but then

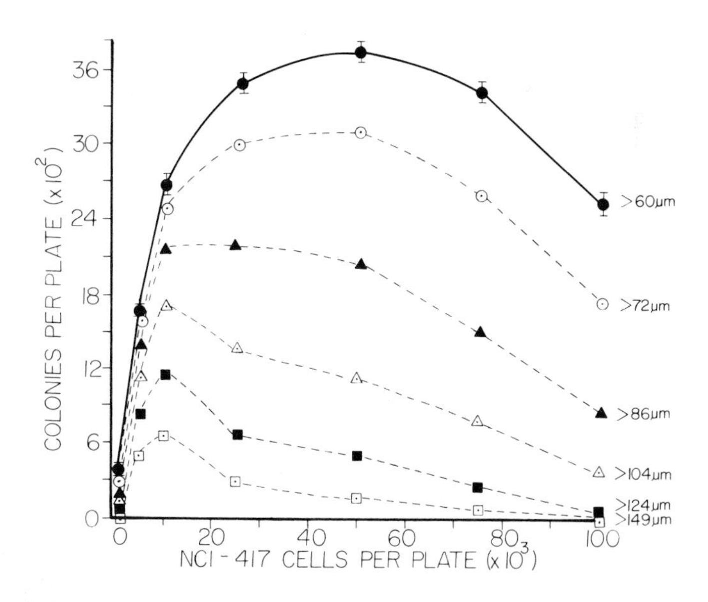

FIGURE 1. Relationship between lung cancer (NCI 417) cell number plated and the number and size of colonies formed in soft agar.

plateaus and eventually decreases with the successive increases in cell numbers plated. The formation of the larger colonies (i.e., > 104 μm diameter) is suppressed markedly at cell concentrations of 25,000 or higher. This finding is most likely due to the fact that we are using a closed non-refed, culture system. This system has a limited supply of nutrients which appears to be inadequate to support colony formation at higher plating densities. For the NCI 417 tumor cells, the optimum plating concentration (OPC) was found to be 10,000 cells per plate, giving a colony forming efficiency (CFE) of 26.7%. The OPC is defined as the cell number plated which results in the highest CFE. Qualitatively similar curves were obtained with the other 5 cell lines. The OPC for each tumor, with the CFE in brackets, is as follows: breast - 50,000 cells per plate (0.9%); melanoma - 10,000 (25%); myeloma - 10,000 (20.9%); colon - 10,000 (36.8%); and ovary - 5,000 (37.9%). These data have important implications particularly when testing the efficacy of drugs that inhibit tumor cell growth. For example, if an anticancer drug is added to a culture plate containing an excessive number of plated cells, the drug may actually stimulate colony formation by inhibiting growth or killing a portion of the plated cells resulting in a greater supply of nutrients available to the remaining cells; presumably, the reverse situation would apply to agents that stimulate cell growth (31; see also Thomson _et al_., this volume). Therefore, caution should be exercised when interpreting drug effects on colony survival curves; this process can be facilitated considerably by using cell concentrations that are close to the optimal plating concentration.

The duration of incubation found to be most suitable for the cell lines used in this study are as follows: 14 days (breast and melanoma); 10 days (lung); 9 days (colon); 8 days (myeloma); and, 7 days (ovary).

Cell Size, Number of Cells in Colonies, and Proliferative Capacity

Recent studies in our laboratory have shown that tumor cells grow in soft agar as disc-ellipsoids (or oblate spheroids) (32). If the cell size and colony size is known, it is possible to determine the cell number in a given colony size using the formula:

$$2.4\ (\text{colony diameter})^{2.38} \div (\text{cell diameter})^{2.80}$$

From these data it is further possible to calculate the proliferation capacity i.e., number of cell population doublings per given colony diameter.

In the present study, the cell sizes, cell numbers per median colony diameter (MCD), and the proliferative capacities were determined for all six human tumor cell lines. The order of cell sizes, ranging from the largest to the smallest was found to be: lung > melanoma > breast > ovary colon > myeloma. The median diameters (μm) of the colonies varied considerably between cell lines [from 83 (breast) to 126 (colon)]. This factor, coupled with differences in cell size resulted in large differences between the cell lines with regard to the number of cells present in the median sized colonies. For example, with breast cancer there was only 27 cells per MCD, whereas with myeloma and colon there were close to 150 cells. The proliferative capacities of the cell lines were as follows: 4-5 for breast and melanoma; 5-6 for lung; 6-7 for ovary; and, 7-8 for myeloma and colon. When testing the efficacy of drugs that inhibit or stimulate cell growth, it would be desirable to work with cells that have a proliferative capacity of at least 3. This is essential particularly when evaluating biological response modifiers, which might not influence cell growth until after 3 or 4 population doublings have taken place. Our studies indicate that all six tumor cell lines had proliferative capacities that were adequate for studies with antiCaM drugs.

Influence of AntiCaM Drugs on Tumor Clonogenicity

Drug dose/response curves were prepared by continuous exposure of each cell line to five graded concentrations of each antiCaM drug. A typical set of dose/response curves for lung cells (NCI 417) are illustrated in FIGURE 2. From this figure it can be seen that each antiCaM drug inhibited the clonogenicity of lung cells in a dose-dependent manner. Since all of the curves are parallel when expressed on a log drug concentration scale, this suggests that the drugs inhibit colony formation through the same mechanism, which presumably is by selective inhibition of CaM. The dose/response curves obtained with the other five cell lines were qualitatively similar to lung.

The order of drug potency for inhibiting clonogenicity of lung cells in soft agar, from the most potent to the least potent, is as follows: calmidazolium > penfluridol > trifluoperazine > W-7 > haloperidol. For the other cell lines

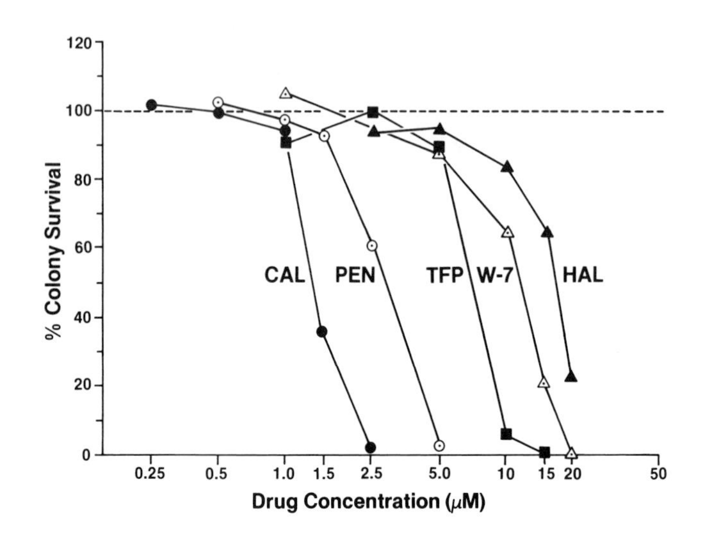

FIGURE 2. Effect of antiCaM drugs on the soft agar clonogenicity of human lung cells (NCI 417).

the order of potency was similar with the exception of colon, in which haloperidol was found to be more potent than W-7. The reason for this difference may be due, at least in part, to a unique difference in the permeability of colon cells to haloperidol since the IC_{50} value for this drug in colon is considerably lower than that of the other cell lines (see TABLE I). It is of interest to note that the order of potency obtained consistently with each cell line is closely related to the relative inhibitory potency of these agents on CaM-dependent enzyme systems such as cyclic nucleotide phosphodiesterases and Ca^{2+}, Mg^{2+}-ATPase. This is additional evidence supporting the view that these drugs are inhibiting clonogenicity by binding to intracellular CaM.

The drug concentrations required to inhibit colony formation by 50%, commonly known as IC_{50}, are summarized for all cell lines in TABLE I. These data suggest that, in general, the sensitivity of the cell lines to growth inhibition by a given antiCaM drug is remarkably similar. The average IC_{50} values (μM) for the six cell lines are: 1.4 (calmidazolium); 2.0 (penfluridol); 7.4 (trifluoperazine); 14.3 (W-7); and 15.1 (haloperidol). This finding is somewhat puzzling, considering the differences between the cell lines with regard to plating efficiencies, cell sizes and growth rates. These results suggest that in spite of these differences, the total amount of CaM required for growth regulation in these cell lines is similar.

TABLE I. IC_{50} values (μM) of antiCaM drugs for inhibiting the clonogenicity of human tumor cells in soft agar

Drug	MDA-MB 231	UACC 81.46A	RPMI 8226	NCI 417	WiDr	HEY
Calmidazolium	1.3	1.3	1.0	1.4	1.3	2.3
Penfluridol	2.3	1.5	1.3	2.9	1.3	2.7
Trifluoperazine	8.0	3.6	11.0	7.0	2.5	12.0
W-7	14.5	12.0	12.0	12.5	17.5	17.5
Haloperidol	14.0	14.5	20.5	17.0	7.5	17.0

Although we have shown that antiCaM agents can inhibit the growth of tumor cells at relatively low concentrations _in vitro_, it remains to be established whether these drug concentrations are achievable _in vivo_. Since these drugs are quite lipophylic, it is likely that their apparent volume of distribution _in vivo_ will be at least as large as the total body water. If this is the case, drug concentrations found to inhibit clonogenicity should be achievable _in vivo_. This is supported by the study of Driscoll _et al._ (12) which showed that phenothiazines such as fluphenazine and butyrophenones such as triperidol significantly inhibited growth of mouse tumor models used by the National Cancer Institute (USA) to screen for anticancer activity, namely L1210 leukemia, B_{16} melanoma and P388.

Effect of AntiCaM Analogs on Clonogenicity

Chemical analogs are available for each type of antiCaM agent used in the above study which have lower affinities for CaM than the parent compound and therefore are less effective in inhibiting CaM-dependent processes. For example: TFP.SO; W-12; R24555 (analog of calmidazolium);

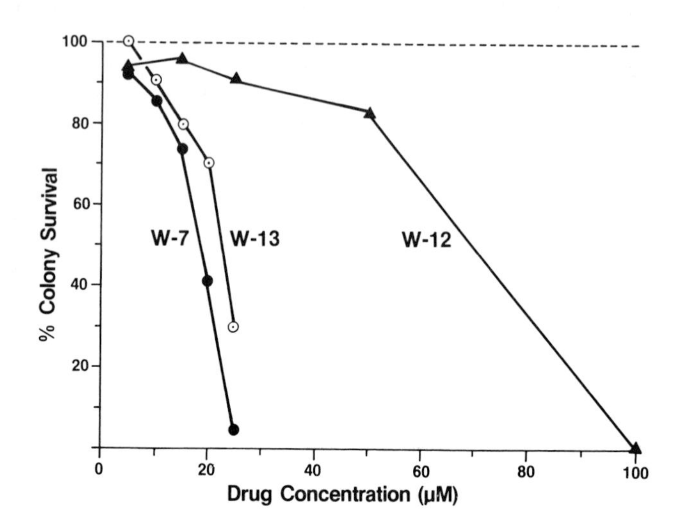

FIGURE 3. Comparative effects of W-12 and the antiCaM agents W-7 and W-13 on the clonogenicity of colon cancer cells (WiDr).

and R1670 (analog of haloperidol and penfluridol). These agents have been used in the present study as tools to determine the extent to which CaM-independent effects play a role in colony formation. The results of an experiment using W-12 and colon cancer cells are shown in FIGURE 3. It is clear that the growth inhibiting activity for W-12 is much lower than W-13 or W-7 i.e., the IC_{50} value for W-12 was 70 μM compared with 20 μM for W-13 and 18 μM for W-7. Similar results were obtained with the other analogs using the other cell lines. This provides additional evidence that CaM is involved in regulating cell replication and proliferation and that antiCaM agents selectively inhibit the influence of CaM on cell growth.

CONCLUSIONS

The following conclusions can be drawn from the present studies:

1. Two parameters to consider when testing the effect of agents on tumor cell clonogenicity in a non-fed semi-solid assay system are the plating concentration and the proliferative capacity of the cells.

The plating concentration used should be close to the cell number giving the highest plating efficiency while the proliferative capacity of the cells should be at least 3-4 population doublings (to reach the median colony diameter).

2. CaM plays a significant role in promoting replication and proliferation of tumor cells. This conclusion is based on the evidence that:

(a) all five chemical types of antiCaM agents tested inhibited the clonogenicity of six different cancer cell types in a dose-dependent manner.

(b) the growth-inhibiting efficacies of antiCaM agents were similar to their respective potencies in inhibiting CaM-dependent enzymes. Further, chemical analogs of antiCaM agents that have weak antiCaM activity are also weak inhibitors of tumor cell clonogenicity.

(c) the log-dose response curves in all cell lines tended to be parallel, suggesting a common mechanism of action, presumably selective CaM inhibition.

3. CaM is a feasible target to exploit in developing a new approach to cancer chemotherapy.

ACKNOWLEDGEMENTS

The authors wish to acknowledge the competent assistance of Jackie Bitz in preparing this manuscript and Linda Kimball for preparation of graphs. We also appreciate the technical assistance of Julie Buckmeier in the proliferative capacity studies. Dr. Ron Buick and Dr. Desmond Carney kindly gave us permission to use their ovary (HEY) and lung (NCI 417) cell lines respectively. Our thanks to Dr. H. Van Belle, Janssen Pharmaceuticals, Beerse, Belgium for providing us with information regarding the phosphodiesterase IC_{50} values for several antiCaM agents.

This work was supported by grants from the Saskatchewan Cancer Foundation, the American Cancer Society (PDT 184) and the National Cancer Institute (U.S.A.) (CA 17094, CA 27502).

REFERENCES

1. Boynton, A.L., Whitfield, J.F., and MacManus, J.P. Calmodulin stimulates DNA synthesis by rat liver cells. Biochem. Biophys. Res. Commun., 95: 745-749, 1980.

2. Bregman, M.D., and Meyskens, F.L., Jr. Inhibition of human malignant melanoma colony-forming cells in vitro

by prostaglandin A_1. Cancer Res., 43: 1642-1645, 1983.
3. Bregman, M.D., Sander, D., and Meyskens, F.L., Jr. Prostaglandin A_1 and E_1 inhibit the plating efficiency and proliferation of murine melanoma cells (Cloudman S-91) in soft agar. Biochem. Biophys. Res. Commun., 104: 1080-1086, 1982.
4. Brinkley, B.R., Beall, P.T., Wible, L.J., Mace, M.L., Turner, D.S. and Cailleau, R.M. Variations in cell form and cytoskeleton in human breast carcinoma cells in vitro. Cancer Res., 40: 3118-3129, 1980.
5. Cailleau, R., Young, R., Olivé, M., and Reeves, W.J., Jr. Breast tumor cell lines from pleural effusions. J. Natl. Cancer Inst. U.S.A., 53: 661-674, 1974.
6. Carney, D.N., Mitchell, J.B., and Kinsella, T.J. In vitro radiation and chemotherapy sensitivity of established cell lines of human small cell lung cancer and its large cell morphological variants. Cancer Res., 43: 2806-2811, 1983.
7. Chafouleas, J.G., Bolton, W.E., Hidaka, H., Boyd, A.E. III, and Means, A.R. Calmodulin and the cell cycle: involvement in regulation of cell-cycle progression. Cell, 28: 41-50, 1982.
8. Chafouleas, J.G., and Means, A.R. Calmodulin is an important regulatory molecule in cell proliferation. In: A.L. Boynton, W.L. McKeehan, and J.F. Whitfield (eds.), Ions, Cell Proliferation and Cancer, pp. 449-464. Toronto: Academic Press, 1982.
9. Chafouleas, J.G., Pardue, R.L., Brinkley, B.R., Dedman, J.R. and Means, A.R. Regulation of intracellular levels of calmodulin and tubulin in normal and transformed cells. Proc. Natl. Acad. Sci. USA, 78: 996-1000, 1981.
10. Cheung, W.Y. Calmodulin. Scient. Amer., 246: 62-70, 1982.
11. Crossin, K.L., and Carney, D.H. Evidence that microtubule depolymerization early in the cell cycle is sufficient to initiate DNA synthesis. Cell, 23: 61-71, 1981.
12. Driscoll, J.S., Melnick, N.R., Quinn, F.R., Lomax, N., Davignon, J.P., Ing, R., Abbott, B.J., Congleton, G., and Dudeck, L. Psychotropic drugs as potential antitumor agents: a selective screening study. Cancer Treatment Rep., 62: 45-74, 1978.
13. Durham, A.C.H., and Walton, J.M. Calcium ions and the control of proliferation in normal and cancer cells. Review. Biosci. Rep., 2: 15-30, 1982.

14. Fozard, J.R., and Koch-Weser, J. Pharmacological consequences of inhibition of polyamine biosynthesis with dl-α-difluoromethylornithine. Trends in Pharmacol. Sci., March, 107-110, 1982.
15. Fukushima, M., Kato, T., Ueda, R., Ota, K., Narumiya, S., and Hayaishi, O. Prostaglandin D_2, a potential antineoplastic agent. Biochem. Biophys. Res. Commun., 105: 956-964, 1982.
16. Gietzen, K., Sadorf, I., Xu, Y.-H., Galla, H.-J., and Bader, H. The influence of drugs on calmodulin-dependent Ca^{2+}-transport ATPase and cyclic nucleotide phosphodiesterase. Acta Physiolog. Lat. Amer., 32: 244-246, 1982.
17. Gietzen, K., Wüthrich, A., and Bader, H. R24571: A new powerful inhibitor of red blood cell Ca^{++}-transport ATPase and of calmodulin-regulated functions. Biochem. Biophys. Res. Commun., 101: 418-425, 1981.
18. Hamburger, A.W., and Salmon, S.E. Primary bioassay of human tumor stem cells. Science, 197: 461-463, 1977.
19. Hickie, R.A. Regulation of cyclic AMP and cyclic GMP in Morris hepatomas and liver. Adv. Exp. Med. Biol., 92: 451-487, 1978.
20. Hickie, R.A., Wei, J.-W., Blyth, L.M., Wong, D.Y.W., and Klaassen, D.J. Cations and calmodulin in normal and neoplastic cell growth regulation. Can. J. Biochem. Cell Biol., 61: 934-941, 1983.
21. Hidaka, H., Asano, M., and Tanaka, T. Activity-structure relationship of calmodulin antagonists naphthalenesulfonamide derivatives. Mol. Pharmacol., 20: 571-578, 1981.
22. Hidaka, H., Sasaki, Y., Tanaka, T., Endo, T., Ohno, S., Fujii, Y., and Nagata, T. N-(6-aminohexyl)-5-chloro-1-naphthalenesulfonamide, a calmodulin antagonist, inhibits cell proliferation. Proc. Natl. Acad. Sci., 78: 4354-4357, 1981.
23. Hidaka, H., Yamaki, T., Naka, M., Tanaka, T., Hayashi, H., and Kobayashi, R. Calcium-regulated modulator protein interacting agents inhibit smooth muscle calcium-stimulated protein kinase and ATPase. Mol. Pharmacol., 17: 66-72, 1980.
24. Horwitz, S.B., Chia, G.H., Harracksingh, C., Orlow, S., Pifko-Hirst, S., Schneck, J., Sorbara, L., Speaker, M., Wilk, E.W., and Rosen, O.M. Trifluoperazine inhibits phagocytosis in a macrophage like cultured cell line. J. Cell Biol., 91: 798-802, 1981.
25. Kakiuchi, S., and Sobue, K. Control of the cytoskeleton by calmodulin and calmodulin-binding proteins. Trends in Biochem. Sci., 8: 59-62, 1983.

26. LaPorte, D.C., Gidwitz, S., Weber, M.J., and Storm, D.R. Relationship between changes in the calcium dependent regulatory protein and adenylate cyclase during viral transformation. Biochem. Biophys. Res. Commun., 86: 1169-1177, 1979.
27. Leof, E.B., Wharton, W., O'Keefe, E., and Pledger, W.J. Elevated intracellular concentrations of cyclic AMP inhibited serum-stimulated, density-arrested BALB/c-3T3 cells in mid G_1. J. Cell Biochem., 19: 93-103, 1982.
28. MacManus, J.P., Braceland, B.M., Rixon, R.H., Whitfield, J.F., and Morris, H.P. An increase in calmodulin during growth of normal and cancerous liver in vivo. FEBS Lett., 133: 99-102, 1981.
29. Marton, L.J., Oredsson, S.M., Hung, D.T., and Deen, D.F. Effects of polyamine depletion on the cytotoxicity of cancer chemotherapeutic agents. Adv. Polyamine Res. 4: 33-40, 1983.
30. Means, A.R., Tash, J.S., and Chafouleas, J.G. Physiological implications of the presence, distribution, and regulation of calmodulin in eukaryotic cells. Physiol. Rev. 62: 1-38, 1982.
31. Meyskens, F.L., Jr., Thomson, S.P., Hickie, R.A., and Sipes,N.J. The size of tumor colonies in semisolid medium is inversely related to the number of cells plated. Br. J. Cancer, 48: 863-868, 1983.
32. Meyskens, F.L., Jr., Thomson, S.P., and Moon, T.E. Quantitation of the number of cells within tumor colonies in semisolid medium and the growth as disc-ellipsoids. Cancer Res., 44: 271-277, 1984.
33. Moskowitz, N., Shapiro, L., Schook, W., and Puszkin, S. Phospholipase A_2 modulation by calmodulin, prostaglandins and cyclic nucleotides. Biochem. Biophys. Res. Commun., 115: 94-99, 1983.
34. Prozialeck, W.C., and Weiss, B. Inhibition of calmodulin by phenothiazines and related drugs: structure-activity relationships. J. Pharmacol. Exper. Ther., 222: 509-516, 1982.
35. Roufogalis, B.D. Specificity of trifluoperazine and related phenothiazines for calcium-binding proteins. In: W.-Y. Cheung (ed.), Calcium and Cell Function, Vol. 3, pp. 129-159. Toronto: Academic Press, 1982.
36. Roufogalis, B.D., Minocherhomjee, A.M., and Al-Jobore, A. Pharmacological antagonism of calmodulin. Can. J. Biochem. Cell Biol., 61: 927-933, 1983.
37. Salmon, S.E. Background and overview. In: S.E. Salmon (ed.), Cloning of Human Tumor Stem Cells, Prog. Clin. Biol. Res., Vol. 48, pp. 3-13. New York: Alan R. Liss, Inc.,1980.

38. Sasaki, Y., and Hidaka, H. Calmodulin and cell proliferation. Biochem. Biophys. Res. Commun., 104: 451-456, 1982.
39. Singer, A.L., Sherwin, R.P., Dunn, A.S., and Appleman, M.M. Cyclic nucleotide phosphodiesterases in neoplastic and non-neoplastic human mammary tissues. Cancer Res., 36: 60-66, 1976.
40. Swierenga, S.H.H., Whitfield, J.F., Boynton, A.L., MacManus, J.P., Rixon, R.H., Sikorska, M., Tsang, B.K., and Walker, P.R. Regulation of proliferation of normal and neoplastic rat liver cells by calcium and cyclic AMP. Ann. N.Y. Acad. Sci., 349: 294-311, 1980.
41. Van Eldik, L.J., and Burgess, W.H. Analytical subcellular distribution of calmodulin and calmodulin-binding proteins in normal and virus-transformed fibroblasts. J. Biol. Chem. 258: 4539-4547, 1983.
42. Veldhuis, J.D., and Hammond, J.M. Role of calcium in modulation of ornithine decarboxylase activity in isolated pig granulosa cells in vitro. Biochem. J., 196: 798-801, 1981.
43. Vincenzi, F.F., Hinds, T.R., and Raess, B.U. Calmodulin and the plasma membrane calcium pump. Ann. N.Y. Acad. Sci., 356: 232-244, 1980.
44. Watanabe, K., and West, W.L. Calmodulin, activated cyclic nucleotide phosphodiesterase, microtubules, and vinca alkaloids. Fed. Proc., 41: 2292-2299, 1982.
45. Watterson, D.M., Van Eldik, L.J., Smith, R.E., and Vanaman, T.C. Calcium-dependent regulatory protein of cyclic nucleotide metabolism in normal and transformed chicken embryo fibroblasts. Proc. Natl. Acad. Sci., U.S.A., 73: 2711-2715, 1976.
46. Wei, J.-W., and Hickie, R.A. Increased content of calmodulin in Morris hepatoma 5123t.c.(h). Biochem. Biophys. Res. Commun., 100: 1562-1568, 1981.
47. Wei, J.-W., Hickie, R.A., and Klaassen, D.J. Inhibition of human breast cancer colony formation by anticalmodulin agents: trifluoperazine, W-7, and W-13. Cancer Chemother. Pharmacol., 11: 86-90, 1983.
48. Wei, J.-W., Morris, H.P., and Hickie, R.A. Positive correlation between calmodulin content and hepatoma growth rates. Cancer Res., 42: 2571-2574, 1982.
49. Weiss, B., Prozialeck, W.C., and Wallace, T.L. Interaction of drugs with calmodulin. Biochemical, pharmacological and clinical implications. Biochem. Pharmacol., 31: 2217-2226, 1982.
50. Whitfield, J.F., MacManus, J.P., Boynton, A.L., Durkin, J., and Jones, A. Futures of calcium, calcium-binding proteins, cyclic AMP and protein kinase in the quest

for an understanding of cell proliferation and cancer. In: R.A. Corradino (ed.), Functional Regulation at the Cellular and Molecular Levels, pp. 61-87. New York: Elsevier North Holland, Inc., 1982.

51. Wolff, D.J., Ross, J.M., Thompson, P.N., and Brostrom, C.D. Interaction of calmodulin with histones. Alteration of histone dephosphorylation. J. Biol. Chem., 256: 1846-1860, 1981.

52. Wong, P.Y.-K., Lee, W.H., Chao, P.H.-W., and Cheung, W.-Y. The role of calmodulin in prostaglandin metabolism. Ann. N.Y. Acad. Sci., 356: 179-189, 1980.

THE SENSITIVITY OF HUMAN LYMPHOBLASTIC CELL LINES AND CFU-C TO 1,10-ORTHOPHENANTHROLINE

Pamela S. Cohen and Stephen D. Smith

Department of Pediatrics
Stanford University School of Medicine,
Stanford, California
Division of Hematology/Oncology
Children's Hospital at Stanford
Palo Alto, California.

INTRODUCTION

Impaired T-lymphocyte function is a well recognized sequela of clinical zinc (Zn) deficiency, seen in patients with acrodermatitis enteropathica (8) or prolonged hyperalimentation without Zn-supplementation (10). In humans and animal models, manifestations of the T-lymphocyte dysfunction include anergy, thymic atrophy, defective cell-mediated immunity and decreased or absent T-cell response to PHA-stimulation (1,5) with relative sparing of B-lymphocyte and granulocyte number and function (7,20). These abnormalities in Zn-deficient patients can be corrected by Zn supplementation.

1,10-Orthophenanthroline (OP) is a Zn-chelating agent which causes inhibition of DNA synthesis in human T-lymphocytes (19), the mouse lymphoblastic cell lines L1210 (2) and the human lymphoblastic cell line CCRF-CEM (4). These observations led us to study whether OP and another Zn-chelator, EDTA, might inhibit T-lymphoblast proliferation as measured by a colony-formation assay, and whether addition of Zn might reverse this effect.

This work was supported in part by grant CH-182A from the American Cancer Society. Pamela S. Cohen is a recipient of a clinical fellowship from the American Cancer Society. Stephen D. Smith is a Scholar of the Leukemia Society of America.

HUMAN TUMOR CLONING
ISBN 0-8089-1671-8

MATERIALS AND METHODS **Experimental Cells.** Three cell lines (K-B-2, K-T-1, SUP-T-1) established in our laboratory were used as the source of malignant lymphoid cells. The K-B-2 cell line was cultured from a child with Burkitt's lymhoma in relapse. This cell line possessed surface immunoglobulin (mu, kappa), lacked lambda light chain, lacked EBV-genome and did not rosette with sheep red blood cells (sRBC). These cells grew in agar (cloning efficiency, C.E., of 20%) and had an average doubling time of 29 hours. The K-T-1 cell line was cultured from a child with T-acute lymphoblastic leukemia (T-ALL) who had failed chemotherapy. These cells rosetted with sRBCs and reacted with anti-Leu 1 (Pan T-cell) and anti Leu 9 (Pan T-cell) monoclonal antibodies (Becton Dickinson, Mountain View, CA). The K-T-1 cells readily grew in agar with a C.E. of 12% and an average doubling time of 50 hours. The SUP-T-1 cell line was cultured from a child with T-non-Hodgkin's lymphoma (T-NHL) at relapse. These cells did not rosette with sRBCs but did react with anti-Leu 1, anti-Leu 9 and anti-Leu 6 (pan-thymocyte) monoclonal antibodies. These cells had a 5% C.E. on agar and had an average doubling time of 35 hours.

Human myeloid/monocytic colonies (CFU-C) were cultured from routine bone marrow aspirates from children with ALL OR NHL in remission. Marrow aspirates were performed at a time when the myelosuppressive effects of maintenance chemotherapy had resolved. Cell suspensions of marrow mononuclear cells were prepared by separation on a Ficoll-Hypaque gradient. The protocol procedures were approved by the Medical Committee for the Use of Human Subjects in Research, and informed consent was obtained.

Experimental Drugs. 1,10-Orthophenanthroline (Sigma Chemical, St. Louis, MO) was suspended in modified McCoy's 5A (11), warmed to $100^{o}C$ for five minutes, filtered and serially diluted in modified McCoy's 5A. CaEDTA (Sigma Chemical) was similarly prepared except for heating. The drugs were used immediately or refrigerated at $4^{o}C$ and used within four days. Stock solution of $ZnSO_4$ (Sigma Chemical) was prepared in sterile water and filtered.

***In vitro* Exposure of Cells to Drugs.** A modification of the semi-solid agar technique of Pike and Robinson (11) was used, employing a feeder layer and overlayer.

Complete medium (CM) contained modified McCoy's 5A with 15% newborn calf serum, penicillin (5 units/cc), streptomycin (0.5 μg/cc) and L-glutamine (40 μM/cc). Neither mitogens, thiols nor antifungal agents were added. In the Zn blockage experiments, stock solutions of Zn were added to CM prior to incubation with cells.

Feeder layer. Heparinized blood was sedimented at 20°C and the WBC-rich plasma (10%) was mixed with agar (0.5%) in CM, and plated onto 35 mm tissue culture dishes.

Overlayer-One Hour Drug Incubation. A 1:1 mix of experimental cells and drug solution, (or modified McCoy's 5A for control) was incubated for one hour at 37°C in 5% CO_2. K-T-1, K-B-2, SUP-T-1 and bone marrow mononuclear cells were incubated with drug at concentrations of 1 X 10^5, 1 X 10^4, 5 X 10^4 and 1 X 10^5 respectively. The drug (or McCoy's 5A in control plates) was removed by dilution with CM and then centrifuged X 500 g for 5 minutes at 37°C. Serial dilutions and centrifugations were repeated twice and the final calculated dilution of test drug was 1:10,000. Cells were then resuspended in an agar-CM mixture (final agar concentration 0.3%) and plated in 35 mm tissue culture dishes containing feeder layers. When incubating bone marrow mononuclear cells, 10% giant cell tumor-colony stimulating factor (GCT) (Gibco Chemical) was added to enhance CFU-C growth (3).

Overlayer-Two Week Incubation. Experimental cells were added to a mixture of CM and 0.3% agar to a final concentration of 1 X 10^5 cells/cc of bone marrow mononuclear cells, 1 X 10^5 cells/cc of K-T-1, 5 X 10^4 cells/cc of SUP-T-1, and 1 X 10^4 of K-B-2. Cells were plated over feeder layers. 50 μl of OP, CaEDTA or modified McCoy's 5A (control) in appropriate dilutions was added to the plated cells. Ten percent GCT was added to the media in experiments using bone marrow mononuclear cells to enhance CFU-C growth.

The Zn blockage experiments were performed in an identical manner except that the cells (SUP-T-1) were incubated in one of 3 different media: 1) standard complete medium (CM) with baseline concentratons of 35 μg/dl of Zn. 2) CM containing Zn 350 μg/dl or 3) CM containing Zn 700 μg/dl. OP was tested at concentrations of 0.6, 1.5 and 2.6 μg/cc cells plated, corresponding to the ID_{50}, ID_{70} and ID_{90} concentrations respectively.

Culture Conditions. Cells were incubated for 14 days at 37° C in 5% CO_2 and colonies which contained 40 or more cells were scored.

Statistical Analysis. One-hour and two-week incubation experiments were performed in duplicate and repeated five times. Zn blockade experiments were performed in duplicate and repeated three times. The results graphed on Figures 1, 2 and 3 were expressed as the percent of colony inhibition after drug treatment as compared to control.

The ID_{50} (dose which inhibited 50% of colony formation when compared to control) was extrapolated by plotting mean values for each dose response experiment as a linear regression. The regression lines of the two-week and one-hour incubation experiments were compared for significance by t-test, at the OP concentrations of 1 μg/cc and 537 μg/cc, respectively. These points were chosen for analysis because of their approximation to the CFU-C ID_{50}. In the Zn blockade experiments, mean values for each dose-response experiment were compared for significance by student t-test at the OP concentrations of 0.6, 1.5 and 2.6 μg/cc.

RESULTS

Figure 1 shows the dose-response curve obtained with CFU-C and the three cell lines after exposure to OP *in vitro* for one hour. The ID_{50}s for K-B-2, K-T-1, SUP-T-1 and CFU-C respectively were, 112, 417, 975 and 537 μg/cc. At 537 μg/cc of OP, only the B-cell line K-B-2 was different from CFU-C in

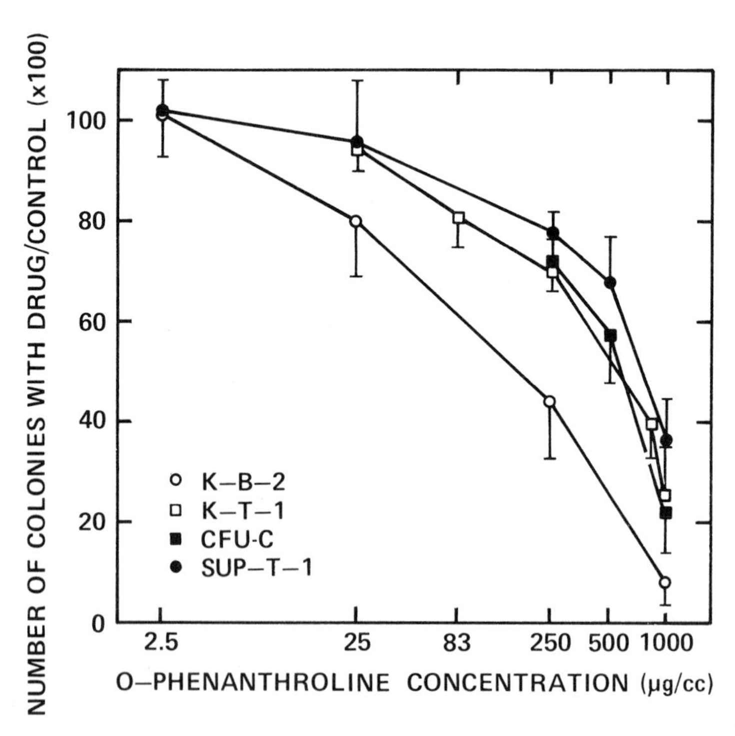

Figure 1. Dose-response curves of cell lines and CFU-C exposed to OP for one hour.
Points, means of quintuplicate assays ± SEM.

OP sensitivity (p=0.009); there was no difference between the sensitivity of the two T-cell lines and CFU-C to OP. In addition, at 537 μg/cc of OP, there was a difference between the sensitivity of the cell lines SUP-T-1 and K-B-2 (p=0.001) but not between K-T-1 and K-B-2 (p=0.08).

Figure 2 shows the dose-response curves obtained with CFU-C and the three cell lines after incubation with OP for two weeks. The ID_{50}s for K-T-1, SUP-T-1, K-B-2 and CFU-C respectively were 0.43, 0.69, 0.835 and 1.3 μg/cc. At 1 μg/cc, all three cell lines demonstrated significantly increased sensitivity to OP when compared to that of CFU-C. The order of OP sensitivity (at 1 μg/cc) was, from greatest to least sensitivity, K-T-1 > SU-T-1 > K-B-2 > CFU-C. Therefore, the T-cell lines were more sensitive than the B-cell line, which in turn was more sensitive than CFU-C at 1 μg/cc of OP after continuous two-week exposure. In addition, at this concentration there was significant difference in sensitivity between K-T-1 and K-B-2 (p=0.02) but not between SUP-T-1 and K-B-2 (p=0.4).

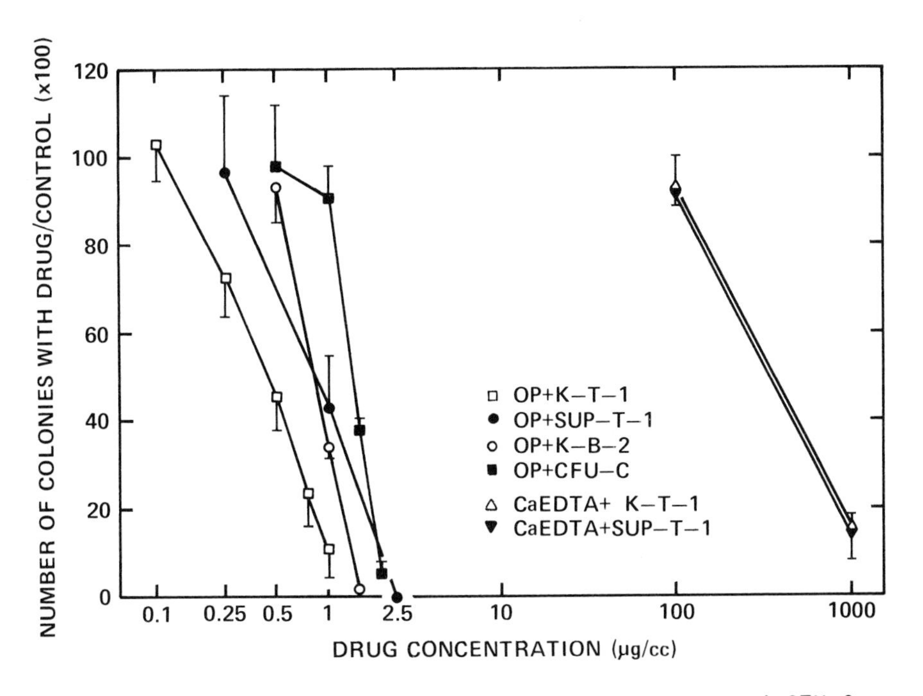

Figure 2. Dose-response curves of cell lines and CFU-C exposured to OP, or EDTA, continuously for 14 days.
Points, means of quntiplicate assay ± SEM.

Figure 2 also shows the dose-response curves obtained after incubating K-T-1 and SUP-T-1 with CaEDTA for two weeks. The ID_{50}s for K-T-1 and SUP-T-1 respectively were, 355 µg/cc and 398 µg/cc. There was no significant difference between them. There was, however, greater than a 400-fold difference between the ID_{50}s of the T-cell lines when exposed to CaEDTA and OP.

Figure 3 shows the dose-response curves obtained after incubating SUP-T-1 with 0.6, 1.5 or 2.5 µg/cc OP continuously for two weeks, in media with and without Zn supplementation. At 0.6 µg/cc OP, the inhibition of colony formation was completely blocked by zinc supplementation. However, at higher OP concentrations of 1.5 and 2.6 µg/cc, the inhibition of colony formation was only partially blocked by the addition of supplemental zinc (p 0.05) when cells treated with 1.5 µg/cc OP were grown in CM containing 700 µg/cc Zn.

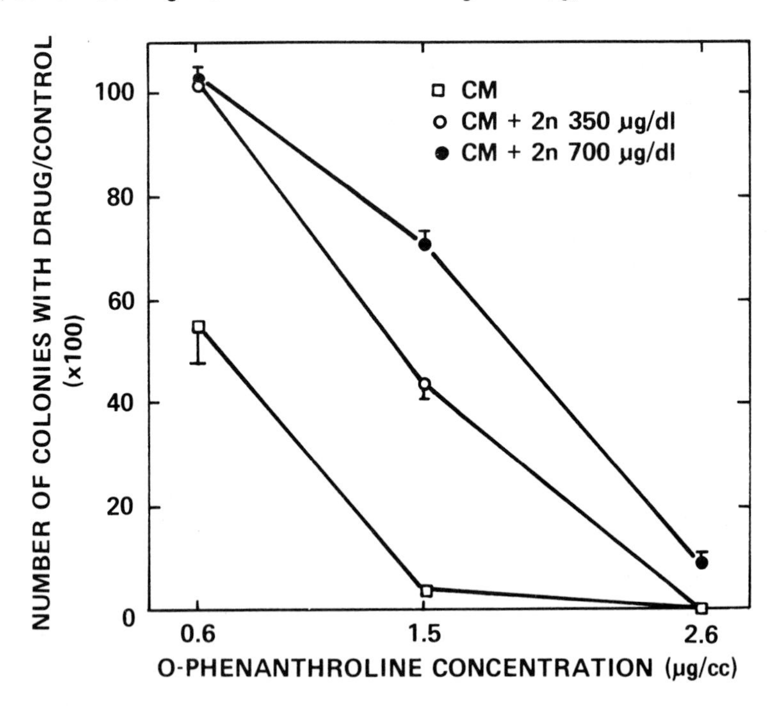

Figure 3. Dose-response curves of SUP-T-1 cells exposed to OP continuously for 14 days, while incubated in □ control media (CM) containing 35 µg/dl of Zn, ○ CM containing 350 µg/dl of Zn or ● CM containing 700 µg/dl of Zn.
Points, means of triplicate assays ± SEM.

DISCUSSION

These studies show that (1) O-phenanthroline inhibited the *in vitro* colony formation of the T- and B-lymphoblastic cell lines which varied depending on the concentration and duration of drug exposure, (2) CFU-C colony formation was similarly affected, (3) no consistent selective effect on T-lymphoblasts was observed, (4) after two weeks exposure, EDTA inhibited the colony-formation of the T-cell lines only at doses at least 400 fold greater than those of OP and, (5) the addition to the medium of Zn at a concentration ten fold higher than that of the control medium partially blocked the colony-inhibiting effect of OP.

There was a 150 to 500-fold difference in the ID_{50}s of short (one hour) versus long-term (two week) OP exposure, for each of the cell types tested. This indicates that the inhibitory effect of OP on colony-formation is a time-dependent phenomenon. By using DNA synthesis as another measure of cell proliferation, Berger et al (2) found that DNA synthesis was also inhibited by OP in a time-dependent manner. *In vitro* incubation of L1210 leukemia cells with OP at a concentration of 5 μg/cc inhibited DNA synthesis by 9% after one hour, but continued incubation resulted in 96% inhibition within 4 hours. Thus, one-hour drug exposure in our experiments might not be sufficient for maximal inhibition of DNA synthesis and subsequent colony formation.

There also appeared to be no consistent preferential effect of OP on the T-lymphoblast lines. The OP sensitivity of both T-cell lines was greater than that of CFU-C or the B-cell line after 14 day exposure at 1 μg/cc of OP. However, this same pattern of sensitivity was not observed after one hour of OP exposure. Thus, the selective effect on malignant T-lymphoblasts could be demonstrated only after prolonged exposure.

Differences in OP sensitivity were also observed between the cell lines, which varied dependent on concentration and duration of drug exposure. The cell lines were plated at varying concentrations to optimize colony counting, since the doubling times of the cell lines differed widely (see Methods). If OP sensitivity of the cells was a function of the plating concentrations (and therefore the amount of OP available per cell), OP sensitivity would parallel plating concentration i.e., K-B-2 SUP-T-1 K-T-1. This was not observed in either one hour or two week incubation experiments.

After two week incubation, there was a greater than 400-fold difference between the ID_{50}s of the two T-cell lines exposed to OP and EDTA. This indicates a relative resistance to EDTA by these cells. There are conflicting reports in the literature concerning the comparative abilities of EDTA and OP to inhibit various indices of cell proliferation. Zanzonico et al (20) studied the PHA and Con-A stimulation of mouse lymphocytes incubated with EDTA (as measured by tritiated thymidine uptake) and found almost complete inhibition of both at 0.1 mM EDTA. This concentration is approximately ten fold less than the observed ID_{50} in our experiments. Rubin (12) also found that EDTA at a concentration of 1 mM, inhibits DNA synthesis about ten-fold in cultures of chick embryo cells, as measured by tritiated thymidine uptake. Conversely, Krishnamurti (6) studied the effects of EDTA and OP on DNA synthesis of Ehrlich ascites cells as measured by tritiated thymidine uptake, and found that the rate of DNA synthesis of EDTA exposed cells (at 3.02 μM) is the same as that of untreated controls, whereas the rate of DNA synthesis of OP-exposed cells is significantly reduced (at 5 μM). Relatively little EDTA entered the cells, but a significant fraction of OP was taken up by the cells. This mechanism may be in operation in the lymphoblastic cells presented here as well.

Zinc can partially or completely block the effect of OP, if added to the medium at the onset of two-week drug exposure. This suggests but does not prove that the mechanism of the OP effect is due to its interaction with Zn. Simultaneous addition of Zn and OP to the culture medium may simply reduce the amount of unbound OP available to cells. Further elucidation of a Zn-OP interaction would require rescue of the OP effect with Zn once cells have been exposed to OP for sufficient time to induce colony inhibition. Moreover, the OP effect may be not due to its Zn chelating ability, but rather to chelation of other divalent cations such as iron or copper. Alternatively, the OP effect may be due to a mechanism entirely unrelated to chelation.

Current methods for the preclinical screening of antitumor agents used by the NCI have largely relied on _in vivo_ models (18). Because of the continuing need for better antitumor drugs, as well as the costly and time-consuming nature of _in vivo_ drug screening, there has been an increased interest in the use of _in vitro_ screening methods at the NCI (15). A frequently used _in vivo_ method, described by Salmon (13) is that of the colony formation assay, which utilizes a short, one-hour drug incubation period. _In vitro_ exposure for one hour may be too brief to reflect the effectiveness of certain cell-cycle specific drugs, e.g. antimetabolites, particularly if some cells are in a resting state, or in a non-sensitive phase of the cell cycle. Indeed, Park (9) demonstrated marked improvement in the ability of a leukemic colony-forming assay to predict clinical response to cytosine arabinoside after prolonged, as opposed to short *in vitro* exposure times. Moreover, the ID_{50}s after prolonged exposure to OP were well within range of currently effective drugs in Salmon's one-hour exposure assay, which ranges between 0.02 μg/cc for Adriamycin to 1.5 μg/cc for Bleomycin (14). The ID_{50}s after one hour OP exposure were not within this range.

In 1974, the National Cancer Institute screened OP for antitumor efficacy against L1210 leukemia *in vivo.* No antitumor effect was observed after giving up to 400 mg/kg/day twice daily intraperitoneally, for four days. However, the dosing schedule used did not take into account the time-dependent nature of the OP effect demonstrated here and elsewhere (2). Moreover, there are many clinically useful drugs which are not effective in the L1210 model, but are effective in other animal tumor models such as P388 leukemia, the current model used for preliminary drug screening. On the basis of the data described here, it would seem possible that OP could demonstrate antitumor effects *in vivo*, if tested by a continuous exposure dosing schedule, or if tested in tumor models other than L1210. These studies are currently underway in our laboratory. Should OP show efficacy in this setting, prolonged or continuous drug exposure should be considered routinely as an additional drug screening method which would help to identify drugs otherwise overlooked by current methods of *in vitro* and *in vivo* drug screening.

REFERENCES

1. Allen JI, Kay NE, McClain CJ: Severe zinc deficiency in humans: Association with a reversible T-lymphocyte dysfunction. Ann Int Med 95:1540-157, 1981.
2. Berger NA, Johnson ES, Skinner AM: OP inhibition of DNA synthesis in mammalian cells. Exp Cell Res 96:145-155, 1975.
3. Dispersio JF, Brennan JK, Lichtman ML, Speiser BL: Human cell lines that elaborate colony-stimulating activity for the marrow cells of man and other species. Blood 51:507-519, 1978.
4. Falchuk KH, Krishan A: 1,10-OP inhibition of lymphoblast cell cycle. Cancer Res 37:2050-2056, 1977.
5. Good RA, Fernandes G, West A: Nutrition, immunity and cancer--A review. Part 1: Influence of protein or protein-calorie malnutrition and zinc deficiency on immunity. Clin Bull 9:3-12, 1979.
6. Krishnamurti C, Saryan LA, Petering DH: Effects of EDTA and 1,10-OP on cell proliferation and DNA synthesis of Ehrlich ascites cells. Cancer Res 40:4092-4099, 1980.

7. Lueker RW, Simonel CE, Fraker PJ: The effect of restricted dietary intake on the antibody mediated response of the zinc-deficient A/J mouse. J Nutr 108:881-887, 1978.
8. Oleske JM, Westphal ML Shore S, et al: Zinc therapy of depressed cellular immunity in acrodermatitis enteropathica: It's correction. Am J Dis Child 133:915-918, 1979.
9. Park CH, Wiernik PH, Morrison FS, et al: Clinical correlations of leukemic clonogenic cell chemosensitivity assessed by in vitro continuous exposure to drugs. Cancer Res 43:2346-2349, 1983.
10. Pekarek RS, Sandstead HH, Jacob RA, et al: Abnormal cellular immune responses during acquired zinc deficiency. Am J Clin Nur 32:1466-1471, 1979.
11. Pike BL, Robinson WA: Human bone marrow colony growth in agar-gel. J Cell Physiol 76:77-84, 1970.
12. Rubin H. Inhibition of DNA synthesis in animal cells by ethylene diamine tetraacetate (EDTA) and its reversal by zinc. PNAS 69:712-716, 1972.
13. Salmon SE, Hamburger AW, Soehnlen BJ et al: Quantitation of differential sensitivity of human tumor stem cells in anticancer drugs. N Engl J Med 298:1321-1327, 1978.
14. Salmon SE (ed). Cloning of Human Tumor Stem Cells. Prog in Clin and Biol Res. Vol 48. 1980. Alan Liss, Inc., New York.
15. Shoemaker RH, Wolpert-DeFilippes MK and Venditti JM: Applications of a human tumor clonogenic assay to screening for new antitumor drugs. In: Proceedings of the 13th International Congress of Chemotherapy, Vienna, Austria. August 28-September 2, 1983. In press.
16. Smith SD, Trueworthy RC, Kisker SE, et al: In vitro sensitivity of normal granulocytic and lymphoma colonies to vinca alkaloids. Cancer 51:417-422, 1983.
17. Venditti JM: The National Cancer Institute Antitumor Drug Discovery Program. Current and future perspectives: A commentary. Cancer Treat Rep 67:767-772, 1983.
18. Weisenthal LM: In vitro assays in preclinical antineoplastic drug screening. Sem Oncol 8:362-376, 1981.
19. Williams RO and Loeb LA: Zinc requirements for DNA replication in stimulated human lymphocytes. J Cell Biol 58:594-601, 1973.
20. Zanzonico P, Fernandes G, Good RA: The differential sensitivity of T-cell and B-cell mitogenesis to in vitro zinc deficiency. Cell Immunol 60:203-211, 1981.

MITOMYCIN C PHARMACOLOGY IN VITRO

Robert T. Dorr
Research Assistant Professor of Medicine
Cancer Center Division
The University of Arizona
Tucson, Arizona

INTRODUCTION

Mitomycin C (MMC) is an antitumor antibiotic which has been in clinical use since the mid-1960s[6]. Despite clinical activity in adenocarcinomas of the breast, lung, stomach, and pancreas,[10] MMC must be conservatively dosed to avoid severe, cumulative bone marrow toxicity.[3] Two mechanisms of MMC cytotoxicity have been suggested for the drug. Iyer and Szybalski originally described alkylation of bacterial DNA by MMC,[14] whereas more recently, DNA strand scission via quinone redox cycling to free radicals has been suggested.[2] Knowledge of the precise mechanism of MMC cytotoxicity in human tumor cells might facilitate the identification of a specific antidote. This has already been accomplished with sodium thiosulfate ($Na_2S_2O_3$) for cisplatin,[12] and with N-acetylcysteine (NAC) for cyclophosphamide[4]. Thus the purpose of the following studies was to elucidate the mechanism of MMC cytotoxicity in mammalian tumor cells and to test putative pharmacologic antidotes in vitro.

METHODS

Four established human tumor cell lines were tested against MMC in vitro using the soft-agar cloning methods of Hamburger and Salmon.[11] The cells were exposed to MMC for one hour at 37° C using 0.1, 1.0, 10.0, and 20 μg/ml·hr^{-1} of drug. Cells were washed twice in complete media (plus 10% fetal calf

HUMAN TUMOR CLONING
ISBN 0-8089-1671-8

serum) prior to plating in 0.3% agar. To test the influence of exogenous sulfur nucleoplhiles on MMC cytotoxicity, either a 1-hr co-incubation or a 24 hr nucleophile pretreatment was used. The concentrations of each sulfur nucleophile were chosen to approximate the maximal molar ratios of MM:nucleophile achievable in vivo. This involved the use of up to 4,000μg/ml NAC[7] and 1,600 μg/ml of $Na_2S_2O_3$[10]. In addition to sulfur nucleophiles, the oxygen free radical scavengers, DMSO (10mM) and mannitol (10mM) were also evaluated in a 1-hr co-incubation experiment. Statistical tests for additive or inhibitory effects on the survival of tumor colony forming units was assessed by the method of Drewinko et al[8].

The influence of concomitant MMC exposure to microsomal enzymes in a reduced environment was also evaluated. For these experiments human tumor cells were exposed in vitro to MMC plus a 0.45 μM-filtered commercial S-9 preparation(Litton Bionetics, Charleston S.C., U.S.A). A freshly-reconstituted NADPH-generating system was also included.[16] Cells were washed and plated in the standard fashion following the 1-hr co-incubation.

The alkaline DNA elution assay of Kohn et al[15] was used to correlate the molecular effects of MMC-DNA interactions with in vitro clonogenic assay results. Mouse leukemia L-1210 cells pre-labeled with ^{14}C-thymidine were exposed to MMC for 1 hour and then post-irradiated on ice (6.0 Gy) prior to lysis and elution through 2.0 μM PVC filters for 16 hrs at pH 12.1. This facilitates the detection of DNA crosslinking by enhanced ^{14}C-DNA retention on the filters compared to the x-ray controls (no MMC). Assays performed with a proteinase K digestion step evaluated the presence of DNA-DNA crosslinks. Assays were also performed without irradiation to assess whether MMC caused DNA single strand breaks (detected as increased ^{14}C-DNA elution over controls). Crosslinking was quantitated using the criteria of Fornace and Kohn[9].

RESULTS

In the four human tumor cell lines, only MCF-7 breast adenocarcinoma was sensitive to MMC at 1/10 the peak achievable plasma MMC concentration of 1.0 μg/ml[18]. This follows the pharmacologic sensitivity criteria of Alberts et al.[1] With the other three human tumor cell lines, MMC exposures of 10µg/ml·hr were required to reduce colony formation to < 30 % of controls. These relatively resistant cell lines included HEC-1A endometrial cancer, WiDr colorectal cells and 8226 myeloma cells. In the HEC-1A cell line, concomitant 1-hr MMO exposure with NAC at concentrations from 10^{-3} mM to 1.0 mM did not reduce or enhance MMC cytotoxicity patterns. Using a 24 hr sulfhydryl pretreatment regimen, neither NAC (50mM) nor glutathione (5mM) altered MMC activity in this cell line. Similar results were obtained with the WiDr cells. These large sulfhydryl concentrations were not directly cytotoxic to the tumor cells as long as the pH was adjusted to neutrality. The effect of $Na_2S_2O_3$ on MMC activity was investigated in 8226 cells. As with NAC, $Na_2S_2O_3$ was not directly toxic to cells over the range of 3-633µM. However, in contrast to NAC, $Na_2S_2O_3$ significantly enhanced MMC cytotoxicty using 1 hr co-incubations at molar ratios of 10:1 ($Na_2S_2O_3$, Table 1-1).

TABLE 1-1
% Survival of 8226 Myeloma Tumor Colony Forming Units Treated with Mitomycin C and $Na_2S_2O_3$ (SD)

MMC ($\mu M \cdot hr^{-1}$)	MMC Control	$Na_2S_2O_3$	MMC Plus $Na_2S_2O_3$
0.3	105(10.7)	109 (1.7)	54 (16.7)*
3.0	99 (12.1)	107 (5.1)	43 (17.9)*
15.0	61 (11.1)	91 (15.2)	7.9 (8.1)*
30.0	5 (7.5)	104 (10.1)	1.1 (2.5)
60.0	< 0.1(0.2)	95 (7.9)	< 0.1 (0.1)

* indicates $p < .05$ compared to controls

With MCF-7 breast cancer cells MMC alone reduced colony survival to about 30% of control at 0.1 µg/ml·hr^{-1}. The addition of NAC (5.0mM) and especially glutathione (0.5mM) significantly reduced MMC cytotoxicity after a 1-hr co-incubation (Figure 1-1).

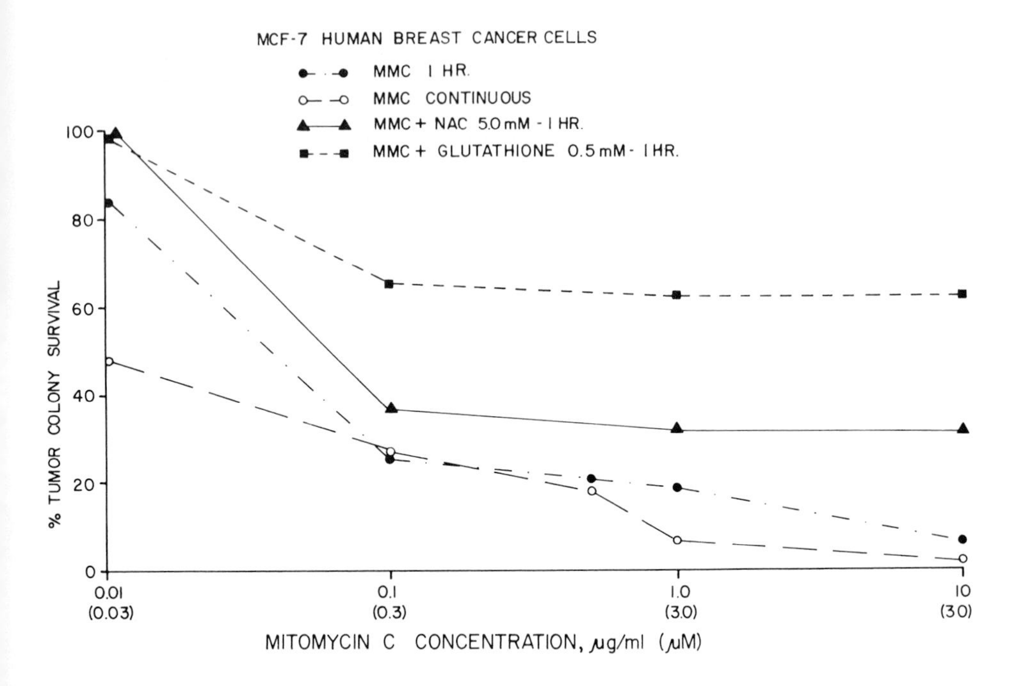

In this same cell line the co-incubation of MMC with the free radical scavengers DMSO (10mM) and mannitol (10mM) showed no inhibition or augmentation of MMC cytotoxicity. These results suggest that MMC cytotoxicity does not depend upon the generation of oxygen free radicals in vitro.

The question of a need for metabolic activation of MMC was also not known. Thus the effect of a 1 hr co-incubation of MMC with S-9 microsomal enzymes and NADPH was evaluated in human tumor cells in vitro. With HEC-1A cells there was no alteration in MMC cytotoxocity when combined with S-9, NADPH or both. If properly filtered the S-9 preparation was not directly cytotoxic, confirming the recent report of Metelmann and Von Hoff[17]. With 8226 cells the combination of MMC plus the S-9/NADPH system significantly reduced MMC activity (Table 1-2). In this case the S-9 preparation significantly reduced tumor colony survival possibly indicating inadequate S-9 filtration.

A final series of experiments used alkaline DNA elution of L-1210 murine leukemia cells to determine the type of DNA lesion responsible for MMC cytotoxicity. In this system MMC produced dose-dependent

TABLE 1-2

Survival of Clonogenic 8226 Myeloma Cells Exposed to Mitomycin C, NADPH, and S-9 Microsomal Enzymes

MMC Exposure ($\mu g/ml \cdot hr^{-1}$)	% Survival (SD)			
	MMC Alone	MMC plus S-9	MMC plus NADPH	MMC plus S-9/NADPH
None	100	44(17.2)*	105(25)	99(29)
0.1	109(25.9)	C	117(31)	104(36)
1.0	119(8.7)	20(4.9)+	90.9(21)	97.9(25)
10.0	0.52(1.2)	15.2(5.7)+	1.5(2.9)	66.3(22)*
20.0	0.79(2.4)	7.5(4.4)+	0.1(1.2)	1.5(4.1)*

C= Sample contaminated
* indicates $p < .05$ compared to control (MMC Alone)
+ Survival adjusted to the S-9 only control

DNA crosslinking following 1 hr MMC exposures in the range of the cytotoxic dose-response curve (Table 1-3). Similar assays performed without the post-irradiation step demonstrated no MMC-induced DNA single strand breaks using pharmacologic doses of MMC. Other assays performed with a proteinase K

TABLE 1-3

Dose-Response of Mitomycin C-Induced Total DNA Crosslinking

MMC Exposure ($\mu g/ml \cdot hr^{-1}$)	Post-Irradiation (6.0 Grays [Gy])	%Remaining ^{14}C-DNA at 12hr	Relative Elution	Cross-link Factor
None	-		-	-
None	+	17.9	0.726	-
20.0	+	89.9	0.025	-
0.1	+	16.8	0.753	0.96
0.5	+	19.3	0.693	1.05
1.0	+	34.4	0.442	1.64
5.0	+	64.6	0.168	4.32
10.0	+	71.0	0.127	5.71
20.0	+	79.1	0.081	8.96

digestion step to remove DNA-protein adducts demonstrated that roughly half of the total crosslinking was due to DNA-DNA crosslinks. Again there was a dynamic range for the DNA-DNA crosslinks which correlated with in vitro cytotoxicity determined using clonogenic assays.

CONCLUSIONS

These results show that MMC activity is limited in vitro as it is in vivo[5]. The sulfur nucleophiles NAC and $Na_2S_2O_3$ were not useful as antagonists to MMC-induced cytotoxicity in 3 of 4 human tumor cell lines. In contrast, in the most sensitive tumor (MCF-7 cells), there was some evidence of reduced MMC activity with glutathione and NAC. This suggests that MMC-sensitive cells may be amenable to pharmacologic antagonism by sulfur nucleophiles. However, the lack of any effects in 3 of the MMC-resistant tumor lines argues strongly against a generic approach to enhancing MMC efficacy using NAC or $Na_2S_2O_3$.

There was also no evidence that MMC requires metabolic activation to produce typical cytotoxicty patterns in vitro. Indeed the results in 8226 cells suggest that MMC-microsomal metabolism leads to inactive species. The DNA alkaline elution studies clearly showed a dose-effect correlation between DNA crosslinking and in vitro activity in the clonogenic assay. Thus there was no evidence of MMC-induced DNA strand scission by alkaline DNA elution and in tumor cells the free radical scavengers did not reduce MMC cytotoxicity.

In summary, MMC cytotoxicity cannot be consistently blocked using the clinically-compatible sulfur nucleophiles NAC and $Na_2S_2O_3$. There does not appear to be a requirement for metabolic activation of MMC. The molecular mechanism for MMC cytotoxicity appears to involve DNA crosslinking exclusively. In this regard, the clonogenic assay appears to have wide applicability to the in vitro evaluation of molecular pharmacologic mechanisms for new and old anticancer drugs.

REFERENCES

1. Alberts DS, Salmon SE, Chen H-SG:Pharmacologic studies of anticancer drugs with the human tumor stem cell assay. Cancer Chemother Pharmacol 6:

253-264, 1981.
2. Bachur NR, Gordon SL, Gee MV: A general mechanism for microsomal activation of quinone anticancer agents to free radicals. Cancer Res 38:1745-1750, 1978.
3. Baker LH, Vaitkevicius VK: Development of an acute intermittent schedule-mitomycin C. In Catre SK and Crooke ST (eds): Mitomycin C: Current Status and New Developments. New York, Academic Press, 1979, pp. 77-82.
4. Brock N, Pohl J, Stekar J: Studies on the urotoxicity of oxazophosphorine cytostatics and its prevention. Comparative study on the uroprotective efficacy of thiols and other sulfur compounds. Europ J Clin Oncol 17:1155-1163,1981.
5. Clark GM, Von Hoff DD: Activity of mitomycin C in a human cloning system in Ogawa M, Rozencweig M, Staquet MJ (eds): Mitomycin C: Current Impact on Cancer Chemotherapy. Amsterdam, Excerpta Medica, 1982, pp. 19-24.
6. Crooke ST, Bradner WT: Mitomycin C: A review. Cancer Treat Rev 3:121-129, 1976.
7. Doroshow JH, Locker CY, Ifrim I, Meyers CE: Prevention of doxorubicin cardiac toxicity in the mouse by N-acetylcysteine. J Clin Invest 65:128-135, 1981.
8. Drewinko B, Loo TL, Brown B, Gottlieb JA, Freireich EJ: Combination chemotherapy in vitro with adriamycin. Observations of additive, antagonistic, and synergistic effects when used in two-drug combinations on cultured human lymphoma cells. Cancer Biochem Biophys 1:187-195, 1976.
9. Fornace AJ, Kohn KW: DNA-protein crosslinking by ultraviolet radiation in normal human and xeroderma pigmentosum fibroblasts. Biochim Biophys ACTA 435;95-103, 1976.
10. Godfrey TE, WIlbur DW: Clinical experience with mitomycin C in large infrequent doses. Cancer 29; 1647-1652, 1972.
11. Hamburger AW, Salmon SE: Primary bioassay of human tumor stem cells. Science 197:461-463, 1977.
12. Howell SB, Pfeifle CL, Wung WE: Intraperitoneal cisplatin with systemic thiosulfate protection Ann Intern Med 97:845-851, 1982.
13. Howell SB, Taetle R: Effects of sodium thiosulfate on cis-dichlorodiammineplatinum (II) toxicity and antitumor activity in L-1210 leukemia Cancer Treat Rep 64:611-616, 1980.

14. Iyer VN, Szybalski W: Mitomycins and profiromycin Chemical mechanism of activation and crosslinking of DNA. Science 145:55-58, 1964.
15. Kohn KW, Erickson LC, Ewig RAG: Fractionation of DNA from mammalian cells by alkaline elution. Biochem 15:4629-4637, 1976.
16. Lieber MM, Ames MM, Powis G, Kovach JS: Anticancer drug testing in vitro: Use of an activating system with the human tumor cell assay. Life Sci 28: 287-293, 1981.
17. Metelman HR, Von Hoff DD: Application of a microsomal drug activation system in a human tumor cloning assay. Invest New Drug 1:27-32, 1983.
18. van Hazel G, Kovach JS: Pharmacokinetics of mitomycin C in rabbits and humans. Cancer Chemother Pharmacol 8:189-192, 1982.

Factors Determining Methotrexate Cytotoxicity in Human Bone Marrow Progenitor Cells: Implications for In Vitro Drug Testing of Human Tumors

GUNTER E. UMBACH, GARY SPITZER, JAFFER A. AJANI,
HOWARD THAMES, VERENA HUG, FRED RUDOLPH, BENJAMIN DREWINKO

Departments of Gynecology (G.E.U), Hematology (G.S.),
Medical Oncology (J.A.A., V.H.), Biomathematics (H.T.),
and Laboratory Medicine (B.D.)
The University of Texas
M. D. Anderson Hospital and Tumor Institute at Houston
Houston, Texas 77030
and
Department of Biochemistry (F.R.)
Rice University, Houston, Texas 77251

INTRODUCTION

The objective of the "human tumor stem cell assay" is to predict the sensitivity of an individual patient's tumor cells to specific anticancer drugs. The in vitro result is determined by the degree of inhibition of colony formation, which depends on the inherent chemosensitivity of the tumor cells, the drug concentration, and the duration of exposure to the drug. In addition, the in vitro cytotoxic activity of agents whose inhibitory effect is mediated by the inhibition of a critical metabolite also depends on the composition of the culture environment. If the culture environment supplies the critical metabolite, the tumor cells may appear resistant to the drug regardless of clinical responsiveness to the drug (3). At present, all drugs, including methotrexate, are tested under similar culture conditions by most tumor cloning laboratories.

This investigation was supported by NIH Grants CA 28153, CA 14528, CA 23272, the Leukemia Society of America, and the Max Kade Foundation.
HUMAN TUMOR CLONING
ISBN 0-8089-1671-8

The culture environment employed usually contains high concentrations of nucleosides. They are constituents of the culture medium (20) or originate from the undialyzed fetal bovine serum (FBS) used as medium supplement (2,10,26). It is well known that nucleosides are capable of protecting cells from methotrexate cytotoxicity (7,16). Since myelosuppression is one of the dose-limiting toxicities in the clinical use of methotrexate, we investigated the influence of drug concentration, exposure duration, and culture conditions on the cyto toxic activity of methotrexate in normal human bone marrow progenitor cells.

MATERIALS AND METHODS

Normal bone marrow cells were obtained by aspiration from 31 patients. Light-density mononuclear bone marrow cells were cultured in a modified bilayer soft agar system (6,25) containing alpha-MEM (K.C. Biological, Kansas City, Kansas), enriched with 10% undialyzed or dialyzed human placental conditioned medium as a source of colony-stimulating factor, and supplemented with 15% serum of different sources. The types of sera used were: undialyzed FBS, dialyzed FBS, and horse serum. Graded dilutions of methotrexate (Lederle Parenterals, Carolina, Puerto Rico) were added to the liquid suspension culture for pulse treatment (pre-incubation) or incorporated in the soft agar system (continuous exposure). Cells that had undergone pulse treatment with methotrexate were centrifuged, resuspended, and plated. In every experiment, duplicate or triplicate cultures were processed for each drug concentration and treatment interval. GM-CFUC (granulocyte-macrophage colony-forming units in culture) formed colonies after 8 to 10 days. Only cultures with 30 colonies (aggregates of at least 40 cells) or more were included in the evaluation of drug cytotoxicity. Survival fractions of GM-CFUC were calculated as the mean number of colonies in experimental dishes divided by the mean number of colonies in concomitant non-drug-treated controls x 100%.

RESULTS

A 1-hour or 2-hour pulse-treatment with 2×10^{-4} M methotrexate in serum-free, nucleoside-free medium failed to produce significant cytotoxic activity on GM-CFUC, killing only 18% (2 SE $\pm$30) and 7% (2 SE $\pm$18) of GM-CFUC, respectively.

We explored the possibility of evaluating cell responsiveness to methotrexate by prolonged pulse treatment in nucleoside-free, serum-free media. However, the cloning efficiencies of untreated GM-CFUC after 24 hours and 48 hours of preincubation in liquid suspension culture were only one third and one fifth of the cloning efficiency observed after a 1-hour pulse treatment, respectively. We therefore judged this avenue to be inadequate due to the loss of most control GM-CFUC.

We investigated serum requirements that would allow the manifestation of methotrexate cytotoxicity at continuous contact of GM-CFUC and drug. The following sera were used: undialyzed FBS, dialyzed FBS, and undialyzed horse serum. Nucleoside concentrations in these sera as determined by high performance liquid chromatography (1) are shown in Table 1. Guanosine, deoxyguanosine, inosine, adenosine, and deoxyadenosine were below detection thresholds (0.1 µM) in all samples. Dialysis of FBS effectively reduced the hypoxanthine concentration from 240 µM to 0.26 µM and the thymidine concentration from 7.4 µM to below the sensitivity limit. There was no major difference in autoclaved versus nonautoclaved dialyzed FBS. Nucleoside concentrations in the horse serum were very low and comparable to those in the dialyzed FBS.

Folic acid concentrations (Table 1) as determined by radioassay (18) were low in all types of sera used compared to the high concentrations present in standard culture media.

TABLE 1: Concentration of Nucleosides and Folic Acid in Sera and Selected Culture Media

	Hypoxanthine (µM)	Thymidine (µM)	Folic acid (ng/ml)
Undialyzed FBS	240	7.4	3.2
Dialyzed FBS	0.26	<0.1	2.0
Horse serum	<5*	<0.1	2.0
Dialyzed FBS after autoclaving	1.2	<0.1	not tested
Medium: Alpha-MEM	0	0	10^6
Medium: CMRL 1066**	0	40	10^4

* Co-eluting substances interfered with more precise measurements.

** In addition, contains 3 other nucleosides.

Data for culture media from manufacturer.

TABLE 2: Effect of Type of Serum Supplementation on Methotrexate Cytotoxicity in Human Granulocyte-Macrophage Colony-Forming Units in Culture (GM-CFUC) at Continuous Exposure. (LD_{50} = Methotrexate Concentration Lethal to 50% of GM-CFUC)

	Type of Serum	LD_{50}
A)	15% Undialyzed FBS	$>10^{-4}$ M
B)	15% Dialyzed FBS	$10^{-6.1}$ M
C)	15% Undialyzed horse serum	$10^{-8.0}$ M

All differences in LD_{50} were significant ($p<0.05$).

The effects of various types of serum supplementation on the cytotoxic activity of methotrexate for continuous exposure are shown in Table 2. There is a greater than 100-fold difference in the LD_{50} of methotrexate on GM-CFUC in (nucleoside-rich) undialyzed versus (nucleoside-depleted) dialyzed FBS. Interestingly, there is still a 100-fold difference in the LD_{50} of methotrexate on GM-CFUC in dialyzed FBS versus horse serum despite similar nucleoside concentrations.

The effect of undialyzed FBS was further investigated by adding graded concentrations to cultures supplemented with dialyzed FBS at a methotrexate concentration of 10^{-4} M. As little as 0.3% and 3% of undialyzed FBS protected 20% and 84% of GM-CFUC, respectively, from methotrexate cytotoxicity; 10% of undialyzed FBS rendered methotrexate totally ineffective.

When adding undialyzed FBS to cultures supplemented with horse serum (methotrexate concentration 10^{-7} M), we found that 10% of undialyzed FBS were able to protect all GM-CFUC from methotrexate cytotoxicity. However, the addition of identical amounts of dialyzed FBS had no significant protective effect on GM-CFUC (data not shown).

DISCUSSION

The absence of methotrexate cytotoxicity at 1-hour and 2-hour pulse-treatment can be explained by the rescue effect of nucleosides (14,17) in the undialyzed FBS-containing media, in which the cells were subsequently plated.

The more than 100-fold difference in the LD_{50} of methotrexate on GM-CFUC in undialyzed versus dialyzed FBS can be explained by the difference in nucleoside concentrations in undialyzed versus dialyzed FBS. This agrees with studies showing that the addition of nucleosides reverses methotrexate cytotoxicity (7,16) and furthermore restores methotrexate-suppressed immune response occurring in dialyzed FBS (5).

However, the 100-fold difference in LD_{50} of methotrexate for GM-CFUC in dialyzed FBS versus horse serum cannot be explained by differences in nucleoside concentration, because both types of sera showed negligible nucleoside concentrations as determined by high performance liquid chromatography. Since this method detects only free nucleosides, we explored the possibility of protein-bound nucleosides in the FBS (Joseph R. Bertino, personal communication) which, unlike free nucleosides, would not have been removed by dialysis. Since dialyzed FBS had the same nucleoside concentrations before and after autoclaving-induced denaturation of serum proteins, we excluded this hypothesis.

Exogenous folic acid cannot account for the described differences (19) since folic acid concentrations were similar in all types of sera used and represent only a fraction of the folic acid present in standard culture media.

The above results and the fact that dialyzed FBS offered no protection of GM-CFUC against methotrexate cytotoxicity in the presence of horse serum are consistent with the following two hypotheses: (1) FBS contains a nondialyzable (or partially dialyzable) non-nucleoside substance capable of protecting human GM-CFUC against methotrexate cytotoxicity; this substance may be inhibited by horse serum. (2) Horse serum sensitizes GM-CFUC to methotrexate by a mechanism other than lack of nucleosides; this mechanism can be overcome by a dialyzable factor in the FBS. The capacity of FBS and horse serum to modulate methotrexate cytotoxicity by factors unrelated to nucleosides and folates does not fit current concepts of methotrexate action but is supported by our data.

We did not investigate how culture conditions affect the methotrexate cytotoxicity in fresh human tumor cells. Other investigators reported that 12% (Vernon K. Sondak and David H. Kern, personal communication) to 15% (Daniel D. Von Hoff, personal communication) and 17% (9) of tested human tumor specimens showed in vitro sensitivity to methotrexate in nucleoside-rich culture environments at drug concentrations ranging from 4×10^{-7} M to 8×10^{-6} M. These findings are not in concordance with our observations in GM-CFUC that failed to show sensitivity to methotrexate at concentrations that were 13 to 250 times higher than the concentrations employed by these investigators for fresh human tumor cells. This wide gap in methotrexate sensitivity as observed in the culture dish does not correlate with clinical experience, where tumoricidal doses of methotrexate are usually associated with myelosuppression.

Selective in vitro toxicity of methotrexate has been reported for certain types of hematologic malignancies. Taetle et al. (23) reported that chronic lymphocytic leukemia colony-forming cells could not be rescued after methotrexate exposure when cultured in nucleoside-rich media, as opposed to normal lymphocyte colony-forming cells that could be rescued. Other investigators (11), however, found no difference in the sensitivity of normal versus malignant hematologic cells to a variety of cytotoxic drugs.

Malignant cells from a variety of cell lines were furthermore not affected by methotrexate in nucleoside-rich culture environments (15,27). No major difference in the inhibition of pyrimidine and purine synthesis has been noted in cells from normal bone marrow versus malignant cell lines (7).

In view of the complex factors influencing methotrexate cytotoxicity in vitro we do not know to which extent the reported in vitro results of tumor specimens for methotrexate actually reflect the in vivo responsiveness of tumors to this agent. Individual in vitro/in vivo correlations are clearly necessary.

We agree with Alberts et al. (3) and Rupniak et al. (19) that current assays using undialyzed FBS or nucleoside-rich culture media are not optimal for evaluating the sensitivity of tumor cells to methotrexate. Since bone marrow progenitor cells are one of the main host tissues affected by the clinical use of methotrexate and since they can be cultured under conditions similar to those used for tumor cells, they appear as a logical frame of reference for selecting appropriate in vitro drug testing conditions for hematologic (22) and solid neoplasms (4,8). This approach may reduce the risk of false in-vitro-sensitive results originating from testing at too high drug doses and of false in-vitro-resistant results from testing at too low doses (8,24). When the chemosensitivity of tumor cells is tested at only one in vitro drug concentration, we suggest continuous exposure to 10^{-8} M (0.005 μg/ml) methotrexate in nucleoside-free media supplemented exclusively with horse serum. These drug testing conditions correspond to the LD_{50} for human bone marrow progenitor cells and lie within the chemotherapy-associated GM-CFUC depression reportedly ranging from approximately 30% to 85% for selected cytotoxic regimens (12,13). Such a modified assay might more effectively test the sensitivity of tumor cells to methotrexate.

Our study reemphasizes the importance of adequate drug concentrations, exposure durations, and culture environments for the successful evaluation of anticancer agents in the stem cell assay (3,19,21,26).

REFERENCES

1. Agarwal RP: Simple and rapid high performance liquid chromatographic method for analysis of nucleosides in biologic fluids. J Chromatogr 231:418-424, 1982
2. Agrez MV, Kovach JS, Beart RW et al.: Human colorectal carcinoma: Patterns of sensitivity to chemotherapeutic agents in the human tumor stem cell assay. J Surg Oncol 20:187-191, 1982
3. Alberts DS, Chen HSG, Salmon SE: In vitro drug assay: Pharmacologic considerations, in Salmon SE (ed): Cloning of Human Tumor Stem Cells. New York, Alan R Liss, 1980, pp 197-207
4. Armstrong RD, Cadman E: 5'-Deoxy-5-fluorouridine selective toxicity for human tumor cells compared to bone marrow. Cancer Res 43:2525-2528, 1983
5. Bogyo D, Mihich E: Reversal of the in vitro methotrexate suppression of cell-mediated immune response by folinic acid and thymidine plus hypoxanthine. Cancer Res 40:650-654, 1979
6. Bradley TR, Metcalf D: The growth of mouse bone marrow cells in vitro. Aust J Exp Biol Med Sci 44:287-300, 1966
7. Howell SB, Mansfield SJ, Taetle R: Thymidine and hypoxanthine requirements of normal and malignant human cells for protection against methotrexate cytotoxicity. Cancer Res 41:945-950, 1981
8. Hug V, Thames H, Spitzer G, et al.: Normalization of in vitro sensitivity testing of human tumor clonogenic cells. Cancer Res (in print, 1984)
9. Jones SE, Salmon SE, Dean JC: In vitro cloning of human breast cancer. In Hofmann, Berens, Martz (eds), Proceedings of the International Conference of Predictive Drug Testing on Human Tumor Cells, Zurich, 1983, Abstract 4
10. Kern DH, Bertelsen CA, Mann BD, et al.: Clinical application of the colonogenic assay. Ann Clin Lab Sci 13:10-15, 1983
11. Kirshner J, Preisler HD: In vitro drug sensitivity studies of colony-forming units in culture in chronic myelocytic leukemia: Lack of specificity between chronic-phase patients and normal donors. Cancer Res 43:6094-6095, 1983
12. Lohrmann HP, Lepp KA, Schreml W: 5-Fluorouracil, 1,3-bis (2-chloroethyl)-1-nitrosurea, and 1-(2-chloroethyl)-3-(4-methylcyclohexyl)-1-nitrosurea: Effect on the human granulopoietic system. JNCI 68:541-547, 1982
13. Lohrmann HP, Schroml W, Lang M, et al.: Changes of granulopoiesis during and after adjuvant chemotherapy of breast cancer. Brit J Haematol 40:369-381, 1978
14. Nederbragt H, Uitendahl MP, van der Grint L, et al.: Re-

versal of methotrexate inhibition of colony growth of L1210 leukemia cells in semisolid medium. Cancer Res 41:1193-1198, 1981
15. Neuman D, Tisman G, Dine M: Attempts to quality control and standardize the human tumor stem cell assay (HTSCA). Proc Amer Soc Clin Oncol 1:31 (Abstract C-123), 1982
16. Pinedo HM, Chabner BA: Role of drug concentration, duration of exposure and endogenous metabolites in determining methotrexate cytotoxicity. Cancer Treat Rep 61:709-715, 1977
17. Pinedo HM, Zaharko DS, Bull J, et al.: The reversal of methotrexate cytotoxicity to mouse bone marrow cells by leucovorin and nucleosides. Cancer Res 36:4418-4424, 1976
18. Rothenburg SP, Da Costa M, Rosenberg Z: A radioassay for serum folate: use of a two-phase sequential-incubation ligand-binding system. N Engl J Med 286:1335-1339, 1972
19. Rupniak T, Whelan RDH, Hill BT: Concentration and time-dependent inter-relationships for antitumour drug cytotoxicities against tumour cells in vitro. Int J Cancer 32:7-12, 1983
20. Salmon SE, Hamburger AW, Soehnlen B, et al.: Quantitation of differential sensitivity of human tumor stem cells to anticancer drugs. N Engl J Med 298:1321-1327, 1978
21. Selby P, Buick RN, Tannock I: A critical appraisal of the "human tumor stem cell assay". N Engl J Med 308:129-134, 1983
22. Spiro TE, Mattelaer MA, Efira A, et al.: Sensitivity of myeloid progenitor cells in healthy subjects and patients with chronic myeloid leukemia to chemotherapeutic agents. JNCI 66:1053-1059, 1981
23. Taetle R, Dong T, Mendelsohn J: In vitro sensitivity to steroid hormones and cytotoxic agents of normal and malignant lymphocyte colony-forming cells. Cancer Res 43:3553-3558, 1983
24. Tisman G, Herbert V, Edlis H: Determination of therapeutic index of drugs by in vitro sensitivity tests using host and tumor cell suspensions. Cancer Chemother Rep 57:11-19, 1973
25. Verma DS, Spitzer G, Beran M, et al.: Colony stimulating factor - Augmentation in human placental conditioned medium. Exp Hematol 8:917-923, 1980
26. Von Hoff DD, Clark GM, Stogdill BJ, et al.: Prospective clinical trial of a human tumor cloning system. Cancer Res 43:1926-1931, 1983
27. Williams CK, Ohnuma T, Holland JF: Inhibition by chemotherapeutic agents of human bone marrow progenitor cells and clonogenic cells of a lymphoblastic cell line. Europ J Cancer 17:519-526, 1981

A NEW BIOASSAY FOR IN VITRO DRUG STABILITY

S.U. HILDEBRAND-ZANKI

The Division of Surgical Oncology
John Wayne Clinic, Jonsson Comprehensive Cancer Center,
UCLA School of Medicine, Los Angeles, CA 90024

D.H. KERN

Surgical Service
Veterans Administration Medical Center
Sepulveda, CA 91343

INTRODUCTION

The accuracy of predicting tumor sensitivity *in vitro* ranges from 60% (Von Hoff et al., 1983) to 86% (Kern et al., 1983). Prediction of resistance is somewhat better and varies between 85% (Von Hoff et al., 1983) and 99% (Alberts et al., 1980). The relatively low accuracy of predicting sensitivity is likely due in part to the difficulty in adjusting the drug doses to reflect *in vivo* conditions. Determination of drug concentrations used in chemosensitivity testing has been based on *in vivo* plasma half-lives and concentration x time products (CXT) (Alberts et al., 1980). However, little data are currently available that show if drugs are stable under *in vitro* assay conditions. This lack of knowledge has made it difficult to determine what mode of drug exposure simulates the clinical situation. One hour exposure times used in some laboratories (Salmon et al., 1978; Von Hoff et al., 1981) can lead to falsely high resis-

Supported by VA Medical Research Service and by Grant CA29605 of the National Cancer Institute.

HUMAN TUMOR CLONING
ISBN 0-8089-1671-8

tance rates for cell cycle specific drugs. On the other hand, continuous exposure (Bertelsen et al., 1984) may give high CXT's and falsely high sensitivity rates. By determining the pharmacokinetics of anti-cancer drugs *in vitro* it may be possible to increase the predictive accuracy of the assay and thus its clinical usefulness.

The little *in vitro* data currently available deal almost exclusively with chemical stabilities as determined by the use of HPLC (Keller and Ensminger, 1982; Poochikian et al., 1981). However, the results from these tests may not necessarily reflect the behavior of a drug under the conditions of the clonogenic assay. Interaction of a drug with the complex culture environment may alter its pharmacokinetics as well as its stability. Therefore, we set out to develop a simple, rapid and sensitive bioassay to determine the half-lives for drugs when used in the clonogenic assay.

MATERIALS AND METHODS

Assay system

Cells were cultured on an underlayer of 0.5% agar which was prepared by mixing 3.5 ml molten 3% agar with 16.5 ml CEM (MA Bioproducts, Walkersville, MD) supplemented with 15% heat-inactivated FCS (Flow Laboratories, McLean, VA), 100 U/ml penicillin (Grand Island Biological Company, Grand Island, NY), 100 µg/ml streptomycin (Grand Island) and 50 µg/ml Fungizone (Grand Island). One ml of this mixture was added to each 35 x 10 mm well of a six-well plate (Linbro 76-247-05, Flow Laboratories). Cells were suspended in 0.6% agar in supplemented CEM and 1 ml of the cell suspension was added to each of 3 hardened underlayers at a final concentration of 3.3×10^4 cells/well. Plates were incubated at 37° in the presence of 6% CO_2. After 48 hours 5 µCi of tritiated thymidine (specific activity 2.0 Ci/mM, New England Nuclear, Boston, MA) were layered over each well and the plates returned to the incubator for an additional 24 hours. Incorporation of thymidine was terminated by transferring the agar layers to 15 ml centrifuge tubes (No. C3051-870, Beral Scientific, Arleta, CA), adjusting the volume to 13 ml with phosphate-buffered saline (PBS) (Grand Island) and boiling the tubes for 40 minutes in a waterbath. The tubes were centrifuged, the pellet washed with PBS and dissolved in 3 ml of 0.85N KOH for 1 hour at 80°. Tubes were cooled to below 4° and the hydrolysate precipitated through the addition of 30 µl of 1% human serum albumin (Cutter Biological, Berkeley, CA) and 2.4 ml of ice cold 30% trichloroacetic acid (TCA). Precipitation was allowed to continue overnight at 4° and then the precipitate was collected by centrifugation. The pellet was washed with 5% TCA, dissolved in 0.3 ml of 0.075N KOH and transferred to liquid scintillation vials containing 5 ml of Liquiscint (National Diagnostics, Somerville, NJ). Radioactivity in each vial was measured in a Beckman liquid scintillation counter (LS230, Beckman Instruments, Irvine, CA). Sodium azide added at a final concentration of 4 x 10^{3} µg/ml to 3 control wells served as the positive control. An assay was considered valid if the average count of the untreated controls was greater than 300 counts per minute and the sodium azide control showed at least 80% inhibition of thymidine uptake when compared to the untreated controls.

Preparation of drugs

Drug solutions were prepared from standard formulations for the following drugs: Adriamycin (Adria), Actinomycin D (Act D), Bleomycin (Bleo), BCNU, Cisplatin (c-Plat), Dacarbazine (DTIC), 5-Fluorouracil (5FU), Melphalan (L-Pam), Mitomycin C (Mit C), Vinblastine (VBL) and Vincristine (VCR). Five experimental drugs were supplied by the Investigational Drug Branch, National Cancer Institute, Bethesda, MD in 50 mg quantities as lyophilized powders. PCNU (NSC 95466), DHAD (NSC 301739), CBDCA (NSC 241240), AZQ (NSC 182986), and PALA (NSC 224131). Drugs supplied by the NCI were reconstituted with PBS to the appropriate concentrations with the exception of AZQ and PCNU wich were dissolved in dimethyl sulfoxide (DMSO) and further diluted with PBS so that the final DMSO concentration did not exceed 3%. The reconstituted drugs were stored in 0.5 ml aliquots at -196°.

Cell lines

Three human melanoma cell lines, UCLA/SOM14 (M14), UCLA/SOM19 (M19) and UCLA/SOM29 (M29), were used to determine *in vitro* drug stabilities for 11 standard and 5 experimental anti-cancer drugs. To determine the half-lives for the 16 drugs chosen, we first tested them against each cell line to find the most sensitive line for each drug. M19 was sensitive to 14 of the 16 drugs, the exceptions being DTIC and PALA. PALA, however, was active against M14 and DTIC showed greatest activity against M26. Due to phenotypic changes and variation in drug susceptibility with increasing passage numbers (Kern et al., 1982), we chose to use frozen aliquots of identical passage to ensure cell stability. For each cell line 50 vials (5 x 10^6 cells/vial) of cells of the same passage were grown and frozen so that all subsequent tests were performed with identical cells.

Drug stability at 37°

At pre-determined times, drug aliquots were thawed and added to the underlayers of 3 wells and preincubated at 37° in a 6% CO_2 atmosphere. Preincubation times were 1,2,3,4,5 and 6 hours and 1,2,4,7 and 14 days. Cells mixed with agar solution were then added to each drug well. The plates were returned to the incubator and assayed as described above.

Determination of drug half-lives

Drug half-lives were determined from two graphs: a plot of percent survival versus drug concentration and a plot of percent survival versus drug preincubation time. From these two graphs a third graph was constructed relating residual drug concentration to the drug preincubation time. Half-lives were determined from this graph as the time required for the drug concentration to decrease by a factor of 0.5.

RESULTS

Dose-response curves were constructed for each drug. For further testing a drug concentration was chosen that reduced survival by 50% to 90% and was on that part of the dose-response curve where changes in survival were maximally sensitive to changes in concentration. Drug concentrations selected are listed in Table 1.

Drug stability under assay conditions was measured by adding the drugs to prepared underlayers and incubating them for up to 14 days before the addition of sensitive cells. Seven of the standard agents (Act D, Bleo, DTIC, 5FU, Mit C, VBL and VCR) were stable for the entire preincubation period and gave $t_{1/2}$ > 14 days. Adria, c-Plat, L-Pam and BCNU were unstable with $t_{1/2}$ of 29, 18.5, 1.8, and 1.0 hours respectively. Two of the experimental drugs, DHAD and PALA had $t_{1/2}$ > 14 days while CBDCA, AZQ and PCNU were unstable with $t_{1/2}$ of 94, 72 and 1 hours respectively. A compilation of $t_{1/2}$ is shown in Table 2.

TABLE I
In Vitro Concentrations of Anti-Cancer Drugs Used For Drug Stability Testing

Standard Drugs	Concentration In μg/ml	Experimental Drugs	Concentration In μg/ml
Act D	0.2	AZQ	10.0
Adria	0.4	CBDCA	400.0
BCNU	2.0	DHAD	10.0
Bleo	2.0	PALA	100.0
c-Plat	3.0	PCNU	400.0
DTIC	10.0		
5FU	10.0		
L-Pam	5.0		
Mit C	3.0		
VBL	5.0		
VCR	0.5		

TABLE II
In Vitro Stability Of Drugs at 37°

Standard Drugs	In Vitro Half-Life	Experimental Drugs	In Vitro Half-Life
Act D	>14 days	DHAD	>14 days
Bleo	>14 days	PALA	>14 days
DTIC	>14 days		
5FU	>14 days	CBDCA	94 hours
Mit C	>14 days	AZQ	72 hours
VBL	>14 days	PCNU	<1 hour
VCR	>14 days		
Adria	29.0 hours		
c-Plat	18.5 hours		
L-Pam	1.8 hours		
BCNU	1.0 hour		

DISCUSSION

The *in vitro* stability of any drug used in the human tumor cloning assay (HTCA) should be established before conclusions as to its efficacy can be drawn. To determine the *in vitro* stability of anti-cancer drugs, we followed these steps. First we had to find a cell line that was sensitive to the drug being tested. The next step was to construct a dose response curve in order to find the appropriate concentration at which the drug should be tested. Once the appropriate concentration was determined, we preincubated the drug on prepared underlayers for up to 14 days before the addition of cells. The great advantage of this assay is that the drug stabilities are tested in the same system subsequently used for the testing of drug chemosensitivities. It is important that culture conditions be simulated in order to take into account drug binding to the agar or serum proteins which may render the agent unavailable for uptake by the cells. Cases in point are c-Plat and CBDCA. When comparing the parent compound with its analogue one has to consider that c-Plat irreversibly binds to plasma protein (Repta and Long, 1980) to a much greater extent than does CBDCA (Curt et al., 1983), so that a substantial portion of the drug is lost to the system.

The assay we developed is rapid for obtaining drug stability data. Results are available in 5 days after plating and thus no long delays in drug testing are incurred. Inhibition of thymidine incorporation has been shown to be an accurate, reproducible and highly sensitive method for determining chemosensitivity.

Pharmacokinetic factors must be taken into account in any *in vitro* chemosensitivity assay. One of the most important parameters is the CXT relating drug concentration to exposure time. The higher the CXT for a given drug *in vitro*, the greater is the probability of cell kill. Knowledge of the clinically achievable CXT as well as a drug's terminal phase plasma half-life has been used to determine drug concentrations *in vitro* (Alberts, 1980). Usually, drug concentrations selected for *in vitro* testing reflect readily achievable plasma levels (Kern et al., 1983; Mann et al., 1982) or some fraction (e.g., one-tenth) of the peak plasma levels (Moon et al., 1981; Salmon et al., 1978; Von Hoff et al., 1983). However, optimal *in vitro* concentrations and exposure times are unknown. One-hour incubation of cells and drugs before plating may not be adequate for drugs that act during a specific phase of the cell cycle (e.g, 5-fluorouracil). On the ohter hand, long-term exposures may present unrealistically high CXT's for other drugs. Some anti-cancer agents, such as BCNU and L-Pam, are very unstable, and it is impossible to calculate their CXT's without accurate *in vitro* stability data.

Knowledge of *in vitro* pharmacokinetics is especially important for drugs that are in Phase I or Phase II trials and for which little or no *in vivo* data are available. *In vitro* dose-response curves, times of exposure and half-lives must be determined in order to relate *in vitro* cytotoxicity data to clinical dosage and scheduling. Such data are also important for drug screening, such as being conducted by the National Cancer Institute. All drugs are tested initially at a single concentration of 10 μg/ml in continuous exposure. However, direct comparison of drug cytotoxicity is difficult without knowledge of *in vitro* drug half-lives. Even though the *in vitro* concentration of two drugs is the same, their CXT's can vary widely.

With an increase in our knowledge of the *in vitro* pharmacokinetics of anti-cancer drugs, the dependability with which chemosensitivity testing can predict clinical response will also surely increase.

REFERENCES

Alberts, D.S., Chen, H.S.G., Soehnlen, B., et al. In Vitro Clonogenic Assay for predicting response of ovarian cancer to chemotherapy. Lancet 2:340-342, 1980.

Alberts, D.S., Chen, H.S.G., Salmon, S.E. In Vitro Drug Assay: Pharmacological considerations. In S.E. Salmon (Ed.), *Cloning of Human Tumor Stem Cells*. New York: Alan R. Liss, 1980, 197-207.

Bertelsen, C.A., Sondak, V.K., Mann, B.D., et al. Chemosensitivity testing of human solid tumors. Cancer, 1984, in press.

Curt, A.C., Grygiel, J.J., Corden, B.J., et al. A Phase I and pharmacokinetic study of diamminecyclobutane-dicarboxylatoplatinum (NSC 241240). Cancer Res. 43:4470-4473, 1983.

Keller, J.H. and Ensminger, W.D. Stability of cancer chemotherapeutic agents in a totally implanted drug delivery system. Am. J. Hosp. Parm. 39:1321-1323, 1982.

Kern, D.H., Bertelsen, C.A., Mann, B.D., et al. Clinical application of the clonogenic assay. Ann. Clin. Lab. Sci. 13:10-15, 1983.

Mann, B.D., Kern, D.H., Giuliano, A.E., et al. Clinical correlations with drug sensitivities in the clonogenic assay. Arch. Sur. 117:33-36, 1982.

Moon, T.E., Salmon, S.E., White, C.S., et al. Quantitative association between in vitro human tumor stem cell assay and clinical response in canacer chemotherapy. Cancer Chemother. Pharm. 6:211-218, 1981.

Poochikian, G.K., Cradock, J.C., and Flora, K.P. Stability of anthracycline agents in four infusion fluids. Am. J. Hosp. Pharm. 38:483-486, 1981.

Repta, A.J. and Long, D.F. Reactions of Cisplatin with human plasma and plasma fractions. In A. Prestayko, S.T. Crooke, and S.K. Carter (Eds.), Cisplatin:Current status and new development.New York:Academic Press Inc., 1980, 285-304.

Salmon, S.E., Hamburger, W., Soehnlen, B., et al. Quantitation of differential sensitivity of human-tumor stem cells to anti-cancer drugs. New Engl. J. Med. 298:1321-1327, 1978.

Von Hoff, D.D., Casper, J., Bradley, E., et al. Association between human tumor colony-forming assay results and response of an individual patient's tumor to chemotherapy. Am. J. Med. 70:1027-1032, 1981.

Von Hoff, D.D., Clark, G.M., Stogdill, J., et al. Prospective clinical trial of a human tumor cloning system. Cancer Res. 43:1926-1931, 1983.

DRUG SENSITIVITY TESTING ON IN VIVO AND IN VITRO PASSAGES OF A HUMAN MELANOBLASTOMA BY USE OF THE HUMAN TUMOR COLONY FORMING ASSAY

K.Schieder, C.Bieglmayer, G.Breitenecker, P.Csaicsich, A.Rogan, and H.Janisch

2nd Dept.Obstet.Gynecol.Univ.Vienna, Spitalgasse 23
1090 Vienna

Abstract

A tumor xenograft of a human melanoblastoma grown in a nude mouse was enzymatically disaggregated in a single cell suspension, which was partly syringed into nude mice again, partly cells were grown in a monolayer culture.
Plating of this melanotic tumor in a soft agar culture showed two different pigmentary variants of colonies, dark melanotic ones and light, Masson's staining negative ones. After several in vitro passages the monolayer culture consisted only of hypomelanotic cells.
Drug sensitivity of cells derived either from a melanotic nude mouse xenograft or from hypomelanotic in vitro passages was tested by use of the Human Tumor Colony Forming Assay. For evaluation of colonies an image analysing system (Omnicon 3000, Bausch and Lomb, Rochester, N.Y.) was utilized. Most tested drugs (Adriamycin, Cis-platinum, 5-Fluorouracil, Methotrexate, Mitomycin C, Hydroxyperoxycyclophosphamide, Vincristine, Bleomycin, Tamoxifen(TMX), and Medroxyprogesterone-acetate(MPA) proved to be more effective against the hypomelanotic cells, whereas the melanotic tumor was affected only by Bleomycin, TMX and MPA. Several in vitro passages later tumor cells became resistant against all tested drugs. These cells were also grown in a nude mouse building up a hypomelanotic tumor. After this animal passage tumor cells remained resistant against TMX and MPA. Surprisingly this hypomelanotic tumor grown in a nude mouse was now sensitive against Adriamycin. This might be discussed as a result of different selection modes acting on clonogenic cells by the nude mouse. Different effects of the used anticancer agents in regard to the plating efficiency and the proliferative capacity of the clonogenic cells could be observed.

HUMAN TUMOR CLONING
ISBN 0-8089-1671-8

Introduction

A good correlation between in vitro prediction of drug sensitivity of an individual tumor by use of the Human Tumor Stem Cell Assay and the clinical response of the tumor has been described by a number of authors (1,2,3,4,5).
The human tumor nude mouse system (6,7,8) has been used for experimental examination of in vitro clonogenic assay data and to compare the inhibition of tumor cell colony growth in vitro with the growth delay of tumor xenografts in vivo.
A correlation between the in vitro drug sensitivity of the different xenografts in nude mice and the in vivo response has been found (9,10,11,12, 13).
In our study we have used the Human Tumor Colony Forming Assay (HTCFA),as described by HAMBURGER and SALMON (14,15) to demonstrate the effect of the most common used anticancer agents on a melanoblastoma xenograft,to find out possible alterations of drug sensitivity after a number of in vitro passages,and to compare drug sensitivity of a well defined in vitro passage with that of a tumor grown in a nude mouse from cells of the same in vitro passage.

Material and Methods

A human nodular melanoblastoma of the vulva had been heterotransplanted into athymic nude mice. The grown tumor xenograft was disaggregated enzymatically,and the resulting tumor cell suspension partly was syringed into nude mice again partly

was grown in a long term culture as an adherent monolayer culture.After several in vitro passages cells of the monolayer culture were syringed into nude mice again building up a tumor.
Drug sensitivity from the second mouse passage of the melanoblastoma xenograft,from cells of the 24th and 27th in vitro passage,and from that tumor grown in a nude mouse from cells of the 27th in vitro passage was studied by use of the colony forming assay.

Establishment of the human nodular melanoblastoma in nude mice

All animals used in this study were 6-10 week-old male,congenitaly athymic,NMRI-nude mice,homozygous for nu/nu allele and bred in our laboratory under aseptical conditions (16).
The fresh tumor sample was minced aseptically in slices measuring 3-4mm in diameter and 0.5mm thick. Tumor pieces were set subcutaneously into the left and right anterior chest wall of the nude mice.
Tumor cell suspensions containing $3x10^6$ cells/ml Mc Coys 5A Medium + 10% foetal calf serum (FCs) were injected subcutaneously in 0.25ml aliquots in the region of the left and right anterior chest wall of the nude mice.

Tumordisaggregation

Tumor pieces were taken aseptically from 1-2cm in diameter holding xenografts and minced with crossed scalpels in HEPES-buffered Mc Coys 5A medium + 10% FCs + 1% penicillin/streptomycin solution (10.000 units/ml,Gibco,Ca.).

Enzymatic digestion was made in the same medium containing 0.4% Collagenase II (Worthington Biochemical Corp.,Freehold,NY) and 0.002% DNase (Sigma Chemical Co.St.Louis,Mo) for two hours at 37°C in a flask with a magnetic stirrer.The resulting cell suspension was filtered through a gauze and thereafter passed through a 20μ nylon monofilament and centrifuged (150xg for 10 minutes, at 10°C).Then the cell pellet was resuspended in medium (17,18).

Cell culture

Tumor cells derived from the in vivo passages were grown in 75sqcm tissue culture flasks,containing 25ml of HEPES-buffered Mc Coys 5A Medium with 10% FCs and 1% Penstrep as adherent monolayer cultures.They were incubated in humidified atmosphere of 5% CO^2/95% Air at 37°C and were passaged every third day following treatment with trypsin (30).

Culture assay for tumor colony forming cells

Cells were cultured essentially as described by HAMBURGER and SALMON,conditioned medium was not required for colony formation in vitro (14,15,16).

Drug sensitivity studies

Stock solutions of intravenous formulations of standard anticancer substances including Adriamycin,Cis-platinum,5-Fluorouracil,Methotrexate, Mitomycin C,Vincristine,and Bleomycin were prepared in sterile buffered saline or water and stored at -70°C aliquots sufficient for an individual assay.Subsequent dilutions for drug incubations were made in 0.9% NaCl (1).

Instead of Cyclophosphamide (Endoxan),which is inactive in vitro,an active metabolite (4-hydroxyperoxy-cyclophosphamide -Asta-Werke,GFR) was used. 10^{-2}m stock solutions of Tamoxifen(TMX) and Medroxyprogesterone-acetate(MPA) were prepared in ethanol.

For all drug studies the cut-off concentrations (an empirically established boundary for calculation of sensitivity in vitro) were used (20,21, 22,23).

Tumor cell suspensions were transferred to tubes and adjusted to a final concentration of $3x10^6$ cells per ml and were incubated in the presence of the appropriate drug dilution or control medium for one hour at 37°C in 5% CO^2 and humidified air. Then cells were centrifuged (150xg for 10 minutes, 10°C),washed twice with FCs containing medium and plated in triplicates as described by HAMBURGER and SALMON (one hour exposure).For Bleomycin,TMX and MPA continuous exposure was utilized by mixing the drugs with the plating media before plating (24,25).

After 10-20 days in culture the grown colonies (greater than 60μ in diameter) were counted and grouped into size classes based on colony diameter (bin to bin ratio 1.2) by an optical image analyser (Omnicon 3000,BAusch and Lomb,N.Y.,USA).

The number of colonies from the control group (3-6 plates) was compared with the drug treated plates and the percent decrease in tumor colony forming units (HTCFUs) was determined.In vitro sensitivity was definded by a greater than 70% reduction of the number of colonies grown in the drug treated group in comparison with the untreated control group(26).

Morphological and cytogenetic studies

For morphological studies air dried slides of the tumor cell suspensions were stained with Papanicolou and Masson's stains.

Permanent slides of stem cell cultures were prepared by the dried slide method (27).

Histological slides of the tumors grown in the nude mice were stained by Hematoxylin-Eosin and Masson's stains.

For cytogenetic analysis of tumor stem cell cultures in soft agar we used the procedures and techniques described by Jeffrey M.TRENT (28).

RESULTS

Colony growth

In stem cell cultures set up from melanoblastoma xenografts in nude mice (fig.1) the number of the grown colonies ranged from 31-161 per plate ($5x10^5$ cells plated).The average number of colonies was 102 (plating efficiency 0.02%).The largest colonies reached a diameter of about 124μ.We were able to appreciate two different types of colony formations in the same plate (fig.2).There were dark colonies,consisting of 30-40 Masson's staining positive cells,and light colonies containing

fig.1:Melanotic and hypomelanotic xenografts of a human melanoblastoma grown in athymic nude nu/nu mice.

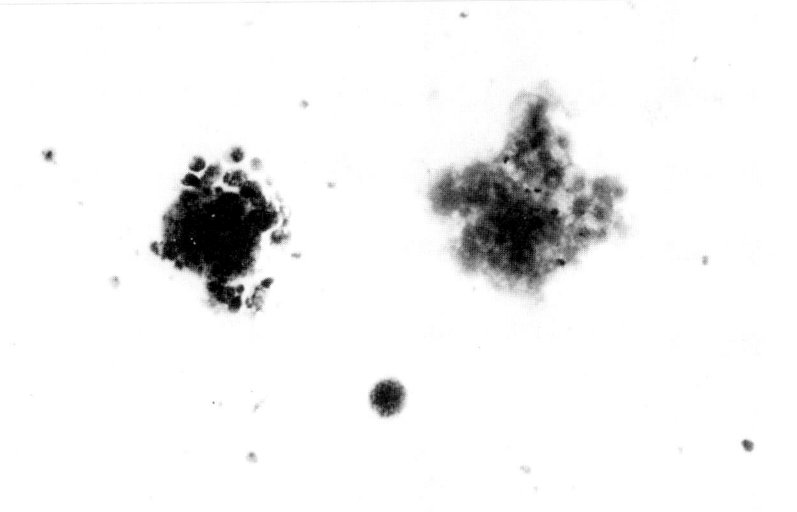

fig.2:Dark and light colony formations grown from cells of a human melanoblastoma xenograft in the same soft agar culture (Masson's stain,x100).

30-50 Masson's staining negative cells. Two different pigmentary variants of melanoma cells, the hypomelanotic large cell colony variant and the melanotic small cell colony variant have been described elsewhere (19). However we were not able to find differences in cell size between the dark, melanotic and the light, hypomelanotic cells.
The ratio between the dark and the light colonies mostly was 9:1, but 1:1 in plates treated with TMX or MPA. These cells of the melanoblastoma xenograft forming two different types of colony formations in the same assay were also grown in a monolayer culture. After several in vitro passages the long term culture only consisted of hypomelanotic cells.
In a HTCFA of the 24th passage of the monolayer culture there were only colonies containing light hypomelanotic cells. Surprisingly we found some isolated melanotic cells in a few colonies.
The average number of colonies was 458 (range from 227-630, plating efficiency 0.09%). The largest colony size was 179μ.
A HTCFA set up from cells of the 27th in vitro passage showed a plating efficiency of 1.17% (average number of colonies 1946, range from 1519-2300, 167.000 cells plated). 13 days after plating the colonies reached a diameter of about 179μ.
The hypomelanotic cells of the 27th in vitro passage were also grown in a nude mouse, building up a hypomelanotic tumor (fig.1). In the colony forming assay of this tumor Melanoma Tumor

Colony Forming Units (MTCFUs) formed 349-773 hypomelanotic colonies per plate (average number 603, plating efficiency0.12%,22 days after plating). The largest colonies reached a diameter of about 372μ.

Chromosome analysis was prepared from a stem cell culture plate of the assay made from cells of the 27th in vitro passage.Whereas there were some polyploid cells,the most metaphases were found to have a chromosome number in the hyperdiploid range (49-51 chromosomes/metaphase).Allmost all cells showed a trisomy 11,an absence of the chromosomes 8 and three small marker chromosomes.

Drug sensitivity studies

Drug sensitivity of the second passage of the melanoblastoma xenograft,of the hypomelanotic cells of the 24th and 27th in vitro passage and of the hypomelanotic tumor grown in a nude mouse from cells of the 27th in vitro passage was studied fig.3).

Most of the tested drugs (Adriamycin,Cis-platinum, 5-Fluorouracil,Methotrexate,Mitomycin C,4-Hydroxy-peroxy-cyclophosphamide,Vincristine,Bleomycin, Tamoxifen (TMX) and Medroxyprogesterone-acetate (MPA) proved to be more effective against the hypomelanotic cells of the 24th in vitro passage of the monolayer culture,whereas the melanotic tumor was only effected intermediately by Bleomycin,TMX and MPA (fig.3).The effect of TMX on MTCFUs of the 24th in vitro passage in comparison to the melanotic tumor was equal,that of MPA was reduced

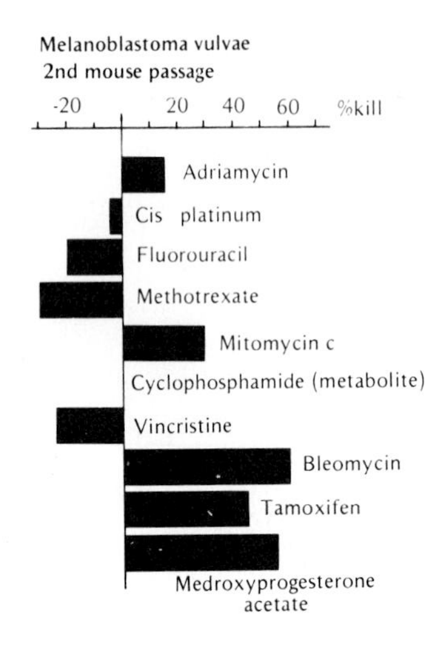

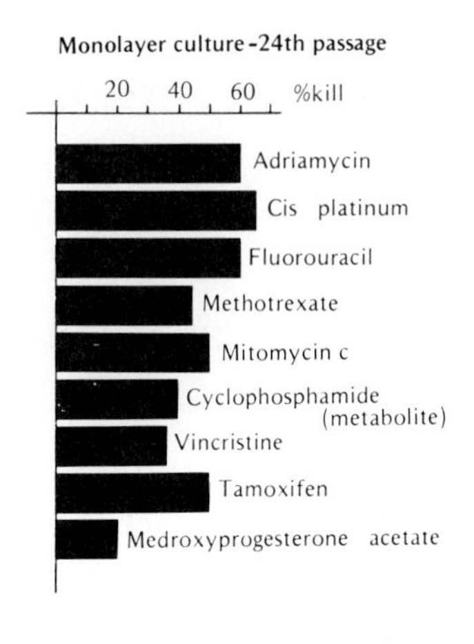

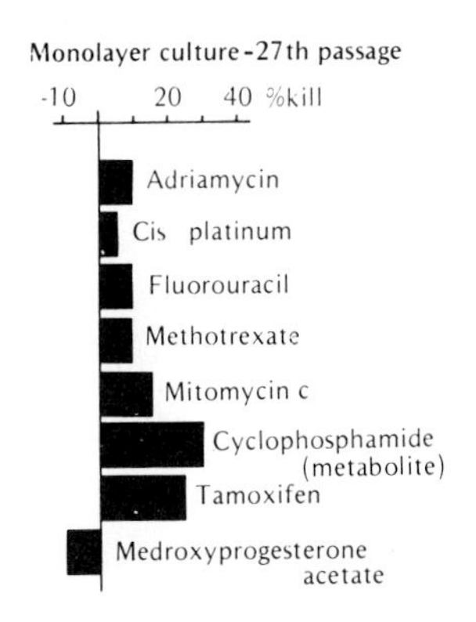

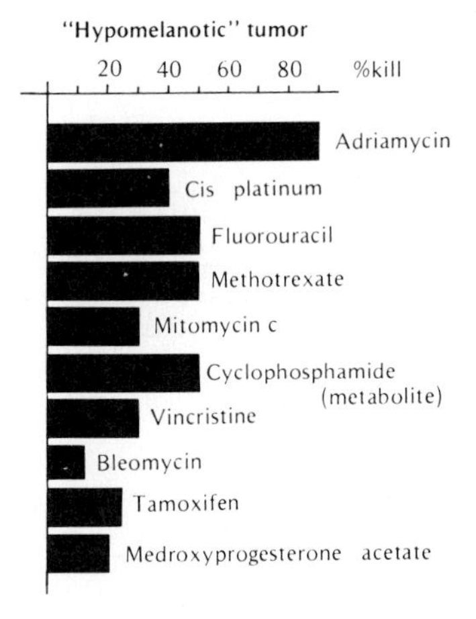

fig.3:Results of the drug sensitivity testing on the different in vivo and in vitro passages of a human melanoblastoma.%kill is the reduction of the number of colonies in comparison to the untreated control group.

by half.Clonogenic cells of the 27th in vitro passage were already resistant against all tested drugs.
MTCFUs of the hypomelanotic tumor grown in the nude mouse were also not inhibited to form colonies by TMX and MPA.Surprisingly this tumor was sensitive against Adriamycin (fig.4).
Because colonies grown from cells of the hypomelanotic tumor were not confluent,we were able to observe the growth of the colonies over three weeks.Using an image analysing system it was possible to measure not only the number (plating efficiency) but also the size (proliferative capacity) of the colonies (25).The used substances influenced the number and size of the grown colonies in a different way (fig.4).Some drugs, like Adriamycin and Cyclophosphamide,had an inhibitory effect on both,the number and size of the colonies.Such substances did not only prevent the formation of colonies,but they also had a decreasing effect on colony size of the remaining MTCFUs.These anticancer substances had a greater inhibitory effect on the proliferative capacity of the MTCFUs,than other agents,like Vincristine or Methotrexate,which had only a decreasing effect on the plating efficiency.Bleomycin and MPA did not inhibite the plating efficiency,but most of the remaining colonies became larger than colonies grown in the untreated control group,meaning an increase in the proliferative capacity of the MTCFUs.

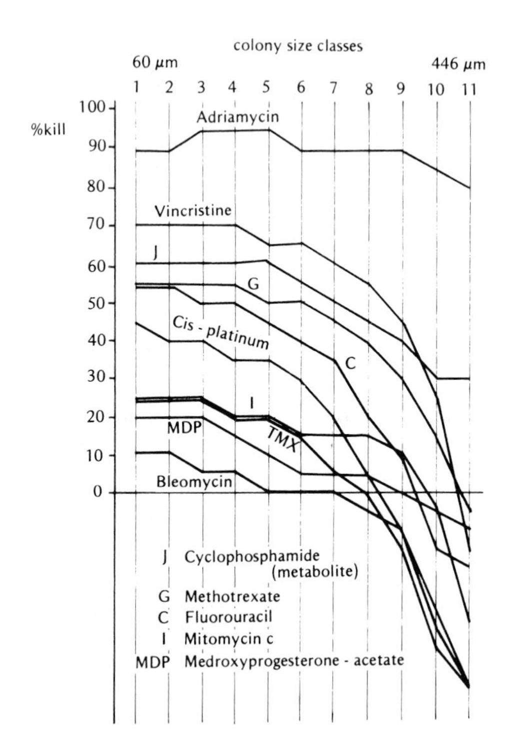

fig.4:Effect of the tested drugs on the number and size of the grown colonies in a HTCFA of the hypo-melanotic tumor grown in a nude mouse.%kill is the reduction of the number of colonies in comparison to the untreated control group.For evaluation of the colonies in regard to their number and size an optical image analyser (Omnicon 3000,Bausch and Lomb,Rochester,N.Y.) was utilized.

DISCUSSION

Chemosensitivitytesting on the melanoblastoma xenograft,which was histological equal to the original tumor,has shown that the most common used anticancer agents are not very effective against this malignant tumor.This result correlates well with the clinical experience with malignant melanoblastoma.

We were able to show the common presence of two different types of colony formations in the same agar culture.The growth of colonies containing melanotic cells was not inhibited by cytostatic drugs,whereas TMX and MPA showed some inhibitory effect on MTCFUs forming melanotic colonies.This could explain,why clonogenic cells of the hypomelanotic monolayer culture and the hypomelanotic tumor had lost all their sensitivity against TMX and MPA.

Several investigators have already described a good correlation between in vitro clonogenic assay data and the in vivo growth delay of human tumor xenografts in nude mice treated with the respective drugs (9,19,11,12,13).

Supposing a greater than 70% decrease of the number of grown colonies in the non control group in comparison to the untreated control group as a definition for drug sensitivity in vitro,for 7 of 8 tested agents a correlation of chemosensitivity between MTCFUs of the 27th in vitro passage and clonogenic cells of the tumor grown in a nude mouse from cells of the same in vitro passage was found.

However this tumor was sensitive against Adriamycin.
Sensitivity testing on the tumor grown in the nude mouse by use of the clonogenic assay may exclude a number of hostfactors modifying the tumor response to drugs,a possible complication of in vivo experiments.Therefore the difference in sensitivity against Adriamycin might be discussed as a result of selection modes acting on clonogenic cells by the nude mouse.Possible reasons could be better growth conditions in the nude mouse for some few Adriamycin sensitive clonogenic cells of the in vitro passage.
The different effects od the used anticancer substances on the MTCFUs in regard to the plating efficiency and the proliferative capacity,respectively,as shown on clonogenic cells of the hypomelanotic tumor,could correlate with the observations of other investigators that anticancer agents at lower dosages can produce an increase in the proliferative capacity of clonogenic cells (25, 29).As we used the cut-off concentrations,this effect of some drugs could mean that a concentration 10% lower than the clinical achievable one produce an increase in the proliferative capacity of the MTCFUs.

Acknowledgments
We thank Ingrid Weinseiss for their dedicated technical assistance.

This paper was supported by grants from:

Bürgermeisterfond der Stadt Wien
Fonds zur Förderung der wissenschaflichen Forschung
Jubiläumsfond der Österreichischen Nationalbank

REFERENCES:

1. Sydney E.Salmon,M.D.,Ann W.Hamburger,PH.D.,Barbara Soehnlen,B.S.,et al.:Quantitation of Differential Sensitivity of Human Tumor Stem Cells to Anticancer Drugs. N.Engl.J.Med.,298,1321-1327(1978)
2. Alberts,D.S..Salmon,S.E.,Chen,H-S.G,et al.:Pharmakologic studies of anticancer drugs with the human tumor stem cell assay.Cancer Chemother.Pharmacol.,6,253-264(1981)
3. Alberts,D.S.,Salmon,S.E.,Chen,H-S.G.,et al.:In vitro clonogenic assay for predicting response of ovarian cancer to chemotherapy.Lancet,2,340-342(1980)
4. Ozols,R.F.,Wilson,J.K.,Weltz,M.D.,et al.:Inhibition of ovarian cancer colony formation by Adriamycin and its metabolites.Cancer Re.,40,4109-4112(1980)
5. Von Hoff,D.D.,Casper,J.,Bradley,E.,et al.:Association between human tumour colony forming assay results and response of an individual patient's tumour to chemotherapy.Amer.J.Med.,70,1027-1032(1981a).
6. Giovanella,B.C.,Stehlin,J.S.,Fogh,J.,et al.:Serial transplantations of human malignant tumours in nude mice and their use in experimental chemotherapy.In: D.P.Houchens and A.A.Ovejera(ed.),Proc.of Symposium of the use of Athymic(nude)Mice in Cancer Research,pp. 163-179,G.Fischer,Stuttgart,New York(1978).
7. Giovanella,B.C.,Stehlin,J.S.,Shepard,R.C.,et al.:Experimental chemotherapy of human malignant tumours serially heterotransplanted in nude mice.Proc.Amer.Ass.Cancer Res.,18,17 (1977).
8. G.Bastert,H.P.Fortmeyer,H.Eichholz,et al.:Human Breast Cancer in Thymusaplastic Nude Mice In:G.B.A.Bastert et al (eds.):Thymusaplastic Nude Mice and Rats in Clinical Oncology Gustav Fischer Verlag Stuttgart New York 1981
9. Bateman,A.E.,Peckham,M.J.,and Steel,G.G.,Assays of drug sensitivity for cells from human tumours:invitro and in vivo tests on xenografted tumours.Brit.J.Cancer 40,81-88(1979)
10. Bateman,A.E.,Selby,P.J.,Steel,G.G.,et al.,In vitro chemosensitivity tests on xenografted human melanomas Brit.J.Cancer,41,189-198(1980)

11. Karimullah A.Zirvi,Hideo Masui,Fernando C.Giuliani,et al. Correlation of Drug Sensitivity on Human Colon Adenocarcinoma Cells grown in soft Agar and in Athymic mice. Int.J.Cancer:32,45-51(1983).
12. P.J. Selbey,V.D.Courtenay,T.J.McElwain,et al.,Colony Growth and Clonogenic Cell Survival in Human Melanoma Xenografts Treated with Chemotherapy.Br.J.Cancer(1980) 42,438.
13. Vicram Guptha,Awtar Krishan,and C.Gordow Zubrod, Correlation of in Vitro Clonogenic Assay Data with in vivo Growth Delays and Cell Cycle Changes of a Human Melanoma Xenograft.Cancer Res.43,2560-2564,June 1983
14. Hamburger,A.W.and Salmon,S.E.,Primary biosaasy of human tumour stem cells.Science,197,461-163(1977)
15. Hamburger,A.W.,Salmon,S.E.,Kim,M.B.,et al.,Direct cloning of human ovarian carcinoma cells in agar. Cancer Res.,38,3438-3444 (1978)
16. H.P.Fortmeyer and G.BAstert,Breeding and Maintenance of nu/nu Mice and Rats,In:G.B.A.Bastert et al(eds.): Thymusaplastic Nude Mice and Rats in clinical Oncology Gustav Fischer Verlag Stuttgart New York 1981
17. Harry K.Slocum,Z.P.Pavelic,and Y.M.Rustum,An enzymatic Method for the Disaggregation of Human Solid Tumors for Studies of Clonogenicity and Biochemical Determinants of Drug Action,In:Cloning of Human Tumor Stem Cells,pages 339-343,Sydney E.Salmon,Editor,1980 Alan R.Liss,Inc., 150 Fifth Avenue,New York,NY 10011
18. Zlatko P.Pavelic,Harry K.Slocum,Youcef M.Rustum,et al., Growth of cell colonies in soft agar from biopsies of different Human solid tumours.Cancer Res.40,4151-4158, Nov.1980
19. Frank L.Meyskens,Jr.,Human Melanoma Colony Formations in Soft Agar.In:Cloning of Human Tumor Stem Cells,pages 85-89,Sydney E.Salmon,Editor,1980 Alan r.Liss,Inc., 150 Fifth Avenue,New York,NY 10011
20. David S.Alberts,H-S.,George Chen,Sydney E.Salmon,In vitro Drug Assay:Pharmacologic Considerations.In: Cloning of Human Tumor Stem Cells,pages 197-207,Sydney E.Salmon,Editor,1980 Alan R.Liss,Inc.,150 Fifth Avenue, New York,NY 10011
21. Thomas E.Moon,Quantitative and Statistical Analysis of the Association Between in Vitro and in Vivo Studies.In: Cloning of HUman Tumor Stem Cells,pages 209-221,Sydney E.Salmon,Editor,1980 Alan R.Liss,Inc.,150 Fifth Avenue, New York,NY 10011

22. David S.Alberts and H-S.,George Chen,Tabular Summary of Pharmacokinetic Parameters Relevant to In Vitro Drug Assay,In:Cloning of Human Tumor Stem Cells,pages 351-359,Sydney E.Salmon,Editor,Alan R.Liss,Inc.,150 Fifth Avenue,New York,NY,10011
23. David S.Alberts,Sydney E.Salmon,H-S.George Chen,et al., Pharmacologic Studies of Anticancer Drugs with the Human Tumor Stem Cell Assay.Cancer Chemother Pharmacol (1981)6:253-264
24. Bernhardt E.Kressner,Roger R.A.Morton,Alexander E.Martens,et al.,Use of an Image Analysis System To Count Colonies In Stem Cell Assays of Human Tumors.In: Cloning of Human Tumor Stem Cells,pages179-193,Sydney E.Salmon,Editor,Alan R.Liss,Inc.,150 Fifth Avenue, New York,NY,10011
25. Marvin D.Bregman and Frank L.Meyskens,Jr.,Inhibition of Human Malignant Melanoma Colony Forming Cells in Vitro by Prostaglandin A_1,Cancer Research 43,1642-1645,April 1983
26. Daniel D.Von Hoff,M.D.,Barbara Forseth,BS,Hans-Robert Metelmann,DDS,et al.,Direct Cloning of Human Malignant Melanoma in Soft Agar Culture.Cancer,Vol.50,No.4, August 15,1982
27. Sydney E. Salmon,Morphologic Studies of Tumor Colonies, In:Cloning of Human Tumor Stem Cells,pages 135-151, Sydney E.Salmon,Editor,Alan R.Liss,Inc.,150 Fifth Avenue,New York,NY 10011
28. Jefrey M.Trent,Protocols of Procedures and Techniques in Chromosome Analysis of Tumor Stem Cell Cultures in Soft Agar,In:Cloning of HUman Tumor Stem Cells,pages 345-349,Sydney E.Salmon,Editor,Alan R.Liss,Inc.,150 Fifth Avenue,New York,NY 10011
29. Fuller B.B.,and Meyskens,F.L.,Jr.,Endocrine responsivness in human melanocytes and melanoma cells in culture, J.Natl.Cancer Inst.,66:799-802,1979
30. John Paul,Cell & Tissue Culture,Fifth Edition Churchil Livingstone,Edinburg,London and New York 1975

Effects of Estrogen, Tamoxifen and Progesterone on a Human Breast Cancer Cell Line in the Soft-Agar Assay

DOMINIC FAN, CHRISTINE SCHNEIDER
SUSAN FAN, HELENE BLANK,
LAURA GILLEN and LEE ROY MORGAN

*Department of Pharmacology and Experimental Therapeutics,
Louisiana State University Medical Center,
1901 Perdido Street, New Orleans, LA 70112*

SUMMARY

An estrogen receptor (ER) and progesterone receptor (PR) positive human breast carcinoma cell line, ZR-75-1, was exposed to 3 log concentrations of antiestrogen Tamoxifen (TAM). Under paired conditions separate groups of cells were pre-treated with 0.1 mM 17-β-estradiol for 30 min prior to exposure of TAM and to growth in soft-agar. TAM inhibited colony formation better in the absence of added estradiol with the greatest inhibitory effect seen in cells exposed to lower than peak plasma levels of TAM (0.01 μg/ml). This inhibitory effect may indicate an uptake of TAM into the cells via the ER and the initiation of a block in the synthesis of macromolecules such as proteins, causing cell damage, or simply a chemical cytotoxic phenomenon. Our observations of cells preloaded with estradiol tend to agree with the former hypothesis because estradiol reversed the inhibitory effects of TAM, indicating that a competitive binding of the compounds to ER might have occurred. Sequential exposure of the soft-agar growth to equimolar progesterone further suppressed colony formation and thus approximates the better clinical responses to progesterone and other hormonal therapies in patients with ER and elevated PR proteins.

Supported in part by a NIH-BRSG #SO-RR-5376, and by a research grant from the Cancer Association of Greater New Orleans.

HUMAN TUMOR CLONING
ISBN 0-8089-1671-8

INTRODUCTION

Tamoxifen (TAM), a non-steroidal estrogen antagonist (Harper 1967), is currently used in treatment of breast cancer (Morgan and Donley 1983; Waseda *et al.* 1981). Approximately 50-60% of breast cancer patients with positive estrogen receptor (ER) has been reported to respond to TAM, whereas only 10% of the ER negative tumors responded to the therapy (Jensen *et al*. 1975; Matsumoto *et al.* 1978). Although it is known that antiestrogens such as TAM may bind to and interact with ER in target cells (Horwitz and McGuire 1978), the mechanism of antiestrogenic action is not well understood. Short-term administration of TAM in breast cancer cell line and cancer patients has been shown to induce progesterone receptor (PR) (Horwitz and McGuire 1979; Mouriguand *et al.* 1983; Namer *et al.* 1980; Waseda *et al.* 1981). Those patients whose tissue contains elevated quantities of PR have been found more responsive to hormone therapies than those lacking PR (Eckert and Katzenellenbogen 1982), and greater than 70% of patients with tumors containing both ER and PR responded to endocrine manipulations (Allegra *et al.* 1979; Osborne and McGuire 1979). In the present study, we used the soft-agar assay to investigate the influence of TAM, estradiol and progesterone on the colony formation of a human breast carcinoma cell line.

CELL CULTURE AND GROWTH CONDITIONS

An ER and PR positive human breast carcinoma cell line ZR-75-1 was obtained from the American Type Culture Collection (Rockville, MD), which was derived from a malignant ascitic effusion (Engel *et al.* 1978). The cells were grown as anchoring-dependent cultures in plastic flasks (*Figure 1*), nourished with RPMI-1640 medium (GIBCO, Grand Island, NY) supplemented with 10-20% fetal calf serum (GIBCO) and antibiotics (100 U/ml penicillin; 100 μg/ml streptomycin). The flasks were incubated at 37° C in a CO_2 incubator, the medium changed every 2-3 days, and the cells diluted by trypsinization at approaching confluency. Under experimental conditions these cells had a population doubling time of 94 hr. The saturation density was 1.92×10^8 cells/cm^2. Cell counts and viability were determined with a hemacytometer and the trypan blue exclusion procedure.

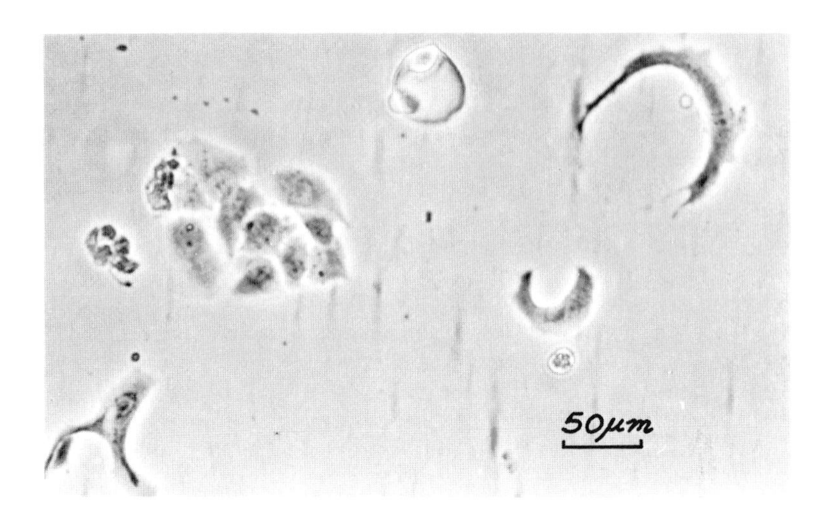

Figure 1. Phase contrast micrograph of a log-phase growth of ZR-75-1 cells in plastic culture flask. Polygonal cells are predominant.

TAMOXIFEN AND ESTRADIOL TREATMENT

Semi-confluent cultures of ZR-75-1 cells were removed from the culture flasks with a 0.05% trypsin/0.02% versene mixture (GIBCO), washed in Hanks' balanced salt solution (GIBCO), and then resuspended in an enriched CMRL medium (GIBCO) containing supplements of nutrients and antibiotics as described by Hamburger and Salmon (1977) except that the spleen macrophage colony stimulating factor was not used, and that $CaCl_2$ and DEAE-dextran were reduced to 1/10 and 1/5 of the quantities, respectively. Half of the cells was exposed to 3 log concentrations (0.01, 0.1, and 1.0 µg/ml) of TAM (tamoxifen citrate, Stuart Pharmaceuticals, Wilmington, DE) at 37°C for 1 hr. Under paired conditions, a separate group of cells was pre-incubated with 0.1 mM 17-β-estradiol (Sigma Chemical Co., St. Louis, MO) for 30-60 min prior to exposure to TAM.

SOFT-AGAR ASSAYS AND PROGESTERONE TREATMENT

After TAM incubation, melted Bacto-agar (Difco Laboratories, Inc., Detroit, MI) was added to the cell suspension to a final concentration of 0.3% agar, and the mixture plated in a 35 mm plastic tissue culture dish on a feeder layer containing an enriched McCoy's 5A medium and 0.5% agar as described by

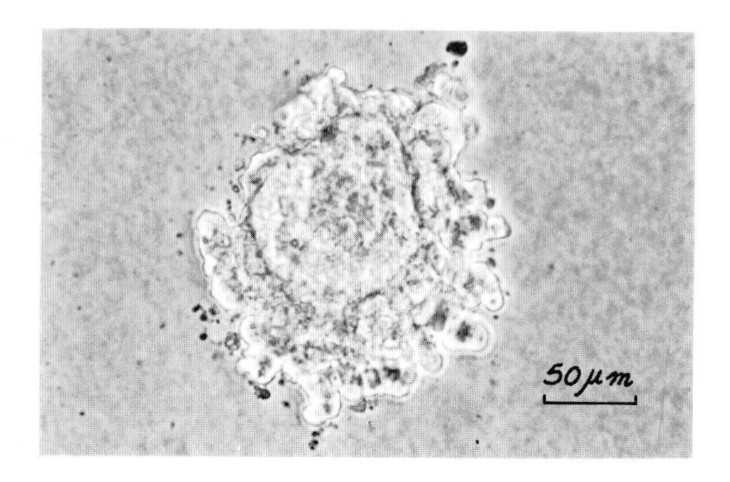

Figure 2. Phase contrast micrograph of a ZR-75-1 colony in soft-agar after 4 weeks of incubation.

Hamburger and Salmon (1977), except for the addition of 2-mercaptoethanol (5 mM) and the reduction of DEAE-dextran to 1/5 of the quantity. The seeding density for ZR-75-1 cells in this study was kept between 1-5x10^4 cells per dish and the plates incubated at 37°C in a CO_2 incubator. After 3-5 days, 0.1 mM progesterone (Sigma Chemical Co., St. Louis, MO) was then added to the growth in an over-layer of the enriched CMRL medium containing 0.3% agar. Incubation was continued for another 3-4 weeks and the plates were examined under an inverted phase contrast microscope. Since these cells grew as tight aggregates in soft-agar (*Figure 2*), it was not possible to determine the number of cells in each growth. An ocular micrometer was instead used to only count colonies of 30 μm or larger in size. Plating efficiency, defined as the number of colonies formed per 100 viable cells plated, was compared between groups. The effects of TAM, estradiol, and progesterone on the development of ZR-75-1 cells in soft-agar were measured as % of inhibition or % of stimulation compared to their independent controls.

EFFECT OF ESTRADIOL AND PROGESTERONE

ZR-75-1 cells have been shown to possess both ER (29 fmol/mg cytoplasmic protein) and PR (43 fmol/mg cytoplasmic protein), as well as other steroid receptors (Engel *et al.* 1979). When plated in soft-agar medium, discrete colonies with a diameter

TABLE I

Plating Efficiency (%) of ZR-75-1 in Soft-Agar

	Plus progesterone	Minus progesterone
Plus estradiol	4%	10%
Minus estradiol	7%	12%

Estradiol and progesterone; each at 0.1 mM.
Plating efficiency (%); defined as the number of colonies formed per 100 viable cells plated.
Numbers are the mean of 3-5 determinations.

of 30-180 μm developed after prolonged incubation (*Figure 2*). Although these cells contain ER, the plating efficiency in soft-agar is not different with or without the addition of 0.1 mM 17-β-estradiol to the medium (TABLE I). It is possible that sufficient amounts of estrogens are present in the growth medium from the supplement of fetal calf serum. However, sequential exposure of soft-agar growth to 0.1 mM progesterone impaired the plating efficiency by 71% in those cells incubated without additional estradiol, and by 150% in cells "primed" with equimolar estradiol (TABLE I). Estradiol, therefore, has significant influence on the responses of ZR-75-1 cells to progesterone in the soft-agar.

Elevated PR in breast cancer cells has been found inducible via estrogen processing and the subsequent synthesis of progesterone binding proteins in ER positive cells (Horwitz and McGuire 1979; Peel and Shih 1975). The interactions between these hormones, their receptors, and TAM are schematically shown in *Figure 3*. Progesterone processing by target cells has been found similar to that of estrogens by which the hormone-receptor complex is translocated into the nucleus (*Figure 3*) (Jensen and DeSombre 1972). Therefore, an increase in PR could result in an increased rate of progesterone processing and thus an enhanced inhibitory effect on the development of breast cancer cells (TABLE I and *Figure 3*). Progesterone has been used in treatment of breast cancer with significant success (Morgan and Donley 1983). The inhibitory effect of progesterone on the colony-forming ability of ZR-75-1 cells in soft-agar reflects such clinical response. Unfortunately, despite extensive research efforts on ER and PR in breast cancer cells, the molecular mechanism from which the inhibitory actions of nuclear hormone-receptors are derived, whether transcriptional or translational, is not clearly delineated.

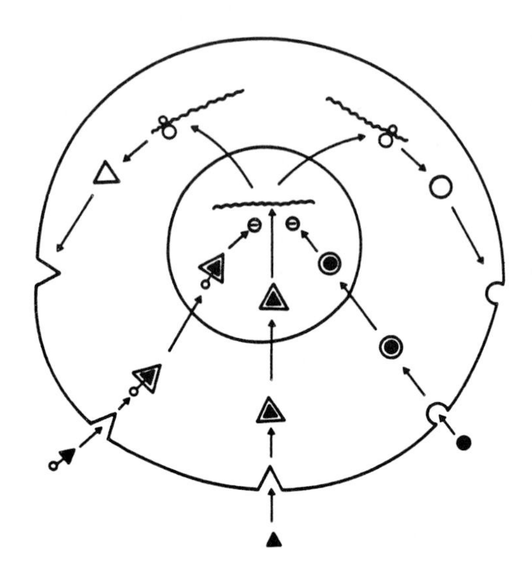

Figure 3. Schematic presentation of the possible interactions of estrogen (▲; ES), tamoxifen (▲̥; TAM), and progesterone (●; PG) in ZR-75-1 cells. The binding of ES to estrogen receptor (△; ER) initiates the processing of ES and the translocation of the ES-ER complex to the nucleus. Specific mRNA (~~~~) are then formed and translated during the synthesis of ER and PG receptor (○; PR) both of which subsequently migrate to the cell surface. Similarly, TAM and PG may also react with ER and PR, respectively, and induce inhibitory effects (⊖) on the functions of ES.

INFLUENCE OF TAMOXIFEN ON COLONY FORMATION

Similar to its inhibitory effects reported by others (Coezy *et al*. 1982; Morgan and Donley 1983; Sutherland *et al*. 1983; Waseda *et al*. 1981), TAM decreases colony formation of ZR-75-1 breast cancer cells in soft-agar under certain conditions (*Figures 4-A* and *4-C*). However, these inhibitory effects were reversed in cells preloaded with 0.1 mM estradiol (*Figures 4-B* and *4-D*). In the absence of both hormones (*Figure 4-A*), TAM suppressed colony formation more at 0.01 μg/ml than at the peak plasma level (0.1 μg/ml). At one log higher than the peak plasma concentration, TAM does not suppress but instead stimulates growth (*Figure 4-A*).

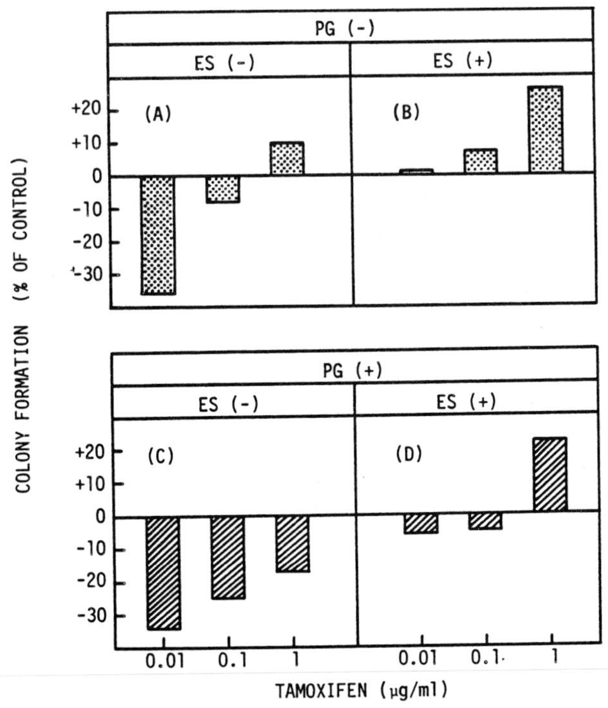

Figure 4. Effects of 0.1 mM progesterone (PG), 0.1 mM estradiol (ES), and tamoxifen on the formation of ZR-75-1 colonies in soft-agar. (▒) minus PG; (▨) plus PG. Values are the mean of 3-4 determinations.

Sequential exposure of the cells to 0.1 mM progesterone appears to amplify the action of TAM. In the absence of equimolar estradiol, TAM inhibited growth at all concentrations, although still in a reversed dose-response fashion (*Figure 4-C*). As expected, the rescue effect of estradiol (*Figure 4D*) was not as prominent as that seen in cells not exposed to progesterone (*Figure 4-B*), indicating the inhibitory influence of progesterone.

Although these observations are somewhat preliminary, it is tempting to interpret further their possible clinical implications relating to menopausal status. The findings in *Figure 4-A* appear to represent an *in vitro* postmenopausal state where TAM is effective in controlling cancer cell growth at low dosages. The stimulatory effect of TAM at higher dose may explain the low clinical responsive rate (50%) in ER positive cancer patients (Jensen *et al.* 1975; Matsumoto *et al.* 1978). This hindrance may be corrected by sequential treatment with progesterone which may also enhance the efficacy of TAM at peak plasma levels in a postmenopausal situation (*Figure 4-C*). The low rate of responses to and/or the estrogenic properties of TAM observed in premenopausal patients are seen in ZR-75-1 cells as shown in *Figure 4-B*. Progesterone administration under this condition does not eliminate the stimulatory effects associated with high concentrations of TAM and produces only slight cytotoxic effects at low dosages of TAM (*Figure 4-D*). These interpretations, however, remain to be clarified with further studies.

CONCLUDING REMARKS

The soft-agar assay developed by Hamburger and Salmon (1977) is a very useful procedure for studying the effects of antiestrogens and their interactions with hormonal manipulations in human breast cancer cells. In the present study, we used this technique to successfully demonstrate (i) the antiestrogenic effect of long-term treatment with TAM (Waseda *et al*. 1981), (ii) the inhibitory effect of progesterone (Morgan and Donley 1983), (iii) the synergistic cytotoxicity of TAM and progesterone in breast cancer cells, and (iv) the comparative observations of the effects of TAM and the hormonal influences at various menopausal stages. The reversed dose-responses and the estrogenic effect associated with high dose of TAM seen in ZR-75-1 cells, however, does not agree with the estrogenic properties observed in cells treated with low dose of TAM (Horwitz *et al*. 1978).

We feel that although the ER and PR assays have become a routine practice in clinical oncology, the response rate can be inconsistent and is only about 50% in patients with positive ER and 70% in patients with both ER and PR. The soft-agar assay, nevertheless, has provided us with an alternative procedure which may be of great value in predicting clinical responses to endocrine therapies.

REFERENCES

Allegra JC, Lippman ME, Thompson EB, Simon R, Barlock A, Green L, Huff KK, Do HMT, Aitken SC, Warren R: Relationship between the progesterone, androgen, and glucocorticoid receptor and response rate to endocrine therapy in metastatic breast cancer. Cancer Res 39:1973-1979, 1979.

Coezy E, Borgna J-L, Rochefort H: Tamoxifen and metabolites in MCF-7 cells: correlation between binding to estrogen receptor and inhibition of cell growth. Cancer Res 42:317-323, 1982.

Eckert RL, Katzenellenbogen: Effects of estrogens and antiestrogens on estrogen receptor dynamic and induction of progesterone receptor in MCF-7 human breast cancer cells. Cancer Res 42:139-144, 1982.

Engel LW, Young NA, Tralka TS, Lippman ME, O'Brien SJ, Joyce MJ: Establishment and characterization of three new contiunous cell lines derived from human breast carcinomas. Cancer Res 38:3352-3364, 1978.

Hamburger AW, Salmon SE: Primary bioassay of human tumor stem cells. Science 197:461-463, 1977.

Harper MJK, Walpole ALA: A new derivative of triphenylethylene: effect on implantation and mode of action in rats. J Reprod Fertil 13:101-119, 1967.

Horwitz KB, Koseki Y, McGuire WL: Estrogen control of progesterone receptor in human breast cancer: role of estradiol and antiestrogen. Endocrinol 103:1742-1751, 1978.

Horwitz KB, McGuire WL: Nuclear mechanisms of estrogen action: effects of estradiol and anti-estrogens on estrogen receptors and nuclear receptor processing. J Biol Chem 253:8185-8191, 1978.

Horwitz KB, McGuire WL: Estrogen control of progesterone receptor induction in human breast cancer: role of nuclear estrogen receptor. Adv Exp Med Biol 117:95-110, 1979.

Jensen EV, DeSombre ER: Mechanism of action of the female sex hormones. AnnRev Biochem 41:203-230, 1972.

Jensen EV, Polley TZ, Smith S, Block GE, Ferguson DJ, DeSombre ER: Prediction of hormonal dependence in human breast cancer, in McGuire WL, Carbone PP, Vollmen EP (eds): Estrogen Receptors in Human Breast Cancer. New York, Raven, 1975, pp 265-275.

Matsumoto K, Ochi H, Nomura Y, Takatani O, Izuo M, Okamoto R, Sugano H: Progesterone and estrogen receptors in Japanese breast cancer, in McGuire WL (ed): Hormones, Receptors, and Breast Cancer. New York, Raven, 1978, pp 43-58.

Morgan LR, Donley PJ: Megesterol acetate versus tamoxifen in postmenopausal women with advanced breast cancer. Mod Med Can 38:855-860, 1983.

Mouriquand J, Jacrot M, Louis J, Mermet M-A, Saez S, Sage J-C, Mouriquand C: Tamoxifen-induced fluorescence as a marker of human breast tumor cell responsiveness to hormonal manipulations: correlation with progesterone receptor content and ultrastructural alterations. Cancer Res 43:3948-3954, 1983.

Namer M, Lalanne C, Boulieu E-E: Increase of progesterone receptors by tamoxifen as a hormonal challenge test in breast cancer. Cancer Res 40:1750-1752, 1980.
Osborne CK, McGuire WL: The use of steroid hormone receptors in the treatment of human breast cancer: a review. Bull Cancer (Paris) 66:203-210, 1979.
Peel JR, Shih Y: Estrogen inducible uterine progesterone receptors characteristics in the ovariectomized immature and adult hamster. Acta Endocrinol 80:344-354, 1975.
Sutherland RL, Hall RE, Taylor IW: Cell proliferation kinetics of MCF-7 human mammary carcinoma cells in culture and effects of tamoxifen on exponentially growing and plateau-phase cells. Cancer Res 43:3998-4006, 1983.
Waseda N, Kato Y, Imura H, Kurata M: Effects of tamoxifen on estrogen and progesterone receptors in human breast cancer. Cancer Res 41:1984-1988, 1981.

The Value of Human Tumor Continuous Cell Lines for Investigating Aspects of the Methodologies Used for *In Vitro* Drug Sensitivity Testing

BRIDGET T. HILL, R. D. H. WHELAN, LOUISE K. HOSKING,
B. G. WARD & ELIZABETH M. GIBBY
Laboratory of Cellular Chemotherapy,
Imperial Cancer Research Fund,
London, WC2A 3PX, England.

INTRODUCTION

In the last six years, laboratory workers have increasingly utilized human tumor material for their experimental studies. The independent development by two groups of clonogenic assay procedures directly applicable to human biopsy specimens (Hamburger & Salmon, 1977; Courtenay & Mills, 1978) provided the major impetus for this work and has led many groups to investigate further or to reinvestigate the potential for human tumor drug sensitivity testing *in vitro*. Data from a number of studies provide evidence that for certain tumors these *in vitro* assays can be used to quantitate differential sensitivities of human tumor 'stem' cells and antitumor agents and hence predict clinical responses (Salmon et al. 1978; Von Hoff et al. 1981; Mann et al. 1982). However, these refer to retrospective clinical studies and although the few results from prospective trials (reviewed in Hill, 1983) tend to agree with these earlier reports, their predictive values are lower. The fact that resistance has been predicted more accurately than sensitivity must also be remembered. Whilst it remains to be established whether the clonogenic assay system is cost effective, providing superior results not only as measured by response rates but also by response durations and survival figures. At the same time, however, it is apparent that the *in vitro* laboratory procedures used in these predictive tests also need improvement. The facts that certain tumor types form colonies more readily than others under currently defined assay conditions and that colony-forming efficiencies (CFEs) are frequently less than 0.1% need to be addressed. In addition, the criteria for *in vitro* drug 'sensitivity' or 'resistance' need to be analysed more rigorously. In our laboratory, we have addressed these problems, before embarking on laboratory/clinical correlation studies.

HUMAN TUMOR CLONING
ISBN 0-8089-1671-8

We have attempted to optimize *in vitro* drug sensitivity evaluations by: (i) improving CFEs obtained with clonogenic assays, (ii) establishing the reproducibility of the assays and hence *in vitro* response criteria, and (iii) examining both time and concentration dependent drug cytotoxicities. Rather than being restricted by a limited supply of tumor cells available from fresh biopsy samples, we have elected to use human tumor continuous cell lines for these comparative studies. This has enabled us to establish, initially in a model system, the reproducibility and reliability of our experimental data, before attempting such comparisons with more heterogeneous tumor cell populations obtainable from biopsy specimens.

RESULTS AND DISCUSSION

In preliminary studies working directly with biopsies from melanomas or ovarian carcinomas, we found that improved CFEs were obtained using the Courtenay procedure as opposed to that of Hamburger & Salmon. Similarly, for a range of human tumor cell lines, as shown in Table 1, the Courtenay assay provided the highest CFEs, representing an improvement of 10 to 40-fold over results using the Hamburger & Salmon method. However, modifying the Hamburger & Salmon assay with additions from the Courtenay assay (red blood cells ± low oxygen tension) significantly enhanced CFE in all the lines tested. These additions also resulted in improved colony quality which greatly facilitated their manual/visual scoring. However, with these higher CFEs obtained using this modified method, we noticed a further problem, the formation of variable numbers of colonies on the plastic dish surface, which could be reduced but not eliminated by using non-tissue culture grade dishes. For example, with HN-1 cells whilst 14.6% of colonies formed in the agar as many as 0.8% were noted on the plastic surface.

In recent studies, we have noted that another critical variable condition of these clonogenic assays is the humidity achieved during the incubation period. With the Courtenay procedure we either packaged the culture tubes in gas-tight polystyrene boxes including tubes of water to maintain high humidity or recently have used a Hereaus incubator which can be controlled to 95% humidity. However, samples assayed by the Hamburger & Salmon method were routinely processed in a standard laboratory incubator also in use for other studies. We have now shown that by increasing the humidity in the standard incubator, CFEs obtained with the Hamburger & Salmon assay can be significantly improved, although with additional 'rbc' further enhancement is still noted (see Table 1.II).

TABLE 1

Comparison of Colony-Forming Efficiencies Obtained*

I % CFEs Obtained Using Different Assay Procedures

Cell Line - Tumor Origin	Courtenay Assay	Hamburger & Salmon Assay	'Modified' Hamburger & Salmon Assay**
COLO205-Colon	30.0 ± 3.0	0.81 ± 0.20	13.3 ± 1.7
LOVO-Colon	29.2 ± 1.3	1.75 ± 0.10	11.4 ± 1.3
DU145-Prostate	10.9 ± 2.2	0.41 ± 0.06	8.5 ± 1.2
PC3mA2-Prostate	41.0 ± 1.3	2.00 ± 0.20	20.0 ± 0.6
HN1-Tongue	14.8 ± 0.8	1.50 ± 0.17	12.9 ± 1.5
SKOV3-Ovary	5.3 ± 0.3	<0.002	<0.002
MCF7-Breast	7.8 ± 0.4	0.08 ± 0.02	1.8 ± 0.1
CHP100-Neuro-blastoma	30.0 ± 2.6	0.80 ± 0.07	18.5 ± 1.7

II % CFEs Obtained Using Various Assay Conditions

Hamburger & Salmon Assay	LOVO Cells	CHP100 Cells	PC3mA2 Cells
'Low' humidity	1.1 ± 0.1	2.0 ± 0.5	2.1 ± 0.3
'High' humidity	4.3 ± 0.9	5.8 ± 0.8	4.1 ± 0.7
'High' humidity + 'rbc' ± low O_2	14.4 ± 0.5	18.8 ± 0.5	14.0 ± 0.5

*In part after Hill & Whelan, 1983.

**+ red blood cells ± low O_2 tension i.e. 5% O_2, 5% CO_2, 90% N_2.

During these studies the following question was raised: does the CFE obtained or the assay procedure adopted influence the *in vitro* drug response? Tveit et al. (1981) reported that sensitivity of melanoma cells to CCNU, vinblastine, abrin and in some cases also DTIC was found to be considerably lower in the Courtenay than in the Hamburger & Salmon assay. In our studies with ovarian tumor biopsy material and using the Courtenay assay to evaluate sensitivities to adriamycin and cisplatin, we consistently observed increased cell kill with increasing drug concentrations and enhanced kill with longer exposure times (Rupniak et al. 1983). These data, therefore, contrast with the plateaus in survival curves reported in some studies where the Hamburger & Salmon assay was employed

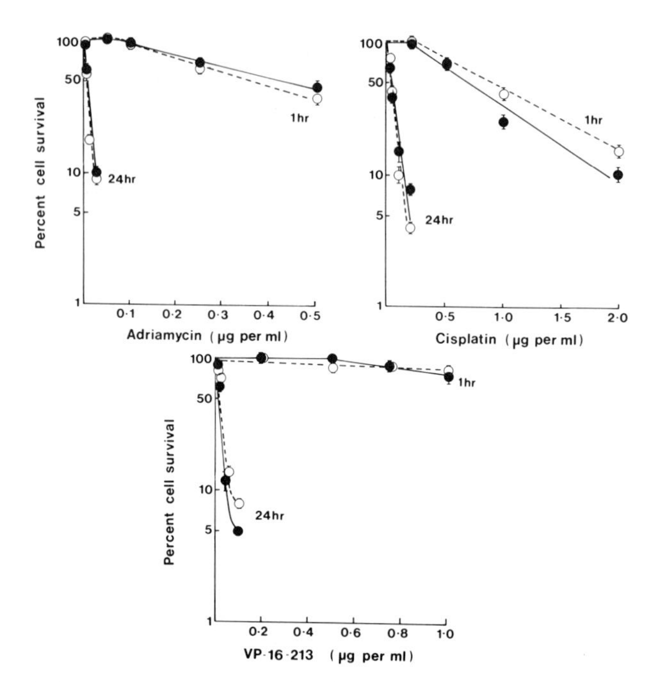

Figure 1. Effects on survival of logarithmically-growing LOVO cells following exposure for 1 or 24 hours to a range of drug concentrations, using the Courtenay assay (●) or the Hamburger and Salmon assay (o).

(Hamburger et al. 1978; Alberts et al. 1981). Therefore, with one cell line the COLO205, we initially estimated the effects of adriamycin using varying drug concentrations and exposure times by the two different assay methods. Our results showed that irrespective of the assay used and the CFE obtained, the resulting survival curves were essentially similar (Hill & Whelan, 1983). We have now extended these observations, using the LOVO cell line and monitoring responses to three drugs, adriamycin, cisplatin and VP-16-213, we again show (Figure 1) that comparable dose response curves were obtained employing either the Courtenay or Hamburger & Salmon assay. Similar data have also been derived for these three drugs using two other human tumor cell lines, PC3mA2 and CHP100. Therefore, these results confirm and extend our earlier observations that: (i) duration of exposure is an important determinant of drug induced cytotoxicity under these assay conditions, (ii) that employing only a 1 hour drug exposure is inadequate for certain agents e.g. VP-16-213, and (iii) irrespective of the type of assay procedure used or the CFE obtained, consistent dose response relationships result.

TABLE 2
Comparison of IC_{50} Values
From Survival Curves Derived From
Clonogenic Assays

I Comparison of IC_{50} Values (μg/ml) Obtained by the Same Operator Repeating Experiments.

Adriamycin on SKOV3 Cells for 24 hr.	Aclacinomycin on COLO205 Cells for 24 hr.	5-Fluorouracil on LAN-1 Cells for 24 hr.	Cisplatin on SKOV3 Cells for 24 hr.
0.027	0.023	1.43	0.40
0.028	0.025	1.50	0.60
0.027	0.028	1.57	0.73
0.025	0.029	1.60	0.85
	0.031		1.00
Range			
(x1.1)	(x1.4)	(x1.1)	(x2.5)

II Comparison of IC_{50} Values (μg/ml) Obtained by Different Operators.

Adriamycin on SKOV3 Cells for 1 hr.	VP-16-213 on SKOV3 Cells for 24 hr.	Methotrexate on LAN-1 Cells for 24 hr.	Cisplatin on SKOV3 Cells for 24 hr.
0.185	0.067	0.06	1.20
0.203	0.081	0.15	0.73
0.215	0.085	0.20	0.43
0.225		0.21	0.40
		0.36	
Range			
(x1.2)	(x1.3)	(x6)	(x3)

For routine clinical in vitro drug sensitivity testing it is essential that the procedure adopted is reproducible, not only within a single laboratory but also between laboratories if multicenter studies are to be considered. Table 2 and Figure 2 show that for three drugs the IC50 values (the drug

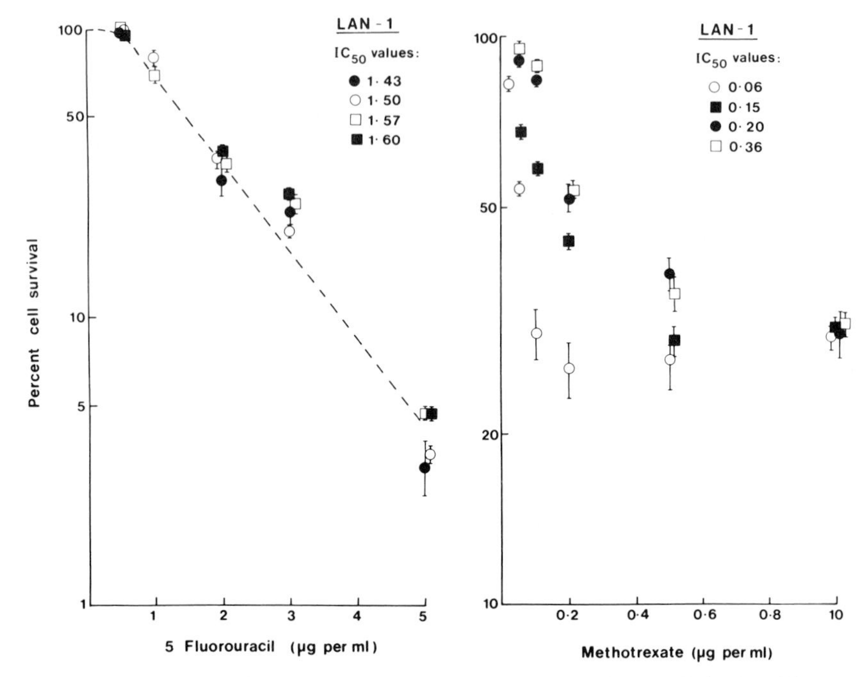

Figure 2. Effects of a 24 hour exposure of logarithmically-growing LAN-1 neuroblastoma cells to a range of methotrexate or 5-fluorouracil concentrations. Reproducibility with 5-fluorouracil was good, but large variations were noted with methotrexate.

concentration required to reduce survival by 50%) obtained by the same skilled operator repeating experiments are very consistent, however, more variation was noted with cisplatin. Similarly, when experiments were carried out with different operators there was remarkable consistency for adriamycin and VP-16-213 (Table 2 and Figure 3) but results with methotrexate and cisplatin (Table 2 and Figure 2) were more variable. These data indicate therefore that the reproducibility of *in vitro* drug sensitivity tests may be drug dependent and suggest that this variability may be associated with: (i) different drug sources, solubility or stability, for example cisplatin is one of the more difficult drugs to solubilize at the relatively high concentrations needed for *in vitro* studies, and (ii) the use of different batches of sera or media which may selectively 'protect' or 'rescue' cells from certain agents e.g. the potential for thymidine to rescue from the cytotoxicity of methotrexate. If further studies confirm our proposal that the extent of reproducibility of these drug

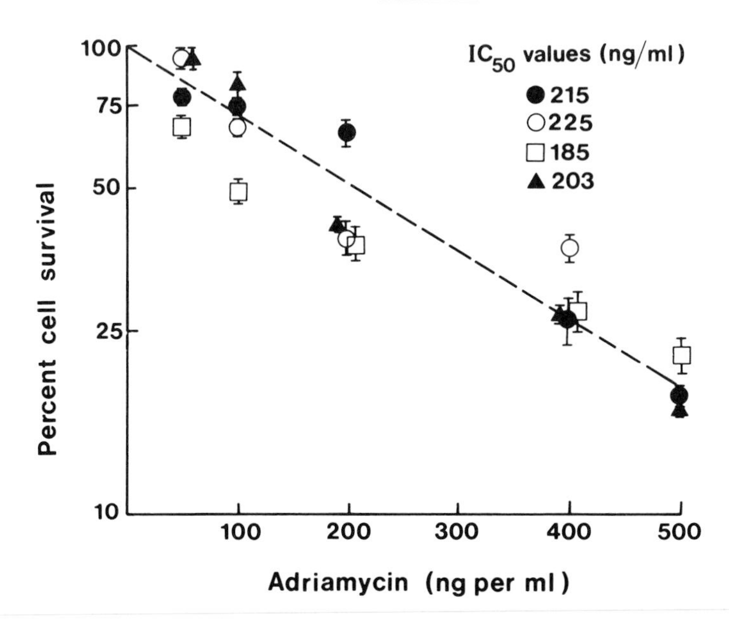

Figure 3. Effects of a 1 hour exposure of logarithmically-growing SKOV3 cells to a range of adriamycin concentrations. The Courtenay assay when used by 4 different operators proved highly reproducible.

sensitivity evaluations depends on the agents being tested, the introduction of an 'internal standard' may prove useful for establishing in vitro criteria for 'sensitivity' or 'resistance'.

It is apparent from earlier studies with biopsy material that depending on the tumor type being investigated different 'cut-off' points have been adopted by various groups. For example, at clinically 'achievable' drug concentrations with ovarian tumors, a ≥ 70% reduction in survival is required for sensitivity (Alberts et al. 1980) whilst only ≥ 49% is the figure accepted for colorectal tumors (Agrez et al. 1982). From our studies with continuous cell lines we have recently considered the possibility that the criteria for in vitro 'sensitivity' or 'resistance' may also have to vary depending on the drug under evaluation. Working with a series of thirty continuous human tumor lines derived from neuroblastoma (4) and tumors of the ovary (4), bladder (6), breast (3) colon (3) prostate (4) and head and neck (6) we have shown that for 24 hour drug exposures the range of IC50 values for certain drugs

TABLE 3
Use of Human Tumor Continuous Cell Lines for In Vitro Drug Sensitivity Evaluations

Drug Tested (24 Hr Exposure)	No. of Lines Tested	Range of IC50 Values (μg/ml)	
Adriamycin	18	0.006 - 0.03	(x5)
mAMSA	9	0.005 - 0.06	(x12)
5-Fluorouracil	15	0.13 - 3.03	(x23)
VP-16-213	8	0.006 - 0.17	(x28)
Cisplatin	17	0.01 - 0.79	(x79)
Vincristine	7	0.0005 - 0.09	(x180)
Methotrexate	22	0.0025 - 45	(x18000)

is very narrow, for example for adriamycin IC50 values differed only by a factor of 5, whilst for other agents the range was very large, for example for cisplatin or methotrexate IC50 values differed by 80 or 18000-fold respectively (see Table 3). It will be interesting to see whether similar data are obtained with tumor cells derived directly from biopsies. If this should prove to be so, we suggest that: (i) for drugs with a narrow range of IC50 values it may be particularly important that accurate and reproducible values are obtained, since any small variability may shift the response between 'sensitive' and 'resistant', and (ii) for other drugs with a very wide range of IC50 values, utilizing a single point on the dose response curve may inadequately reflect the overall tumor response and it may be important to focus attention on other features of the survival curve, for example the presence or absence of a 'resistant' tail or plateau region.

OVERALL CONCLUSIONS

Using a series of human tumor continuous cell lines: (i) the Courtenay assay consistently yielded significantly higher CFEs than the Hamburger & Salmon assay, (ii) addition of 'red blood cells' + 'high' humidity ± low oxygen tension to the Hamburger & Salmon assay significantly enhanced the CFEs obtained, (iii) *in vitro* drug sensitivity evaluations (1 or 24 hour exposure durations) were not markedly influenced either by the CFE obtained or the assay procedure adopted, (iv) the reproducibility of IC50 values derived from survival curves resulting from repeat experiments by the same operator or by different skilled operators was dependent on the drug being tested: e.g. IC50 values for adriamycin and 5-fluourouracil rarely differed by a factor greater than 2 but with cisplatin and methotrexate more variation was noted, (v) the range of IC50 values for certain drugs was very narrow e.g. IC50 values for adriamycin differed only by a factor of 5 in 18 lines tested whilst for other agents the range was very large e.g. for cisplatin or methotrexate IC50 values differed, amongst 22 or 17 lines tested, by 80 or 18000-fold respectively, (vi) employing a single set of standard criteria for *in vitro* 'sensitivity' or 'resistance' for all antitumor drugs may be inappropriate.

REFERENCES

AGREZ MV, KOVACH JS, BEART RW et al. : Human colorectal carcinoma: patterns of sensitivity to chemotherapeutic agents in the human tumor stem cell assay. J Surg Oncol 20:187-191, 1982.

ALBERTS DS, SALMON SE, CHEN HSG et al. : *In vitro* clonogenic assay for predicting response of ovarian cancer to chemotherapy. Lancet ii:340-342, 1980.

ALBERTS DS, SALMON SE, CHEN HSG et al. : Pharmacologic studies of anticancer drugs with the human tumor stem cell assay. Cancer Chemother Pharmacol 6:253-264, 1981.

COURTENAY VD, MILLS J : An *in vitro* colony assay for human tumours grown in immune-suppressed mice and treated *in vivo* with cytotoxic agents. Br J Cancer 37:261-268, 1978.

HAMBURGER AW, SALMON SE : Primary bioassay of human tumor stem cells. Science 197:461-463, 1977.

HAMBURGER AW, SALMON SE, KIM MB et al. : Direct cloning of human ovarian carcinoma cells in agar. Cancer Res 38: 3438-3443, 1978.

HILL BT : An overview of correlations between laboratory tests and clinical responses, in Dendy PP, Hill BT (eds) : Human Tumour Drug Sensitivity Testing *In Vitro*. London, Academic Press, 1983, pp235-249.

HILL BT, WHELAN RDH : Attempts to optimise colony-forming efficiencies using three different survival assays and a range of human tumour continuous cell lines. Cell Biol Int Rep 7:617-624,1983.

MANN BD, KERN DH, GIULIANO AE et al. : Clinical correlations with drug sensitivities in the clonogenic assay. Arch Surg 117:33-36,1982.

RUPNIAK HT, WHELAN RD, HILL BT : Concentration and time - dependent inter-relationships for antitumour drug cytotoxicities against tumour cells *in vitro*. Int J Cancer 32:7-12, 1983.

SALMON SE, HAMBURGER AW, SOEHNLEN BJ et al. : Quantitation of differential sensitivities of human tumor stem cells to anti-cancer drugs. New Engl J Med 298 : 1321-1327, 1978.

TVEIT KM, ENDRESEN L, RUGSTAD HE et al. : Comparison of two soft agar methods for assaying chemosensitivity of human tumours in vitro : malignant melanomas. Br J Cancer 44 : 539-544, 1981.

VON HOFF DD, CASPER J, BRADLEY E et al. : Association between human tumor colony-forming assay results and response of an individual patient's tumor to chemotherapy. Am J Med 70 : 1027-1032,1981.

V. CLINICAL TRIALS

Preclinical and Clinical Applications of Chemosensitivity Testing with a Human Tumor Colony Assay

Sydney E. Salmon
Professor of Medicine, Department of Internal Medicine
Director, Cancer Center
University of Arizona College of Medicine
Tucson, Arizona 85724

SUMMARY

Using the human tumor colony assay (HTCA or tumor stem cell assay), the growth and chemosensitivity of clonogenic tumor cells present in fresh biopsies of human tumors has been investigated. Clinical correlations have been made between in vitro chemosensitivity and the response of patients with metastatic cancer to chemotherapy. In a series of trials, HTCA has had a 71% true positive rate and a 91% true negative rate for predicting drug sensitivity and resistance respectively of cancer patients to specific chemotherapeutic agents. The impact of sensitivity on survival is also being assessed. HTCA has also had several major areas of application to new drug development and screening. Care is needed in experimental design to assure that results are clearly interpretable.

INTRODUCTION

A major objective in cancer research has been to develop simple techniques to predict drug sensitivity of human cancers. In the mid-1970's, Dr. Hamburger and I developed an assay system which proved capable of supporting tumor colony formation from fresh biopsies of human cancers (1). An analogous assay procedure was developed independently by Courtenay and Mills (2) and applied preclinically to human tumor xenografts. Such assays have proven useful for studies of cancer biology, prediction of response and survival in cancer patients, and to aid in screening and assessment of new anticancer drugs. This paper briefly reviews progress and limitations in these areas. Details concerning

HUMAN TUMOR CLONING
ISBN 0-8089-1671-8

methodology and statistical approaches are summarized in other sections of this volume and previously (3-5).

RESULTS AND DISCUSSION

Tumor Colony Growth

Although not all tumor specimens grow in HTCA, most human tumor types have been grown with the assay system (6). Some of the recognized limitations in HTCA in its current stage of development include difficulties in obtaining true single cell suspensions, and low cloning efficiencies observed for many tumor types. The best growth rates have been observed in ovarian, lung cancer and melanoma. More than 50% of specimens for various types of common cancers give rise to at least 30 colonies per plate. A linear relationship is generally observed between the number of cells plated and the number of tumor colonies formed. Evidence that the colonies are comprised of tumor cells has been obtained with morphologic, cytogenetic and biomarker studies (7). Tumor colonies (>30 cells) arising from single tumor cells by definition arise from clonogenic tumor cells. Evidence that tumor colonies contain cells with stem cell characteristics has been obtained by growth of tumors in nude mice from tumor colonies (e.g., in lung cancer) (8), and by "self-renewal" studies in vitro (9,10). Thus, it is apparent that at least some of the cells giving rise to colonies in HTCA are closely related to tumor stem cells. The low cloning efficiencies observed may in fact reflect tumor stem cell characteristics (11).

Drug Sensitivity

Survival curves for tumors exposed to cycle non-specific drugs for one hour prior to plating generally show progressively increasing lethality with increasing drug dosage. However, substantial heterogeneity in sensitivity is observed from patient to patient. Many of the apparently flat survival curves observed in early experiments appear to have been artifactual and were likely due to the presence of aggregates in the cell suspensions. Use of day 1 subtraction counts, and a viability stain, as well as the addition of a positive control have substantially reduced interpretive problems. Drugs which are schedule-dependent often manifest flat survival concentration curves with one hour exposure, but steep inhibition curves at substantially lower doses when the drug is added to the agar for continuous exposure studies. Schedule dependent curves have been observed with bleomycin, VP-16 and vinblastine, but not with doxorubicin, cis-platinum or alkylating agents (12). Serial in vitro studies have also been evaluated in a series of patients before and after initiation of therapy (13). Tumors which initially exhibited sensitivity to a specific drug in vitro prior to therapy frequently showed resistance in vitro on re-biopsy after the patient had relapsed (p=.03) (3). Additionally, in ovarian cancer, the frequency of in vitro sensitivity to common agents such as cis-platinum, doxorubicin and bleomycin was greater in patients who had received no prior chemotherapy than in those who were in relapse (14).

Correlation of In Vitro Drug Sensitivity, Clinical Response and Survival

Correlation between in vitro chemosensitivity in HTCA and clinical response was first evaluated at the University of Arizona Cancer Center in a series of patients with multiple myeloma or ovarian cancer (4). A highly significant correlation was observed between in vitro tumor resistance to specific drugs and failure of the patient to respond to the same drugs clinically. In most instances, when patients exhibited in vitro sensitivity, they responded to the drug clinically. However, some patients, whose cells were sensitive in vitro, failed to respond (false positives) (4). A number of subsequent studies (summarized in Table 1) have independently confirmed and extended the initial report.

TABLE 1

Published Series of Clinical Correlations with Human Tumor Colony Assays

Tumor Types	No. Trials	True + (%)	True - (%)	Ref. (#)
Ovarian, Myeloma	32	11/12 (92)	20/20 (100)	(4)
Ovarian	44	8/11 (73)	33/33 (100)	(15)
Melanoma	48	12/19 (63)*	25/29 (86)	(16)
Multiple	123	15/21 (71)	100/102 (98)	(17)
Melanoma	49	10/11 (91)**	38/38 (100)	(18)
Multiple	36	9/11 (82)	24/25 (96)	(19)
Lung	24	8/12 (67)	12/12 (100)	(20)
Myeloma	33	11/15 (73)	15/18 (83)	(21)
Multiple	246	26/43 (60)	172/203 (85)	(5)
TOTALS:	635	110/155 (71)	439/480 (91)	

True + = Responses/Total
in vivo /sensitive
in vitro

True - = Treatment/Total
failure /resistant
in vivo in vitro

Studies summarized all used method of Hamburger and Salmon (1) with the exception of Tveit et al. (18) who used the method of Courtenay and Mills (2).

* Mixed responses included as responses.
**Mixed responses and stable disease included as responses.

Overall, HTCA had a 71% true positive rate and a 91% true negative rate. Only preliminary information is currently available on the relationship between in vitro chemosensitivity and overall survival. However, data on myeloma (21) and ovarian cancer (this volume, Alberts et al.) suggest that in vitro chemosensitivity to available agents used for treatment in these tumor types is a favorable prognostic factor for survival. For most tumor types, it is important to underscore that the limitations in growth rate and cloning efficiency, and the lack of effective drugs as well as clinical decisions for selection of alternative chemotherapy irrespective of assay results, currently preclude routine application of HTCA. Other potential problems such as intrinsic tumor heterogeneity (e.g., between primaries and metastases) and difficulties in predicting combination chemotherapy effects (either in vitro or in vivo) remain to be resolved. As discussed elsewhere in this volume, major efforts at present are directed towards enhancing in vitro growth conditions by the use of specific growth or conditioning factors for specific tumor types such as lung (8) or breast cancer (22), or by use of an altered microenvironment (23).

In Vitro Phase II Trials

The correlations of in vitro drug sensitivity and clinical response have provided evidence for validity of HTCA and added rationale for its application to new drug development. One application has been described as the "in vitro phase II trial" (24,25), which has two key objectives: 1. to determine concentrations of a new agent which exhibits significant cytotoxic activity against TCFU's, and 2. to obtain a preliminary idea of the "response rate" by tumor type for specific new agents. Such information can potentially be of value in prioritizing tumor types for clinical trial which may be anticipated to exhibit clinical sensitivity to a given new agent. Figure 1 illustrates an in vitro phase II study of 4'deoxydoxorubicin (esorubicin).

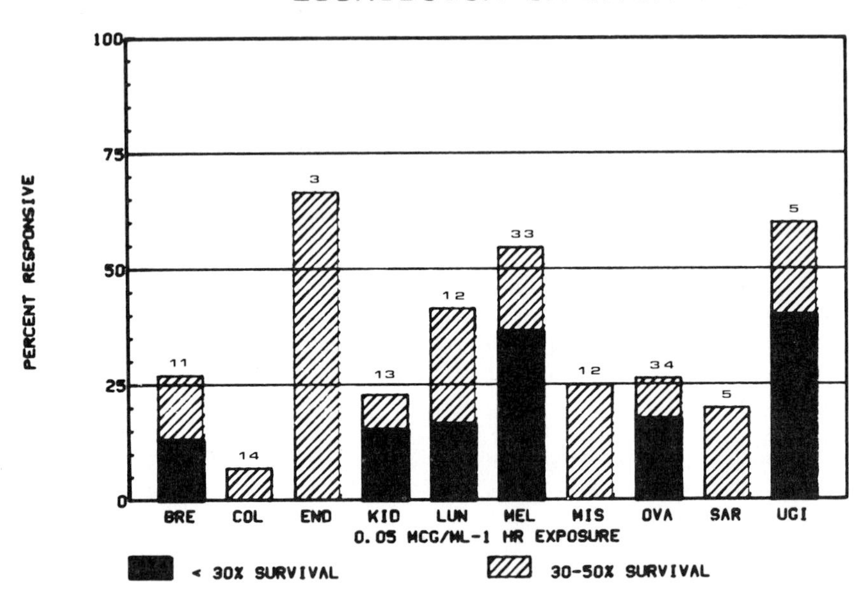

Figure 1. Effects of a one-hour exposure of esorubicin on the growth of human tumor colony forming units. The number of tumors of each type is shown above each bar. Tumor types: BRE=breast, COL=colon, END=endometrial, KID=kidney, LUN=lung, MEL=melanoma, MIS=miscellaneous, OVA=ovarian, SAR=sarcoma, and UGI=upper gastrointestinal neoplasms (3 stomach, 1 pancreatic and 1 hepatic carcinoma)(28). © 1983 by the ASCO.

Similar studies have been published on mitoxantrone (26), bisantrene (27) and recombinant leukocyte interferon (28).

Preclinical Screening

Another important application of HTCA for new drug development is in the initial search for new anticancer drugs (3,29). The NCI has recently initiated a major effort with HTCA as a means for identifying novel antitumor agents which may have been missed with the leukemia prescreen. Initial results from this study are summarized elsewhere (this volume, Shoemaker et al.) and appear to be very encouraging. The long term goal of the NCI study is to select novel agents identified with HTCA screening for advancement to clinical trial to establish whether an in vitro screen can identify unique and effective compounds.

CONCLUDING REMARKS

HTCA provides a useful tool with which to assess in vitro chemosensitivity in relation to cancer chemotherapy as well as serving as an aid in drug development. The relative infrequency of marked in vitro sensitivity of common tumor types to many of the currently available anticancer drugs is consistent with clinical experience. Nonetheless, available correlations provide strong evidence that HTCA is able to predict clinical chemosensitivity or resistance with reasonable accuracy. It is of importance to realize that tumor cells in a petri dish are not a tumor in the patient, and that many major differences exist including the lack of inherent means to activate or inactivate drugs which undergo metabolism. Additionally, there are a number of difficulties in routinely applying HTCA colony assay for chemosensitivity testing of human tumors. Improved methods are also needed to prepare better single cell suspensions from solid tumors and to enhance tumor colony growth. Criteria for drug sensitivity are empirical and subject to change. After growth is improved, randomized clinical studies will be needed to assess the potential clinical impact of routine chemosensitivity testing. Effective use of HTCA is highly dependent on the availability of effective anticancer drugs. Currently, available agents are "first generation" drugs, and lack efficacy for a number of common tumor types. Therefore, application of HTCA for new drug screening and development may prove to be its most useful application.

ACKNOWLEDGEMENTS

These studies were supported by grants CA-21839, CA-17094 and CA-23074, and contract NO1-CM-17497, from the National Institutes of Health, Bethesda, Maryland 20205.

REFERENCES

1. Hamburger AW, Salmon SE: Primary bioassay of human tumor stem cells. Science 197:461-463, 1977.

2. Courtenay FD, Mills J: An in vitro colony assay for human tumors grown in immune-suppressed mice and treated in vivo with cytotoxic agents. Br J Cancer 37:261-268, 1978.

3. Salmon SE: Cloning of Human Tumor Stem Cells, (ed). New York, Alan Liss, 1980.

4. Salmon SE, Hamburger AW, Soehnlen B, Durie BGM, Alberts DS, Moon TE: Quantitation of differential sensitivity of human tumor stem cells to anticancer drugs. New Engl J Med 298:1321-1327, 1978.

5. Von Hoff DD, Clark GM, Stogdill BJ, Sarosdy MF, O'Brien MT, Casper JT, Mattox DE, et al.: Prospective clinical trial of a human tumor cloning system. Cancer Res 43: 1926-1931, 1983.

6. Salmon SE: Background and overview. In: Cloning of Human Tumor Stem Cells, Salmon SE (ed), New York, Alan Liss, 1980, p 3-14.

7. Salmon SE: Morphologic studies of tumor colonies. In: Cloning of Human Tumor Stem Cells, Salmon SE (ed), New York, Alan Liss, 1980, p 135-152.

8. Carney DN, Broder L, Edelstein M, Gazdar AF, Hansen M, et al.: Experimental studies of the biology of human small cell lung cancer. Cancer Treat Rep 67(1):21-26, 1983.

9. Buick RN, MacKillop WJ: Measurement of self-renewal in culture of clonogenic cells from human ovarian carcinoma. Br J Cancer 44:349-355, 1981.

10. Thomson SP, Meyskens FL Jr: Method for measurement of self-renewal capacity of clonogenic cells from biopsies of metastatic human malignant melanoma. Cancer Res 42: 4606-4613, 1982.

11. MacKillop WJ, Ciampi A, Till JE, Buick RN: A stem cell model of human tumor growth: implications for tumor cell clonogenic assays. J Natl Cancer Inst 70:9-16, 1983.

12. Ludwig R, Alberts DS, Miller TP, Salmon SE: Evaluation of anticancer drugs schedule dependency using an in vitro human tumor clonogenic assay. Cancer Chemother and Pharmacol, in press, 1984.

13. Salmon SE, Alberts DS, Meyskens FL Jr, Durie BGM, Jones SE, Soehnlen B, Young L, et al.: Clinical correlations of in vitro drug sensitivity. In: Cloning of Human Tumor Stem Cells, Salmon SE (ed), New York, Alan Liss, 1980, p 223-245.

14. Alberts DS, Chen HSG, Salmon SE, Surwit EA, Young L, Moon TE, Meyskens FL Jr, et al.: Chemotherapy of ovarian cancer directed by the human tumor stem cell assay. Cancer Chemother and Pharmacol 6:279-285., 1981.

15. Alberts DS, Salmon SE, Chen HSG, Surwit EA, Soehnlen B, Young L, Moon TE: Predictive chemotherapy of ovarian cancer using an in vitro clonogenic assay. Lancet 2:340-342, 1980.

16. Meyskens FL Jr, Moon TE, Dana B, Gilmartin E, Casey WJ, Chen HSG, Franks DH, Young L, Salmon SE: Quantitation of drug sensitivity by human metastatic melanoma colony forming units. Br J Cancer 44:787-797, 1981.

17. Von Hoff DD, Casper J, Bradley E, Sandbach J, Jones D, Makuch R: Association between human tumor colony forming assay results and response of an individual patient's tumor to chemotherapy. Amer J Med 70:1027-1032, 1981.

18. Tveit KM, Fodstad O, Lotsberg J, Vaage S, Pihl A: Colony growth and chemosensitivity in vitro of human melanoma biopsies. Relationship to clinical parameters. Int J Cancer 29:533-538, 1982.

19. Mann BD, Kern DH, Giuliano AE, Burk MW, Campbell MA, Kaiser LR, Morton DL: Clinical correlations with drug sensitivities in the clonogenic assay. Archives of Surgery 117:33-36, 1982.

20. Carney DN, Broder L, Edelstein M, Gazdar AF, Hansen M, et al.: Experimental studies of the biology of human small cell lung cancer. Cancer Treat Rep 67(1):21-26, 1983.

21. Durie BGM, Young L, Salmon SE: Human myeloma in vitro colony growth: interrelationships between drug sensitivity, cell kinetics and patient survival duration. Blood 61:929-934, 1983.

22. Hug V, Spitzer G, Drewinko B, Blumenschein G, Haidle C: Improved culture conditions for the in vitro growth of human breast tumors. In: Proceedings Amer Assoc for Cancer Res, San Diego, 1983, 24:35, abstract #138.

23. Von Hoff DD, Hoang M: A new perfused capillary cloning system to improve cloning of human tumors. In: Proceedings Amer Assoc for Cancer Res, San Diego, 1983, 24: 310, abstract #1225.

24. Salmon SE: Applications of the human tumor stem cell assay to new drug evaluation and screening. In: Cloning of Human Tumor Stem Cells, Salmon SE (ed), New York, Alan Liss, 1980, p 291-312.

25. Salmon SE, Meyskens FL Jr, Alberts DS, Soehnlen B, Young L: New drugs in ovarian cancer and malignant melanoma: in vitro phase II screening with the human tumor stem cell assay. Cancer Treat Rep 65(1-2):1-12, 1981.

26. Von Hoff DD, Coltman CA, Forseth B: Activity of mitoxantrone in a human tumor cloning system. Cancer Res 41:1853-1855, 1981.

27. Von Hoff DD, Coltman CA, Forseth B: Activity 9-10 anthracenedicarboxaldehyde bis ((4,5-dihydro-1H-imidazol-2yl) hydrazone)dihydrochloride (CL-216,942) in a human tumor cloning system. Cancer Chemother and Pharmacol 6:141-144, 1981.

28. Salmon SE, Durie BGM, Young L, Liu R, Trown PW, Stebbing N: Effects of cloned human leukocyte interferons in the human tumor stem cell assay. JCO 1(3): 217-225, 1983.

29. Salmon SE: Application of the human tumor stem cell assay in the development of anticancer therapy. In: Cancer: Achievements, Challenges and Prospects for the 1980's; Burchenal JH, Oettgen HF (eds), New York, Grune and Stratton, 1980, p 2:33-43.

Improved Survival of Patients with Relapsing Ovarian Cancer Treated on the Basis of Drug Selection Following Human Tumor Clonogenic Assay

David S. Alberts
Professor of Medicine
Department of Internal Medicine, Department of Pharmacology

Susan Leigh
Research Associate, Cancer Center

Earl A. Surwit
Associate Professor of Medicine
Department of Obstetrics and Gynecology

Ruth Serokman
Research Assistant, Cancer Center

Thomas E. Moon
Research Professor of Medicine
Department of Internal Medicine
Assistant Director, Cancer Center

Sydney E. Salmon
Professor of Medicine, Department of Internal Medicine
Director, Cancer Center

University of Arizona College of Medicine
Tucson, Arizona 85724

INTRODUCTION

Ovarian cancer patients who experience relapse with measurable disease following treatment with combination chemotherapy have a very poor prognosis. Objective response rates to subsequent anticancer drug treatment are generally <25% and median survival durations are, in most

HUMAN TUMOR CLONING
ISBN 0-8089-1671-8

series, less than six months (1-4). We (5,6) and others (7,8) have previously reported that the human tumor clonogenic assay (HTCA) is associated with an accuracy of 60%-70% in the prediction of objective clinical response and a >95% accuracy in the prediction of clinical resistance in patients with relapsing ovarian cancer. Despite these positive results, there has been no evidence that selection of drug therapy, based on in vitro sensitivity data, leads to improved patient survival in this poor prognosis group of patients.

Over the last five years, we have had the opportunity to follow the clinical course of 69 patients with relapsing ovarian cancer whose tumors grew adequately in the HTCA. These patients have been evaluable for both response to chemotherapy and survival following tumor biopsy and successful in vitro cell culture. The following report represents an evaluation of the in vitro assay results and clinical courses of these 69 patients, including survival duration analysis following HTCA.

METHODS

Patients

Eligibility criteria for patient entry into this clinical trial included the following: 1. histopathologically proven ovarian cancer of epithelial type; 2. clinically measurable disease, either by palpation, roentgenograms, ultrasound or CT scanning; 3. relapsing disease following prior multiple anticancer drug regimens; 4. residual disease biopsied for histopathological evaluation and HTCA; 5. adequate in vitro colony growth (i.e., ≥30 colonies per 35 mm dish) with evaluation of tumor colony sensitivity to at least three anticancer agents; 6. evaluable for survival duration and objective clinical response to chemotherapy instituted immediately following HTCA; 7. chemotherapy either selected on the basis of HTCA data or empirically selected by the clinician. Following HTCA, chemosensitivity data were supplied to the referring physician by the cloning laboratory with specific suggestions for therapy, using the most active 1,2 or 3 agents. The referring physician was given the choice to select chemotherapy according to the drug assay results or multiple agent therapy selected empirically.

Standard criteria were used to evaluate response. Complete clinical response required disappearance of all measurable disease for greater than one month. Partial response required a >50% reduction in the sum of the products of the largest perpendicular diameters of all measurable tumor masses for longer than one month. Disappearance of cytologically positive malignant effusions for greater than one month was also included in this response category. Non-response indicated <50% reduction in measurable disease, partial clearance of malignant effusions or no evidence of change in disease status.

Human Tumor Clonogenic Assay

All tumor biopsies and malignant effusions were obtained for pathologic evaluation and HTCA in accord with guidelines approved by the Human Subjects Committee of the University of Arizona Health Sciences Center. The HTCA was carried out as described by Hamburger and Salmon for epithelial cancers of the ovary (9,10). Tumor samples for culture were obtained from solid tissues and malignant pleural and peritoneal effusions (collected in sterile containers with 10,000 units preservative-free heparin/liter of fluid). Solid tumor biopsies as well as malignant effusions were disaggregated into single cell suspensions, using mechanical or enzymatic methods. The resulting tumor cell suspensions were exposed for one hour to various anticancer drugs prior to plating in the HTCA (11). The anticancer drugs tested were obtained from the National Cancer Institute, Bethesda, Maryland, or the pharmaceutical industry. When available, the clinical formulation of the drug was utilized and reconstituted in the appropriate diluents, followed by dilution into complete tissue culture media. Stable anticancer agents were stored in replicate single use tubes at -80°C. Aliquots of cells were exposed at 37°C to at least two concentrations of the drugs, so that a one log dose response curve could be constructed. Drugs were studied at low concentrations (each in triplicate), generally ranging between .01 and 1.0 µg/ml. Procedures for cell plating were carried out as has been detailed previously (9,10).

All 35 mm culture plates were monitored for cellular aggregation by inverted microscopy on the day of plating.

Assays with more than 20 aggregates in the controls were discarded. Culture plates were scanned every few days by inverted microscopy and colonies counted on days 10-16, using a Bausch and Lomb FAS II image analysis system. Currently acceptable colonies are those proliferative cellular units observed by the FAS II to have relatively circular shape with a diameter of at least 60μ (12).

The definition of adequate in vitro growth for evaluation of drug effect was a minimum of 30 colonies per 35 mm petri dish in the controls. The median number of tumor colonies from a large series of countable ovarian cancer experiments was >100 colonies/control dish (500,000 nucleated cells plated). While cloning efficiencies in this system are low, there has been sufficient growth for the evaluation of drug effects in about 60% of ovarian tumors.

Criteria for defining in vitro sensitivity for specific drugs were based on calculation of the percentage survival of tumor colony forming units (TCFUs) in drug-exposed plates relative to those in control plates at "boundary" drug concentrations which are readily achievable (i.e., approximately 10% of the in vivo plasma concentration.time products) after standard clinical drug doses. In our initial drug studies using the HTCA, the in vitro TCFU survival-drug concentration curves were ranked for sensitivity using an "area-under-the-curve technique" (11). Currently, a mathematically simpler ranking of the percent survival data up to empirically derived "cutoff" drug concentrations is being used to evaluate drug sensitivity (13). In each instance, correlations are drawn between the in vitro and in vivo data in order to form a "training set" for each drug to allow prospective application of HTCA results to clinical outcome. For ovarian tumors, we have classified a tumor as "sensitive" to a specific drug if survival of TCFUs is reduced to 30% of control or less, and as "intermediately sensitive" if survival of TCFUs is in the range of >30%-50%. When survival of TCFUs is >50% of control at the "cutoff" concentration, the ovarian tumor is considered to be "resistant" to the specific drug tested.

Analysis of Correlations Between In Vitro HTCA Data and Clinical Response

Our prospective clinical trials in ovarian cancer have been of two types: A. prospective correlative trials; or B. prospective decision-aiding trials. In the prospective correlative trial, the laboratory tests the specific drug or drugs the patient is scheduled to receive after biopsy is performed, and treatment is independent of the laboratory result and is carried out without knowledge of it. In the prospective decision-aiding trial, the patient's specific treatment is predicted on the basis of in vitro tumor sensitivity data to a specific drug or drugs. In all instances, the laboratory results are obtained by technical staff without influence from the physicians responsible for patient treatment, who independently assess the patient's clinical response.

RESULTS

Patient Characteristics

Sixty-nine patients with relapsing ovarian cancer met all eligibility criteria for this study. They could be split into three separate groups according to in vitro drug chemosensitivity results and subsequent selection of anticancer drug treatment. In 24 patients, treatment was selected on the basis of sensitivity in vitro to the best 1,2 or 3 agents. A second group of 20 patients received empirically selected chemotherapy, although their tumors showed sensitivity in vitro to at least one drug. Finally, a third group of 25 patients was characterized by a lack of tumor sensitivity to any tested anticancer agent. There were no significant differences between the three groups of patients with respect to median age, median number of months from initial tumor diagnosis to in vitro HTCA, number of drugs included in prior treatment regimens, percent of patients having received cis-platin in prior therapy, median number of drugs tested in vitro, median number of drugs associated with in vitro chemosensitivity, and median number of drugs selected for therapy following HTCA.

The objective response rates to chemotherapy following the HTCA were significantly different (p=.015, Kruskal-Wallis) between the three groups of patients.

Patients treated according to in vitro chemosensitivity results had a 50% objective response rate (one complete response plus 11 partial responses in 24 patients) and a median duration of response of 5.1 $\pm$ 1.8 months. Thirteen (54%) of the 24 patients in this group received single agent therapy. There were only four (20%) objective responses in the 20 patients whose tumors showed sensitivity in vitro to at least one anticancer drug, but who were treated by their physician with empirically selected therapy. The lowest response rate (8%) to subsequent drug therapy was observed in the 25 patients whose tumors showed complete resistance to all drugs tested in vitro.

Shown in figure 1 are the actuarial survival curves for the three groups of patients entered into this study. Note that the outermost curve represents the actuarial survival data for the 24 patients treated according to HTCA results.

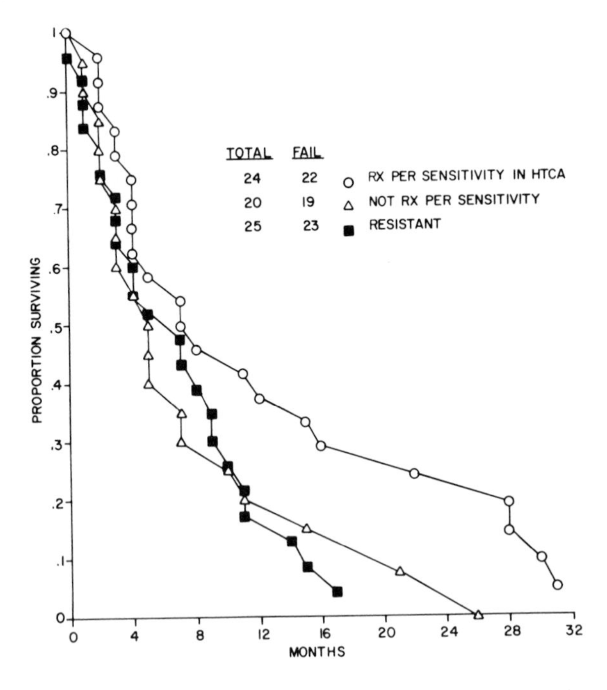

Figure 1. Actuarial survival curves for 69 ovarian cancer patients entered into prospective chemotherapy trial following successful HTCA.

This curve was significantly different ($p<.05$, Log Rank Test) from the survival curves of the other two groups of patients.

In order to evaluate the effect on survival duration of in vitro sensitivity to one or more drugs, we have compared median survival durations of patients, regardless of subsequent treatment, whose tumors showed sensitivity (i.e., <30% survival of TCFUs) or intermediate sensitivity (i.e., <50% survival of TCFUs) to no anticancer drug, one drug, or more than one drug. There were no significant differences between the median survivals of these three categories of patients.

DISCUSSION

Use of the HTCA to select individualized chemotherapy for patients with ovarian cancer and other solid tumors has a predictive accuracy of >60% for clinical response and >85% for clinical resistance (5,6,10,14,15,16). Until now, there has been no evidence that enhanced response rates, resulting from use of HTCA data to guide therapy, have been associated with improved survival durations. In this study, we have determined that patients with relapsing ovarian cancer treated on the basis of chemosensitivity results experienced a significant improvement in survival over those patients whose therapy was selected on an empirical basis. The improved response rate in patients treated according to HTCA results appears to be the major factor responsible for improved survival. No other important prognostic factor (i.e., median age, median duration of survival between diagnosis and in vitro HTCA, number of drugs used in prior chemotherapy, response to prior chemotherapy, number of drugs associated with in vitro tumor sensitivity, or number of drugs used in therapy following HTCA) had an uneven distribution between the three groups of patients included in this study. Obviously, ultimate proof that selection of chemotherapy, based on HTCA results, leads to improved patient survival will require completion of prosepctive clinical trials wherein patients, whose tumors show in vitro chemosensitivity, are randomly assigned to treatment according to assay data or based on empirical choice by the clinician.

Even though this trial suggests a survival advantage for those relapsing ovarian cancer patients treated according to HTCA results, median survival was still quite short (i.e., seven months), and it will be necessary to prove that previously untreated ovarian cancer patients will also benefit with improved survival if therapy is selected on the basis of assay results. One such study (discussed elsewhere in this volume) is being carried out by the Piedmont Oncology Group (17), but final results from this study will not be available for several years.

Despite the significantly improved survival for those patients treated according to in vitro chemosensitivity results, we still cannot recommend the HTCA for routine clinical use in patients with ovarian cancer for the following reasons: 1. only approximately 60 of ovarian tumors obtained for cell culture yield adequate growth for drug assay (i.e., $\geq$30 colonies/35mm petri dish). Thus, the assay cannot be routinely applied to all ovarian cancer patients, who require chemotherapy; and 2. significant quality control requirements have limited effective use of HTCA primarily to the research laboratory rather than in routine clinical laboratories.

A major question remains concerning whether the improved survival observed in our patients treated according to HTCA data was associated with non-specific in vitro chemosensitivity or the assay's ability to select a specific agent for treatment. When survival durations for all 69 patients included in the study were evaluated according to degree of chemosensitivity (i.e., $\leq$50% or $<$30% survival of TCFUs to no drug versus one or more than one drug), there was no evidence that inherent tumor sensitivity itself was associated with a survival advantage. Such data suggest that the improved survival in patients treated according to assay results was related to the assay's selection of specific agents for treatment.

The majority of patients treated according to assay results received single agents. Since combination chemotherapy has been associated with the highest response rates in ovarian cancer (18), the HTCA must prove useful in the selection of combination chemotherapy. Thus far, one specific two-drug combination, cis-platin plus doxorubicin, appears to be accurately tested in the HTCA

(11). However, other combinations may not be successfully tested because of either in vitro chemical incompatibilities or inability to simulate in vivo conditions in an in vitro assay. Future studies must address the problem of drug combination testing in vitro before the assay can have a major impact on ovarian cancer therapy.

ACKNOWLEDGEMENTS

This work was supported by grants CA-17094, CA-23074, CA-28139 from the National Cancer Institute, Department of Health and Human Services, Bethesda, Maryland, 20205.

REFERENCES

1. Stanhope CR, Smith JP, Rutledge F: Second trial drugs in ovarian cancer. Gynecol Oncol 5:52, 1977.

2. Young RC, Von Hoff DD, Gormley P, et al.: Cis-dichlorodiammineplatinum (II) for the treatment of advanced ovarian cancer. Cancer Treat Rep 63:1539, 1979.

3. Alberts DS, Hilgers RD, Moon TE, et al.: Cisplatinum combination chemotherapy for drug resistant ovarian carcinoma. In: Cisplatin: Current Status and New Developments; Carter SK and Crook SR (eds), New York, Academic Press, 1980, p393-401.

4. Surwit EA, Alberts DS, Crisp W, et al.: Multiagent chemotherapy in relapsing ovarian cancer. Amer J Obstet and Gynecol 146(6):613-616, 1983.

5. Alberts DS, Salmon SE, Chen HSG, et al.: In vitro clonogenic assay for predicting response of ovarian cancer to chemotherapy. Lancet 2:340, 1980.

6. Alberts DS, Chen HSG, Salmon SE, et al.: Chemotherapy of ovarian cancer directed by the human tumor stem cell assay. Cancer Chemother and Pharmacol 6:279, 1981.

7. Ozols RF, Willison JKV, Grotzinger KR, et al.: Cloning of human ovarian cancer cells in soft agar from malignant effusions and peritoneal washings. Cancer Res 40:2743, 1980.

8. Welander CE, Homesley HD, Jobson VW: In vitro chemotherapy testing of gynecologic tumors: basic for planning therapy? Am J Obstet and Gynecol 147:188–195, 1983.

9. Hamburger AW, Salmon SE, Kim MB, et al.: Direct cloning of human ovarian carcinoma cells in agar. Cancer Res 38:3438, 1978.

10. Salmon SE, Hamburger AW, Soehnlen B, et al.: Quantitation of differential sensitivity of human tumor stem cells to anticancer drugs. New Engl J Med 298:1321, 1978.

11. Alberts DS, Salmon SE, Chen HSG, et al.: Pharmacologic studies of anticancer drugs with the human tumor stem cell assay. Cancer Chemother Pharmacol 6:253, 1981.

12. Salmon SE, Young L, Lebowitz J, Thomson S, et al.: Evaluation of an image analysis system for counting human tumor colonies. In: Proceedings, Fourth Conference on Human Tumor Cloning, Tucson, Arizona, January 8–10, 1984, abstract no. 51.

13. Moon TE, Salmon SE, Chen HSG, Alberts DS: Quantitative association between in vitro and in vivo studies. Cancer Chemother Pharmacol 6:211–218, 1981.

14. Von Hoff DD, Casper J, Bradley E, et al.: Association between human tumor colony forming assay results and response of an individual patient's tumor to chemotherapy. Amer J Med 70:1027–1032, 1981.

15. Von Hoff DD, Clark GM, Stogdill BJ, et al.: Prospective clinical trial of a human tumor cloning system. Cancer Res 43:1926–1931, 1983.

16. Mann BD, Kern DH, Giuliano AE, et al.: Clinical correlations with drug sensitivities in the clonogenic assay. Archives of Surgery 117:33–36, 1982.

17. Welander CE, Homesley HD, Jobson VW: Multiple factors predicting responses to combination chemotherapy in patients with ovarian cancers. In: Proceedings, Fourth Conference on Human Tumor Cloning, Tucson, Arizona, January 8–10, 1984, abstract no. 29.

18. Decker DG, Fleming TR, Malkasian GD, et al.: Cyclophosphamide plus cis-platinum in combination: treatment program for stage III or IV ovarian carcinoma. Obstet Gynecol 60:481–487, 1982.

Multiple Factors Predicting Responses to Combination Chemotherapy in Patients with Ovarian Cancers

CHARLES E. WELANDER

Assistant Professor
Department of Obstetrics and Gynecology

TIMOTHY M. MORGAN

Assistant Professor
Section on Community Medicine (Biostatistics)

HOWARD D. HOMESLEY

Professor
Department of Obstetrics and Gynecology
Bowman Gray School of Medicine of Wake Forest University
Winston-Salem, North Carolina 27103

INTRODUCTION

During the last decade, the treatment of ovarian cancer has changed remarkably. It remains apparent that surgical cytoreduction (debulking) procedures are useful, particularly in those cases where the largest mass of residual tumor can be reduced to less than 1 cm in size.[9] However, even more important than the degree of surgical debulking which can be achieved is the apparent improvement in clinical response rates seen with new innovations in chemotherapy. The Gynecologic Oncology Group protocol #47 showed that the addition of cisplatin to a previous regimen of cyclophosphamide and doxorubicin improved response rates

HUMAN TUMOR CLONING
ISBN 0-8089-1671-8

from 46% to 71%, with 50% being clinical complete responders. The median duration of survival was increased from 9.5 months to 15.0 months.[14] However, second look operations done on clinical complete responders still reveal that approximately two-thirds of the patients have persistent disease. The fact also becomes apparent that a survival duration of 15 months is not a cure, and the good benefits which cisplatin bestowed on patients with ovarian cancers are limited.

As the biology of tumor cell growth and responses to therapy becomes better understood, it is clear that tumors have heterogeneous subpopulations of cells with different growth rates and degrees of drug sensitivity.[6] The empiric clinical means to deal with that problem is to administer multiple drugs to patients, either simultaneously or in an alternating fashion, in an attempt to avoid the selection of a resistant subpopulation of tumor cells. The empiric choice of chemotherapeutic agents, however, does not necessarily provide maximal benefit to each patient with ovarian cancer.

One of the exciting applications of new laboratory technology to clinical oncology has been the human tumor clonogenic assay (HTCA) as described by Hamburger and Salmon.[10] The majority of the early studies correlating _in vitro_ predictions with clinical responses evaluated only one drug at a time.[2,16] The results of single drugs tested _in vitro_ do provide the clearest correlations with patient responses, and are necessary in order to develop a foundation of credibility for the new HTCA. However, based upon our understanding of the heterogeneity of tumor cell populations and of the limited cytotoxic activity of each drug by itself, it is no longer standard clinical practice to treat patients who have newly diagnosed ovarian cancers with single agent therapy. Therefore, an important question is raised, whether the new HTCA can be used to create individualized multiple agent treatment regimens for ovarian cancer patients.

When several drugs are combined _in vitro_ in the HTCA there are interactions of the combination that are not always predictable. One published report studied combinations of two drugs _in vitro_, using cells from 15 ovarian, 5 uterine cancers, and 1 testicular cancer.[4] In that series, combined cisplatin/vinblastine and combined doxorubicin/cisplatin showed added effects in 53% and 67% of the _in vitro_ trials, respectively. Another report studied combinations of three drugs, including cisplatin, doxorubicin, and cyclophosphamide tested with ovarian cancer cells. The overall colony reduction of this drug combination was similar to that seen when the best agent in the combination was tested alone.[17] The latter study seems to suggest that clear drug synergy in vitro may be difficult to demonstrate.

Since the scant literature concerning drug combinations studied in the HTCA gives mixed results, the clearest endpoint can probably be reached by testing multiple drugs, each as a single agent, with ovarian tumor specimens. Using the degree of in vitro cytotoxicity from the HTCA, the "best drugs" can be selected and administered to a patient.

Another question is important to consider, concerning multiple host factors which impact on patient responses to chemotherapy. Published reports have shown that patient performance status,[18] residual tumor bulk following surgery,[9] and tumor grade[7,15] each contribute to the ultimate outcome of a patient's therapy.

In this chapter data are presented from HTCA testing of single chemotherapeutic agents and corresponding clinical trials, treating patients with combination chemotherapy. Host factors, including tumor grade, bulk of tumor and patient performance status are analyzed, each correlated with ultimate responses to therapy and survival.

MATERIALS AND METHODS

Tumor Procurement and Processing

Tumor specimens from 45 patients with advanced previously untreated ovarian cancers were tested by the Gynecologic Oncology Laboratory. The clonogenic assay methodology used in this study has been described in the literature.[10,11] A few essential features of the methodology are noteworthy. Viable tumor cells must be obtained from the patient and transported to the laboratory as quickly as possible. There is decreasing viability of cells in a tumor nodule when there is prolonged transit time from the operating room to the laboratory, particularly apparent when overnight transport of tumor specimens is necessary.

Cells are disaggregated into a single cell suspension using both mechanical mincing and collagenase 0.8% (Worthington) with DNAse 0.03% (Cal Biochem). The cell suspension is then placed on to a Ficoll-Paque® gradient in order to remove the erythrocytes, dead cells, and cellular debris.[8] The cells from the gradient interface are counted, adjusted in volume and concentration, and prepared for drug exposure.

The optimal methods of drug exposure are understood only partially. Whether short-term (1 hour) incubation prior to plating in agar or long-term drug exposure achieved by incorporating the drug with the cells into the agar is better is not clear.[1] We are testing a group of 10 cytotoxic drugs in this study, based upon our previous HTCA experience, identifying these as the 10 most active agents with ovarian cancer cells. The group of drugs includes cyclophosphamide, doxorubicin, cisplatin, 5-fluorouracil, mitomycin C, vinblastine, bleomycin, methotrexate, vincristine, and hexamethylmelamine. Based upon our laboratory experience during the past two years, the HTCA results correlate best with clinical responses when continuous *in vitro* drug exposure is used for all except mitomycin C. Substantial overestimation of tumor cell sensitivity is found when continuous *in vitro* exposure of mitomycin C is used; therefore the shorter one hour exposure method is substituted for that drug only. Because of solubility problems with hexamethylmelamine, an analog of the drug, pentamethylmelamine has been substituted (provided by the National Cancer Institute, Drug Synthesis and Chemistry Branch). At least two concentrations of each drug are optimally tested, using an achievable peak plasma level (PPL) and a lower concentration at one-tenth PPL. It is not yet clear which drug concentration will correlate better with clinical responses, but both are recorded.

Plating in agar is done following drug exposure. Tumor cells are plated at a density of 500,000 cells per well, in triplicate wells for each drug concentration studied. Six control wells are used for each experiment. A feeder layer is routinely used in each experiment, prepared with xenogeneic adherent peritoneal cells (primarily macrophages) harvested from adult BDF_1 mice stimulated with Pristine (2,6,10,14-tetramethylpentadecane, Aldridge Chemical Company, Milwaukee, Wisconsin).[19] The other nutrient additives to the medium are similar to those described originally by Hamburger and Salmon.[10,11] Culture plates are incubated at 37°C in 7.5% CO_2 until colony growth is adequate for counting. Colonies are defined as aggregates of >50 cells. At least 30 colonies must be present in a control well for the assay to be valid. Counting is done with an image analyzer, the FAS-II (Bausch & Lomb). In the experience of the Gynecologic Oncology Laboratory growing primary ovarian cancer specimens, the range of days growth in culture prior to being ready for counting is 7-32 days; 40.5% of assays are reported in under 14 days and 83.4% are reported in less than 21 days.

Statistical Methods

Tests for the equality of two proportions were performed using the one-sided Fisher's exact test. Estimates of survival distributions were computed using Kaplan-Meier estimates;[12] differences between survival distributions were tested using the Breslow-Wilcoxon test.[5]

RESULTS

Of the 45 patients studied with the HTCA and treated with combination chemotherapy, 20 were given the "best three drugs", based on the HTCA results of the drugs tested (Table I). The other 25 received "state-of-the-art" chemotherapy, namely, cyclophosphamide, doxorubicin, and cisplatin (CAP), regardless which drugs were determined to be the most active in the assay. Each patient received eight courses of therapy, given ideally at three-week intervals. Clinically determined responses are recorded in Table II. The patient group receiving the "best three drugs" had 6/20 (30%) complete responses and 17/20 (85%) complete and partial responses. Of those receiving CAP chemotherapy, there were 3/25 (12%) complete responses and 17/25 (68%) complete and partial responders. These results are suggestive of the utility of prospective drug selection based on HTCA results; however, at this point the results are not statistically significant.

The predictive variables associated with clinical responses examined in this study are the HTCA data, patient

TABLE I
Chemotherapeutic Agents Tested in the HTCA with Ovarian Cancer Cells from Untreated Patients

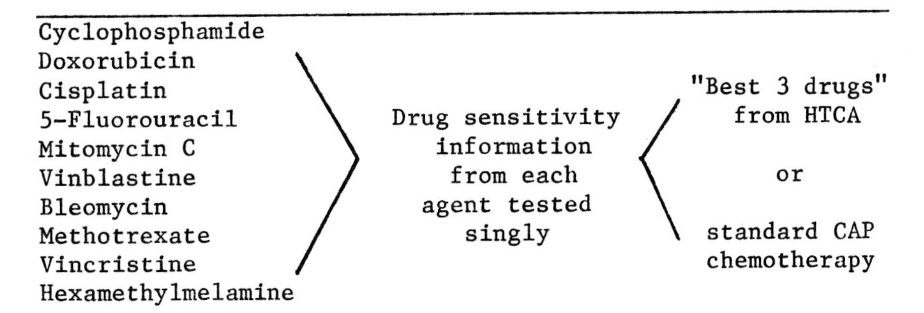

TABLE II
Clinically Determined Responses to Multiple Agent Chemotherapy of Untreated Ovarian Cancer Patients

Regimen	No. of Patients	CR	PR	Stable + Progression
"Best 3 drugs"	20	6 (30%)	11 (55%)	3 (15%)
CAP	25	3 (12%)	14 (56%)	8 (32%)

performance status, grade of tumor, and residual tumor bulk following cytoreductive surgery. Looking first at correlations between HTCA data and patient responses, the total group of responders in general and the complete responders in particular increases as the number of drugs showing activity against the patient's tumor increases (Table III). If there are no drugs which, as single agents, show _in vitro_ activity (less than 50% colony survival at 0.1 PPL) in a particular patient's case, there still are clinical responders when the "best three drugs" of the available choices are combined into a treatment regimen. However, the probability of a complete response increases as the number of active drugs increases, from 3/24 (12.5%) for patients who have no drugs with less than 50% colony survival to 5/11 (45.5%) for patients with two or more active drugs.

Patient performance status is correlated with response rates as shown in Table IV. Patients who have performance status of 3 were significantly more likely to be non-responders (83%) as compared to patients with better performance status levels 0-2 (15%) (p=0.002).

TABLE III
Predictive Correlation of HTCA Data with Responses to Multiple Agent Chemotherapy

Number of Drugs Showing <50% Colony Survival	Number of Patients with Complete Response	Number of Patients with Partial Response	Number of Patients Not Responding (Stable + Progression)
No drugs (0.1 PPL)	3/24 (12.5%)	15/24 (62.5%)	6/24 (25%)
1 drug (0.1 PPL)	2/10 (20%)	4/10 (40%)	4/10 (40%)
2 or more drugs (0.1 PPL)	5/11 (45.5%)	5/11 (45.5%)	1/11 (9.1%)

TABLE IV
Predictive Clinical Features Correlated with Responses in Advanced Ovarian Cancers

Performance Status (GOG Classification)

Perf. Status	CR + PR	Stable + Progression
0	1/1 (100%)	0/1
1	16/18 (89%)	2/18 (11%)
2	16/20 (80%)	4/20 (20%)
3	1/6 (17%)	5/6 (83%)

Residual Tumor Bulk (Following Cytoreductive Surgery)

Residual Tumor	CR + PR	Stable + Progression
<1 cm (optimal)	8/8 (100%)	0/8
>1 cm (suboptimal)	26/37 (70%)	11/37 (30%)

Tumor Grade (FIGO Classification)

Grade	CR + PR	Stable + Progression
I	4/4 (100%)	0/4
II	6/7 (86%)	1/7 (14%)
III	21/28 (75%)	7/28 (25%)
Unknown	3/6 (50%)	3/6 (50%)

Patients with residual tumor less than 1 cm (optimal) were more likely to have responses in general and complete responses in particular than patients with suboptimal tumor bulk (Table IV). The optimal residual bulk group had 8/8 (100%) overall responses, as compared to 26/37 (70%) responses in the suboptimal group ($p=0.084$), and 5/8 (62.5%) complete responses compared to 4/37 complete responses for the optimal and suboptimal groups, respectively ($p=0.002$). Patients who had good performance status 0-2 or optimal residual tumor bulk, had 33/39 (85%) responses, compared to patients with performance status of 3 and suboptimal tumor bulk, who had 1/6 (17%) responses.

Grade of tumor is less well correlated with responses, although there is a trend toward poorer responsiveness as tumor grade increases (Table IV).

The proportion of responders to therapy was 17/20 (85%) for patients given the "best three drugs" selected by the HTCA results, and 17/25 (68%) for patients given CAP ($p=0.17$). As shown in Table IV, the individual factors tumor bulk and performance status can predict clinical response in most patients. For example, patients with optimal tumor bulk or performance status 0-1 had 90% responses, while patients with performance status 3 had 83% non-responses. The intermediate group of 19 patients with suboptimal tumor bulk and performance status 2 is not well predicted by the individual factors (79% responses). Of this group 10 were given the "best three drugs" and 9 given CAP. Responders to therapy were the 10/10 (100%) given the "best three drugs" and the 5/9 (56%) given CAP ($p=0.033$). Although sample sizes are small, these results suggest the clinical benefit of using prospective HTCA multiple drug selection on patients whose response to therapy and overall prognosis is not easily determined by pretreatment host factors.

Apart from responses to therapy, survival is also correlated with these predictive parameters. In the brief duration of this study, the median survival time has not yet been reached (Figure 1). Therefore, it is difficult to establish firm relationships between the predictive parameters and survival. Some trends, however, can be observed with these preliminary data.

Clinical responses to therapy can be correlated with increased survival (Figure 2). The distributions of survival times for complete, partial, and non-responders are statistically significant ($p=0.0027$). The patients having complete clinical responses are all surviving at this point in time with a median time to follow-up of 16 months (range 5-22 months). The distribution of survival times is larger for partial responders than non-responders. As seen in Figure 3, the distribution of survival time was highly associated with performance status ($p=0.001$). Tumor bulk was also associated with survival time ($p=0.046$). All eight patients with optimal tumor bulk are surviving with a median length of follow-up of 16 months (range 9-22 months).

DISCUSSION

Physicians practicing clinical oncology are always searching for means to predict the outcome of cancer therapy. Discoveries of the past 30 years have fallen into two categories, one of specific predictive tests which give therapeutically useful information about which treatment to select, and the category of non-specific parameters which are correlated, in a general manner, with results of standard therapy. The HTCA has been particularly intriguing as a specific way for clinicians to obtain information aiding in the selection of active chemotherapeutic agents for an

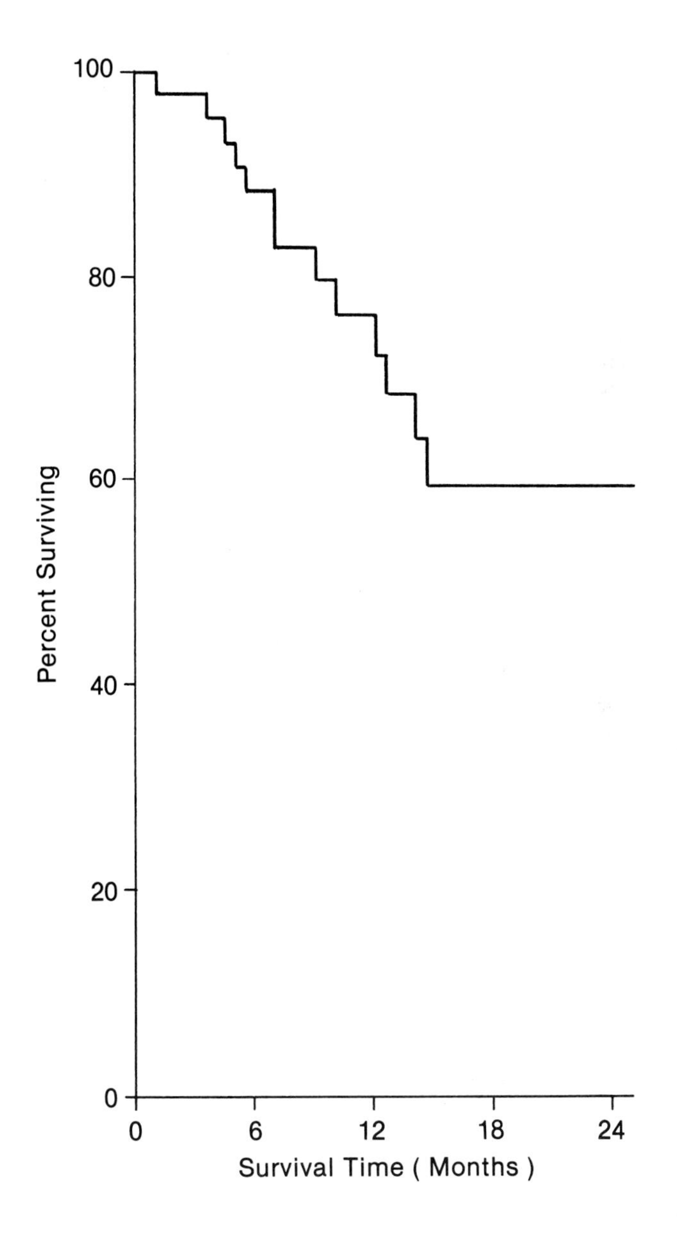

Figure 1. Overall predicted probability of survival for previously untreated patients (Kaplan-Meier estimates)

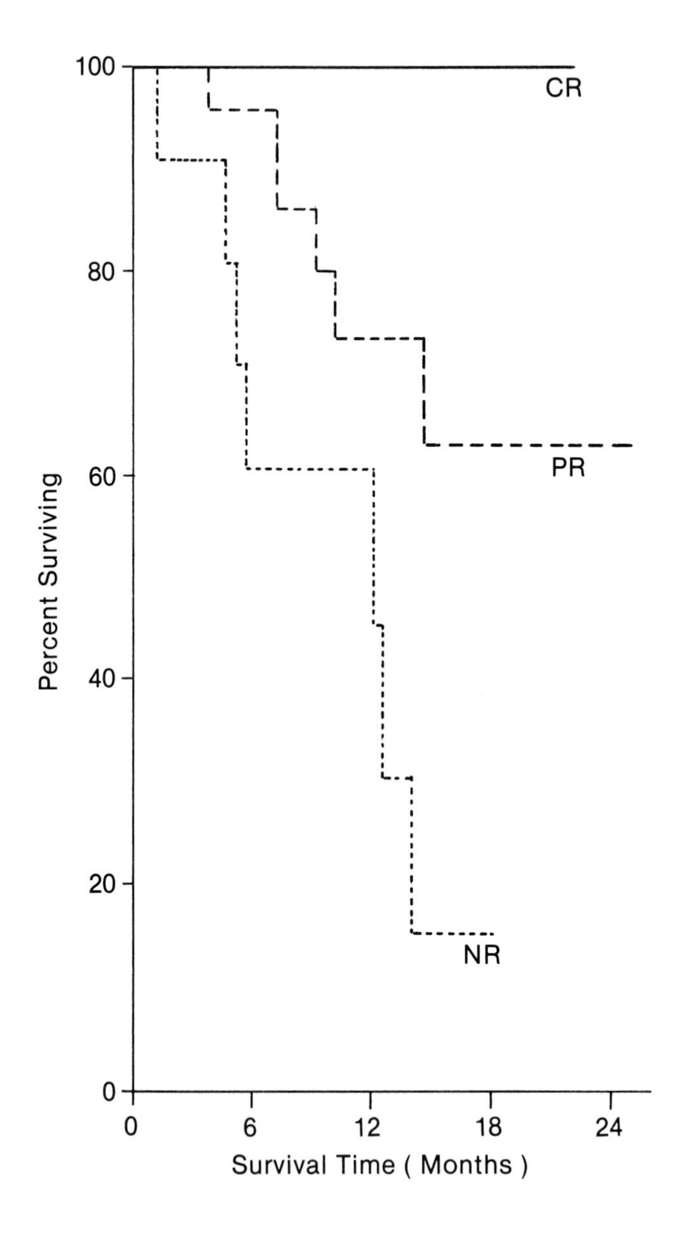

Figure 2. Predicted probability of survival for previously untreated patients by response (Kaplan-Meier estimates) (generalized Wilcoxon test for equality of survival curves, p=0.0027)

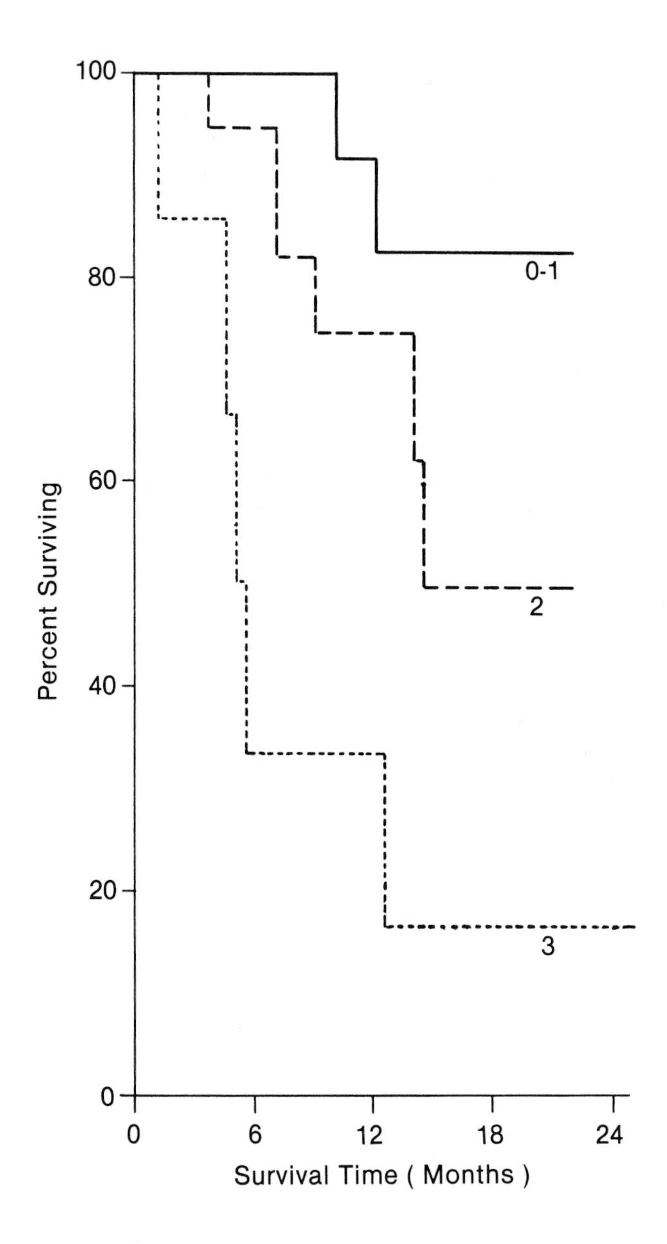

<u>Figure 3</u>. Predicted probability of survival for previously untreated patients by performance status (Kaplan-Meier estimates) (generalized Wilcoxon test for equality of survival curves, $p<0.001$)

individual patient. The early retrospective and correlative trials testing the potential usefulness of the HTCA used only single agents, both in the laboratory and in patient treatment. On the basis of clinical correlations using these single agent trials, empiric definitions of in vitro responsiveness have been determined. Colony survival in the HTCA to less than 30% of control values has been defined as highly predictive of a clinical response, and colony survival between 31% and 50% has been correlated to a lesser degree with clinical responses.[13] When drugs failed to reduce colony survival to less than 50% of control values, there was only the remote probability of a clinical response. Indeed, the HTCA trials have shown greatest accuracy in predicting clinical non-responsiveness.

This study is looking at HTCA data testing 10 drugs individually but treating patients with a combination of three drugs. Fortunately, clinical responses are seen in patients treated with the "best three drugs" even when none of the drugs alone would predictably be active. This raises the possibility of in vivo drug synergy as a mechanism of enhanced clinical activity. However, in vitro attempts to demonstrate drug synergy have been met with mixed results. As a practical matter, it would be difficult to determine in advance for an individual patient which drugs might have additive or synergistic combined effects. A screening assay would first be required, followed by a second in vitro assay to study combinations of the most active single agents. The availability of fresh tumor cells for repeated testing and the time required to perform such studies would preclude this method as a routine procedure. We are, therefore, left with this somewhat indirect means to determine the best drugs for an individual patient, by testing agents individually and then administering the best combination to each patient.

One of the distressing points of confusion in specific predictive testing is the impact of host factors on patient responses to therapy and survival. The individual factors such as performance status, bulk of disease, and tumor grade each correlate with treatment outcome. An unfavorable combination of these three can outweigh favorable specific predictive HTCA data. Multivariate analysis of these factors correlated with responses and with survival are not yet possible, due to small numbers of patients in each sub-category analyzed. This will be accomplished when the study has accessioned more patients.

As the HTCA begins to assume a place in selection of treatment for individual patients, the justification for the assay's credibility is based upon demonstrated correlations between the laboratory and the clinic. However, the inability of the published HTCA-clinical correlations to reach greater than 90% accuracy for predicting both responses and non-responsiveness to therapy may be related to host factors which have not been controlled for in studies published heretofore.

The method of endpoint evaluation in predictive testing is crucial. Classical chemotherapy trials have been designed to detect "responses" to therapy, whether partial or complete. Some partial responses are of such brief duration as to be of inconsequential value to a patient. If the HTCA is to have clinical benefit for patient treatment planning, then "responses" noted must be significant in terms of palliation and/or survival. A study published by Alberts, _et al_. from the University of Arizona showed that survival duration can be increased among patients treated with drugs prospectively chosen using HTCA data.[3] The study reported in this chapter shows in a preliminary fashion that responders do survive longer than non-responders. The ability to predict which patients will achieve a complete response is less accurate, but remains a goal toward which we are striving. This study is now being expanded in a prospectively randomized fashion in order to accrue large enough patient groups to answer more conclusively the questions posed in the introduction.

REFERENCES

1. Alberts DS, Chen HSG, Salmon SE: In vitro drug assay; pharmacologic considerations, in Salmon SE (ed): Cloning of human tumor stem cells. New York, Alan R. Liss, Inc., 1980, pp 197-207.
2. Alberts DS, Chen HSG, Soehnlen B, Salmon S, Surwit EA, Young L: In vitro clonogenic assay for predicting response of ovarian cancer to chemotherapy. Lancet Aug 16:340, 1980.
3. Alberts DS, Chen HSG, Young L, _et al_.: Improved survival for relapsing ovarian cancer patients using the human tumor stem cell assay to select chemotherapy. Proc Amer Soc Clin Oncol 22:462, 1981.
4. Alberts DS, Salmon SE, Surwit EA, Chen HSG, Moon TE, Meyskens FL: Combination chemotherapy in vitro with the human tumor stem cell assay. Proc Am Assoc Cancer Res 22:153, 1981.
5. Breslow W: A generalized Kruskal-Wallis test for comparing K samples subject to unequal patterns of censorship. Biometrika 57:579, 1970.

6. Coldman AJ, Goldie JH: A mathematical model of drug resistance in neoplasms, in Bruchovsky N, Goldie JH (eds): Drug and hormone resistance in neoplasia, Vol. 1. Boca Raton, Florida, CRC Press, Inc., 1982, pp 55-78.
7. Day TG, Gallager HS, Rutledge FN: Epithelial carcinoma of the ovary: prognostic importance of histologic grade. Natl Cancer Inst Monograph 42:15, 1975.
8. Gaines JT, Welander CE, Homesley HD: Improved cloning efficiency in the human tumor stem cell assay using Ficoll gradient separation. Proc Am Assoc Cancer Res 24:311, 1983.
9. Griffiths CT: Surgical resection of tumor bulk in the primary treatment of ovarian carcinoma. Natl Cancer Inst Monograph 42:101, 1975.
10. Hamburger AW, Salmon SE: Primary bioassay of human tumor stem cells. Science 197:461, 1977.
11. Hamburger AW, Salmon SE, Kim MB, et al.: Direct cloning of human ovarian carcinoma cells in agar. Cancer Res 38:3438, 1978.
12. Kaplan EL, Meier P: Nonparametric estimation from incomplete observations. J Am Stat Assoc 53:457, 1958.
13. Moon TE: Quantitative and statistical analysis of the association between in vitro and in vivo studies, in Salmon SE (ed): Cloning of human tumor stem cells. New York, Alan R. Liss, Inc., 1980, pp 209-221.
14. Omura GA, Ehrlich CE, Blessing JA: A randomized trial of cyclophosphamide plus adriamycin with or without cisplatinum in ovarian carcinoma. Proc Am Soc Clin Oncol 1:104, 1982.
15. Ozols RF, Garvin AJ, Costa J, Simon RM, Young RC: Histologic grade in advanced ovarian cancer. Cancer Treatment Reports 63:255, 1979.
16. Salmon SE, Alberts DS, Meyskens FL, et al.: Clinical correlations of in vitro drug sensitivity, in Salmon SE (ed): Cloning of human tumor stem cells. New York, Alan R. Liss, Inc., 1980, pp 223-245.
17. Sondak VK, Korn EL, Morton DL, Kern DH: Absence of in vitro synergy for chemotherapeutic combinations tested in the clonogenic assay. Proc Am Assoc Cancer Res 24:316, 1983.
18. Stanley KE: Prognostic factors for survival in patients with inoperable lung cancer. J Natl Cancer Inst 65:25, 1980.
19. Welander CE, Natale RB, Lewis JL Jr: In vitro growth stimulation of human ovarian cancer cells by xenogeneic peritoneal macrophages. J Natl Cancer Inst 69:1039, 1982.

Feasibility of a Prospective Randomized Correlative Trial of Advanced Non-Small Cell Lung Cancer Using The Human Tumor Clonogenic Assay

THOMAS P. MILLER, M.D.
Assistant Professor of Medicine

LAURIE A. YOUNG, B.S.
Research Associate

LINDA PERROT, M.D.
Research Associate

SYDNEY E. SALMON, M.D.
Professor of Medicine and Director, Cancer Center

University of Arizona Cancer Center and
Veterans Administration Hospital, Tucson, Arizona

INTRODUCTION

New approaches to the treatment of patients with non-small cell lung cancer (NSCLC) are needed. In the United States approximately 130,000 patients developed lung cancer in 1983 and only a small fraction (10%) can expect long term survival following modern surgical approaches. The vast majority of patients have metastatic disease at the time of diagnosis (either clinically apparent or microscopic) and will remain incurable until effective systemic therapy is developed. For the past 15 years patients with NSCLC have received empiric trials of single agents or drug combinations without any consistent evidence of improved survival or palliation of symptoms. This traditional approach to developmental therapeutics has led to inconsistent and contradictory results. For example, patients with advanced NSCLC treated with MACC (methotrexate, adriamycin, cyclophosphamide, and CCNU) were initially reported to have a 44% objective response rate and a median survival of 29-35 weeks (1). A subsequent confirmatory trial, however, noted a 12% response rate and a median survival of 15 weeks (2). These differences in survival duration and response rate correlate most closely

HUMAN TUMOR CLONING
ISBN 0-8089-1671-8

with several pretreatment clinical prognostic features and not with therapy (3,4). In view of this discouraging track record using traditional empiric trials of drugs in patients with NSCLC, it seems appropriate to investigate other methods of identifying active drugs for patients with this disease. Today we report our early laboratory experience of NSCLC in the Human Tumor Clonogenic Assay (HTCA). The results of that experience are then extrapolated to determine if a prospective randomized clinical correlative trial in patients with NSCLC is feasible.

PATIENTS AND METHODS

Between April, 1976 and October, 1983 the HTCA laboratory at the University of Arizona received 200 NSCLC specimens from patients with lung cancer. Single cell suspensions, *in vitro* exposure to pharmacologically-achievable doses of drugs, and plating in a bi-layer soft agar system was performed with modifications of the method of Hamburger and Salmon (5). All NSCLC specimens received by the HTCA laboratory and identified as lung cancer were retrospectively reviewed with regard to histology, site of tumor biopsy, condition of specimen on arrival at the laboratory, and whether or not adequate growth occured to determine *in vitro* drug sensitivity information. Growth of 30 colonies/control plate was required to determine drug sensitivity data. More recently, 30 colonies/ control plate is still required but after baseline (day 1) numbers of viable 60 micron-images are subtracted from the final colony count using the FAS II image analyzer, and a positive control (abrin) is utilized to assure that true proliferative growth from single cells has occurred. Differences in growth rates are tested for significance using the chi-squared test.

RESULTS

In vitro drug sensitivity/resistance information was obtained in 48 of 200 NSCLC specimens (24%) submitted to the HTCA laboratory (Table 1). All samples submitted as NSCLC are included. Adenocarcinoma of the lung was most frequently biopsied and gave the highest rate of drug sensitivity/resistance information (39%). Specimens submitted without histologic information resulted in a very low rate of drug information (7%) which is probably a

TABLE 1

Results by Histology Using HTCA

Histology	No.	No. (%) Studies	No. Inadequate	Adjusted Studies*
Adeno	79	31 (39)	19	31 (48)
Squamous	32	5 (16)	9	5 (17)
Large Cell	27	4 (15)	9	4 (22)
NSCLC	18	5 (28)	3	5 (33)
Unknown	44	3 (7)	25	3 (16)
Totals	200	48 (24)	65	48 (36)

*No. (%) of studies resulting in drug information after subtracting the number of unacceptable specimens from the total number submitted for HTCA.

reflection of the overall care given to the acquisition and submission of this group of specimens (25 of 44 (57%) were thought to be unacceptable for HTCA at the time of arrival in the laboratory). If inadequate specimens are removed from analysis, the rate of obtaining drug sensitivity/ resistance information increases to 36%. The difference in the rate of obtaining successful drug tests by histology is significant p=0.001.

Six factors were identified which predicted a very low likelihood of growth sufficient to obtain drug sensitivity data. These factors are all readily apparent prior to plating of the specimen and include 1) a clinically negative cytology, 2) including preservative with the specimen, 3) contamination, 4) a delay in arrival to the laboratory of greater than 24 hours, 5) insufficient tumor cells (generally less than 6 million), and 6) viable tumor cells comprising less than 25% of the sample submitted using the tryphan blue die exclusion technique (Table 2).

The frequency that drug sensitivity data was obtained is compared to the site of tumor biopsy in Table 3. There is no significant difference between these rates. A majority of specimens came from 4 sites among these patients with NSCLC. Three of the 4 sites (pleural fluid, peripheral

TABLE 2

Unacceptable Specimens Defined

Reason	No. Specimens Received	No. Drug Studies (%)
Cytology negative	45	1
Preservative incl.	3	1
Contaminated	8	0
Delay > 24^0	6	1
Insufficient cells	13	0
< 25 % viable	22	3
Total	97	6 (6)

All lung cancer histologies included.

lymph nodes, and subcutaneous nodules) are readily accessible for biopsy with a minimum of discomfort for the patient.

Fifty-eight of 242 (24%) lung cancer samples submitted between 1976 and 1982 gave adequate growth to evaluate drug

TABLE 3

Rate of Obtaining Drug Sensitivity Information by Tumor Site

Tumor Site	No. Specimens	No. (%) Inadequate	Adjusted Drug Studies (%)
Pleural fluid	72	20 (28)	17 (33)
Lymph nodes	36	19 (53)	11 (65)
Primary	35	10 (29)	7 (28)
Skin & S.C.	24	6 (25)	6 (33)
Unknown	17	6 (35)	4 (36)
Brain	11	3 (27)	1 (12)
Pericardial f.	3	1 (33)	1 (50)
Bone marrow	2	0	1 (50)

sensitivity/resistance. More recently (1983) drug sensitivity/resistance information was obtained in 13 of 36 specimens (36%) without exclusions due to unacceptable specimens. The proportion of unacceptable specimens decreased from 37% (1976-1982) to 25% (1983). This apparent increase in the rate of obtaining drug sensitivity data is not due to patient selection since 36 samples obtained over 9 months in 1983 is similar to 242 samples collected over a 6 year period.

To determine if 36% is an adequate rate for obtaining drug sensitivity data in order to perform a prospective randomized trial of the HTCA in patients with NSCLC, we reviewed our experience in 30 patients entered on an empiric trial of chemotherapy at the University of Arizona over the past year for the availability of tumor for biopsy (Table 4). Eighteen patients (60%) had tumor metastases which were biopsied or had tumor in locations which could have been biopsied. Sixteen biopsies were actually obtained from 12 patients which included 13 different sites of disease. Pleural effusions were collected most commonly and several patients had multiple thoracenteses to obtain samples. Drug studies were obtained in 6 of these 12 patients (50% growth rate). An additional 5 patients had tumor accessible which could have been biopsied.

TABLE 4

Accessible Tumor in 30 Patients
With Metastatic NSCLC Treated on Phase II Drug Studies

Tumor Site	No. Patients	No. Samples	Accessible No Biopsy	No. Drug Studies
Pleural f.	6	9	0	3
Pericardial f.	2	1	1	1
Neck nodes	5	5	0	2
Skin/S.C.	4	0	4	
Primary	1	1	0	
Totals	18	16*	5	6

*16 biopsies from 12 patients (13 sites)

DISCUSSION

Growth in culture was adequate for in vitro testing, and subsequently drug sensitivity/resistance data was obtained in 24% of all NSCLC specimens submitted from 1976 to 1983 and in 36% of specimens submitted during 1983. During 1983, 30 patients were entered on phase II empiric trials of chemotherapy for metastatic NSCLC, and 16 patients (60%) had tumor accessible for biopsy for HTCA. To determine if a 36% growth rate in 60% of patients with metastatic NSCLC is adequate for randomized correlative trial, we estimated the numbers of patients required to detect a difference in the response rate at a significance level of p=0.05 and a power of 0.8 (Table 5). The estimate requires information regarding the anticipated response rate to empiric (standard) treatment and HTCA-directed therapy. First, the response rate to empiric single agent therapy is estimated to be 15% based on past results of single agents in patients with NSCLC (6). The response rate for a large number of patients treated with adriamycin, vinblastine, vindesine and cis-platinum varies between 14%-18% (6). Second, the response rate in NSCLC to HTCA-directed therapy is unknown. However, from experience in other solid tumors it is estimated to be between 45% (melanoma) and 60% (ovarian carcinoma) (7,8). If patients are stratified for known important prognostic features, and if the randomization is weighted to reflect current laboratory growth rates (36% in this study), then a randomized correlative study to compare

TABLE 5
Proposed HTCA Randomized Trial
Statistical Considerations

Response Rate - Assumptions		No. Patients	
Empiric	HTCA Directed	ARM	Total
15%	45%	39	245
15%	60%	20	127

*To detect differences in response rate p=.05, power = 0.8.

assay selected therapy to empiric therapy is feasible (Table 5). The numbers of patients with NSCLC required to detect a significant difference in the response rates given in Table 5 is readily available even at relatively small centers.

A number of other factors will influence the statistical considerations. The histologic subtype of NSCLC is a significant determinant of the proportion of samples which provide drug sensitivity/resistance information. Adenocarcinoma is associated with a 39% yield of drug information (1976-1983) compared to a 16% yield for squamous histologies. At the University of Arizona approximately 66% of patients entering phase II treatment trials of advanced disease have had adenocarcinomas which would allow a decrease in the estimated number of patients required for study. The rate of obtaining drug sensitivity information (as opposed to drug resistance information) also influences the estimates of the numbers of patients required to reach significance. This proportion is not accurately known for NSCLC due to several problems in a retrospective review of a developmental research tool. First, most patients had been previously treated which might be expected to reduce the frequency of finding an _in vitro_ active drug. Secondly, there are very few correlations with single agents to determine the degree of colony inhibition which correlates with clinical response. Thus, what is meant by "sensitive" _in vitro_ is unknown. Third, the proportion of drug tests performed resulting in significant growth inhibition depends on the drugs available (investigational) and actually tested _in vitro_. In our experience, considering only previously untreated patients, using 70% growth inhibition at pharmacologically achievable doses of drugs, and considering only samples that were tested with drugs having a known response rate in NSCLC, approximately 80% of specimens with growth of 30 colonies/plate result in identifying an active drug. This very important proportion needs to be further defined and more accurately measured before exact estimates of patient numbers can be determined for a randomized correlative trial.

Several important variables remain to be defined before exact numbers of patients required to perform a randomized, correlative test of HTCA-directed therapy in NSCLC can be determined. However, the high incidence of the disease, the large number of patients with tumor metastases in accessible locations, and an adequate _in vitro_ growth rate make such a study feasible. The inability of traditional phase II trials to produce consistent results free of investigator bias based on patient selection makes such a study desirable.

REFERENCES

1. Chahinian AP, Mandel EM, Holland JF, et al: MACC (methotrexate, adriamycin, cyclophosphamide and CCNU) in advanced lung cancer. Cancer 43: 1590-1597, 1979.
2. Vogl SE, Mehta CR, Cohen MH: MACC chemotherapy for adenocarcinoma and epidermoid carcinoma of the lung. Low response rate in a cooperative group study. Cancer 44: 864-868, 1979.
3. Stanley KE: Prognostic factors for survival in patients with inoperable lung cancer, JNCI 65: 25-32, 1980.
4. Miller TP, Vance RB, Tong TC: Treatment of advanced non-small cell lung cancer with mitomycin C, cis-platinum, and vindesine: Comparisons to other mitomycin C, vinca-containing combinations. Proc Amer Soc Clinical Oncology 3: in press, 1984.
5. Hamburger AW, Salmon SE: Primary bioassay of human tumor stem cells. Science 197: 461-463, 1977.
6. Takasugi BJ, Miller TP: Chemotherapy of advanced non-small cell lung cancer: A review. Invest New Drugs, in press, 1984.
7. Meyskens FL, Moon TE, Dana, et al: Quantitation of drug sensitivity by human metastatic melanoma colony forming units. Br J Cancer 44: 787-797, 1981.
8. Alberts DS, Chen HSG, Salmon SE et al: Chemotherapy of ovarian cancer directed by the human tumor stem cell assay. Cancer Chemother Pharmacol 6: 279-285, 1981.

ACKNOWLEDGEMENTS

Supported in part by Public Health Service grant CA-17094 and CA-21839 from the National Cancer Institute, National Institute of Health, Department of Health and Human Services, Bethesda, MD 20205.

COLONY GROWTH AND PATIENT SURVIVAL IN PRIMARY OR METASTATIC BREAST CANCER A PRELIMINARY ANALYSIS

MATTI S. AAPRO
Division of onco-hematology
PETER SCHAEFER
Clinic of gynecology
RETO ABELE
Division of onco-hematology
FELIX KRAUER
Clinic of gynecology
PIERRE ALBERTO
Division of onco-hematology
University Hospital of Geneva
Switzerland

C. CILLO
N. ODARTCHENKO
Department of Cell Biology
Swiss Institute for Cancer Research
Lausanne, Switzerland

Clonogenic assays provide a tool to study the biology of fresh human tumor samples. Interest in these techniques has increased since Hamburger and Salmon reported an improved method to grow fresh tumor colonies (4) Several types of clonogenic assays are presently available (2, 3, 6). Since 1980 our group has used a methylcellulose based assay which supports growth of tumor colonies originating from several types of sources (1). In an effort to gain some understanding of the significance of growth versus non-growth, we have analyzed the possibility that such _in vitro_ features might correlate to the survival of breast cancer patients.

MATERIALS AND METHODS

Solid tumor samples were minced after operation into

HUMAN TUMOR CLONING
ISBN 0-8089-1671-8

Iscove's Modified Dulbecco's Medium (IMDM, Gibco) with 10% foetal calf serum (FCS) screened for optimal hemopoietic colony growth support. After transportation to the laboratory these samples were minced further, passed through needles, treated with 1% trypsin and passed through needles again.

Malignant effusions collected with preservative-free heparin were centrifuged, and the cells resuspended in IMDM + 10% FCS. Centrifugation over a Ficoll-Hypaque gradient was used and the cells washed twice (9).

10^5 cells were mixed with one milliliter of a final concentration of 0.8% methylcellulose in IMDM + 15% FCS and plated in 35-millimeter plastic Petri (non-tissue culture) dishes. Cultures were incubated at 37°C in humidified air/5% CO_2.

Colonies are defined as aggregates of 50 or more cells counted at 21 to 28 days after plating using an inverted microscope.

Control of the tumoral nature of the colonies was done by random cytocentrifuge preparation of individual colonies.

Patient's records were examined to obtain data on the surgeon's and pathologist's postsurgical pTNM (UICC) classification, exact nodal status, estrogen receptor status, time to relapse and date of last follow-up or death. All this data had been recorded by observers unaware of the in vitro results.

RESULTS

Fourty samples have been sent for culture between November 1980 and May 1982. Three solid tumor specimens and 3 pleural effusions were not plated because of insufficient cell number (or no tumor cells in effusion) or contamination. One patient has been lost to follow-up.

Twenty-six solid tumor samples and 7 pleural effusions were plated. Distribution of colony growth is shown on table 1.

Table 1
Breast cancer in a methylcellulose based clonogenic assay

SOURCE	colonies/10^5 viable cells 0	1-9 (median 5)	10-210 (median 44)
PRIMARY	12	7	7
PLEURAL EFF.	2	2	3

Table 2
Solid tumor patients

	Nb samples	no growth	growth	(months) survival (median)
T_{1-3}	18	8	10	21
T_4	8	4	4	18.5
N_0	9	3	6	21
N_{1-3}	11	8	3	21
N_{4+}	6	1	5	12
M_0	20	10	10	21
M_1	6	2	4	12
ER_+	17	9	8	21
ER_-	9	3	6	21

T = size of tumor and clinical characteristics per UICC criteria
N_0 = negative nodes N_{1-3} = 1 to 3 positives nodes
N_{4+} = 4 or more nodes
M_0 = no metastatis at preoperative evaluation
M_1 = metastatic at preoperative evaluation
ER = estrogen receptor + = positive or unknown
− = negative

In table 2 T, N, M and ER status are related to colony growth and patient survival.

Patients with T_4, N_{4+}, M_1 status seem to fare worse in terms of survival as compared to patients with T_{1-3}, N_{0-3}, M_0 status. ER receptor status does not seem to make any significant difference in terms of survival in this small series of breast cancer patients.

Tumor colony growth does not correlate to either T, M or ER status. Samples of the 6 patients with 4 or more positive nodes show better *in vitro* growth when compared to N_{1-3} patients (but p=0.17) and the difference is not significant when compared to N_0 patients (p=0.87)(Fisher's exact test).

Table 3
Survival (absolute and median duration, months) relative to "in vitro" growth of breast cancer

	dead/alive	relapse free survival	survival
		SOLID TUMOR SAMPLES	
No growth	2/10	17+	21+
1-9 colonies	5/2	12	16
10 or more	4/3	9	16
		PLEURAL EFFUSIONS	
No growth	2/0	--	17
1-9 colonies	0/2	--	29+
10 or more	3/0	--	13

Table 3 shows that there is a trend for a shorter relapse-free survival for patients whose tumor sample showed significant growth in vitro. Overall survival is not different in terms of median duration.

DISCUSSION

In this small series we have observed that 58 percent of the tumor samples obtained from patients having breast cancer will show growth of tumor colonies (in vitro). Significant growth (more than 10 colonies/10^5 viable cells) has been seen in 30 percent of the samples. These growth characteristics are not directly comparable to those reported by other authors (7) but do not seem significantly different. In a parallel study done at the same time as this one, we have observed that in our hands the soft-agar assay gave results similar to the methylcellulose assay (2).

We have observed the expected correlation between tumor size, node and metastatic status with patient survival. This small series does not show a difference of survival between patients with ER positive and ER negative tumors. The short median follow-up, many other variables and specially the small number of tests might explain the apparent lack of predictiveness of ER receptors.

Tumor colony growth does not seem to be statistically correlated with lymph-node status. However, the small number of samples received for each subset of patients does not allow for a definitive conclusion. The other parameters studied (T, M, ER) do not seem to predict for tumor colony growth "in vitro". We have yet to examine if our series reproduces the reported influence of histological grade on the success of cultures (10).

Breast cancer growth "in vitro" might thus be an independent variable and therefore we studied if it would predict for the clinical behaviour of the tumor in terms of relapse-free and overall survival of the patients. Patients whose solid tumor samples did not grow in vitro have a relapse-free survival of 17+ months and those whose sample grew 10 or more colonies have a relapse-free survival of 9 months. Overall survival of patients with primary breast cancer samples is not different between those whose sample grew or didn't grow. Longer follow-up is needed as it is clear from table 3 that most patients (10 of 12) whose sample didn't grow are still alive whereas 9 of 14 patients whose sample grew are already dead.

The very small number of pleural effusion samples analysed does not allow any conclusion to be drawn.

Our results imply that in vitro growth of tumor colonies might predict the aggressiveness of the patient's particular breast cancer. The small number of patients does however not allow any conclusion as to the independence of this factor from several other predictive parameters. Another group has recently shown that primary stage IV breast cancer patients whose tumor grew in vitro have a poorer prognosis whereas in recurrent stage IV disease colony growth had no correlation with survival (9).

Larger series where node negative patients can be studied should be analyzed to study if this supplementary information can have any clinical usefulness. Also, badly needed improvements of the clonogenic assays for human tumor cells (4, 8) might modify in vitro growth characteristics.

These new growth conditions will require an independent analysis to see how they affect the predictive value of this assay.

REFERENCES

1. Cillo C, Aapro MS, Eliason JF, et al: Methylcellulose clonogenic assay for solid human tumors. Proc AACR 23:312, 1983.
2. Cillo C, Abele R, Alberto P, Odartchenko N: Culture of colony-forming cells from human solid tumors. Proc 1st European Conf Clin Oncol, UICC, Lausanne, p 114, 1981.
3. Courtenay VD, Selby PJ, Smith IE et al: Growth of human tumor cell colonies from biopsies using two-soft agar techniques. Br J Cancer 38:77-81, 1978.
4. Eliason JF, Fekete A, Odartchenko N: Improving techniques for clonogenic assays. In: Hofmann V, Berens M, Martz G (eds): Predictive drug testing on human cells. Berlin, Springer Verlag (in press, 1984).

5. Hamburger AW, Salmon SE: Primary bioassay of human tumor stem cells. Science 197:461-463, 1977.
6. Salmon SE, Hamburger AW, Soehnlen B et al: Quantitation of differential sensitivity of human-tumor stem cells to anticancer drugs. N Engl J Med 298:1321-1327, 1978.
7. Sandbach J, Von Hoff DD, Clark G et al: Direct cloning of human breast cancer in soft agar culture. Cancer 50:1315-1321, 1982.
8. Selby P, Buick RN, Tannock I: A critical appraisal of the "human tumor stem cell assay". N Engl J Med 308:129-134, 1983.
9. Sutherland CM, Mather FJ, Carter RD et al: Breast cancer as analyzed by the human tumor stem cell assay. Surgery 94:370-375, 1983.
10. Touzet C, Ruse F, Chassagne J et al: In vitro cloning of human breast tumor stem cells: influence of histological grade on the success of culture. Br J Cancer 46:668-669, 1982.

TESTING OF MAMMARY CANCER IN THE HUMAN TUMOR STEM CELL ASSAY

DITTRICH,Ch.[1], SATTELHAK,E.[1], JAKESZ,R.[2], KOLB,R.[2], HOLZNER,H.[3], HAVELEC,L.[4], LENZHOFER,R.[1], STEININGER,R.[2], VETTERLEIN,M.[5], MOSER,K.[1], SPITZY,K.H.[1]

[1]Department of Chemotherapy
[2]Ist Department of Surgery
[3]Institute of Pathology
[4]Institute of Biostatistics
[5]Institute of Cancer Research
of the University of Vienna, Austria
A-1090 VIENNA, Lazarettgasse 14

In the attempt to shift cytotoxic treatment of tumor patients from a purely empirical or statistical approach to a more individual, patient-oriented management, different methods have been applied. Numerous tests have been developed, designed to discover before the onset of treatment whether or not a cytostatic agent would be effective. Thus tumor-resistant substances may be eliminated a priori and only tumor-sensitive ones administered. Both from a theoretical and clinical point of view, the Human Tumor Stem Cell Assay (HTSCA) has turned out to be most appropriate for this purpose (4,13).

The stem cell concept is the theoretical background for the human tumor cloning assay, postulating that the so-called tumor stem cell is the origin of any tumor; this cell produces both cells which are identical to the original stem cell in their proliferative capacity as well as cells of all levels of differentiation (15). It is therefore responsible for tumor growth and metastasizing, success of treatment depending on the eradication of these stem cells. The HTSCA adequately reflects action and effect of cytostatic therapy on these cells; therefore it seems to represent a most suitable approach.

For some tumor entities correlation was good between the test results obtained by the stem cell assay and the prognosticated clinical development (9,11,16,17,18,19). Moreover, in myeloma (3) and head and neck cancer (10),

HUMAN TUMOR CLONING
ISBN 0-8089-1671-8

HTSCA turned out to be a very useful prognostic indicator. Treatment of ovarian cancer based on the test proved for the first time that the results were superior to those obtained by standard methods (Alberts, DS., personal communication).

Cases of breast cancer seem to be a special challenge to the test system since even in advanced stages - though no cure - alleviation may be attained. Resistance and clone selection are the factors responsible for failure of action in breast cancer that had been treated effectively before. HTSCA seems adequate in terms of coping with resistance and clone selection, indicating conditions for appropriate treatment, especially after the onset of chemoresistance. Contrary to hypothetical expectations however, practical results are modest indeed (1,12,14). Only the South Texas Central Human Tumor Cloning Group has reported about a larger number of test series (14), other authors confining themselves to a few cases only (1,12). Percentages for tumor growth vary because of differing definitions of the term "sufficient growth" - ranging from five to 50 colonies per plate. The median number of colonies below 50 and cloning efficiency between 0.001 and 0.6% must be considered low.

The following constitutes a brief survey on our own data obtained with the HTSCA in mammary carcinoma (2), using the technique developed by Hamburger and Salmon (4,13).

Out of 77 tumor specimen, 39 (51%) were suitable for test purposes. 45% of the specimen that had to be excluded produced negative cytological results, did not have sufficient viability (below 30%), in 13% the yield of tumor material was too low and 3% of the specimen were contaminated. 56% of the evaluated specimen were obtained by biopsy - 17 primary tumors, two local and three lymph node recurrences. The 17 effusions comprised 11 pleural effusions, 5 ascites and one pericardial effusion with a median viability of 91% - clearly above thatof the biopsies (55%).

Clonal growth was identified in 44% (17/39) of our specimen; biopsies were positive in 18% and effusions in 77%. Growth was sufficient in half of the biopsies and two thirds of the effusions - provided we consider a minimum of 20 colonies per plate as a standard. In the biopsies the median number of colonies with growth was 18, in the effusions 67. Median plating efficiency of all tests was at a level of 0.01%.

Next we studied the effect of autologous serum on the plating efficiency in primary breast cancer, since it was reported that tumor growth might be stimulated by adding serum antibodies (5,7,8). Studies using the colony inhibition technique showed that sera from human and animal - bearing tumors were occasionally effective in blocking tumor cell destruction by lymphocytes in vitro (efferent immune response enhancement effect) (5,6). For this purpose the serum of the patients whose specimen were being tested was added to the upper layer in a final concentration of 25%. Irrespective of whether autologous serum was added or not, there was no statistical significant difference in colony growth, implying in particular that growth was not stimulated.

For technical reasons it was only possible to use five of the 11 tumor specimen with sufficient growth for a comparison between clinical and test results. Expectations regarding treatment on the basis of these test results were confirmed in both cases where sensitivity and in the three cases where resistance was predicted.

Our own results - in line with international experiences - clearly show that the HTSCA is not yet as successfully applicable to mammary carcinoma as to other tumor entities. A number of questions of technique in particular still have to be resolved. Often the tumor cell yield in mammary cancer which is sometimes scirrhous in texture, is not adequate. Also, growth rates and cloning efficiency are frequently insufficient - mostly due to lack of suitable media. Simulation of polychemotherapy in vitro is problematic as such since not enough is known as yet about drug interaction in vivo.

On the one hand clone selection and resistance of solid tumors are important indications for applying the HTSCA. However, heterogeneity of such tumors - causing clone selection and resistance - on the other hand constitutes a general limiting factor in treating solid tumors and therefore a specific one in the case of the HTSCA. The test might indicate effective treatment which is rendered ineffective in vivo with the onset of resistance.

Even though the test as it is now carried out is too sophisticated for routine use, we may expect clinically relevant results from it for the benefit of patients suffering from cancer of the breast - particularly after some questions of methodology and tumor biology in general have been solved.

LITERATURE

1.) Benard, J., Da Silva, J. et al.: Culture of clonogenic cells from various human tumors: drug sensitivity assay. Eur.J.Cancer Clin.Oncol.,19,1,65-72, 1983.

2.) Dittrich, Ch., Holzner, H., et al.:First results of the "Viennese tumor sensitivity testing group" with the human tumor cloning assay. Proc.13th Int.Congress of Chemotherapy, Vienna 1983. K.H.Spitzy, K.Karrer (eds.). Publ.: H.Egerman.

3.) Durie, B.G.M., Young, L. et al.: Human myeloma stem cell culture:relationship between in vitro drug sensitivity and kinetics and patient survival duration. Blood 61, 929-934, 1983.

4.) Hamburger, A.W., Salmon, S.E.: Primary bioassay of human tumor stem cells. Science 197, 461-463, 1977.

5.) Hellström, K.E., Hellström, I.: Cellular immunity against tumor antigens. Adv.Cancer Res.12:167, 1969.

6.) Hellström, K.E., Hellström, I.: Lymphocyte mediated cytotoxicity and blocking serum activity to tumor antigens. Adv.Immunol.18: 209, 1974.

7.) Kaliss, N.: Immunologic enhancement. Int.Rev.Exp.Pathol.8:241, 1969.

8.) Kaliss, N.: Dynamics of immunologic enhancement. Transplant Proc.2: 59, 1970.

9.) Mann, B.D., Kern, D.H., et al.: Clinical correlations with drug sensitivities in the clonogenic assay . Arch.Surg.117, 33-36, 1982.

10.) Mattox, D.E., Von Hoff, D.D.: In vitro stem cell assay in head and neck squamous carcinoma. Amer.J.Surgery, 140, 527-530, 1980.

11.) Meyskens, F.L., Moon, T.E. et al.:Quantitation of drug sensitivity by human metastatic melanoma colony-forming units. Br.J.Cancer 44, 787-797, 1981.

12.) Pavelic, Z.P., Slocum, H.K. et al.:Growth of cell colonies in soft agar from biopsies of different human solid tumors. Cancer Res.40, 4151-4158, 1980.

13.) Salmon, S.E., Hamburger, A.W. et al: Quantitation of differential sensitivity of human tumor stem cells to anticancer drugs. N.Engl.J.Med., 298, 24, 1321-1327, 1978.

14.) Sandbach, J., Von Hoff, D.D. et al. and the South Central Texas Human Tumor Cloning Group.: Direct cloning of human breast cancer in soft agar culture. Cancer, 50, 1315-1321, 1982.

15.) Steel, G.G.: Growth kinetics of tumors: cell population kinetics in relation to the growth and treatment of cancer. Oxford: Clarendon Press, 1977.

16.) Tveit, K.M., Fodstad, O. et al: Colony growth and chemosensitivity in vitro of human melanoma biopsies: relationship to clinical parameters. Int.J.Cancer 29, 533-538, 1982.

17.) Von Hoff, D.D., Cowan, J.et al.: Human tumor cloning: feasibility and clinical correlations. Cancer Chemother.Pharmacol. 6, 265-271, 1981.

18.) Von Hoff, D.D., Casper, J. et al:Association between human tumor colony forming assay results and response of an individual patient's tumor to chemotherapy. Am.J.Med.70 1027-1032, 1981.

19.) Von Hoff, D.D., Clark, G.M. et al: Prospective clinical trial of a human tumor cloning system. Cancer Res.43, 1926-1931, 1983.

The Clinical Usefulness of the Human Tumor Clonogenic Assay (HTCA) in Breast Cancer

Stephen E. Jones
Professor of Medicine, Department of Internal Medicine
Chief, Section of Hematology - Medical Oncology

Judith C. Dean
Laurie Young
Research Associates, Cancer Center

Sydney E. Salmon
Professor of Medicine, Department of Internal Medicine
Director, Cancer Center

University of Arizona Cancer Center
Tucson, Arizona 85724

SUMMARY

The human tumor clonogenic assay (HTCA) was used to evaluate 337 samples of human breast cancer for clonal growth and chemosensitivity to a variety of standard and experimental anticancer agents. Tumor colony forming units (TCFU's) were observed in 44% of studies and 27% were adequate for drug sensitivity studies. Clinical correlations to in vitro drug activity were evaluable in 59 patients. For single agent therapy, response was predicted in 56% and resistance in 89% of 25 cases. For combination chemotherapy, response was predicted in 50% and resistance in 65% of 34 cases. Future research with the HTCA in breast cancer needs to define the conditions necessary to enhance clonal growth in order to exploit the apparent usefulness of the HTCA in predicting clinical activity of single agents for treatment.

Research supported in part by grants CA 17094 and CA 21839 from the National Cancer Institute, DHEW, Bethesda, MD.

HUMAN TUMOR CLONING
ISBN 0-8089-1671-8

INTRODUCTION

The human tumor clonogenic assay (HTCA) has been used to extensively study the clonal growth and chemosensitivity of human cancer and has been the subject of three international conferences at the University of Arizona Cancer Center (1). The HTCA has been reported to predict responsiveness for a variety of human cancers to single agent chemotherapy as well as to accurately predicting resistance (2). The purpose of this paper is to report in a preliminary fashion our data on the use of the HTCA in breast cancer. A more detailed analysis is currently in preparation.

PATIENTS AND METHODS

Between 1978 and 1983, 337 adequate single cell suspensions of human breast cancer were studied by the clonogenic assay laboratory. Tumor tissue was obtained from a variety of sources with the most common being malignant pleural effusions, ascites, metastatic lymph nodes, primary tumors and skin metastases. The assay technique has been previously described (1,2) and is discussed in many chapters in this volume. The clinical records of patients on whom in vitro drug sensitivity information was obtained were reviewed with respect to the type of chemotherapy they received. A majority of patients were included in prospective trials with the intent of correlating in vitro and clinical response. Patients were treated with either single agent or combination chemotherapy programs for metastatic breast cancer. Clinical responses were defined as follows: complete response (CR) = complete disappearance of all symptoms and signs of breast cancer for at least one month; partial response (PR) = 50% or greater decrease in all measurable tumor for at least one month; improved = decrease in tumor not easily measured (for example, ascites pleural effusions, diffuse skin disease) for at least one month. In vitro sensitivity for each anticancer drug was defined as a 50% or greater inhibition of tumor colony forming units (TCFU's) using a one hour incubation of the agent at 10% of the clinically achievable concentration as previously described (1,2,3). In vitro sensitivity for drug combinations was defined as a 50% or greater inhibition of TCFU's based upon the product of individual drug sensitivities (4).

RESULTS

Among the 337 adequate single cell suspensions, colony growth was obtained in 44%. Ninety-one samples (27%) yielded adequate colony growth (greater than 30 colonies per plate) for drug testing and 85 were actually tested against variety of anticancer drugs. Virtually all anticancer drugs, whether standard or experimental, showed evidence of activity against breast cancer ranging from 14% for doxorubicin to 54% for bisantrene. In general, the in vitro response rates were similar to what has been observed for clinical activity of the various anticancer drugs. However there were some exceptions including the low response rate to doxorubicin.

Fifty-nine correlations of clinical response with the results of the HTCA were available including 25 prospective single agent correlations and 34 combination chemotherapy correlations. Among the 59 patients treated with either single agent or combination chemotherapy, 25 clinical responses were noted: 7 patients had improvement, 16 had partial responses, and 2 had complete responses. With respect to the single agent correlations, 16 patients were predicted to be sensitive to a variety of specific single agents and 9 (56%) responded to treatment with the appropriate single agent. Likewise, 9 patients were predicted to be resistant and only one of these patients responded. Therefore, resistance to single agents was correctly predicted in 89% of cases.

Thirty-four patients were evaluable for combination chemotherapy correlations. Among 14 patients whose tumors were predicted to be sensitive to the drug combinations, 7 (50%) of these patients responded to the appropriate treatment. Resistance to combination chemtherapy was correctly predicted in 13 (65%) of 20 additional cases.

DISCUSSION

In this study we were able to demonstrate that colony growth of breast cancer could be achieved from 44% of adequate single cell suspensions prepared from fresh biopsy samples of human breast cancer. Sufficient colony growth for in vitro drug testing occurred in only 27% of cases. With adequate colony growth, in vitro sensitivity to a variety of standard and experimental chemotherapy agents could be readily demonstrated and was generally in the approximate range of activity as would be predicted from known clinical activity of these agents. Thus, the HTCA appears to identify active chemotherapy agents in breast cancer as it has for other types of human cancers (1,2).

There is relatively little data on clinical correlations with the HTCA in human breast cancer (3). We were able to evaluate 59 clinical trials of either single agent or combination chemotherapy in relationship to in vitro results with the HTCA. Published results in other types of cancer treated with single agents indicate that approximately 70% of patients manifesting evidence of in vitro sensitivity to a particular chemotherapeutic agent will respond to that agent (1,2,5). Conversely drug resistance appears to be correctly identified in 85-90% of cases (1,2,5). Our experience with single agent predictivity in breast cancer using the HTCA is very similar with a 56% true response rate for activity and an 89% prediction of resistance.

There is a paucity of data about using the HTCA to predict the effect of drug combinations in any type of human cancer. In this paper we are reporting 34 correlations of clinical response with in vitro activity to the appropriate drug combinations. We used the product of surviving fractions of TCFU's as one indication of potential in vitro activity for the drug combination (4). We were able to successfully predict clinical response in 50% of 14 patients and to predict resistance in 65% of 20 patients. It is not surprising that resistance was not identified as accurately for combination chemotherapy as for single agents, because it fails to take in account the possibility of drug synergism, optimal doses and schedules etc. Additionally, clinical doses of drugs are often reduced in drug combinations whereas they are tested in full dose in vitro. In the future, testing of two or more agents together (as well as individually) in HTCA may prove to be helpful. However, there are major logistic and interpretive problems in the assessment of drug combinations in vitro. Further research is needed to enhance breast cancer growth in vitro and to perfect a suitable means to assess combination chemotherapy in vitro.

REFERENCES

1. Salmon, S. Cloning of human tumor stem cells. New York: A.R. Liss, Inc, 1980.

2. Salmon SE, Hamburger AW, Soehnlen B et al: Quantitation of differential sensitivity of human tumor stem cells to anticancer drugs. N Engl J Med 298: 1321-1327, 1978.

3. Sandbach J, Von Hoff DD, Clark G et al: Direct cloning of human breast cancer in soft agar culture. Cancer 50: 1315-1321, 1982.

4. Alberts DS, Salmon SE, Chen HSG et al: Pharmacologic studies of anticancer drugs using the human tumor stem cell assay. Cancer Chemother and Pharmacol 6:253-264, 1981.

5. Von Hoff DD, Clark GM, Stogdill BJ et al: Prospective clinical trial of a human tumor cloning system. Ca Res 43: 1926-1931, 1983.

Use of Normalized Drug Sensitivities of Human Breast Tumors for Clinical Correlations and For Comparisons of Drug Activities

Verena Hug, M.D.

Assistant Professor of Medicine
Medical Breast Department

Howard Thames, Ph.D.

Professor of Biomathematics
Department of Biomathematics

Margot Haynes

Research Assistant II
Medical Breast Department

George Blumenschein, M.D.

Professor of Medicine
Medical Breast Department

Benjamin Drewinko, M.D., Ph.D.

Professor of Medicine
Department of Laboratory Medicine

Gary Spitzer, M.D.

Associate Professor of Medicine
Department of Hematology
M. D. Anderson Hospital and Tumor Institute
Houston, Texas

HUMAN TUMOR CLONING
ISBN 0-8089-1671-8

INTRODUCTION

We have investigated inherent drug sensitivities of breast tumor cells, to clarify some aspects of clinical treatment response. The clinical treatment response is determined by the patient's performance status, the tumor bulk and the proliferative activity and inherent drug sensitivity of tumor cells; but the contributory role of each of these factors is unknown.

A comparison of the cytotoxic effects of four anthraquinone derivatives of ten patients' bone marrows and tumors revealed that the drug effects were variable and unpredictable on the progenitor cells of the tumors, but were similar on those of the bone marrows (1). Based on this observation, we began to normalize the in vitro activity of anticancer agents by their effects on bone marrow progenitors, the dose-limiting host tissue cells for many anticancer agents, thereby permitting comparison of drug sensitivities of tumor cells to different drugs at equitoxic drug concentrations. We determined the drug sensitivity patterns of individual tumors and of groups of tumors with determinants on disease course. We also defined the relative antitumor effects of agents used for the primary and for the secondary treatment of breast tumors, including that of drug analogs.

MATERIALS AND METHODS

One hundred and twenty-five breast tumors were cultured over a period of 21 months. Double-layer agar cultures were used, as described previously (2-5). Eighty-six percent of these tumors formed more than 30 colonies and 66% (82) more than 50 colonies. These latter 82 tumors were used for the analysis. Twenty-six percent of these tumors were derived from untreated patients and 74% from patients that had received chemotherapy prior to the study. Thirty-four percent of the tumors were estrogen receptor-positive, 37% were estrogen-receptor-negative, with no determination in the remainder of the tumors. The median clonogenic efficiency of all controls was 0.024, with an average number of colonies of 191 per 500,000 cells seeded. The average number of colonies was twice as high in previously treated tumors as it was in untreated tumors (225 vs 120). Two hundred and thirty-four drug sensitivity determinations were performed, 60 on untreated tumors and 174 on treated tumors.

TABLE 1-1

Drugs Compared For Their Cytotoxic Effects On Breast Tumors

	Drug Concentration Ranges: (μg/ml)
Primary Agents For The Treatment Of Breast Carcinoma	
Adriamycin	0.005 - 0.025
Cyclophosphamide	1.0 - 11.0
5-Fluorouracil	0.75 - 2.10
Secondary Agents For The Treatment Of Breast Carcinoma	
Cis-Platinum	0.01 - 0.60
Vinblastine	0.0001 - 0.0100
Elliptinium-Acetate	0.001 - 0.40
VP-16	0.0001 - 0.0600
Anthraquinone Derivatives	
Adriamycin	0.005 - 0.025
4'-Epi-Doxorubicin	0.00125 - 0.02000
Mitoxantrone	0.00008 - 0.004

Table 1 lists the type and concentration ranges of the drugs that were assayed. Continuous drug exposure was used. Drugs that were assayed included primary agents for the treatment of breast carcinoma, doxorubicin, cyclophosphamide and 5-fluorouracil; secondary agents for the treatment of breast carcinoma, including that of the investigational agent, elliptinium acetate, and four anthraquinone derivatives that are currently in clinical trials.

To normalize in vitro drug activities, comparison was made of the sensitivities of tumor progenitors and bone marrow progenitors. The model developed for this purpose is summarized in Figure 1, and has been described previously (6). Briefly, we found that within a cell-kill range of 0-60%, any one drug investigated killed similar proportions of granulocyte macrophage colony-forming units (GM-CFU) from bone marrows of different donors. The effect of any one drug could

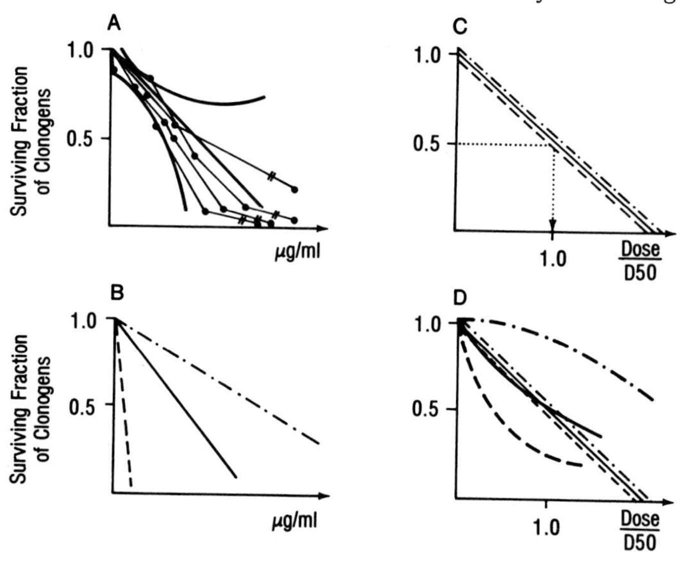

Figure 1: Development of model for normalization of in vitro drug activities. Chart A represents dose-responses of GM-CFU from four donors to a given drug. The thick lines indicate the regression line for all observations (log-linear line) and its confidence bands (curved line). Chart B illustrates the hypothetical regression lines for three different drugs. Chart C shows the same three lines after scaling of the abscissa to doses in units of D50 of each drug. Chart D shows dose-responses to these three drugs of both tumor and bone marrow progenitors. Reproduced with permission of Cancer Research.

TABLE 1-2

Distribution Pattern Of Tumor Sensitivities

(62 Tumors)

Drug-Sensitivity Distribution Of Tumors	Percent of Tumors
Confined To One Class of Sensitivity	29
Confined To Two Adjacent Classes of Sensitivity	52
Distributed Over Two Nonadjacent Or All Three Classes Of Sensitivity	19

therefore be represented by a regression line. By scaling of the abscissa to doses in units of D50 of each drug, these lines could be brought to overlap, thereby permitting a better representation of differences of inherent drug sensitivities of host and tumor cells (Chart B).

Figure 2 shows the measures of sensitivity used to compare inherent drug sensitivities of individual tumors and of groups of tumors with different biologic properties. We have previously reported that sensitivities determined in this way are in reasonable agreement with the clinical drug responsiveness (6) in that 67% of tumors from patients responding to doxorubicin were also sensitive to that drug in vitro, and 53% of tumors from patients resistant to doxorubicin were also resistant to doxorubicin in vitro.

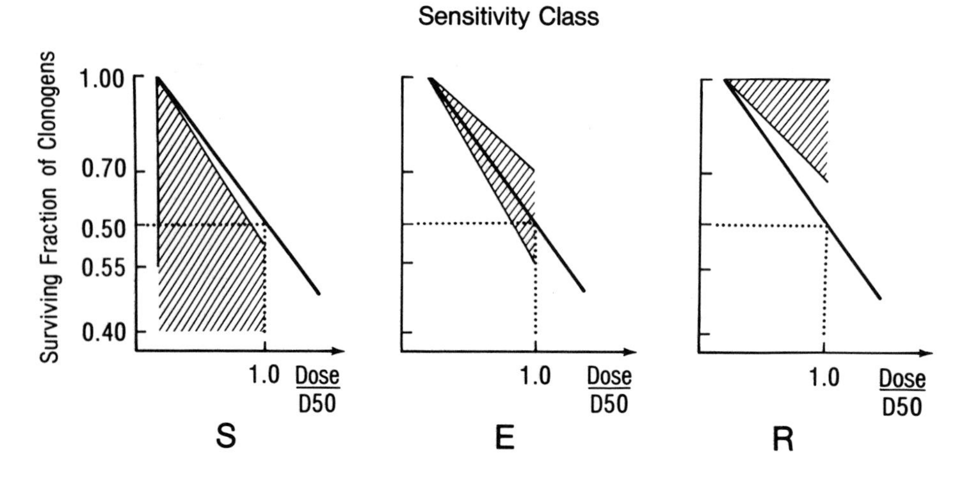

Figure 2: Measures of sensitivity used to compare inherent sensitivities of individual tumors and of groups of tumors that bear prognostic significance. The thick log-linear line represents the GM-CFU regression-line, normalized for all drugs investigated. The shaded areas outline the three classes of tumor response relative to that of bone marrow. E refers to tumors that are equally sensitive to bone marrow. They have dose-responses within the area overlying the bone marrow regression line that encompasses the cell-kill range of 30-55% at the LD50; S refers to tumors that are more sensitive than bone marrow. Their dose-responses are within the area to the left; R refers to the tumors less sensitive than bone marrow. Their dose-responses are within the area to the right.

To compare the antitumor effects of different cytotoxic drugs, the areas enclosed by the tumor and bone marrow survival curves were calculated and their relative size was used for the ranking of the drug effects. With this criterion, the concordance of in vitro and in vivo findings for the tumors from 21 patients treated with single agents was 71%.

RESULTS

The pattern of drug sensitivity distribution of 62 tumors tested on two to five drugs is shown in Table 2. Less than 20% of tumors showed drug sensitivities distributed over the entire scale. However, in 44% of all cases, the assay selected an agent with higher activity on tumor than on bone marrow cells. Comparisons for drug sensitivity of 21 untreated tumors with that of 61 treated tumors revealed that drug treatment led to an only modest increase of in vitro drug resistance. Untreated tumor cells were more resistant than bone marrow cells to 17% of the drugs assayed as opposed to 28% in the case of treated tumors. After exclusion of the tumors from patients that had received adjuvant treatment only, this percentage increased to 40%. But, even these clinically treatment-refractory tumors were still sensitive to 20% of the drugs tested. Comparison for drug sensitivities of 28 estrogen-receptor-positive tumors with that of 30 estrogen-receptor-negative tumors revealed that the estrogen-receptor-positive cells were less drug-sensitive: estrogen-receptor-positive tumor cells were more sensitive than bone marrow cells to only 21% of the drugs assayed, as opposed to 35% in the case of the estrogen-receptor-negative tumors. These lower drug sensitivities of estrogen-receptor-positive tumors was set off by a higher proportion of drugs with equal toxicity to tumor and host cells and not by an increase in the proportion of drugs with lesser antitumor activity.

To compare the antitumor effects of the components of the FAC treatment regimen, these three drugs were assayed on the tumors of 12 patients. 4-Hydroperoxycyclophosphamide, an in vitro active metabolite, was substituted for cyclophosphamide. The antitumor activity was comparable for all three agents. More variation in antitumor activity was observed for the second line agents, when compared to that of doxorubicin. All six tumors tested were less sensitive to cis-platinum, but five of eight were more sensitive to elliptinium. Figure 3 illustrates the relative effects of four anthraquinone derivatives, observed on the breast tumors of

25 patients. Seven patients were untreated and 17 patients had received prior treatment. Adriamycin and 4'epi-doxorubicin predominantly occupied the first two ranks, whereas mitoxantrone and bisantrene primarily occupied the last two ranks. This was true for the untreated and for the treated tumors. When anthracyclines were compared to anthraquinones, the higher efficacy of the former became more apparent.

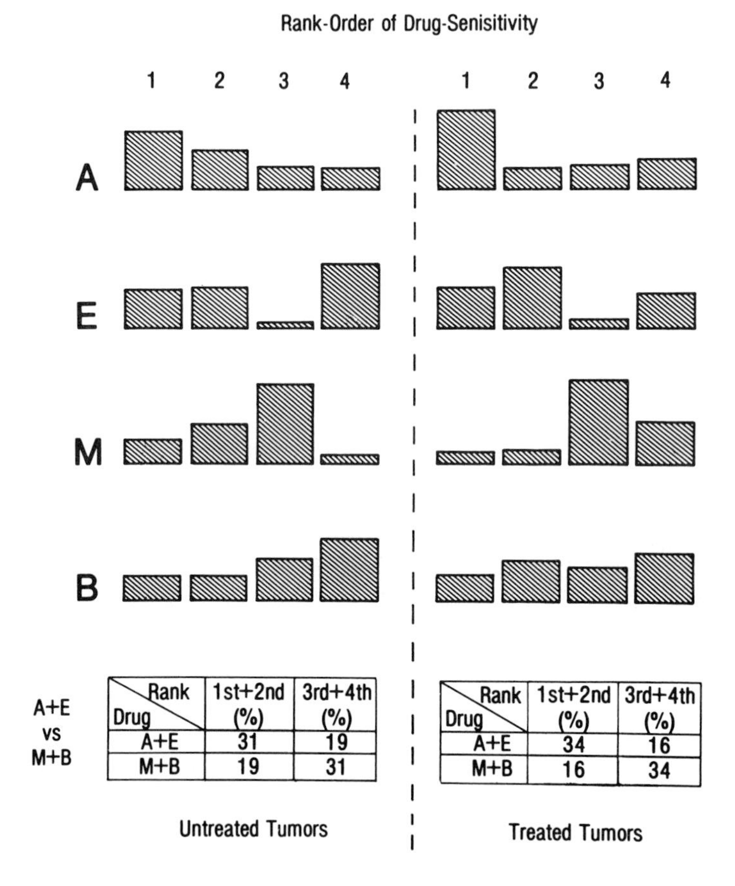

Untreated Tumors

Rank / Drug	1st+2nd (%)	3rd+4th (%)
A+E	31	19
M+B	19	31

Treated Tumors

Rank / Drug	1st+2nd (%)	3rd+4th (%)
A+E	34	16
M+B	16	34

Figure 3: Comparison of the cytotoxic effects of four anthraquinone derivatives. Tumors from 25 patients with breast carcinoma were evaluated. The height of the bars underneath each rank-order of the drug-sensitivity is proportional to the frequency distribution in that rank of the drug that is listed to the left. A comparison of anthracyclines with anthraquinones, made with the same data, is shown in the tables at the bottom of the graph. A = doxorubicin, E = 4'-epi-doxorubicin, M = mitoxantrone and B = bisantrene.

DISCUSSION

Drug normalization can provide a way to compare the effects of many different drugs and tumor cells at equitoxic doses. Bone marrow is the dose-limiting host tissue for many anticancer agents, and the progenitor cells can be recovered under culture conditions similar to those used for the identification of tumor progenitor cells. Normalization of in vitro drug activities by their effects on GM-CFU, therefore provides a biologically valid approach. Using this approach, we found that drug sensitivities of only a small proportion of tumors were distributed over the entire sensitivity scale. However, the assay identified a drug with higher cytotoxicity on tumor cells than on bone marrow cells for 36 of the 82 tumors tested for 234 drugs. We also found that in vitro drug sensitivities of estrogen-receptor-positive tumors were lower than those of tumors that did not contain this protein. This finding is of particular interest, as no clinical study has yet provided a definite statement in this regard. It may reflect a longer generation time of these cells, that may make them less sensitive to cell-cycle-active anticancer drugs.

Of similar interest is the finding that previous drug treatment led to only an insignificant increase in inherent in vitro drug resistance. In fact, tumors of patients that had received adjuvant treatment only, by and large retained their inherent drug sensitivities. It therefore appears that, at least at the advanced disease stages, loss of inherent drug sensitivity is not the only factor resulting in development of resistance to treatment. Other factors, such as reduction of drug exposure secondary to increase of tumor load or to a decrease of bone marrow tolerance, may be equal or more important determinants of development of clinical resistance.

The finding that all three agents used for primary treatment of breast carcinoma revealed similar degrees of in vitro antitumor activities argues for the validity of this approach. The fact that the system was capable of discerning differences in activity of secondary agents is also encouraging. Of particular interest is the fact that the cytotoxic potential of the two anthracyclines is higher than that of the more remotely related anthraquinones (bisantrene and mitoxantrone) and indicates that the cytotoxicity, at least for breast tumors, is not only mediated by the quinone, but also by the sugar-moiety.

We conclude that drug normalization can provide useful information in studying inherent drug sensitivities of groups of tumors with diverse biologic properties and in the comparison of the antitumor effects of cytotoxic agents.

References

1. Hug V, unpublished.
2. Hamburger AW and Salmon SE: Primary bioassay of human tumor stem cells. Science 197: 461-463, 1977.
3. Hamburger AW and Salmon SE: Primary bioassay of human myeloma stem cells. Clin. Invest. 60:846-854, 1977.
4. Hug V, Spitzer G, Drewinko G, Blumenschein GR, and Haidle C: Improved culture conditions for the in vitro growth of human breast tumors. Proc. AACR, (Abstract #138) May 25-28, 1983.
5. Hug V, Spitzer G, Blumenschein GR, Drewinko B, Hortobagyi GN, and Freireich EJ: Improved culture conditions for clonogenic growth of primary human breast tumors. Submitted for publication.
6. Hug V, Thames H, Spitzer G, Blumenschein GR, Drewinko B: Normalization of in vitro sensitivity testing of human tumor changes. In press.

ASSOCIATION BETWEEN HISTOPATHOLOGIC FACTORS AND SENSITIVITY OF HUMAN COLON CANCER TO 5-FLUOROURACIL

D.H. KERN

Surgical Service
Veterans Administration Medical Center
Sepulveda, CA 91343

M.B. LEE

Department of Biomathematics And Division of Surgical Oncology
John Wayne Clinic, Jonsson Comprehensive Cancer Center,
UCLA School of Medicine, Los Angeles, CA 90024

INTRODUCTION

It is generally believed that the least differentiated tumors are the most aggressive and have the most rapid growth (shortest tumor-doubling times). This should be reflected in greater colony formation and/or greater growth rate *in vitro*. Also, tumors which divide more frequently should be more sensitive to cycle specific drugs (e.g. 5-FU) than to cycle non-specific drugs (e.g. Adriamycin). More effective chemotherapy may be possible if histopathologic factors can be used as a guide to selecting effective anti-cancer agents for particular patients. This also may help explain why presently used anti-cancer drugs are of benefit in only a few (20-25%) of colon patients, since a heterogeneous population is being treated. Identification of sub-populations (based on histopathologic factors) selectively sensitive to certain drugs may increase the overall clinical response rates.

Supported by VA Medical Research Service and by Grant CA29605 of the National Cancer Institute.

HUMAN TUMOR CLONING
ISBN 0-8089-1671-8

MATERIALS AND METHODS

Preparation of tumors

All specimens used in this study were histologically confirmed as colon carcinomas. The tumor was minced into pieces less than 2 mm in diameter in the presence of CEM (MA Bioproducts, Walkersville, MD) containing 15% heat-inactivated FCS (Flow Laboratories, McLean, VA), 100 U/ml penicillin (Grand Island Biological Company, Grand Island, NY), 100 μg/ml streptomycin (Grand Island) and 50 μg/ml Fungizone (Grand Island).

Enzymatic dissociation of solid tumors

Ten ml of CEM containing 0.03% DNase (500 Kunitz units/ml) and 0.14% collagenase Type I (Sigma Chemical Company, St. Louis, MO) were used for each gram of tissue. The tumor fragments were added to the enzyme medium, and mixtures were stirred for 90 min at 37° in the presence of 6% CO_2. The free cells were resuspended in 10 ml of supplemented CEM (CEM with 15% heat-inactivated FCS and 100 μg/ml streptomycin and 100 U/ml penicillin). Cell yield and viability were determined by trypan blue exclusion.

Preparation of agar plates

An underlayer of agar was prepared as follows: 16.5 ml of supplemented CEM was warmed to 50° and added to 3.5 ml of 3% agar, also prewarmed to 50°. The solution was rapidly mixed and 1 ml of the solution at 50° was dispersed to each 35 x 10 mm well of a 6-well plate (Linbro 76-247-05, Flow Laboratories). The agar was allowed to set at room temperature. Next, 1.2×10^6 viable tumor cells were suspended in 1.0 ml of supplemented CEM at 37°. After 2.0 ml of 0.6% agar in supplemented CEM was added, the cells were rapidly and evenly suspended. One ml of cell suspension was added to each of three wells over the feeder layer, yielding 4×10^5 viable tumor cells per well. The plates then were placed in a humidified incubator at 37° in the presence of 6% CO_2.

Colony counting

The cells in the agar plates were examined weekly under an inverted microscope at 4X and 10X. The number of colonies (>20 cells) usually reached a maximum within 2-4 weeks. Colony counts were made with an Omnicon FAS II (Bausch and Lomb, Rochester, NY). Minimum colony size was 60μ.

TABLE I

CONCENTRATIONS OF ANTICANCER AGENTS USED FOR CHEMOSENSITIVITY TESTING

Adriamycin	0.4 μg/ml
BCNU	2.0
Melphalan	1.0
Mitomycin C	3.0
Methotrexate	4.0
Vincristine	0.5
5-Fluorouracil	10.0

Inhibition of colony formation by anticancer drugs

Drug solutions were prepared from standard intravenous formulations in continuous exposure. Concentrations used are shown in Table I. Drugs were prepared as 60 X stock solutions and were aliquoted and stored at -70°. Drugs were added to the tumor cell suspension (the upper agar layer) immediately prior to plating; 100 μl of drug were added for each 1.2 x 10^6 cells. For no-treatment controls, 100 μl of CEM was added to the tumor cell suspension. Drug effects were calculated as percent inhibition of colony formation relative to the controls (no drug). Control wells with an average count of less than 30 colonies were considered to be "no growth" and were not evaluated. All drug tests were performed in triplicate. Six control (no drug) wells were included in each experiment.

Histologic characteristics

Permanent slides of tumor colonies were prepared and stained with hematoxylin and eosin. Slides were prepared as follows: two ml of saline were added to each well and excess proteins were eluted off for 15 min at 37°. This step was repeated and excess saline was drawn off. Two ml of 4% formaldehyde was added, and the plates were allowed to set overnight at room temperature. The formaldehyde was removed after 18 h, and the plates were submerged in distilled water for 30 min. The agar layers usually separated at this point; however, when they did not, the top agar layer was gently dissociated mechanically from the feeder layer. The top layer was floated onto a glass slide, slowly dried in a humidified atmosphere overnight, and then stained. Slides were prepared from every tenth culture. Cells from all examined clones showed nuclear pleomorphism, high nucleo-

TABLE II

OVERALL GROWTH OF COLORECTAL CARCINOMAS IN THE CLONOGENIC ASSAY

Number Tumors Received	204
Number with Growth	144 (70%)
Number with No Growth	60 (30%)
Reasons for No Growth	
Small Specimen Size	7 (4%)
Poor Yield of Viable Cells	22 (11%)
Contamination	22 (11%)
Other	9 (4%)

cytoplasmic ratios, nuclear prominence, aminocytosis and cytoplasmic basophilia, features consistent with malignant cells.

RESULTS

During the course of this study, 204 tumor specimens from patients with pathologically confirmed colorectal carcinoma were received. Colony growth was observed in 144 (70%) tumors, while no growth was seen in 60 (30%). For 29 (15%) tumors poor yields of viable tumor cells were obtained (see Table II). Contamination of 22 (11%) primary colon carcinomas precluded evaluation of these assays.

Copies of pathology reports were obtained for 193 patients and form the basis for our evaluation of the relation between histopathologic factors and growth and chemosensitivity in vitro. The histopathologic factors reviewed were:

1) Primary vs. metastatic tumor
2) Site of metastasis
3) Whether or not the patient had chemotherapy prior to surgery
4) Age
5) Sex
6) Degree of differentiation of the tumor
7) Depth of invasion as determined by Dukes classification

TABLE III

ASSOCIATION BETWEEN HISTOPATHOLOGIC FACTORS AND GROWTH IN THE CLONOGENIC ASSAY

Histopathologic Factor	Level	#Grew / #Tested	(%)
Primary vs. Metastatic	M	60/78	76.9
	P	80/115	69.6
Site of Metastasis	Colon	15/20	75.0
	Liver	19/24	79.2
	Nodes	7/10	70.0
	Lung	4/6	66.7
	Other	8/11	72.7
Previous Chemotherapy	No	97/120	80.8
	Yes	13/17	76.5
Age	<50	18/23	78.3
	50-65	56/78	71.8
	>65	55/72	76.4
Sex	F	59/77	76.6
	M	84/124	67.7
Degree of Diff.	Well/Mod	83/101	82.2
	Poor/Undiff	15/20	75.0
Depth of Invasion (Dukes)	A/B	32/41	78.0
	C/D	92/117	79.0

This data, as well as patient demographic data and results of chemosensitivity assays were entered into the UCLA Hospital Computing Center computer. To determine whether growth in vitro was influenced by histopathologic factors, a 2-sample t test was used for factors with two levels, while a 1-way ANOVA was used for factors with more than 2 levels.

The association between histopathologic factors and growth in vitro is shown in Table III. Only minor, statistically insignificant differences in growth were noted for all

TABLE IV

OVERALL SENSITIVITY OF COLORECTAL CARCINOMA TO ANTI-CANCER DRUGS

Drug	Number Tested (%)	Number with 50% Kill or Greater (%)
Adriamycin	68 (53.5%)	10 (14.7%)
BCNU	102 (80.3%)	8 (7.8%)
Melphalan	53 (41.7%)	7 (13.2%)
Mitomycin C	102 (80.3%)	36 (35.3%)
Methotrexate	61 (48.0%)	6 (9.8%)
Vincristine	68 (53.5%)	8 (11.8%)
5-Fluorouracil	118 (92.9%)	28 (23.7%)

histopathologic factors. The mean colony number for all specimens that grew was 105, and did not vary according to histopathologic factors.

The sensitivity of colorectal carcinomas to standard anticancer drugs is shown in Table IV. The two most active anticancer drugs were Mitomycin C and 5-Fluorouracil (5-FU), with in vitro response rates of 35% and 24%, respectively. All other single agents had response rates less than 20%.

Since 5-FU has been the drug most commonly used to treat patients with colorectal carcinoma, we looked for associations between in vitro response to 5-FU and histopathologic factors. These are shown in Table V. There was no relation between age or sex and sensitivity to 5-FU. Patients who had chemotherapy with 5-FU prior to surgery were less sensitive than those patients who had not received chemotherapy at all. Since 90% of the tumors we received were from previously untreated patients, we had data on only 8 patients treated with 5-FU prior to surgery. Substantial differences in chemosensitivity response were observed for 4 histopathologic factors. Metastatic tumors were more sensitive to 5-FU than primaries, and distant metastases were more sensitive than local recurrance. Tumors that were moderately to poorly differentiated were more sensitive than well-differentiated tumors. The more invasive tumors, as determined by Dukes classification of C or D were more sensitive to 5-FU than were tumors classified as A or B.

TABLE V

ASSOCIATION OF HISTOPATHOLOGIC FACTORS WITH SENSITIVITY OF COLORECTAL CARCINOMA TO 5-FLUOROURACIL

Histopathologic Factor	Level	#Sens / #Tested	(%)
Primary vs	P	12/66	18.2
Metastatic	M	15/49	30.6
Previous	No	19/83	22.9
Chemotherapy	Yes	1/8	12.5
Age	<50	4/14	28.6
	50-65	11/47	23.4
	>65	11/47	23.4
Sex	F	11/47	23.4
	M	16/70	22.7
Degree of	Well	3/29	10.3
Differentiation	Mod/Poor	15/55	27.3
Depth of	A/B	4/29	13.8
Invasion (Dukes)	C/D	19/75	25.3
Site of	Colon	4/17	23.5
Metastasis	Nodes	2/6	33.3
	Liver	8/18	44.4
	Lung	1/2	50.0

* Greater than 50% inhibition of colony formation was defined as in vitro sensitivity.

DISCUSSION

Colon cancer in the United States are second in incidence only to skin cancer. Over 100,000 new cases are diagnosed each year, and about 50,000 deaths are expected. Chemotherapy has not been particularly effective against colorectal carcinomas. Most anticancer drugs are ineffective (Moertel, 1977). 5-FU remains the first drug of choice, but objective response rates are only about 20% (Wasserman, 1975). The clonogenic assay has raised hopes that response rates can be increased for some human malignancies to chemotherapy (Hamburger and Salmon, 1977). In correlative trials, a high association between chemosensitivity and clinical course has been demonstrated (Von Hoff, et. al., 1983; Bertelsen, et al., 1984). We have received over 200 specimens from patients with colorectal carcinoma for chemosensitivity testing. We reviewed our data to determine if tumor growth and sensitivity to 5-FU in vitro was associated with histopathologic factors.

Overall, 70% of specimens received in our laboratory formed at least 30 colonies in soft agar. Growth was not associated with histopathologic factors. A major reason for no growth was low yield of viable cells, primarily because some specimens were very small (less than 2 gm) and/or necrotic. Contamination also caused some assays to be unevaluable. Using techniques reported here and elsewhere (Kern, et al., 1982), it is apparent that most colorectal cancers can be successfully tested for chemosensitivity to 5-FU and other drugs.

We found 24% of the colorectal specimens tested to be sensitive to 5-FU. The only other drug with activity was Mitomycin C. The in vitro response rates shown in Table IV closely parallel clinical response rates for these drugs (Moertel, 1977; Crooke and Bradner, 1976; Wasserman, et al., 1975; Bullen et al., 1976).

The sensitivity of colorectal cancers to 5-FU was associated with biological agressiveness (Table V). Poorly differentiated tumors were more sensitive than well differentiated ones. Depth of invasion (Dukes classification) was directly related to sensitivity to 5-FU. Distant metastases were more sensitive than local recurrances. Although none of these associations individually reached statistical significance, the evidence definitely indicates greater sensitivity to 5-FU is directly related to loss of growth control in the cell. More aggressive tumors, with shorter doubling times, are more likely to have their growth arrested by 5-FU. It is important to remember that 5-FU in these experiments was added directly to the cells in the agar in continuous exposure. It is doubtful that associations between histopathologic factors and sensitivity to 5-FU would be apparent with one hour preincubation of drug, as used by some other laboratories (Von Hoff, et al., 1981; Salmon, et al., 1978).

If the associations between histopathology and sensitivity to 5-FU reported here hold true in vivo, pathological staging might be useful for prognosticating responses of colorectal carcinomas to chemotherapy. These data also suggest that cycle non-specific drugs may be more effective on well differentiated or less aggressive tumors than 5-FU. Finally, in vitro chemosensitivity tests may be most useful for screening new anticancer drugs for activity against colorectal carcinoma (Shoemaker, 1983).

SUMMARY

Although adenocarcinoma of the colon is a single histologic type, several histopathologic factors reflect tumor heterogeneity. We analyzed results of clonogenic assays performed on 204 histologically confirmed colon carcinomas for an association of HPF to growth in vitro and sensitivity to 5-Fluorouracil (5-FU). Overall, 144/204 (70.6%) specimens formed at least 30 colonies (> 20 cells) in soft agar. Growth was independent of age, sex, Dukes classification, degree of differentiation, site of lesion, or if the tumor was a primary (P) or metastasis (M). Tumors were tested against 10 μg/ml 5-FU in continuous exposure. We defined sensitivity as at least 50% inhibition of colony formation; 28/118 (23.7%) specimens were sensitive. Chemosensitivity was not related to age or sex, but was associated with several histopathologic factors: 12/66 (18.2%) P vs. 15/49 (30.6%) M, 3/29 (10.3%) well differentiated vs. 14/54 (25.9%) moderate to poorly differentiated, and 4/29 (13.8%) Dukes A/B vs. 19/75 (25.3%) Dukes C/D tumors were sensitive. Distant metastases were more sensitive than locally recurrent tumors. Increased sensitivity to 5-FU was associated with biological aggressiveness and high proliferative potential, suggesting that histopathologic factors are prognostic for responsiveness to chemotherapy.

REFERENCES

Bertelsen, C.A., Sondak, V.K., Mann, B.D., Korn, E.L., and Kern, D.H., Chemosensitivity Testing of Human Solid Tumors, Cancer, 1984, in press.

Bullen, B.R., et al, Randomized comparison of melphalan and 5-Fluorouracil in the treatment of advanced gastrointestinal cancer, Cancer Treat. Rep., 1976, 60, 1267-1273.

Crooke, S.T. and Bradner, W.T., Mitomycin C, a review. Cancer Treat. Rev., 1976, 3, 121-130.

Hamburger, A.W. and Salmon, S.E., Primary bioassay of human tumor stem cells, Science, 1977, 197, 461-463.

Kern, D.H., Campbell, M.A., Cochran, A.J., Burk, M.W., Morton, D.L., Cloning of human solid tumors in soft agar. Int. J. Cancer, 1982, 30, 725-729.

Moertel, C.G., Chemotherapy of gastrointestinal cancer. In H.J. Tagnon and M.J. Staguet (Eds.), Recent Advances in Cancer Treatment. New York: Raven Press, 1977, 311-320.

Salmon, S.E., et al., Quantitation of differential sensitivity of human tumor stem cells to anticancer drugs, N. Engl. J. Med., 1978, 298, 1321-1327.

Shoemaker, R.H., Wolpert-DeFilippes, M.K., Makuch, R.W., and Venditti, J.M., Use of the human tumor clonogenic assay for new drug screening. Proc. Am. Assoc. Cancer Res., 1983, 24, 311.

Von Hoff, D.D., et al., Association between human tumor colony-forming assay and response of an individual patient's tumor to chemotherapy. Am. J. Med., 1981, 70, 1027-1032.

Von Hoff, D.D., et al., Prospective clinical trial of a human tumor cloning system. Cancer Res., 1983, 43, 1926-1931.

Wasserman, T.H., et al., Tabular analysis of the clinical chemotherapy of solid tumors, Cancer Chemother. Rep., 1975, (Part 3) 6, 399-407.

Sensitivity of Renal Cell Carcinoma to Leukocyte Interferon in the Human Tumor Clonogenic Assay and Clinical Correlations

DAVID R. STRAYER
JAN WEISBAND
WILLIAM A. CARTER
ISADORE BRODSKY

Institute for Cancer and Blood Diseases
Barry Ashbee Leukemia Research Laboratories
Hahnemann University
Broad and Vine Streets
Philadelphia, Pennsylvania 19102

INTRODUCTION

Renal cell carcinoma is the most common malignancy involving the kidney with more than 15,000 new cases diagnosed each year (Silverberg, 1983). Although several cytotoxic drugs have activity, no consistently effective systemic chemotherapy exists. Recently, however, preliminary evidence for antitumor activity of leukocyte interferon (α-IFN) was reported in patients with metastatic renal cell carcinoma (Quesada et al., 1983).

The purpose of these studies was 1) to examine the usefulness of a soft agar system for growing renal cell

This work was supported in part by grant P01 CA29545 from the NIH.
The authors express their appreciation to E. Pequignot for statistical analysis and to D. Valley for editorial aid.

HUMAN TUMOR CLONING
ISBN 0-8089-1671-8

carcinoma, 2) to determine the sensitivities of individual renal cell tumors to α-IFN, and 3) to decide if this approach has utility for increasing the response rate of patients with renal cell carcinoma to interferon.

Colony formation, sufficient to evaluate α-IFN effects, was obtained in two-thirds of the tumor samples plated. The tumor type with greatest sensitivity to α-IFN was renal cell carcinoma. An excellent correlation was found between _in vitro_ sensitivity to α-IFN and clinical response. Our results suggest that the human tumor clonogenic assay will be useful for the identification of interferon responders.

MATERIALS AND METHODS

Patient Samples

An informed consent was obtained from all patients prior to the collection of the tumor specimens. Nine different human neoplasms were studied. Thirty-three percent of the tumor samples were from patients who had received anticancer drugs. Chemotherapy was terminated for at least 28 days prior to obtaining the tumor specimen. Each tumor sample was examined microscopically to confirm its malignant nature.

Preparation of Single Cell Suspensions and Exposure to α-IFN

Pleural effusions and ascitic fluid, collected in preservative-free heparinized vacuum bottles, were centrifuged (150xg, 20 min), cell pellets were washed and resuspended in 10 ml McCoy's 5A medium containing 10% fetal calf serum (FCS) and 1% penicillin/streptomycin (P/S). Solid tumors were placed in the same solution and mechanically dissociated by slicing into 1mm x 1mm pieces followed by enzymatic dissociation of cells using the methods described previously (Salmon et al., 1980; Slocum et al., 1981; Von Hoff et al., 1983). Briefly, 20 ml of a sterile solution containing 0.8% collagenase II (Sigma) and 0.002% deoxyribonuclease I (Sigma) in McCoy's 5A medium with 10% FCS was added to 10-20 gm of tissue and incubated 1-2 hours at 37°C. Subsequently, the single cells were separated from any remaining tumor fragments by passing them through a sheet of sterile gauze. The single cells were then centrifuged (150xg for 10 min) and the gauze discarded. The cell pellet was washed twice in 10 ml of Hank's balanced salt solution and resuspended in 10 ml McCoy's 5A medium containing 10% FCS and 1% P/S. Tumor cell suspensions were adjusted to a final concentration of 1.5×10^6 cells/ml.

Assay of Tumor Colony-Forming Units (TCFUs)

The culture system has been described (Salmon et al., 1980; Von Hoff et al., 1983). Briefly, cells in each tube were suspended in CMRL 1066 medium (supplemented with 15% horse serum, 2% insulin, 30 mM vitamin C, 2% FCS, 2% glutamine, and 1% P/S) and 3% agar was added to yield a final concentration of 500,000 cells/ml in .3% agar. One ml of this suspension was pipetted into each 35 mm Petri dish containing a 1 ml bottom layer of 0.5% agar and enriched McCoy's 5A medium containing 10% FCS, 5% horse serum, 2.2% sodium pyruvate, 21 mg L-serine, 2% glutamine and 1% P/S. α-IFN at 100, 300, 1000, and 3000 international reference units (IRU)/ml was added at the time of plating for a continuous exposure. The α-IFN used in these *in vitro* studies and in our clinical trials was prepared by the Cantell method and was obtained from the New York Blood Center. Cultures were maintained at 37°C and 5% CO_2 in a humidified incubator. Each experiment included 9 control plates plus 3 replicate plates for each drug concentration. Colonies (>40 cells) were counted at 10-21 d; ≥30 TCFUs/control plate were required for measurement of possible α-IFN effect. Permanent slides, prepared with Papanicolau's stain, as well as karyologic analysis, were used to establish the malignant nature of the colonies.

Clinical Protocol

Seven patients with renal cell carcinoma were entered into a clinical trial of α-IFN. Each patient was treated at an initial dosage level of 3×10^6 IRU daily, IM, 5 days per week. After 40 doses, two patients (samples 5 and 12 in Table 1) had achieved partial responses and the frequency of α-IFN administration was decreased to three times weekly. A minor response (MR) was defined as a greater than 25%, but less than 50% decrease in tumor size. A partial response required a 50% or greater decrease in tumor size.

RESULTS

Fresh tumor cells from 9 different histologic types of human malignancies were assayed for the effect of α-IFN on their clonogenicity in soft agar. Adequate colony formation to evaluate α-IFN sensitivity (≥30 colonies/ control plate) was obtained in 66% of all tumors and in 80% of renal cell carcinomas analyzed. The mean cloning efficiency for renal cell carcinoma was .050%. Ovarian carcinoma was the only tumor type with a higher cloning efficiency (.099%). Following exposure to 1000 IRU/ml α-IFN, thirty-three percent of the tumor specimens showed a 60% or greater inhibition of TCFUs (Table 1). The tumor type most sensitive to α-IFN was renal cell carcinoma accounting for 69% (9/13) of tumors showing a 60% or greater decrease in colony formation at 1000 IRU/ml and 100% (3/3) of tumors at 100 and 300 IRU/ml. Other tumor types which showed sensitivity (>60% decrease in colony formation) to α-IFN at 1000 IRU/ml were carcinoid 1/3, breast carcinoma 1/4, melanoma 1/3, and ovarian carcinoma 1/4. Of the 16 renal cell carcinomas evaluated, three (19%) showed high *in vitro* sensitivity (<20% colony survival); six (38%) showed moderate sensitivity (20-40% survival); while seven (44%) were resistant (>40% colony survival). The three renal tumors (samples 1-3 in Table 1) which exhibited high sensitivity were the only tumors sensitive to α-IFN at 100 and 300 IRU/ml. There was no significant difference in the sensitivity of renal primary tumors versus metastatic tumors. Whether or not a patient had received prior chemotherapy also did not appear to influence sensitivity to α-IFN.

Significant stimulation ($p<.05$) in TCFUs by α-IFN (1000 IRU/ml) was observed in 7 tumors (samples 33-39 in Table 1) and has been reported by others (Ludwig et al., 1983). No α-IFN stimulation of TCFUs was observed in any renal cell carcinomas at 1000 IRU/ml (0/16) or 300 IRU/ml (0/15). However, at 100 IRU/ml TCFU's from two (2/13) renal tumors (samples 23 and 29) were stimulated ($p<.05$).

In order to determine the clinical utility of this assay system for increasing response rates to IFN, we are entering patients with tumors studied *in vitro* for α-IFN sensitivity into clinical trials of α-IFN. Currently, seven patients with metastatic renal cell carcinoma, for which *in vitro* α-IFN sensitivity data were available, have been treated with α-IFN. Figure 1 shows the *in vitro* dose-response data for these 7 patients.

Of the four patients whose tumors (samples 2, 5, 9, and 12) were sensitive to α-IFN (>60% decrease in TCFU's at 1000 IRU/ml), all four have had clinical responses. Two of the responses have been minor (MR) with both patients (samples 2 and 9) continuing to receive α-IFN 5 days per week. The other two responses were partial (PR) and both

TABLE 1

Antiproliferative Activity of α-Interferon Against Human Tumors Studied in the Clonogenic Assay

Sample Number	Histologic Type	Specimen Source	% Change in TCFUs*
1	Renal	kidney (primary)	97% decrease
2	Renal	skin metastasis	96% decrease
3	Renal	kidney (primary)	84% decrease
4	Carcinoid (rectal)	liver biopsy	77% decrease
5	Renal	kidney (primary)	74% decrease
6	Breast	pleural effusion	70% decrease
7	Renal	bone metastasis	68% decrease
8	Renal	kidney (primary)	66% decrease
9	Renal	kidney metastasis	66% decrease
10	Ovarian	ascites	65% decrease
11	Renal	kidney (primary)	64% decrease
12	Renal	kidney (primary)	61% decrease
13	Melanoma	lymph node	60% decrease
14	Breast	pleural effusion	57% decrease
15	Renal	bone metastasis	56% decrease
16	Breast	skin metastasis	43% decrease
17	Mesothelioma	lymph node	33% decrease
18	Renal	lung metastasis	33% decrease
19	Renal	kidney (primary)	29% decrease
20	Renal	kidney (primary)	26% decrease
21	Carcinoid	ovarian (primary)	24% decrease
22	Leiomyosarcoma	lymph node	24% decrease
23	Renal	abdominal metastasis	21% decrease
24	Melanoma	skin recurrence	17% decrease
25	Carcinoid (rectal)	liver biopsy	17% decrease
26	Renal	kidney (primary)	15% decrease
27	Carcinoid (rectal)	rectal (primary)	15% decrease
28	Colon	liver biopsy	11% decrease
29	Renal	lung metastasis	10% decrease
30	Carcinoid (small bowel)	liver biopsy	6% decrease
31	Melanoma	lung metastasis	5% decrease
32	Leiomyosarcoma	primary (thigh)	2% increase
33	Colon	colon (primary)	9% increase
34	Ovarian	ovarian (primary)	19% increase
35	Endometrial Stromal Sarcoma	lymph node	20% increase
36	Colon	liver biopsy	25% increase
37	Ovarian	pelvic recurrence	38% increase
38	Ovarian	pelvic recurrence	43% increase
39	Breast	lymph node	73% increase

*At an α-IFN concentration of 1000 IRU/ml

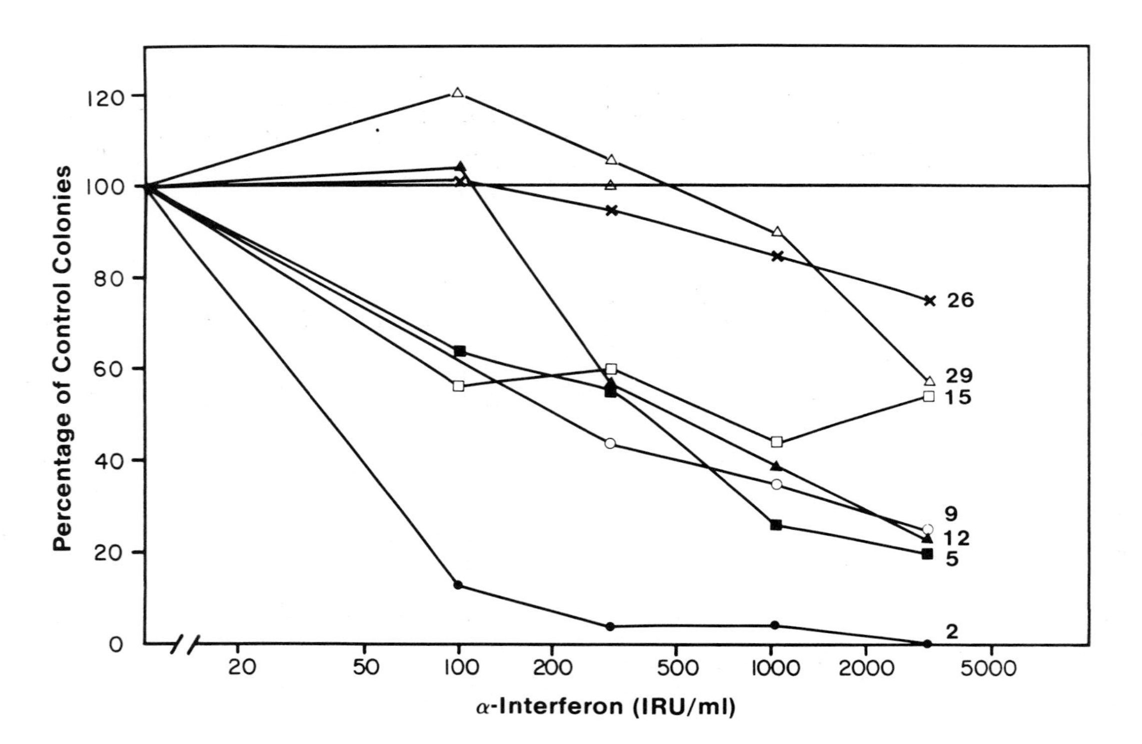

Figure 1. Sensitivity of Renal Cell Carcinoma to α-Interferon in the Clonogenic Assay. Sample numbers are indicated adjacent to each dose-response curve.

patients (samples 5 and 12) are continuing on study now over 8 and 5 months, respectively.

Of the three patients who had tumors resistant to α-IFN (<60% decrease), all three have had progressive disease. Despite the small sample size, the correlation between clinical response and _in vitro_ sensitivity in the human tumor clonogenic assay was significant ($p<.03$).

DISCUSSION

This study demonstrates that the cloning efficiency of renal cell carcinoma in the human tumor clonogenic assay is high (0.050%) and useful sensitivity data can be obtained in the majority (80%) of cases. Nine of 16 (56%) renal tumors were sensitive to α-IFN. Several renal tumors were very sensitive to α-IFN with 78-96% decreases in TCFU's observed at α-IFN concentrations of 100 and 300 IRU/ml. Other renal tumors were resistant with only 10-15% decreases in TCFU's seen at 1000 IRU/ml. The tumor type found most sensitive to α-IFN was renal cell carcinoma. In contrast, the sensitivity of renal cell carcinoma to cloned interferon, αA-IFN, was reported to be less than the sensitivity of eight other human tumors including melanoma, sarcoma, breast, and ovarian cancer (Salmon et al., 1983). Since these four tumor types showed only minimal (0-33%) sensitivity to α-IFN in our study, it appears likely that αA-IFN and α-IFN differ in tumor or tissue specificity. This same possibility was recently raised regarding αA-IFN and αD-IFN specificity (Salmon et al., 1983).

The response rate of renal cell carcinoma to α-IFN was recently reported to be near 52% including partial (26%), minor (10.5%), and mixed (16%) responses (Quesada et al., 1983). Our results suggest that IFN sensitivity data obtained using the human tumor clonogenic assay will be useful for increasing this response rate. Using a >60% decrease in TCFU's at 1000 IRU/ml α-IFN to define a tumor sensitive to α-IFN, we found an excellent correlation between _in vitro_ sensitivity to α-IFN and clinical response in seven patients followed for 3-8 months. The four responding patients continue to receive α-IFN and maximum responses have not yet been reached.

Recently, a clinical trial of α-IFN in patients with carcinoid tumors was reported (Oberg et al., 1983). Although, symptomatic relief of carcinoid syndrome was seen, no tumor responses were documented. It is of interest that one of our patients with a rectal carcinoid had three tumor specimens (samples 4, 25, and 27) analyzed for α-IFN sensitivity. A liver biopsy specimen (sample 4) was sensitive to both α-IFN (Table 1) and β-IFN (β-IFN data not shown). The patient received natural β-IFN for 6.5 months with complete regression of three superficial tumor nodules documented at laparotomy (Strayer et al., 1984). At that time a site of residual carcinoid tumor was biopsied (sample 25) and following the β-IFN treatment the tumor was found to be resistant to both α and β-IFN. A rectal biopsy (sample 27) obtained after β-IFN therapy was also resistant to α and β-IFN. Tumor specimens from two other carcinoid patients were resistant to α-IFN.

Stimulation of TCFU's by the α-IFN preparation was observed in 18% (7/39) of tumors tested at 1000 IRU/ml (Table 1). These results are similar to the findings of Ludwig et al. (1983), who observed stimulation of TCFU's in 26.3% of tumor samples treated with α-IFN. How IFN-induced stimulation of TCFU's *in vitro* might relate to tumor stimulation *in vivo* is unknown. However, we are reluctant to treat patients with IFN, if stimulation of TCFU's has been observed in the human tumor clonogenic assay.

Clearly, follow up of this clinical study will be required before the full impact of this approach on response rates and survival will be known. Nonetheless, our results to date suggest that the human tumor clonogenic assay will be useful for the identification of interferon responders.

REFERENCES

Ludwig CU, Durie BGM, Salmon SE, Moon TE: Tumor growth stimulation in vitro by interferons. Eur J Cancer Clin Oncol 19:1625-1632, 1983.

Öberg K, Funa K, Alm G: Effects of leukocyte interferon on clinical symptoms and hormone levels in patients with mid-gut carcinoid tumors and carcinoid syndrome. N Engl J Med 309:129-133, 1983.

Quesada JR, Swanson DA, Trindade A, Gutterman JU: Renal cell carcinoma: antitumor effects of leukocyte interferon. Cancer Res 43: 940-947, 1983.

Salmon SE, Alberts DS, Durie BGM, et al.: Clinical correlations of drug sensitivity in the human tumor stem cell assay. Recent Results in Cancer Res 74:300-305, 1980.

Salmon SE, Durie BGM, Young L, et al.: Effects of cloned human leukocyte interferons in the human tumor stem cell assay. J Clin Onco 1:217-225, 1983.

Silverberg E: Cancer Statistics, 1983. CA 33: 9-25, 1983.

Slocum HK, Pavelic ZP, Rustum YM, et al.: Characterization of cells obtained by mechanical and enzymatic means from human melanoma, sarcoma, and lung tumors. Cancer Res 41: 1428-1434, 1981.

Strayer DR, Carter WA, Brodsky I, Curley RM, Gain T: Carcinoid tumor response to fibroblast interferon. JAMA, in press, 1984.

Von Hoff DD, Clark GM, Stogdill BJ, et al.: Prospective clinical trial of a human tumor cloning system. Cancer Res 43:1926-1931, 1983.

Human Tumor Cloning Assay With Prospective Clinical Correlations And Association Between In Vitro Sensitivity And The Number Of Drugs Tested

JAFFER A. AJANI, GARY SPITZER, FRASER BAKER, GUNTER E. UMBACH, BARBARA TOMASOVIC, SHAIL K. SAHU, VERENA M. HUG.

Division of Medicine and Department of Gynecology (G.V.), The University of Texas, M. D. Anderson Hospital and Tumor Institute at Houston, Houston, Texas 77030

INTRODUCTION

A test predictive test for response to cancer chemotherapy can have major clinical applications. To develop such a test the human tumor cloning assay developed by Hamburger and Salmon (2) has been widely explored in the past few years (6-8,11,12). Efficacy of this assay has been reported in studies performed retrospectively (2,3,6), and more recently in a large prospective study by Von Hoff (12). Consistent true positive rates of 60-65% and true negative rates of 85-95% have been reported in these studies. Several problems with the assay have been discussed (4,7,12). Some of these problems including: [1] difficulties with obtaining single cell preparations, [2] low plating efficiency of most tumors, [3] questions regarding in vitro drug dosing and duration of drug exposure and [4] criteria for in vitro response. The need for prospective randomized trials in patients treated with the drug selected by the assay has been proposed (9), however, difficulties in

This investigation was supported in part by NIH Grants CA 28153, CA 14528 and Grant 174208 Allotment for the Solid Tumor Cloning Laboratory and the Max Kade Foundation.

HUMAN TUMOR CLONING
ISBN 0-8089-1671-8

implementing such trails have also been encountered. The efforts to resolve these difficulties are under way and several centers continue to study this system to examine its clinical validity. The preliminary results of our ongoing study are reported in this paper. Another objective of this analysis was to examine the relationship between the number of drugs tested and in vitro sensitivity of a miscellaneous group of tumors. Utilizing one-hour drug exposure to tumors from 50 patients with malignant melanoma, Myeskens et al. demonstrated a linear relationship between the number of drugs tested and in vitro sensitivity (5). In that study, when one or two drugs were tested, one active drug was found in only 13% of the tumors as opposed to in 100%, when seven or eight drugs were tested, the difference is statistically significant ($P = <0.0001$). This means that if one could test seven or more drugs on malignant melanoma specimens, the possibility of finding at least one sensitive drug would be extremely high. We analyzed 67 in vitro tests to ascertain if the same phenomenon existed in a miscellaneous group of malignant tumors.

MATERIALS AND METHODS

Collection of tumor cells: Our soft agar human cloning method is based on the techniques described by Hamburger and Salmon (2,8), and has been previously reported (1). In brief, solid tumor specimens, malignant effusions, ascites and bone marrow containing tumor cells were collected from 67 patients undergoing diagnostic or therapeutic procedures, or both, for their malignancy. The solid tissue specimens, transported under sterile conditions, were sliced to 1 mm cubes then further dissociated by using an enzyme cocktail containing type III collagenase (Worthington Biochemical Corporation, Freehold, New Jersey) and DNAse (Sigma Chemical Corporation, St. Louis, Missouri) in a culture medium at a final strengths of 0.75% and 0.005% respectively, for 12 to 18 hours under continuous agitation at 37° C. The fluid specimens containing malignant cells were transported to the laboratory under sterile conditions, mixed with the preservative-free heparin. Cells were then centrifuged at 150 g for 10 minutes, and the concentrated cells were subjected to treatment with the enzyme cocktail, as described for the solid specimens. If necessary, the tumor cells were separated from the excess of red cells by density-gradient Ficoll-Hypaque centrifugation. The bone marrow specimens were diluted several-fold and mononuclear cells from the bone marrow specimens were treated in the same manner as were fluid specimens except the enzyme dissociation was not used. The cell suspension, thus prepared was washed free of enzyme using Hank's Balanced Salt Solution and was allowed to recover in the culture medium and then passed through 18 to 25 guage needles. The viability of cells derived was determined by the trypan blue dye exclusion test.

In vitro exposure of tumor cells: Stock solutions of both standard and investigational drugs were prepared in advance in sterile buffered 0.9% NaCl solution or in water and were stored at -70°C in aliquots sufficient for individual assays. Subsequent desired concentrations were obtained by further dilutions. Each drug was exposed at various concentrations to the tumor cells. At least three drug concentrations currently are being utilized in our laboratory. These three concentrations correspond to a 40%, 78% and 99% inhibition of kill from lethal activity of these drugs against GM-CFUC (LD40, LD78 and LD99) of normal human bone marrow incubated with these drugs. An additional concentration above GM-CFUC (Granulocyte Macrophage-Colony Forming Units) LD99 by a factor of three was also utilized for some specimens to study dose response and to assess true in vitro resistance.

For one-hour drug exposure, the tumor cells were transferred to tubes and adjusted to a concentration of 6 x 10^6 cells in the presence of an appropriate drug concentration. Cells were incubated with and without the drug for one hour at 37°C in Hank's Balanced Salt Solution plus 10% heat-inactivated fetal bovine serum. Cells were then centrifuged at 150 g for 10 minutes, washed twice with Hank's Balanced Salt Solution and cultured. For continuous drug exposure, the drug was mixed in the upper layer with tumor cells and adjusted to the desired concentration for the volume of the upper layer. Control and drug treated tumor cells were plated in 35-mm plastic petri dishes in triplicate. The bottom layer was made up of agar and culture medium F12 (Ham) (K. C. Biological, Lenexa, Kansas) supplemented with 10% heat-inactivated fetal bovine serum to give a final concentration of 0.5% agar. The bottom layer was allowed to gel at room temperature. The top layer containing 5 x 10^5 mononuclear cells and culture medium alpha Minimal Essential Medium (K. C. Biological, Lenexa, Kansas), supplemented with 15% heat inactivated fetal bovine serum with a final agar concentration of 0.32%, was prepared immediately before plating. After preparation of both bottom and upper layers, the plates were examined under an inverted microscope to ensure satisfactory cell preparation, clumps were counted and recorded to be subtracted from the final colony counts. Plates were placed in a high humidity container and gased with a mixture of 12% oxygen, 5% carbon dioxide and 83% nitrogen. The plates were then incubated at 37° C for 14-21 days. The number of colonies containing at least 30 cells(measuring at least 60 um in diameter) were counted on an Olympus inverted stage microscope at 40X magnification. At least 20 tumor cell colonies in the control plates were considered adequate for chemosensitivity testing. The mean colony counts were recorded for triplicate control and treated plates. Tumor specimens were considered sensitive if the survival of colonies was 35% or less compared with that of controls. Standard criteria were utilized for assessing response in vivo. A complete remission was defined as complete disappearance of all measurable tumor for a period of 4 weeks; partial remission, as $\geq$50% decrease in measurable tumor; progressive disease, as any measurable increase in tumor, and no change, as stabilization of previously progressive disease for a period of 6 weeks.

RESULTS

Sixty-seven patients' tumor specimens had sufficient growth for completion of the in vitro chemosensitivity testing. Twenty-six of 67 tumors were tested at one hour drug exposure, whereas 41 specimens were tested at continuous drug exposure. The median number of drugs tested was three (range 1-5). The median number of concentrations per drug was three (range 1-7).No "cut-off" concentrations were utilized in this study and highest degree of colony inhibition by the drug was scored to determine in vitro respone. Overall, 36 of 67 tumors were found to be sensitive to at least one drug in vitro. We examined the relationship between the number of drugs tested and the probability of detecting at least one sensitive drug. Only 18 of 44 tumors (41%) were sensitive in vitro when 1 to 3 drugs were tested, whereas 18 of 23 tumors (78%) were sensitive when 4 to 5 drugs were tested (p=0.008). The likelihood of detecting two or more sensitive drugs in vitro increased with increasing number of drugs tested. Nine of

Table 2.

Patient characteristics and clinical correlations with the human tumor cloning assay.

Tumor Type	# of Pts.	In vitro / In vivo	Correlations S/S	R/R	S/R	R/S	N/C
Melanoma	25		3	2	0	3	17
Ovarian	16		1	5	1	1	8
Lung	8		1	1	0	0	6
Sarcoma	8		0	5	1	0	2
GI	3		0	2	0	0	1
Misc.	7		0	2	0	0	5
Totals	67		5	17	2	2	39

S = Sensitive
R = Resistant
N/C = No Correlations
GI = Gastrointestinal

36 tumors (25%) were sensitive to two drugs, when 2 to 3 drugs were tested compared with 9 of 23 tumors (39%) when 4 to 5 drugs were tested (p=0.38) (Table 1).

TABLE 1. Association of the number drugs tested to in vitro sensitivity.

# of drugs tested	Tumors with	
	one sensitive drug	two sensitive drugs
1	3/8	-
2	5/14	1/14
3	10/22	2/22
4	10/12	6/12
5	8/11	3/11

All tumor specimens were sent to our laboratory with a list of drugs. Physicians were requested to treat patients according to the assay and patient charts were analysed to assess clinical correlations. All correlations in this study are prospective. Clinical correlations are available in 28 of 67 patients. The lack of correlation in the remaining 39 patients resulted for the following reasons: [1] 7 patients were lost to follow-up; [2] 9 patients were ineligible due to a state of no evidence of disease or general poor condition and early death in two patients, [3] 5 patients were not treated according to the assay because of physicians' unwillingness and [4] 18 patients were too early for analysis. The number of each tumor types and the various clinical correlation categories are shown in Table 2.

TABLE 3. Clinical correlations with human tumor cloning assay.

In Vitro / In Vivo	S/S	S/R	R/R	R/S
No. of patients	7/5	7/2	23/17	23/4
% Correlations	71	29	74	26

S = Sensitive
R = Resistant

Table 4. Clinical features of the patients with true positive association.

Tumor type	Prior therapy (# of regimens)	In vitro sensitive to	In vivo therapy with	Response site	Response
Melanoma	Yes (2)	velban+bleomycin	velban, bleomycin +cis-platinum	Soft tissue	PR
Melanoma	No	DTIC	DTIC+vindesine	Lung	<PR
Melanoma	Yes (2)	velban	velban, bleomycin +cis-platinum	Soft tissue	PR
Lung (small cell)	No	VP-16	ECHO*	Lung	<PR
Ovarian	Yes (1)	cis-platinum	cis-platinum+ cyclophosphamide	Abdomen	PR

*VP-16, cyclophosphamide, doxorubicin and vincristine.

Patients who had progressive disease or no change in measurable disease were classified as non-responders. Patients who had partial remissions and minor responses, were classified as responders, since any objective response indicated anti-tumor activity of the drugs utilized. By this criteria 2 patients were assigned to the S/S (sensitive in vitro and in vivo) category, and 2 patients to the R/S (resistant in vitro but sensitive in vivo). One patient with lung cancer had more than 40% reduction of serum CEA after the first treatment with aclacinomycin, but then developed disseminated intravascular coagulation and expired immediately following second treatment with aclacinomycin. This patient was involved in the only single-agent trial and was classified as inevaluable. The details on 5 patients categorized as sensitive in vitro and in vivo are shown in Table 4.

DISCUSSION

Our first objective was to examine the relationship between the number of drugs tested and the probability of detecting an active drug in vitro. More than 50% of the tumors were found to be sensitive in vitro to at least one chemotherapeutic agent. When the number of drugs in vitro was increased from 1 to 3 to 4 to 5 the possibility of finding at least one sensitive drug increased significantly, from 43% to 78%. Two or more sensitive drugs were also found in more instances as the number of drugs used in vitro increased. Our data on a miscellaneous group of tumors concur with that of Meyskens et al. (5) on malignant melanoma. Another important determinant of in vitro sensitivity of tumors appears to be the drug concentration. The dose responsiveness of a variety of malignant cell lines has been amply demonstrated (2,10). A similar dose responsiveness of fresh human tumors in monolayer cultures has also been observed (F. Baker: Unpublished data) utilizing a variety of standard and investigational drugs.The dose response suggests that an inappropriately high dose will result in increased in vitro activity of various drugs and high false positive clinical correlations may result. For one-hour drug exposures, a "cut-off" concentration of 10% of peak plasma drug levels has been arbitrarily chosen and utilized, however, no "cut off" has been proposed for continuous drug exposure.

Continuous drug exposures appear more appropriate than one-hour drug exposures, since cycle specific as well as non-specific drugs can be tested effectively by continuous exposure. In order to assess a biologic reference, lethal concentrations of various anti-cancer drugs on human bone marrow, GM-CFUC (LD40, LD78 and LD99) are utilized in our laboratory. However, the "cut-off" concentration relevant for clinical correlation is yet to be determined and perhaps other biologic references (normal cell lines or sensitive malignant cell lines) are necessary.

A number of clinical trials have demonstrated an association between the clinical course of the tumor and in vitro drug sensitivity (2,3,6,12). The results of our ongoing study confirm that contention. Our data lack single-agent trials to prove true-positive associations, however, difficulities with obtaining single-agent trials have been realized and reported (12). One of the objectives was to demonstrate validity of this assay in the clinical setting. Although a large data base is necessary,our initial results are encouraging.

CONCLUSION

Our data suggest that for a miscellaneous group of tumors, 5 or more drugs should be tested in vitro, so that a sensitive drug may be identified in a higher percentage of patients. This may improve the efficacy of the human tumor cloning assay in clinical use. For continuous drug exposure a biologic reference may be appropriate to determine biologically meaningful "cut-off" concentrations which may result in improved in vitro / in vivo conclusions.

ACKNOWLEDGMENT

The authors appreciate the diligence of Ms. Brenda Benavides in preparation of this manuscript.

REFERENCES:

1. Ajani J.A., Sahu S.K., Spitzer G., Hug V.M., Bodey G.P.: Cloning of human tumor stem cells in soft agar - An overview. Cancer Bulletin 35:16-19, 1983.

2. Hamburger A.W., Salmon S.E.: Primary bioassay of human tumor stem cells. Science 197:461-463, 1977.

3. Hill B.T., Whatley S.A., Bellamy A.S., et al.: Cytotoxic effects and biological activity of 2-Aza-8-germanspiro (4,5) decane-2-propranamine-8, 8-diethyl-N, N-dimethyl dichloride (NSC 192965, spirogermanium) in vitro. Cancer Res 42: 2852-2856, 1982.

4. MacKintosh F.R., Evans T.L., Sikic B.I.: Methodolgic problems in clonogenic assays of spontaneous human tumors. Cancer Chemother Pharmacol 6:205-210, 1981.

5. Meyskens F. L., Moon T. E., Dana B., et al.: Quantitation of drug sensitivity by human metastatic melanoma colony-forming units. Br J Cancer 44:787-797, 1981.

6. Meyskens F.L.: Individualized patient chemotherapy: Use of a human tumor colony formation technique. Front Radiat Ther Oncol 16:55-61, 1982.

7. Salmon S.E., Alberts D.S., Meyskens F.L.: Clinical correlations of in vitro drug sensitivity in human cloning; in Salmon S.E. (ed): Cloning of Human Tumor Stem Cells. New York, Alan Liss Co.: pp 223-245, 1980.

8. Salmon S.E., Hamburger A.W., Soehnlein B., et al.: Quantitation of differential sensitivity of human tumor stem cells to anticancer drugs. N Engl J Med 298:1321-1327, 1978.

9. Selby P., Buick R.N., Tannock I.: A critical appraisal of the "human tumor stem-cell assay." N Engl J Med 308:129-134, 1983.

10. Skipper H.E., Schabel F.M., Lloyd H.H.: Dose-response and tumor cell repopulation rate in chemotherapeutic

trials; in Rosowsky A (ed): Advance in Cancer Chemotherapy. New York and Basel, Marcel Dekker Inc. Vol. 1, pp205-253, 1979.

11. Von Hoff D.D., Casper J., Bradley E., et al.: Association between human tumor colony-forming assay results and response of an individual patient's tumor to chemotherapy. Am J Med 70:1027-1032, 1981.

12. Von Hoff D.D., Clark G.M., Stogdill B.J., et al.: Prospective clinical trial of human tumor cloning system. Cancer Res 43:1926-1931, 1983.

13. Von Hoff D.D., Cowan J., Harris J., Reisdorf G.: Human tumor cloning: Feasibility and clinical correlations. Cancer Chemother Pharmacol 6:265-271, 1981.

Comparison of In Vitro Prediction and Clinical Outcome for Two Anthracene Derivatives: Mitoxantrone and Bisantrene

BERND LATHAN, M.D.

Research Fellow
Medicine/Oncology
University of Texas Health Science Center at San Antonio
San Antonio, Texas

DANIEL D. VON HOFF, M.D.

Associate Professor of Medicine
Medicine/Oncology
University of Texas Health Science Center at San Antonio
San Antonio, Texas

GARY M. CLARK, Ph.D.

Research Associate Professor of Medicine
Medicine/Oncology
University of Texas Health Science Center at San Antonio
San Antonio, Texas

INTRODUCTION

In a variety of clinical studies, the Human Tumor Cloning System has shown some promise for predicting for response or resistance of an individual patient's tumor (3, 5, 7). Based on this experience, the system was utilized to screen for cytotoxic activity of the two new anthracene derivatives, Bisantrene and Mitoxantrone, prior to their entrance into clinical trials (8, 9). It was hoped that the Human Tumor Cloning System has the potential for pinpointing the tumor types against which the drugs would be most useful in clinical trials. With both drugs, various clinical trials have been completed by now. The aim of the present study was to compare the clinical activity of both Bisantrene and Mitoxantrone in Phase II trials with their updated results in the Human Tumor Cloning System.

HUMAN TUMOR CLONING
ISBN 0-8089-1671-8

MATERIALS AND METHODS

Collection of Cells

Either solid tumors, ascites, bone marrow or pleural fluid were utilized for cloning in soft agar. Effusions were collected in preservative-free heparinized vacuum bottles, centrifuged at 150 x g for ten minutes, and washed twice in Hank's balanced salt solution with ten percent heat inactivated fetal calf serum and one percent penicillin and streptomycin solutions (all materials, Grand Island Biological Company, Grand Island, N.Y.). Bone marrow specimens were collected in heparinized syringes and processed in the same manner as were the effusions except that after centrifugation only the buffy coat was removed and processed. Solid tumors removed by biopsy or at operation were immediately placed in McCoy's Medium 5A plus ten percent newborn calf serum plus one percent penicillin and streptomycin in the operating room and transported to the laboratory within 24 hours, where they were mechanically dissociated by forcing through a wire mesh gauze into Hank's balanced salt solution plus ten percent fetal calf serum. They were then passed through progressively smaller needles and processed in the same manner as were the effusions.

In Vitro Drug Exposure

In vitro drug testing was performed at about one-tenth the average peak plasma concentration of the drugs, as determined in earlier pharmacokinetic studies (4, 10). Bisantrene was tested at a final concentration of 0.5 μg/ml and Mitoxantrone at a final concentration of 0.05 μg/ml.

Tumor cell suspensions were transferred to tubes and adjusted to a final concentration of 1.0×10^6 cells per ml in the presence of the appropriate drug dilution or control medium. Cells were incubated with or without drug for one hour at 37° in Hank's balanced salt solution. The cells were then washed twice with Hank's balanced salt solution and prepared for culture.

Assay for Tumor Colony Forming Units (TCFU's)

The culture system utilized in the study has been previously described (1, 6). In brief, cells to be tested were suspended in 0.3 percent agar in enriched CMRL 1066 medium supplemented with 15 percent horse serum to yield a final concentration of cells of 5×10^5 per ml. One ml of this mixture was pipetted into each of three 35 ml petri dishes containing one ml 0.5 percent agar in enriched McCoy's Medium 5A but without conditioned medium. Cultures were incubated at 37° in seven percent CO_2 in humidified air. All assays were set up in triplicate. With a Bausch and Lomb FAS II automatic scanner the plates were then screened for the presence of cell clusters in order to assure the plating of only single cells. Colonies ($\geq$50 cells) usually appeared in ten to 15 days, and the number of colonies on control and drug-treated plates was determined with the automatic counter. At least 20 tumor colonies per control plate were required for a drug experiment to be considered evaluable for measurement of drug effect.

Data Analysis

Colony counts of the three plates for a particular drug concentration were averaged to obtain one data point. For determination of sensitivity to a particular drug, the ratio between the number of colonies surviving at each drug concentration and the number of colonies growing in control plates was calculated.

For purposes of this analysis, three different definitions of *in vitro* activity were utilized, i.e., $\leq$50 percent, $\leq$40 percent, and $\leq$30 percent survival of TCFU's.

Comparison with Phase II Clinical Trials

Data of *in vivo* activity were obtained from published Phase II clinical trials with both drugs. A total of 16 clinical trials were evaluated for Bisantrene and a total of 20 clinical trials for Mitoxantrone (for a complete list, see references). For purposes of this analysis, a partial remission (PR) or a complete remission (CR) in patients was considered as *in vivo* activity of the drugs (Southwest Oncology Group criteria).

Tumor types with less than ten evaluable specimens or patients were not analyzed in this study, and only tumor types with sufficient data for both drugs were evaluated.

RESULTS

Seven different tumor types with sufficient clinical data were available for *in vitro-in vivo* comparison of Bisantrene. The data are detailed in Table 1.

The data of Table 1 are graphically illustrated in Figure 1, for a $\leq$50 percent survival of TCFU's as definition of *in vitro* cytotoxic activity.

For breast cancer and lymphoma (both moderate to good activity) and for ovarian and colorectal cancer (no activity), the activity of Bisantrene in the Human Tumor Cloning System correlated well with its activity in Phase II trials. Lung, kidney cancer and melanoma were outliers. In these tumor types, Bisantrene showed some *in vitro* activity, whereas no *in vivo* activity in Phase II trials was seen. Utilizing a $\leq$30 percent survival of TCFU's as a definition of *in vitro* cytotoxic activity, Bisantrene's *in vitro* activity against lung and

TABLE 1
In Vitro-In Vivo Correlation of Bisantrene

Tumor Type	In Vitro # Sensit./# Eval. Spec. ≤50% Surv. TCFU's (%)	In Vitro # Sensit./# Eval. Spec. ≤30% Surv. TCFU's (%)	In Vivo # CR + PR/ # Eval. Pts. (%)
Breast	37/197 (19)	12/197 (6)	20/131 (15)
Lung	18/94 (19)	3/94 (3)	1/112 (1)
Ovary	15/98 (15)	7/98 (7)	1/14 (7)
Kidney	11/50 (22)	1/50 (2)	2/84 (2)
Melanoma	10/31 (32)	1/31 (3)	0/24 (0)
Lymphoma	8/26 (31)	2/26 (8)	4/18 (22)
Colorectal	1/14 (7)	0/14 (0)	3/55 (6)
TOTAL	100/510 (20)	53/510 (10)	31/438 (7)

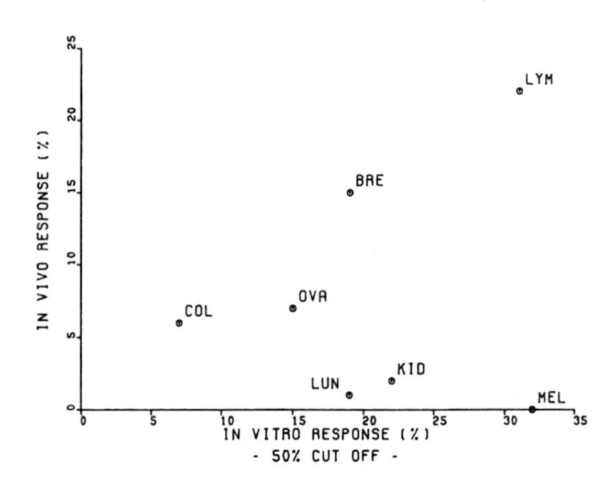

Figure 1. In Vitro-In Vivo Correlation of Bisantrene

kidney cancer, as well as melanoma, dropped to two to three percent, resulting in a better correlation of in vitro and in vivo activity in those tumor types (Table 1). But, on the other hand, the drug's in vitro activity against breast cancer and lymphoma also decreased and was lower than its in vivo activity.

The in vitro-in vivo comparison of the antitumor activity of Mitoxantrone is detailed in Table 2.

For a ≤50 percent survival of TCFU's as definition of in vitro activity, the results are graphically illustrated in Figure 2.

There was a good in vitro-in vivo correlation of the cytotoxic activity against breast cancer and lymphoma as well as for absence of activity against ovarian and colorectal cancer. Again, lung and kidney

TABLE 2
In Vitro-In Vivo Correlation of Mitoxantrone

Tumor Type	In Vitro # Sensit./# Eval. Spec. ≤50% Surv. TCFU's (%)	In Vitro # Sensit./# Eval. Spec. ≤30% Surv. TCFU's (%)	In Vivo # CR + PR/ # Eval. Pts. (%)
Breast	19/123 (15)	7/123 (6)	69/393 (17)
Lung	33/153 (21)	15/153 (10)	11/293 (4)
Ovary	12/105 (11)	5/105 (5)	0/49 (0)
Kidney	5/23 (22)	0/23 (0)	3/79 (4)
Melanoma	9/31 (29)	6/31 (19)	2/81 (23)
Lymphoma	9/35 (26)	4/35 (11)	19/83 (23)
Colorectal	4/37 (11)	2/37 (5)	2/114 (2)
TOTAL	91/507 (18)	39/507 (8)	106/1092 (10)

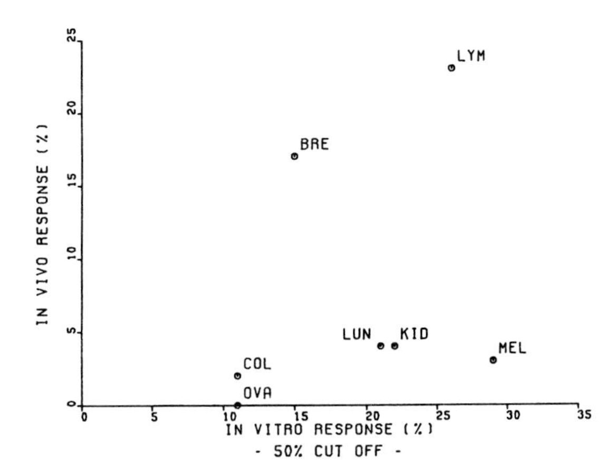

Figure 2. In Vitro-In Vivo Correlation of Mitoxantrone

cancer as well as melanoma were outliers. Mitoxantrone was active in vitro in more than 20 percent of these specimens, but essentially inactive in Phase II clinical trials (Figure 2). When a ≤30 percent survival of TCFU's was utilized for definition of activity in the Human Tumor Cloning System, Mitoxantrone had minor in vitro activity against all tumor types with the exception of melanoma (Table 2).

For quantitation of the in vitro-in vivo correlations for both drugs, the Spearman Correlation Coefficients were calculated for a ≤30 percent, ≤40 percent, and ≤50 percent survival of TCFU's as definition of in vitro activity (Table 3).

In addition, the Spearman Correlation Coefficient of the in vitro-in vivo activity was calculated for both drugs excluding the three outliing tumor types. The Correlation Coefficients for breast, ovarian, colorectal cancer and lymphoma alone are shown in Table 4.

TABLE 3
Spearman Correlation Coefficient
In Vitro vs. In Vivo

Drug	≤50%	≤40%	≤30%
	Survival of Tumor Colony Forming Units		
Bisantrene	-.22	.19	.63
Mitoxantrone	.5	.41	.35

TABLE 4
Spearman Correlation Coefficient for
Breast, Ovary, Colon, Lymphoma
In Vitro vs. In Vivo

Drug	≤50%	≤40%	≤30%
	Survival of Tumor Colony Forming Units		
Bisantrene	1.0	1.0	.8
Mitoxantrone	.95	.8	.94

Overall, more than 80 percent of the patients in the Phase II trials had had prior chemotherapy. Since some trials with Mitoxantrone in breast cancer had been performed in non-pretreated patients, an evaluation of the influence of prior chemotherapy on this tumor type can be assessed (Table 5).

TABLE 5
Mitoxantrone in Breast Cancer

	In Vitro	In Vivo
	# Sensit./# Eval. (%)	# CR + PR/# Eval.(%)
With Prior Chemotherapy	14/61 (23)	39/309 (13)
W/O Prior Chemotherapy	4/25 (16)	30/84 (36)
	p=0.47	p<.0001

With a p-value of 0.47, there was no significant difference in the *in vitro* activity of Mitoxantrone between specimens from patients with or without prior chemotherapy. In contrast, the *in vivo* activity of Mitoxantrone was significantly higher in patients without prior chemotherapy than in patients with prior chemotherapy.

DISCUSSION

The object of the study was to evaluate the capability of the Human Tumor Cloning System to predict for clinical activity of new anticancer agents prior to their entrance into clinical trials. For this purpose, the *in vitro* activity of Bisantrene in 510 evaluable specimens was compared to its clinical activity in 16 published Phase II trials with 438 patients. A similar comparison was performed with Mitoxantrone, utilizing 507 specimens and 1,092 patients from 20 clinical trials. A comparison of this kind has several caveats that should be considered.

First, the patient population for the in vitro and in vivo investigations were different, which makes a precise evaluation difficult.

Second, the patients in the Phase II trials were more often and more heavily pretreated than the patients whose tumors were tested in vitro. Over 80 percent of the patients in the evaluated Phase II trials had had prior chemotherapy, but only 50 to 60 percent of patients that had their tumors tested in vitro had prior treatment. As it could be shown for breast cancer, prior chemotherapy did significantly influence the in vivo cytotoxic activity of Mitoxantrone (Table 5).

The third caveat is that for the in vitro and in vivo investigations different parameters were measured for evaluation of antitumor activity (i.e., survival of TCFU's versus shrinkage of tumor mass).

It is interesting that for both drugs, there was a similar pattern of activity in vitro and in vivo for the individual tumor types. The overall correlation of in vitro with in vivo activity was poor for both Bisantrene and Mitoxantrone. For both anthracene derivatives the Human Tumor Cloning System overpredicted for activity against melanoma, lung and kidney cancer. It could be, that due to unknown mechanisms these tumor types are much more sensitive in vitro than in vivo. In this connection, it is interesting that the new drug, AMSA, had also shown good activity against melanoma in the Human Tumor Cloning System, but its activity against melanoma in published clinical trials was only minor (2). After the exclusion of the three outliing tumor types, the correlation of in vitro-in vivo activity for breast, ovarian, colon cancer and lymphoma was good (Table 4).

It could be argued that the Human Tumor Cloning System in general indicates higher activity in tumor types, that are known for their chemoresistance in vivo, as is the case with lung, kidney cancer and melanoma. In contrast, the minor in vitro activity of both drugs against colorectal cancer with a corresponding low in vivo response rate argues against the hypothesis.

The results of the present study suggest that in future studies with new drugs in the Human Tumor Cloning System, cytotoxic activity or lack of activity against breast, ovarian, colorectal cancer and lymphoma could predict for the outcome of the drugs in subsequent clinical trials. In contrast, in vitro activity against melanoma, lung and kidney cancer would have to be interpreted with caution.

As this paper investigated only seven tumor types, it will be of interest to see how other tumor types correlate for in vitro-in vivo activity when more clinical trials, especially with Bisantrene, are completed.

REFERENCES

1 Hamburger AW, Salmon SE. Primary bioassay of human tumor stem cells. Science 197: 461-463, 1977.

2 Lathan B, Von Hoff DD, Elslager EF. Comparison of Amsacrine and the new analog CI-921 in the human tumor cloning system. IND in press, 1984.

3 Salmon SE, Alberts DS, Meyskens FL Jr, et al. Clinical correlations of in vitro drug sensitivity. Salmon SE, ed. Cloning of Human Tumor Stem Cells. New York: Alan R Liss Inc 223-245, 1980.

4 Stewart JA, McCormack JJ, Krakhoff IH. Clinical and clinical pharmacologic studies of Mitoxantrone. Cancer Treat Rep 66: 1377-1331, 1982

5 Von Hoff DD, Casper J, Bradley E, et al. Association between human tumor colony-forming assay results and response of an individual patient's tumor to chemotherapy. Am J Med 70: 1027-1041, 1981.

6 Von Hoff DD, Casper J, Bradley E, et al. Direct cloning of human neuroblastoma cells in soft agar culture. Cancer Res 40: 3591-3597, 1980.

7 Von Hoff DD, Clark GM, Stogdill BJ, et al. Prospective clinical trial of a human tumor cloning system. Cancer Res 43: 1926-1931, 1983.

8 Von Hoff DD, Coltman CA Jr, Forseth B. Activity of Mitoxantrone in a human tumor cloning system. Cancer Res 41: 1853-1855, 1981.

9 Von Hoff DD, Coltman CA Jr, Forseth B. Activity of 9-10 anthracenedicarboxaldehyde bis ((4,5-dihydro-1 H-imidazol-2-yl)hydrazone)dihydrochloride (CL216,942) in a human tumor cloning system. Leads for Phase II trials in man. Cancer Chemother Pharmacol 1981; 6: 141-144.

10 Von Hoff DD, Myers JW, Kuhn J, et al. Phase I clinical evaluation of 9-10 anthracenedicarboxaldehyde bis ((4,5-dihydro-1 H-imidazol-2-yl)hydrazone) dihydrochloride (CL-216, 942). Cancer Res 41: 3118-3121, 1981.

Phase II Trials With Bisantrene

11 Ahmed T, Kemeny NE, Michaelson RA, Harper HD. Phase II trial of Bisantrene in Advanced Colorectal Carcinoma. Cancer Treat Rep 67: 307-308, 1983.

12 Cavalli F, Clarysse A, Bokkel Huinink WT, et al. Phase II study of Bisantrene in advanced breast cancer. Proc Am Soc Clin Oncol 2: 113, 1983.

13 Chang, AYC, Falkson G, Kaplan BH, et al. Phase II study of Bisantrene (9, 10 Anthracenedicarboxaldehyde CL 216, 942) in

patients with refractory breast cancer. Proc Am Soc Clin Oncol 2: 99, 1983.
14 Coltman CA, Cowan D, Von Hoff DD, et al. Southwest Oncology Group studies with Bisantrene. Proc 13th Int Cong Chemother 213: 1-4, 1983.
15 Fuks JZ, Van Echo DA, Garbino C, Kasdorf H, Aisner J. Treatment of advanced non-small lung cancer with Bisantrene. Cancer Treat Rep 67: 597-598, 1983.
16 Gams RA, Ostroy F, Pocelinko R. Clinical studies of Bisantrene (Cl 216, 942: NSC-33776) administered on a 5-day schedule. Proc 13th Int Cong Chemother 213: 53-57, 1983.
17 McGuire WP, Blough RR, Cobleigh MA, et al. Phase II trial of Bisantrene for unresectable non-small cell bronchogenic carcinoma: An Illinois Cancer Council study. Cancer Treat Rep 67: 841-842, 1983.
18 Murphy WK, Yap BS, Farha P, et al. Proc Am Soc Clin Oncol 2: 205, 1983.
19 Myers JW, Von Hoff DD, Clark GM, Coltman CA. A phase II clinical trial with Bisantrene in patients with advanced refractory small cell lung cancer directed by a cloning assay. Cancer Treat Rep, submitted.
20 Myers JW, Von Hoff DD, Coltman CA Jr, et al. Phase II evaluation of Bisantrene in patients with renal cell carcinoma. Cancer Treat Rep 66: 1869-1871, 1982.
21 Osborne CK, Von Hoff DD, Cowan JD, Sandbach J. Cancer Treat Rep, in press.
22 Perry MC, Forastiere AA, Richards F II, Weiss RB, Anbar D. Phase II trial of Bisantrene in advanced colorectal cancer: A Cancer and Leukemia Group B study. Cancer Treat Rep 66: 1997-1998, 1982.
23 Scher H, Schwartz S, Yagoda A, et al. Phase II trial of Bisantrene for advanced hypernephroma. Cancer Treat Rep 66: 1653-1655, 1982.
24 Spiegel RJ, Levin M, Muggia FM. Clinical trials of Bisantrene in melanoma and renal cell carcinoma. Proc 13th Ind Cong Chemother 213: 45-48, 1983.
25 Yap BS, Yap HY, Murphy WK, Bodney GP. M.D. Anderson studies using Bisantrene. Proc 13th Int Cong Chemother 213: 49-52, 1983.
26 Yap HY, Yap BS, Blumenschein GR, et al. Bisantrene, an active new drug in the treatment of metastatic breast cancer. Cancer Res 43: 1402-1404, 1983.

Phase II Trials with Mitoxantrone

27 Anderson KC, Cohen GI, Garnick MB. Phase II trial of Mitoxantrone. Cancer Treat Rep 66: 1929-1931, 1982.

28 Bonnem EM, Mitchell EP, Woolley PV, et al. Phase II trial of Mitoxantrone in advanced colorectal cancer. Cancer Treat Rep 66: 1995-1996, 1982.
29 Coltman CA Jr, McDaniel TM, Balcerzak SP, et al. Mitoxantrone hydrochloride (NSC-310739) in lymphoma: A Southwest Oncology Group study. IND 1: 65-70, 1983.
30 Cowan JD, Von Hoff DD, McDonald B, et al. Phase II trial of Mitoxantrone in previously untreated patients with colorectal adenocarcinoma: A Southwest Oncology Group study. Cancer Treat Rep 66: 1779-1780, 1982.
31 De Jager R, Cappelaere P, Earl H, et al. Phase II clinical trial of Mitoxantrone: 1,4-Dihydroxy-5, 8-Bis(((2-((2-Hydroxyethyl)Amino) Ethyl)Amino)) 9, 10-Anthracenedione Dihydrochloride in solid tumors and lymphomas. Third NCI-EORTC Symposium on New Drugs in Cancer Therapy. Brussels 1981.
32 Gams RA, Keller J, Golomb HM, Steinberg J, Dukart G. Mitoxantrone in malignant lymphomas. Proc 13th Int Cong Chemother 212: 47-51, 1983.
33 Hilgers RD, Von Hoff DD, Rivkin SE, Alberts DS. Mitoxantrone in epithelial carcinoma of the ovary. Am J Clin Oncol, submitted.
34 Knight WA III, Von Hoff DD, Neidhart JA, et al. Mitoxantrone in advanced breast cancer: A phase II trial of the Southwest Oncology Group. IND 1983; 1: 181-184.
35 Levin M. Pandya K, Khandekar J, et al. ECOG trials of Mitoxantrone (DHAD) in advanced breast cancer. Proc Am Soc Clin Oncol 2: 102, 1983.
36 Mouridsen HT, Van Oosterom AT, Rose C, Nooi MA. A phase II study of Mitoxantrone as first line cytotoxic therapy in advanced breast cancer. Proc 13th Int Cong Chemother 212: 30-34, 1983.
37 Neidhart JA, Roach RW. A randomized study of Mitoxantrone and Adriamycin in breast cancer patients failing primary therapy. Proc Am Soc Clin Oncol 1: 86, 1982.
38 Raghavan D, Bishop J, Mann G, et al. Phase II study of Mitoxantrone (NSC 301739). Proc Am Assoc Cancer Res 24: 140, 1983.
39 Smalley R, Gams R. Phase II study of Mitoxantrone in patients with metastatic breast carcinoma: A Southeastern Cancer Study Group project. Cancer Treat Rep 67: 1039-1040, 1983.
40 Stuart-Harris RC, Smith IE. Mitoxantrone: A phase II study in the treatment of patients with advanced breast carcinoma and other solid tumors. Cancer Chemother Pharmacol 8: 179-182, 1982.
41 Stuart-Harris RC, Smith IE, Cornbleet MA, Smyth JF, Rubens RD. Mitoxantrone, an active well tolerated new drug in advanced breast cancer: A phase 2 study in patients receiving no previous chemotherapy. Proc Am Soc Clin Oncol 2: 100, 1983.
42 Taylor SA. Phase II trial of dihydroxyanthracenedione (DHAD) in advanced renal cell carcinoma: A Southwest Oncology Group

study. Proc Am Soc Clin Oncol 1: 118, 1982.
43 Taylor SA, Von Hoff DD. Phase II trial of dihydroxyanthracenedione in advanced malignant melanoma: A Southwest Oncology Group study. Proc Am Soc Clin Oncol 2: 225, 1983.
44 Valdivieso M, Umsawasdi T, Spitzer G, Bodey GP. Phase II clinical study of Dihydroxanthracenedione (DHAD) in patients with advanced lung cancer. Proc Am Assoc Cancer Res 23: 145, 1983.
45 Von Hoff DD, Chen T, Clark GM, et al. Mitoxantrone for treatment of patients with refractory small cell carcinoma of the lung: A Southwest Oncology Group study. Cancer Treat Rep 67: 403-404, 1983.
46 Yap HY, Blumenschein GR, Schell F, Buzdar AU, Valdivieso M. Dihydroxanthracenedione: A promising new drug in the treatment of metastatic breast cancer. Annals Int Med 95: 694-697, 1981.

In Vitro Drug Sensitivity of Leukemia Colony-Forming Cells in Acute Nonlymphocytic Leukemia: Clinical Correlation Update with Various Drug Exposure Methods.

C. H. Park, M. Amare
University of Kansas Medical Center
and Southwest Oncology Group,
Kansas City, KS

P. H. Wiernik, J. P. Dutcher
Albert Einstein College of Medicine
Bronx, NY

F. S. Morrison
University of Mississippi Medical Center
and Southwest Oncology Group,
Jackson, MS

T. R. Maloney
Wilford Hall Medical Center
Lackland AFB, TX

An in vitro colony assay was developed for leukemic progenitor cells in the bone marrow of patients with acute nonlymphocytic leukemia (4). Chemotherapy sensitivity of these cells was assessed initially by in vitro pulse exposure to the single drugs separately which were administered to the patients in combination. The sensitivity index (SI) thus obtained correlated significantly with clinical responses (2). However, there were a number of false correlations. In an effort to find an improved method with better correlation, a series of experiments was performed varying the methods of exposing cells to chemotherapy drugs. In the first variation (1) cells were exposed to the drug mixture rather than single drugs separately, and in the second variation (5) cells were exposed to single drugs continuously over a prolonged period throughout the entire culture period rather than for a short

HUMAN TUMOR CLONING
ISBN 0-8089-1671-8

pulse duration. There appear to be merits for each of these approaches, and we have now combined both approaches, i.e. the continuous exposure of cells to the drug mixture. The results of this combined method are the major subject of this report. In addition the results of all 4 methods which have been studied are summarized, and a new approach is discussed.

MATERIALS AND METHODS

Patients with newly diagnosed acute nonlymphocytic leukemia without prior treatment were candidates for this study. Intensive combination chemotherapy with ara-C infusion and an anthracycline with or without Vincristine and Prednisone was given to the patients to induce complete remission. All patients for whom both clinical chemotherapy response is evaluable and in vitro chemotherapy sensitivity is available by one of 4 methods (vide infra) are included. Some clinical characteristics of these patients are shown in this (Table 1) and prior publications (1,2,5). Bone marrows were aspirated before chemotherapy, for routine morphological evaluation and this in vitro chemotherapy study, usually from the posterior iliac crest. Details on the marrow handling were described previously (5). Informed consent was obtained from all patients prior to marrow aspiration.

The culture system consisted of 2 layers of 0.3% agar in a 35-mm plastic Petri dish perforated at the bottom by 5 small holes. The top agar layer contained twice-washed buffy coat cells separated from bone marrow aspirates. The details of cell culture method with the unique feeding technique as an essential component have been described (4).

Four different methods have been developed and used to expose cells to drugs in order to assess chemotherapy sensitivity (Table 2). Details of the methods were published for the pulse exposure technique for single drugs (2) and for drug mixtures (1); and continuous exposure for single drugs (5). Continuous exposure with drug mixtures has now been performed in identical fashion as for single drugs (drugs were added once to culture with first feeding) except that the 2 single drug solutions were pooled in a one to one ratio. Therefore the final concentration of each drug was half that of the standard concentration used for single drug testing. As in the case of single drugs, 2 concentration levels, high and low, were studied for drug mixture exper-

TABLE 1

In vitro chemotherapy sensitivity asssessed by continuous exposure of cells to drug mixture

Case number New / old[a]	Treat- ment[c]	Complete Remission[c]	Sensitivity Index[d] Mixture	Single
1 / 1	Ad	Yes	+2.42	+2.29
2[b]	Ad	Yes	+2.00	ND
3 / 10	Ad	Yes	+0.45	-0.67
4 / 3	Rz	Yes	+1.99	+1.41
5 / 2	Rz	Yes	+1.70	+1.39
6 / 4	Dm	Yes	+1.77	+1.25
7 / 9	Dm	Yes	+1.42	-1.07
8 / 7	Dm	Yes	+0.82	+0.02
9 / 8	Dm	Yes	+0.86	+0.29
10 / 11	Dm	Yes	-0.47	-0.24
11 / 13	Ad	No	-0.91	-1.22
12 / 19	Ad	No	+0.22	-0.36
13 / 14	Rz	No	-0.54	-1.86
14 / 15	Rz	No	-0.43	-0.68
15 / 16	Rz	No	-0.62	-2.07
16 / 12	Dm	No	-1.58	-1.58
17 / 17	Dm	No	-1.02	-2.35
18 / 18	Dm	No	-0.46	-1.47
19 / 20	Dm	No	-0.40	+0.08
20 / 21	Dm	No	+1.27	-0.68

a Case numbers of patients for whom chemotherapy sensitivities with single drugs were reported, along with clinical characteristics (5).

b 60 year-old male with acute myelocytic leukemia.

c All patients received a combination chemotherapy including an anthracycline, either Adriamycin (Ad), Rubidazone (Rz), or daunorubicin (DR), and ara-C. For more details on the treatment and response see the prior publication (5).

d In vitro chemotherapy sensitivity indices were obtained by continuously exposing cells to either mixture of an anthracycline and ara-C (Mixture) or 2 drugs separately. The 2 single drug sensitivity indices for individual patients thus obtained have been reported separately (5), and averaged to be shown here (Single). The sensitivity indices with mixture in patients treated with daunorubicin are shown added with a value of 1.5

Table 2

Four Methods of Exposing Cells to Drugs

	Pulse[a]	Continuous[b]
Single-drugs[c]	I	III
Drug-mixture[d]	II	IV

a Cells were exposed to drug for 1 hr at 37°, washed free of drug, and plated in culture.
b Cells were plated in culture and drug added onto the culture on the day of culture.
c Cells were exposed to 2 drugs (ara-C and an anthracycline) separately, and the 2 SI averaged.
d Cells were exposed to a mixture of 2 drugs.

iments. The standard high and low concentrations of individual drugs were previously shown (5). The chemotherapy sensitivity of leukemic colony forming cells was expressed in relation to that of normal bone marrow colony forming cells. A simple ratio of surviving fractions (normal over leukemic colony forming cells) was used to calculate chemotherapy SI for pulse exposure studies (1,2), and a log odd ratio was used instead for the continuous exposure studies (5). The neutral point for SI based on the simple ratio is "1" and that based on the log odd ratio is "0". The values above the neutral points indicate sensitivity; those below it, resistance. Depending on the method used, the magnitude of the SI can be different; but whether a leukemic marrow is sensitive or resistant, and the relative order of the sensitivity or resistance, will remain same.

RESULTS

SI obtained by the continuous exposure of cells to the drug mixture are shown in Table 1 along with single drug SI where the values are available for comparison. Patients treated with a daunorubicin combination had generally low SI, and an arbitrary correction factor of 1.5 had to be added to the SI of these patients to obtain the best separation of patients achieving and not achieving complete remission. In 19 patients single drug SI are available for comparison with drug mixture SI. Four false correlations with single drug

Table 3

Clinical Correlation of in vitro Chemotherapy Sensitivity Obtained with 4 Exposure Methods

Method[a]	No. of Correlations[b]	In vitro/Clinical Corr.[c]				Good Corr.[d]
		S/S	S/R	R/S	R/R	
I	23	7	5	2	9	70
II	14	4	1	1	8	86
III	21	8	3	3	7	71
IV	20	9	2	1	8	85
Total	78	28	11	7	32	77

a See Table 2.
b The number of patients involved was 34. More than one method was studied in 24 patients.
c S, sensitive; R, resistant. (S/R is false positive and R/S false negative correlations).
d Percentage of good correlations (S/S and R/R) over all trials.

SI, 3 false negative and 1 false positive, are corrected by the use of drug mixture SI. However 2 patients not achieving complete remission with true negative single drug SI turned to have false positives according to drug mixture SI. All the results of in vitro/clinical correlations with the 4 different methods (Table 2) are summarized in Table 3. This shows a prediction accuracy of at least 70% regardless of which method is used; and an overall accuracy of 77% in the 78 correlations available. A trend is noted that the drug mixture technique is better than the single drug method for either pulse exposure (86% vs. 70%) or continuous exposure (85% vs 71%).

DISCUSSION

An in vitro/clinical correlation of 85% has been achieved by the use of a continuous exposure of the cells to drug mixtures. This method does not entail major technical problems, and appears to be the most preferable among the 4 methods studied. The use of the drug mixture appears to be better than the use of single drugs for both the continuous and pulse exposures (1). This may be the reflection of synergistic, or antagonistic, effects of component single drugs operative in vitro and in vivo. There is also a practical advantage that the number of cultures to be set up, and therefore the number of cells needed, is reduced. This can be of critical importance for patients from whom a limited number of cells was obtained. The result of pulse exposure to drug mixture has shown a good clinical correlation comparable to that of continuous exposure to drug mixture. However, the previous study with single drugs has shown that the continuous exposure is better than the pulse exposure (5). Also a one hour pulse exposure may not be sufficiently long for phase-specific drugs, such as Ara-C, because S-phase duration of leukemic cells is a minor fraction of the generation time which may be as long as several days. In determining the final value of SI for drug mixture continuous exposure method, an arbitrary value of 1.5 had to be added for patients receiving the daunorubicin combination. In other words, the neutral point for these patients is "-1.5" instead of "0". We have no ready explanation for this requirement. It is conceivable that some drugs may exert their cytotoxic effect in vitro with different efficiency than in vivo, and this can be reflected by the "correction factor" that must be added or subtracted. In so far as this number was determined with preliminary experiments and then applied uniformly to all patients receiving the same drug, the predictive value of this assay will not be compromised.

Although there are differences among the 4 methods, the predictability of any one of these is equal to or over 70% with the overall average of 77% on 78 correlations available. This general approach for prediction of chemotherapy response is therefore mature and can now be considered established. We are however continuing our effort to improve the system further. We do not believe we can ever reach a 100% correlation considering the enormous complexities of this disease from the standpoints of both its basic pathophysiology and clinical aspects. However, we do not think it unrealistic to improve the predictability to at least 90%.

A preliminary result obtained with a new method exposing cells to drugs for an intermediate duration, longer than pulse and shorter than continuous exposures, appears promising (3). This method allows drug exposure of variable durations, which may make it possible to determine the most optimal schedule of drug administration in addition to simple selection of drug.

SUMMARY

In 20 patients with acute nonlymphocytic leukemia treated with combination chemotherapy including ara-C and an anthracycline, the in vitro chemotherapy sensitivity of colony forming cells was assessed by exposing cells to mixtures of these 2 drugs continuously throughout the cell culture period. There was 1 negative in vitro result in 10 patients achieving complete remission (false negative), 2 false positives among the other 10 patients not achieving complete remission, and an overall correlation rate of 85%. This method appears to be the best method of in vitro drug exposure among all 4 methods studied thus far. When the results from all 4 methods are combined, the overall accuracy is 77% for 78 correlations, indicating a maturity of this general approach for chemotherapy response prediction.

ACKNOWLEDGEMENT

The technical assistance of S. O'Brien and M. Reeder, and expert typing of J. Stika are gratefully appreciated. Special thanks are extended to Dr. B.F. Kimler who has read this manuscript critically and offered helpful comments. This work was supported by Grants CA20717, CA32107, CA16385, CA12644, CA12014, and CA32102 from the National Cancer Institute, Department of Health and Human Services. C.H. Park is the recipient of a Research Career Development Award KO4 CA 00534 from the Department of Health and Human Services.

REFERENCES

1. Park, C.H., Amare, M., Morrison, F.S., Maloney, T.R., Goodwin, J.W.: Chemotherapy sensitivity assessment of leukemic colony-forming cells with in vitro simultaneous exposure to multiple drugs: Clinical correlations in acute nonlymphocytic leukemia. Cancer Treat. Rep. 66:1257-1261, 1982.
2. Park, C.H., Amare, M., Savin, M.A., Goodwin, J.W., Newcomb, M.M., Hoogstraten, B.: Prediction of chemotherapy response in human leukemia using an in vitro chemotherapy sensitivity test on the leukemic colony-forming cells. Blood 55:595-601, 1980.

3. Park, C.H., Amare, M., Wiernik, P.H., Morrison, F.S., Dutcher, J.P., Myers, J.A.: Versatile drug exposure times for improved clinical chemotherapy correlation by cloning assay in acute nonlymphocytic leukemia. Proc. Am. Soc. Clin. Oncol. 2:28, 1983.
4. Park, C.H., Savin, M.A., Hoogstraten, B., Amare, M., Hathaway, P.: Improved growth of in vitro colonies in human acute leukemia with the feeding culture method. Cancer Research 37:4595-4601, 1977.
5. Park, C.H., Wiernik, P.H., Morrison, F.S., Amare, M., Van Sloten, K., Maloney, T.R.: Clinical correlations of leukemic clonogenic cell chemosensitivity assessed by in vitro continuous exposure to drugs. Cancer Res. 43: 2346-2349, 1983.

VI. OVERVIEW AND SUMMARY

SUMMARY PERSPECTIVES

Bridget T. Hill, Ph.D.
Head, Laboratory of Cellular Chemotherapy
Imperial Cancer Research Fund
Lincoln's Inn Fields
London WC2A 3PX, England

This Fourth Conference on Human Tumor Cloning provided a forum for presenting and discussing evidence of the advances that have been made in this research area during the last three years. These types of meetings are particularly valuable since they afford a unique opportunity for laboratory research workers and clinicians to meet together to discuss their real problems and, in particular, to lay or renew the foundations for worthwhile collaborative studies. These involve each individual providing their own particular expertize and it is in this way that the overall problems can be tackled and, with time, I am sure, resolved.

There were five main components of the Conference which I shall review briefly in turn, highlighting some of the studies with emphasis on those which have particularly fired my interest. For clarity, this discussion attempts to synthesize both oral and poster presentations and does not precisely parallel the organization of chapters in this volume.

THE BIOLOGY OF CLONOGENIC TUMOR CELLS

Self-Renewal Capacity

A stem cell, by definition, is capable of infinite cell proliferation. Evidence is now available that both melanomas and ovarian carcinomas can give rise to colonies with self-renewal capacity. However, these studies have been possible with only a few highly selected tumor specimens amenable to yielding a truly single cell suspension. In malignant ovarian ascites cell populations, it has been demonstrated that self-renewal potential is restricted to cells with low density which form large colonies, but in melanomas this distinction is not apparent. The majority of clonogenic melanoma cells have self-renewal capacity and have more proliferative capacity than that expressed during primary colony formation. This observation may be of particular

HUMAN TUMOR CLONING
ISBN 0-8089-1671-8

interest in view of the general lack of clinical responsiveness noted in melanomas.

Modulation of Tumor Colony Growth

Numerous attempts have been made in many laboratories to enhance anchorage independent tumor cell growth, employing for example, different media, sera, growth factors and other supplements, or various conditions of hypoxia. In general, low oxygen tension has provided a more favorable environment for colony growth from biopsies of a range of tumor types and certain continuous human tumor cell lines; there is no evidence to suggest that its use results in any reduction in colony-forming efficiencies. Similarly, where tested, the ability of human tumor cells to form colonies in soft agar is enhanced by the presence of autologous phagocytic/adherent cells and this ability is further increased if these cells receive prior relatively low doses of irradiation. The value of epidermal growth factor or conditioned media remains controversial. In one study, no major colony-stimulatory activity was reported in 41 specimens of various types of malignancy by the addition of 25 ng/ml of epidermal growth factor, while another group found it was selectively beneficial in stimulating the growth of well-differentiated as opposed to undifferentiated cells. However, introducing a word of caution, this latter group also pointed out that the acid extracted serum-free conditioned medium from cultured human malignant cells contained not only stimulatory but also inhibitory factors for certain cell types. Conditioned media from certain cell lines was reported as a more effective stimulant for fresh human tumors than the addition of a number of tissue culture supplements. Using conditioned media from either three particular breast tumor cell lines or that from an ovarian tumor cell line, increased colony growth occurred in approximately 50% or more of a number of different types of tumor specimens.

It is, however, probably now time at least to attempt to tune culture conditions more precisely for specific tumor types or even for definite subpopulations within an individual tumor. In this respect, the preliminary demonstration is informative that the effects of antiestrogens and their interactions with hormonal manipulations in human breast cancer cells, using the ZR-75-1 continuous cell line, can be investigated using the soft agar assay.

Improved methods of disaggregation of "solid" tumors is vital if clonogenicity is to be markedly enhanced. The increased cell yields per gram of tumor and enhanced clonogenicity resulting from the use of a hypoosmolar medium appear significant in this respect, although the influence of this prior manipulation on subsequent drug sensitivities needs to be established. Data were also presented indicating that there was a marked variation of clonogenic growth between incubators and a significant loss of water from the clonogenic system during incubation. The importance of maintaining conditions of "high" humidity for increased colony forming efficiencies were also stressed in another presentation. Finally, the demonstration that colony size, linearity of colony formation and drug survival curves depend on the number of cells plated in the bilayer agar clonogenic assay is of considerable significance, with major implications for data interpretation.

Biological Heterogeneity

Heterogeneity with respect to differentiation and proliferative potential has been defined in certain ovarian carcinomas. The concept of a proliferation-differentiation hierarchy is now suggested for human breast cancer, as a result of studies with the MCF-7 cell line, with estrogen receptors not being expressed in tumor stem cells but increasingly expressed in the transit and end-cell compartments, and thus functioning as a marker of differentiation. To confirm and extend this observation, additional studies are now being considered with fresh tumor biopsies.

Heterogeneity has also been identified in a series of cell lines derived from patients with small cell lung cancer. On the basis of certain biological properties, three groups are apparent, namely multipotent cell lines, classic cell lines or variant cell lines. This latter group is characterized by higher colony forming efficiencies, shorter latency periods in nude mice, radioresistance, and amplified levels of the c-myc-oncogene. Furthermore, this study raises the exciting possibility that this biological heterogeneity has prognostic implications since, in patients from whom variant cell lines were derived, either no response or poor responses to systemic chemotherapy were noted.

Correlation of Tumor Characteristics with In Vitro Clonal Growth

A retrospective analysis of 88 ascites or pleural effusions indicated that the following factors predicted for successful colony growth: high tumor cell numbers (>5%), high macrophage counts (>20%), high labeling index (>7%) and low lymphocyte count (<25%). This raises the possibility that manipulation of cellular composition could improve colony forming efficiencies.

Attempts to grow colonies from squamous cell carcinomas of the head and neck in soft agar have been notoriously unsuccessful. Thus, encouragement is provided by the demonstration that the following parameters appear to predict for adequate colony growth: little or no morphologic evidence of cytoplasmic keratinization and 1-5 mitosis per high power field. Interestingly, neither prior therapy received by the patients, extent of nuclear differentiation nor pattern of invasion appeared relevant. In a study of primary breast tumors, higher histological grading gave enhanced colony growth and enhanced potential for growth. Studies of this type with other specific "solid" tumors should be encouraged.

Cytogenetic Studies

These in vitro cloning procedures appear to have greatly facilitated karyotypic analyses which have now provided evidence in support of the concept that tumor-associated chromosomal alterations are frequently observed in human "solid" tumors. It is now possible to use molecular biological techniques to determine whether these sites of chromosome change represent the location of cellular oncogenes. In vitro cytogenetic studies in which the stem cell compartment in chronic myeloid leukemia was evaluated were also reported. Of particular importance too is the increased role that cytogenetic analysis is now playing in studies of drug resistance in human tumors. Investigations of the amplified cellular genes associated with methotrexate resistance in the KB cell line were reported and evidence was provided that in gliomas, cellular resistance to BCNU was associated with near diploid chromosome numbers, while BCNU sensitivity was found in hyperploid or hypoploid cells.

CLONING OF HEMATOPOIETIC MALIGNANCIES

Enhancement of In Vitro Growth

For bone marrow myeloma cells, a critical determinant of improved growth was rigorous T-lymphocyte depletion of plated cells using neuraminidase-treated sheep red blood cells plus non-proliferating human fetal lung fibroblasts (MRC-5). Furthermore, it was reported that continuous in vitro growth of T-cell acute lymphoblastic leukemia of childhood was facilitated by a hypoxic environment with nutritional support which proved inadequate for sufficient growth of CFU-C and normal unstimulated T and B lymphocytes.

Effects of Culture Conditions and Exposure Duration on Drug Sensitivities

It is already recognized that the successful evaluation of antitumor agents in the human tumor clonogenic assays requires the testing of adequate drug concentrations under otpimal exposure durations and culture environments. Confirming a number of studies which have highlighted the complex factors influencing methotrexate cytotoxicity in vitro, it has been recommended here that a "continuous" methotrexate concentration of 10^{8}M in a nucleoside-free culture media supplemented with undialyzed horse serum be adopted since, under these culture conditions, this corresponds to the LD50 for human normal granulocyte-macrophage colony forming units. Continuous drug exposure was also favored for assessing in vitro sensitivity of leukemia colony-forming cells in acute non-lymphocytic leukemia to a combination of cytosine arabinoside and an anthracycline. Under these exposure conditions, the overall clinical correlation rate with patients achieving complete remission was 85%.

The time-dependent inhibitory effect of a zinc-chelating agent, orthophenanthrolene, on colony-formation in malignant lymphoid and normal myeloid colony forming cells was also noted, with superior effects seen after prolonged drug exposure in vitro. Of significance too was the observation that, after prolonged exposure, preferential sensitivity to this agent was noted against the malignant T and B lymphoblastic cell lines as opposed to the normal granulocyte colony-forming cells.

CLINICAL PHARMACOLOGY

Modulation of Drug Resistance or Sensitivity

Two rather specific areas were discussed predominantly and the initiation of further such studies should be encouraged by the fact that both approaches, if confirmed, have definite clinical application.

Recently published studies with the P388 murine leukemia have suggested that the acquired resistance to anthracyclines and vinca alkaloids, which is associated with altered drug transport properties, is reversible, at least in part by concomitant exposure to verapamil, a calcium transport inhibitor. At this meeting, somewhat contradictory data were presented when these types of studies were repeated using human tumor material. One study involving nine fresh tumor samples, three from patients who had failed to respond to adriamycin administration, showed no significant inhibition of growth with either vinblastine or adriamycin at clinically achievable drug concentrations for a one-hour exposure, and the concomitant addition of 1 μg/ml of verapamil failed to modify this lack of response. However, concomitant verapamil treatment did result in increased doxorubicin sensitivity under comparable exposure conditions, in four out of seven tumors evaluated in a second study, including two tumors with acquired resistance to the drug. This later finding also appears to be confirmed in a study using a series of human ovarian cancer cell lines where pharmacologic modulation of adriamycin resistance was observed in some cases by the addition of verapamil, and this has led to an ongoing trial of the combination in relapsed ovarian cancer patients.

The initial demonstration that melphalan cytotoxicity could be modulated in the L1210 murine model system has now been confirmed in human ovarian tumor cell lines. Depletion of glutathione levels by nutritional deprivation of cystine or by buthionine sulfoximine resulted in successful potentiation of melphalan cytotoxicity. This was reported in tumor cell lines derived from both untreated patients or those clinically resistant to alkylating agents. The toxicity of buthionine sulfoximine in large animals is now being evaluated since this demonstration, that pharmacologic manipulation of glutathione levels resulted in an increased cytotoxicity, had potential clinical importance in treating ovarian cancer.

In another report investigating mitomycin C pharmacology, some evidence was provided that mitomycin-sensitive cells, exemplified by the MCF-7 continuous human tumor breast cell line, may be amenable to pharmacologic antagonism by sulfur nucleophiles including glutathione and N-acetyl cysteine. However, this effect was, rather disappointingly, not observed in three other human tumor cell lines which exhibited resistance to mitomycin C.

Drug Sensitivity Profiles of Primary Tumors Versus Metastases

Studies aimed at determining whether drug sensitivity profiles were the same or different in a primary tumor and one of its metastases provided evidence of more heterogeneity of response when metastases were evaluated. From one investigation, involving a miscellaneous group of tumors from 46 patients, it was concluded that while one sampling of the primary may provide an accurate test, one sampling of a metastases will not necessarily reflect the sensitivity of the primary tumor. This serves to introduce a word of caution in selecting tumors for clinical correlation studies, particularly to those investigators who have found tumors from metastatic sites easier to grow in vitro than some primary specimens. In a second smaller study involving 14 patients' tumors, however, considerably more heterogeneity of response to chemotherapy was reported, not only among different tumors from the same patient, but even within the same tumor. Large degrees of discordance were reported which were not considered specific for any particular tumor type or antitumor agent being evaluated. However, in the discussion period, the possibility was raised that these findings might be an artifact of the experimental protocol employed. Clearly, further studies of this type are necessary.

Bioassay for In Vitro Drug Stability

The development of this type of methodology is long overdue and hence a most welcome advance in this area of in vitro drug sensitivity testing. A knowledge of the half-lives of these antitumor agents in vitro is clearly important when attempting to select the optimal exposure duration for the in vitro test. Furthermore, this knowledge is crucial for drugs in phase I or phase II trials for which little or no in vivo data are available.

Drug Metabolism Studies

The questionable requirement for metabolic activation of mitomycin C was investigated using S-9 microsomal enzymes and NADPH in human tumor cells. No conclusive evidence was obtained that activation was necessary to produce typical in vitro cytotoxicity patterns. A novel approach to assessing the relevance of hepatic metabolism to the activity or inactivity of new drugs was suggested by employing co-cultures of rat hepatocytes with human tumor cell lines.

TUMOR TYPES, NEW ASSAY PROCEDURES AND METHODOLOGIES

Comparison of In Vitro Assay Procedures

Consistent evidence was provided by two groups that the

Courtenay and Mills clonogenic procedure provided significantly increased colony forming efficiencies compared with that of Hamburger and Salmon and in a small study of lung cancer specimens this same methodology was also considered superior to the assay described by Carney. The important additional factors appear to be the inclusion of rat erythrocytes, a hypoxic environment and a replenishable liquid top medium. The improved colony quality was also significant and proved particularly valuable if colonies had to be scored manually. However, divergent data were presented in attempting to answer the question as to whether the colony forming efficiency obtained or the clonogenic assay procedure adopted influenced the drug response recorded. One group reported comparable drug sensitivity irrespective of the methodology employed while the other group claimed enhanced chemosensitivity with the Hamburger and Salmon assay as opposed to the Courtenay and Mills method. Further studies to resolve these questions are essential and emphasis was placed on the need not only to tailor methodologies for specific tumor types but also to establish individual "training sets" of drug sensitivity patterns for any altered or new assay methodology before attempting clinical correlation studies.

Four alternative methodologies were also described: (i) the use of 100 μl capillary tubes, considered especially suitable for testing smaller tumor cell samples, was claimed to have improved colony forming efficiencies over those obtained in the 35 mm petri dish system; (ii) the use of tritiated thymidine incorporation in soft agar culture to substitute for manual colony counting in agar; (iii) culture on soft agar containing sheep red blood cells, when results equivalent to the double agar layer technique were reported, although it may be difficult to distinguish between cell aggregates and true colonies under such conditions, and (iv) an alternative rapid screening radiometric system which detects ^{14}Carbon dioxide produced by tumor cells from addition of ^{14}C-glucose. This was not proposed as an alternative to the cloning assay but rather as a first line screen for eliminating inactive compounds.

Quality Control and Data Reproducibility

It was encouraging that a number of groups were addressing these very important aspects. An evaluation of the image analysis system for counting human tumor colonies versus counting by experienced technicians provided evidence that the colony counter proved sufficiently reliable and was considerably faster. Unfortunately, such sophisticated instrumentation is thus far only available to a few groups. The introduction of a number of technical improvements into the assay procedures appear to have improved the reliability of the assay and the quality of the dose response data obtained. These procedures include: (i) background subtraction; (ii) a positive control to avoid clumping artifacts, e.g., abrin, mercuric chloride or sodium azide; (iii) application of the drug to culture surfaces following cell inoculation, hence avoiding mechanical manipulation of cells in the presence of the drug, and (iv) viability staining of colonies with INT dye.

It is particularly important that any lack of reproducibility of assay data is not confused with or sought to be explained by invoking the term heterogeneity. One group working with cell lines provided evidence that when the cloning procedure was adhered to, variation within one laboratory or within different laboratories was acceptable, although this may depend to some extent on the degree of standardization of media, sera and/or drug dilutions. However, another group working with "solid" tumors taken from a nude mouse xenograft established from a cell line, found it difficult to standardize results because of subjectivity inherent to the methodology, especially the problems associated with cell clumping and reliable cell viability estimations.

In Vitro and In Vivo Model Systems

Novel in vitro and in vivo models were described for studying ovarian cancer and in particular addressing the problem of drug resistance. A series of cell lines were initiated from both untreated patients and from patients with disease refractory to clinically used drug combinations. Resistance was also induced in vitro in lines from untreated patients by step-wise drug exposure. In addition, a unique transplantable intraperitoneal in vivo model of human ovarian cancer in athymic nude mice was described. Drug sensitivity testing with the colony assay on in vivo and in vitro passages of a human melanoblastoma were also reported. Differences in sensitivity to adriamycin were noted which might be explained by variable selection modes acting on clonogenic cells in the nude mouse tumor or in monolayer culture, thus introducing a word of caution when attempting to compare data from these two very different model systems. However, in a study with a human squamous cell carcinoma xenograft, a positive correlation was noted between the colony forming assay data and growth inhibition data caused by adriamycin and cis-platin.

Finally, preliminary experiments designed to overcome contamination of tissue samples with microorganisms from oral cavity tumors were described involving prior implantation into the retroperitoneal space of mice after which successful cloning was possible in three of seven tumors.

Use of Cell Lines for Monitoring Drug and Radiation Sensitivities

The feasibility of this approach was definitely established by studies presented at this conference. Studies described provided clear evidence of reproducible dose response curves, although emphasis was placed on the need to consider carefully the colony size "cut offs" for assessing radiation responses. These model systems can also be readily used to establish the influence of duration of exposure on "new" or "old" drug cytotoxicities. Again, it was reported that, for certain agents, for example, methotrexate, hydroxyurea or VP-16-213, a one hour exposure may be inadequate and hence inappropriate for evaluating their in vitro responses. One study using a panel of ten human bladder cell lines attempted to predict drug activity in this tumor type. The authors concluded that while there was some concordance between in vitro and clinical activity, the in vitro assay tended to over-predict for the anthracyclines and underpredict for methotrexate. In future conferences, it is hoped that further studies of this type will be reported now that a number of cell lines are available derived from various different human tumor cell types.

Selectivity of the Clonogenic Assay for Tumor Growth

Evidence was presented confirming the fact that some histologically non-malignant tissues may proliferate in soft agar. Most of the examples quoted related to the use of infiltrated bone marrow samples, e.g., in studies of prostate or lung tumors when a significant degree of granulocyte-macrophage colony formation was noted. However, the fact that benign prostatic hypertrophy cells are clonogenic was considered of value since it provided an in vitro model for studying this common, even if non-malignant, condition. However, one paper presented at this conference provided evidence which helped to put this problem of selectivity into perspective. It was reported that from over 1900 specimens received for chemotherapy sensitivity testing, 38 were classified as non-malignant on pathological examination but only 12 "grew" as colonies, representing approximately 0.5% of the total.

CORRELATIVE CLINICAL TRIALS AND

IN VITRO CHEMOSENSITIVITY STUDIES

Predictive Use of Human Tumor Colony Assay Data in Clinical Correlation Studies

Clinical correlations have been made between in vitro chemosensitivity and the response of patients with metastatic cancer to chemotherapy. Surprisingly few new data were presented at this conference. Initially, results from a series of trials were reviewed showing that the human tumor colony assay has had an overall 71% true positive rate and a 91% true negative rate for predicting drug sensitivity and resistance respectively of patients to specific antitumor agents. Then two positive studies were reported for ovarian cancer, employing the standard Hamburger and Salmon assay procedure. In one case, the in vitro data were used to design multiple drug protocols and in the other, the impact of in vitro sensitivity on survival was recorded (discussed below). Data from a large study of human breast cancers were presented involving 407 biopsies from 288 patients. Unfortunately, adequate in vitro growth for drug testing (>30 colonies per plate) was obtained only in 91 (27%) of the 337 viable samples, thus highlighting the major remaining problems, in dealing with this tumor type, including inadequate sampling and inadequate growth. Nevertheless, the assay appeared to detect in vitro activity as well as resistance to a variety of agents and to predict the clinical responsiveness to standard as well as investigational single agents. These general overall conclusions could also be drawn from a smaller study involving 67 tumor samples taken from 64 patients.

In melanoma, an assay based on the clonogenic procedure devised by Courtenay and Mills was used to predict individual clinical responses. In a largely retrospective analysis of 74 patients, there was a positive correlation between in vitro sensitivity (quantitated as expected growth delay in human melanoma xenografts grown in athymic nude mice) and the in vivo response. This calibrated assay thus is able to select patients with tumors that are resistant and who should not have standard chemotherapy and, in many cases, is also capable of selecting an effective alternative agent.

Two rather different approaches to improving the accuracy for predicting clinical sensitivity to chemotherapy were presented. It was first suggested that by modifying the definition of survival fraction so that it reflects the number of cell doublings and possibly the specific tumor types being investigated, a better indication of drug reduced proliferative capacity would be provided which may yield a better index of clinical response. The second study related to primary breast tumors, where with specific improvements of the soft agar assay methodology, 83% evaluable assays were successfully performed on 60 separate specimens. These authors conclude that a comparison of in vitro sensitivity of tumor and normal bone marrow cells, together with stratification of patients according to tumor load, provides a more accurate method for predicting clinical sensitivity to chemotherapy. Indeed, this same group further recommended that Do values (the drug concentration required to kill 63% of granulocyte-macrophage colony forming units) should be used as a reference for selecting biologically meaningful in vitro drug concentrations. In this way, the risk of false in vitro sensitive results originating from testing too high drug doses and of false in vitro resistance from testing at too low a dose could be minimized.

New Agent Screening

The human tumor colony assay is now being increasingly widely used for new drug development and indeed this may be its most useful application. Data were presented on certain anthracycline analogs, the anthracene derivatives, mitoxantrone and bisantrene, the newer vinca alkaloid, vinzolidine, interferon and mis-matched double stranded RNA (with interesting sensitivity reported for these two latter materials in renal cell carcinomas) and a number of NCI potential anticancer drugs tested in primary screening for new agents. Other applications of the clonogenic assay methodology presented included: the relative efficacy of anti-calmodulin agents, suggesting calmodulin as a promising target to exploit in new drug development; evaluation of some "natural drugs"; the design and testing of biomodular therapy for human melanomas; and, correlations of in vitro tumor cell radiosensitivity with radiotherapy response in carcinoma of the cervix.

In general, however, it remains to be seen whether these results will correlate with future Phase II clinical trials and it remains to be established whether this initial screening really needs to be carried out on fresh tumor biopsy material rather than employing continuous human tumor cell lines.

Evaluation of Drug Combinations

It was particularly encouraging in view of the wide clinical usage of drug combinations that the possibility of using the in vitro assay to select or evaluate combinations of agents has now been considered by several groups. In one study, using 11 tumors and testing 39 combinations by continuous in vitro drug exposure, no difference was noted in the response to single versus multiple agents. However, in brain tumors where single agents versus an eight drug combination are currently being evaluated clinically, while the eight drug combination and "best" drug for each tumor selected by the assay yielded comparable results, the combination proved significantly better than the least effective drug. Therefore, this indicates that the combination has an equal chance of being successful in a given brain tumor population compared to the random choice of a single agent. In a further study of advanced ovarian cancer, however, ten individual drugs were tested and the patients were randomized to receive either the standard cyclophosphamide-doxorubicin-cisplatin (CAP) regimen or the "best three drugs" chosen by the in vitro assay. In terms of response rate, comparable figures were obtained using either the empirically chosen CAP regimen or that combination of "best drugs" selected by the assay methodology. Survival data from this particular study are eagerly awaited. In breast cancer, however, the human tumor colony assay proved less predictive for combinations than for single agents, but in the discussion section the difficult question of how accurately to evaluate combinations was raised. This particular problem clearly needs to be addressed in future studies.

Prognostic Factors

We were reminded in ovarian cancer that poor prognostic host factors can exert a negative influence on response and sometimes outweigh favorable human tumor colony assay predictive data. In colon cancer, increased sensitivity to 5-fluorouracil was associated with biological aggressiveness and high proliferative potential, suggesting that histopathological factors are prognostic for responsiveness to chemotherapy. Additionally, in a small series of 26 "solid" tumor samples and 7 pleural effusions from patients with breast cancer, it was reported that positive colony growth in a methyl-cellulose based assay was associated with poor patient survival.

Prospective Clinical Studies and Survival Data

There were very few reports of prospective studies at this meeting and only one study presented survival data. A prospective study on tumors from 49 patients with miscellaneous tumor types suggested the clinical usefulness of the human tumor colony assay indicating that at least five drugs should be evaluated to enhance the success rate of in vitro selection of sensitive drugs. I hope that this will not deter groups from setting up prospective, randomized, controlled clinical trials which are clearly essential for validation of the true potential of these clonogenic assays for accurately selecting effective agents. Indeed, the feasibility of conducting such a trial now in non-small cell lung cancer was described at this conference.

In terms of survival data, long term follow-up is obviously needed and at future meetings, one would hope to hear of more results from such studies. In this respect we should be encouraged by the evidence of the one study presented that chemotherapy selected by the human tumor colony assay can be used successfully to enhance survival of ovarian cancer patients who have relapsed following standard chemotherapy regimens.

CONCLUDING REMARKS

After reviewing these presentations, I consider that we might take as the overall theme of this meeting the term "encouragement":

- encouragement to those who initiated this work which now has sufficiently firm foundations to continue;
- encouragement to those who are starting out in this research area, since so much remains to be done, and
- encouragement to continue to submit grant proposals, even if they have been rejected in the past.

I am aware that many problems associated with the use and interpretation of the current clonogenic assay procedures remain, some of which are listed in Table 1, and they must be tackled by the laboratory researcher.

TABLE 1

PROBLEMS ASSOCIATED WITH THE USE AND INTERPRETATION OF CURRENT CLONOGENIC ASSAYS

1. Not all tumor types grow under these conditions.
2. Low colony forming efficiencies.
3. Small proportion of tumor specimens suitable for drug testing.
4. Specificity: "normal" versus "malignant".
5. Difficulties in preparing single cell suspensions.
6. Quantitation of data and dose response survival curves.
7. Colony size: ? clumps ? clusters.
8. Uncertainty of in vitro criteria for drug sensitivity.
9. Inadequate in vitro pharmacologic data on antitumor agents.
10. Choice of clinical criteria for assessing drug responses.
11. Description of in vitro-in vivo correlations.

I was rather disappointed that at this conference there were so few reports of advances in these areas, but I trust that these problems are being addressed and I feel confident that many can be overcome, although probably not in the short term. In particular I would hope and indeed recommend that groups concentrate their endeavors on one or two particular tumor types, since in this way it is most likely that the specific requirements necessary for optimal growth can be delineated for each individual tumor type. The question of definitely identifying the cells which form colonies in soft agar as "normal", "benign", or "malignant" also remains to be answered. With the latest reports at this conference of successful clonogenic growth from breast tumors and squamous cell carcinomas of the head and neck, it becomes increasingly important that this distinction can be made accurately. Criteria for drug sensitivity must also be rigorously defined, not only in the laboratory but also in assessing clinical responses. It is now becoming apparent that such criteria may have to vary not only according to the tumor type being studied but also with the drug being tested. With the development of new or improved methodologies, a careful and extensive calibration of the in vitro data has to be carried out for each drug and each tumor type.

There is now definite evidence to show that these cloning procedures are already proving their value in enlarging our knowledge of the biology and cytogenetics of human tumors and in particular certain of the "solid" tumors. Of major interest to me, however, is their potential clinical value, as listed in Table 2, although I hasten to point out that I consider the first proposal for individualizing chemotherapy programs quite impractical, at least at the present time.

TABLE 2

POTENTIAL CLINICAL VALUE OF CLONOGENIC ASSAY PROCEDURES

1. Tailoring individual chemotherapy regimens.
2. Identification of patterns of chemosensitivity for patients with unknown primary carcinomas.
3. In vitro Phase II studies of new agents.
4. Initial screening of analogs.
5. Establishing patterns of cross resistance and sensitivity in relapsing patients.
6. Monitoring development of drug resistance with serial samples.
7. Tumor detection or staging.
8. Relationship between clonogenicity and prognostic factors.

However, if the full potential of these studies is to be realized, what we need is more evidence. It is up to us all to provide this in the next decade, fortified and encouraged by these types of conferences and workshops. Although it should be appreciated that even then there can be no guarantee that these approaches will be generally accepted or more widely adopted since people can react to evidence in two very different ways. The first is exemplified by Bertrand Russell, when he stated that "the degree of probable validity attached to a proposition should be proportional only to the evidence in its favor." I trust that many would agree with him. However, there is an alternative view, shared perhaps by the many skeptics who have not participated in this conference, exemplified by Malcolm Muggeridge, "One of the stupidest theories of Western life has been that of evolution. It couldn't be true. Certainly there is no evidence for it. And even if there were evidence for it, I wouldn't accept it".

VII. ABSTRACTS

1 PROSTAGLANDINS, MACROPHAGES AND HUMAN TUMOR CLONING. M.E.Berens and S.E.Salmon. Cancer Center, U.of Arizona, Tucson,USA

We studied the role of autologous macrophages (MØ) on human tumor cloning by selective depletion of these host cells, and by reconstitution with MØs as feeder cells. Subpopulations of cells were investigated for their abilities to produce prostaglandin (PG) E2. In separate experiments, gas chromatography was used to measure the profile of the different PGs. In 5/6 MØ-depletion experiments, tumor cloning decreased following elimination of MØs (mean: 34% control). Reconstitution with MØs enhanced cloning in a dose response fashion ($n=6$, $r^2=0.517$). PG production in 17 different samples was highly variable; however PGs were found to be of MØ origin and not from the fresh tumor cells ($p=0.02$). Cloning experiments with tumor cell lines have shown that PGF2a promotes clonal growth. The present studies reveal that cells from fresh human tumors produce very little of this PG. PGE1 is also minimally present. Tumor MØs produce primarily PGE2 and PGI2, both of which lack effects on tumor cell proliferation, but demonstrate potent actions on immune and hemostatic mechanisms. We conclude that tumor MØs support clonal growth of human epithelial tumor cells, and are also the dominant source of PGs (mainly E2 and I2) in tumors. Furthermore, the profile of PGs in tumors suggests that these compounds are more important to indirect mechanisms of tumor survival than for proliferation of tumor cells.
[Supported by NCI Grants CA 21839 and CA 17094.]

2 ANALYSIS OF CLONOGENIC SURVIVAL USING A MODIFIED COULTER PARTICLE COUNTER. Bonni J. Hazelton and Peter J. Houghton. St. Jude Children's Research Hospital, Memphis, TN

A modified Coulter particle counter (CPC) has been employed to rapidly and reproducibly determine the cloning efficiency of a human rhabdomyosarcoma cell line, RD, grown in agarose. The CPC method is applicable to systems in which tightly packed colonies are formed. In order to establish the adaptability of this method to other cell systems, the integrity of fixed colonies produced by a series of human colon carcinomas was examined. HCT-15 forms colonies which are stable, with constant stirring, for up to one hour. Two other lines, WiDr and DLD-1, exhibited approximately a 20% loss of detectable colonies within 10 minutes after dissolving the agarose. No further loss was seen over the subsequent 50 minutes, suggesting that the initial decrease may be a loss of loosely bound peripheral cells. The response of RD cells to a 24 hour exposure to vincristine (VCR) or cis-dichlorodiamminoplatinum (cis-DDP) was evaluated using the CPC method and correlates well with the results obtained using the conventional visual method of enumeration. Using a C1000 Channelyzer it was observed that survival was related to colony volume up to 2.5×10^4 μm^3 above which the response was independent of volume. With proper validation this technique may prove valuable for examining drug sensitivity and drug interactions in established cell lines. The method will eliminate subjectivity with visual counting and allow for larger assays, thus enhancing sensitivity. Supported by American Cancer Society grant CH-156 and by ALSAC.

3 The Inhibitory Effects of Actinomycin D on Malignant Melanoma Cells Cultured on Basement Membranes. M.J.C. Hendrix, J.L. Brailey, H.N. Wagner, Jr., K.R. Gehlsen, and B. Persky, Departments of Anatomy, Univ. of Ariz., College of Medicine, Tucson, AZ., and Loyola Univ. Stritch School of Medicine, Chicago, ILL.

In order to study the invasive characteristics of a patient-derived malignant melanoma cell line (81-46c), interactions of tumor cells with a basement membrane (BM) were studied in vitro with TEM. From fresh placentae, the amnion was dissected away, and the amniotic epithelium was removed chemically. 81-46c cells were seeded onto membranes within chambers and incubated at 37°C for time periods ranging from 2 through 14 days. In 7 trial runs, the membranes within chambers contained media with serum; 4 trials were in serum-free media. After 14 days in culture, 81-46c tumor cells, in the presence of serum, never penetrated the collagenous stroma underlying the BM. Local degradation of the BM was seen at focal points of interaction between tumor cells laden with secondary lysosomes, and the broken membrane. Tumor cells seeded onto BM's in the absence of serum penetrated the BM and the stroma within 3 days. In 4 trials, 81-46c cells were allowed to interact with the BM in the absence of serum for 24 hours, and were treated with Actinomycin D at a final concentration of 0.1 μg/ml. This drug was selected based on the sensitivity demonstrated by the patient's tumor cells to continuous drug exposure. The action of this rRNA inhibitor prevented a significant number of 81-46c cells from invading the BM. Antibodies to Actinomycin D were used to stain the 81-46c cells as they interacted with the BM. Immunohistochemical studies revealed an intense nuclear staining pattern in a majority of the tumor cells on the BM. Those tumor cells which invaded the BM initially are considered to be drug resistant and can penetrate the BM in a shorter period of time in subsequent trials.

4 Variation in Interferon-Induced Antiproliferation: Direct Effects of Target Cells. L.J. Krueger, P. J. Andryuk, and D.R. Strayer, Cancer Institute, Hahnemann University, Philadelphia, PA

The interferons, a family of related proteins, presently are in clinical trials to determine conditions for maximum effectiveness in cancer therapy. Factors which influence interferon-induced antitumor action are being identified in several rapid in vitro systems. We have coupled standard tissue culture techniques with the clonogenic assay system to define important intermediate stages in interferon action. Our studies, using tumor cells of diverse histological origin, have defined separable phases of the interferon induced antiproliferative state. These phases include: 1) potentiation 2) memory 3) initiation 4) maintenance and 5) escape. Interferon requirements for maximum effectiveness differ among cell types from 2.0 IRU/ml α-IFN to $>$10,000 IRU/ml (IFN resistance) and between dose schedules. Our results show that through an understanding of the five phases involved in the antiproliferative response, a 5-15 fold reduction can be obtained in the concentration of interferon required for the antiproliferative response.

In addition, an intermediately sensitive cell line, the human fibrosarcoma HT-1080 has been studied in vitro. Initial results indicate that a subpopulation(s) exists demonstrating different interferon sensitivities. These observations indicate that the subpopulation(s) result either from: 1) genetic events, 2) phenotypic modulations, or a combination of these mechanisms. Studies of clonogenicity in soft agar following interferon exposure indicate the presence of stem cell subpopulations with different sensitivities to interferon. Analysis of individual HT-1080 clones is currently underway.

5 COMPARISON OF TUMORS CONTAINING MUCH AND FEW STROMA IN THE CLONOGENIC ASSAY. G. Link, G. Rauthe, and J. Mussmann. Dept. of Gynecology and Obstetrics, Univ. of Giessen, Giessen, F.R.G.

As often reported in literature colonies of tumor cells are generally collected easier from tumors containing few stroma than from those containing plenty of stroma. At ovarian tumors we were able to demonstrate that colonies - grown in soft agar following the method of HAMBURGER and SALMON - of tumors with a dense cell pattern are exhibiting a similar morphological pattern as the original tumor at light microscopy. Therefore, it has been of interest if such a morphological analogy between original tumor and colony could be found at neoplasms containing only few tumor cells, investigated in the mammary cancer as a typical example.

The tumor colonies grown in the clonogenic assay were mixed with agar containing graphite following formaldehyde fixation and centrifuged. After embedding in paraffin histological sections of the tip of the agar enriched with the colonies were produced for staining.

Since proliferation of stromapart of the tumor in soft agar culture is impossible lacking of mesenchymal surroundings, the present investigation shall demonstrate whether morphological qualities of tumors with much stroma can be imitated in the clonogenic assay, too.

Comparing colonies of tumors containing much and little stroma we hope to enlarge our knowledge concerning the morphological realization of cell growing in agar culture.

6 A HYPOOSMOLAR MEDIUM TO DISAGGREGATE TUMOR CELL CLUMPS INTO VIABLE AND CLONOGENIC SINGLE CELLS FOR THE HUMAN TUMOR STEM CELL CLONOGENIC ASSAY.

Leibovitz, A., Liu, R., Hayes, C., Salmon, S.E. University of Arizona Cancer Center, Tucson, AZ

A hypoosmolar medium and tissue processing technique is described which is useful for disaggregation of residual human tumor cell clumps persisting after mechanical or enzymatic treatment of solid tumors and malignant effusions. The addition of the hypoosmolar procedure to the standard methods for disaggregation of human solid tumor specimens increased the viable single cell yield by 47% and of malignant effusions by 67%. In 5 of the 26 solid tumor specimens tested in the human tumor stem cell assay, clonogenic single cells were obtained with the hypoosmolar procedure, whereas no growth was observed using standard methods. Overall, the success rate for clonogenicity increased from 46% to 65% for the 26 solid tumors in this study with the major improvement occurring in ovarian cancer. Clonogenicity was obtained in 80% of malignant effusions both by standard methods and the hypoosmolar techniques. Recently, we have directly compared the hypotonic method to standard techniques as the initial mode of disaggregation and found it to enhance both yield (cells/gram) and clonogenicity. The increased total yield of clonogenic cells obtained with this procedure enhances the opportunity for experimental versatility and in vitro drug testing.

7 Immunohistochemical Characterization of Malignant Melanoma Cells Cultured in Soft Agar Following Drug Perturbation.

B. Persky, F.L. Meyskens, Jr., and M.J.C. Hendrix, Department of Anatomy, Loyola University Stritch School of Medicine, Maywood, IL., and Departments of Internal Medicine and Anatomy, College of Medicine, Tucson, AZ.

In order to evaluate the immunohistochemical characterization of drug-treated colonies and their extracellular matrices, a patient's malignant cells were plated in triplicate in soft agar and incubated with retinoic acid ($1x10^{-6}$M), Bleomycin (0.1 ug/ml), and Melphalan (0.075, 0.15, and 0.3 ug/ml). Plates were refed on day 14 with 0.5 ml of F10 media for retinoic acid and Bleomycin-treated colonies, and with 0.5 ml F10 media plus Melphalan in concentrations of 0.075, 0.15, and 0.3 ug/ml. Seven days later colonies were automatically scored. The percent survival of drug-treated cells verses controls was 85% for retinoic acid, 2% for Bleomycin, and 44%, 18%, and 3% for concentrations of 0.075, 0.15, and 0.3 ug/ml Melphalan respectively. Immunohistochemical analysis of untreated colonies utilizing the peroxidase anti-peroxidase (PAP) technique revealed a positive reaction product for fibronectin, which was localized primarily on the tumor cell surface and the associated extracellular matrix (ECM). Trace amounts of type IV collagen and laminin were noted in the ECM. A strong positive reaction for vinculin was seen primarily to be cell surface associated and extracellular in nature. Immunological probing of drug-treated colonies revealed similar intensities and patterns as the control colonies with one exception. Specifically, colonies treated with 0.3 ug/ml of Melphalan and stained with anti-laminin showed a demonstrable increase in laminin localization. This observation suggests that tumor cells treated with Melphalan may preferentially utilize laminin for attachment to a substrate during metastasis.

8 CORRELATION OF IN VITRO GROWTH CHARACTERISTICS OF HUMAN BREAST CARCINOMA WITH THEIR HISTOLOGICAL GRADING. R. Rashid, V. Hug, G. Spitzer, G. Blumenschein, B. Drewinko, and D. Johnston. M.D. Anderson Hospital and Tumor Institute, Houston, Texas

We compared histological grading using the criteria of Bloom and Richardson (Br J Ca 11:359-377, 1957) and in vitro growth properties of 31 primary breast carcinomas. One aliquot of each tumor sample was processed for routine histological examination and the remainder was used for clonogenic growth in semisolid agar cultures. Seven tumors did not form colonies and 24 formed between 2-482 colonies (mean 86). The difference in histological grading between tumors that produced clonogenic growth and those that did not was the following:

In Vitro Growth	No. of Specimens	No. of Tumors with Histological Grade	
		I	II or III
Absent	7	6	1
Present	24	6	17

p=0.0086 by Fischer's Exact Test

A higher histological grading was also associated with a higher potential for clonogenic growth:

Total No. of Tumors	No. of Colonies	Histological Tumor Grade I(n)	II(n)	III(n)
7	0	6	1	0
18	1-100	5	9	4
6	100	1	2	3

Tubular formation (glandular differentiation) was a histological feature of all tumors that did not form colonies in vitro, but of only 58% of tumors that yielded clonogenic growth (p=0.044).

These results suggest that clonogenic breast tumor cells have less differentiated morphological features.

9 THE KINETICS, EXTENT, AND LIMITS OF CELLULAR PROLIFERATION WITHIN THE CLONOGENIC ASSAY.

Stephen P. Thomson, Frank L. Meyskens, Jr., and Thomas E. Moon. Cancer Center, University of Arizona, Tucson, Arizona

We developed techniques to quantify the number of cells within tumor colonies of different sizes and quantitated the proliferation of clonogenic cells in agar. Daily observations of cells in agar and serial photography showed that there can be a 3 to 4 day delay in the onset of proliferation in agar followed by rapid growth and then abrupt cessation of proliferation. We quantified the extent of proliferation of cells from melanoma biopsies of 7 patients and 11 cell lines derived from various tissues after they were allowed to proliferate in agar until they stopped. Approximately 10% of cells divided 1 to 5 times while only 0.01% divided 6 to 9 times. The total number of cells within the colonies at the end of growth was different while the total volume of cells within the colonies per plate was similar, approximately $10^9 \mu m^3$. Refeeding increased the total volume of cells within colonies in a dose-dependent manner. Plating fewer cells increased the proportion of larger colonies but the total volume and number of cells within the colonies was similar. These data suggest that $10^9 \mu m^3$ cellular volume/plate represents a maximal cellular volume for proliferation in the closed, non-refed agar system. Replating studies using the same biopsy cells have shown that clonogenic melanoma cells can self-renew and have more proliferative capacity than that expressed during primary colony formation. Thus, clonogenic assays only measure the initial proliferative capacity.

10 POSSIBLE NON-RANDOM CHROMOSOME ALTERATION IN RHABDOMYOSARCOMA. J.M. Trent*, F. Thompson*, J.Casper+, and J. Fogh°. *Cancer Center, Univ. of Arizona, Tucson, AZ; +Midwest Children's Cancer Center, Medical College of Wisconsin, Milwaukee WI; °Memorial Sloan Kettering Cancer Center, Rye, NY.

We have examined tumor colony forming units (TCFUs) and/or monolayer cultures from 7 cases of rhabdomyosarcoma. Detailed chromosome banding analysis was performed using G-, C-, and Q-banding techniques. Results revealed a wide range of chromosome numbers with the majority of specimens displaying highly aneuploid karyotypes. The single chromosome most consistently involved in either numeric or structural alterations was chromosome 1. Marker chromosomes involving either the long (q) or short (p) arm of chromosome 1 were observed in all evaluable samples. The most consistent chromosome alteration observed was the deletion or translocation of the short arm of chromosome 3 at band p21. All samples studied from TCFUs, and 3 of 4 established cell lines examined displayed alteration of 3p21. Our results support the concept that tumor-associated chromosome alterations are frequently observed in human solid tumors.

Research support in part by CA-29476.

11 USE OF A RADIOMETRIC SYSTEM TO SCREEN FOR NEW ANTINEOPLASTIC AGENTS. D.D. Von Hoff, B. Forseth, L. Warfel. University of Texas Health Science Center, San Antonio, Texas, and Johnston Laboratories, Towson, Maryland.

As now constituted, the human tumor cloning system has several problems which makes it difficult to utilize for screening thousands of compounds to select new antineoplastic agents. The BACTEC System represents an instrument which measures $^{14}CO_2$ produced by tumor cells from ^{14}C glucose or other radiolabel containing media. Although the system has a number of theoretical limitations (metabolism of malignant and nonmalignant cells is measured and it does not distinguish between injured and reproductively dead cells), it does have potential for rapidly screening large numbers of agents against a variety of human tumor cell lines. To test that ability, four human cell lines, including two breast cancer lines (MCF-7 and ZR-75), Hep 2 (laryngeal carcinoma) and HL60 (promyelocytic leukemia line) were studied. For all four lines there was a direct relationship between the number of cells seeded in a vial and the amount of $^{14}CO_2$ produced. The optimal time for making the $^{14}CO_2$ measurements was six days. For all four classes of antineoplastics tested (Adriamycin, vinblastime, methotreate, and cis-platinum) there were excellent dose response effects with decreases in $^{14}CO_2$ production with increasing concentration of drugs. Comparison of percent survival of these cell lines measured by the BACTEC versus the conventional two-layer agar system for continuous drug exposure showed a correlation coefficient of $r=0.87$ ($p=<0.0001$; 35 data points). For the one hour comparison the correlation was not as good $r=0.354$ ($p<0.03$; 36 data points). While in no way a replacement for the more biologically attractive cloning assays, the BACTEC system does provide an alternative first line screen for eliminating inactive compounds as well as for detecting compounds with some cytotostatic or cytocidal activity.

12 SURVIVAL CURVES OF CLONOGENIC HUMAN MELANOMA CELLS IN SOFT AGAR ARE CONSISTENT WITH SURVIVAL CURVES OF OTHER MAMMALIAN CELL LINES. Karin H. Yohem, Marvin D. Bregman, and Frank L. Meyskens, Jr., Department of Internal Medicine, University of Arizona Health Sciences Center, Tucson, Arizona

Clonally derived human melanoma cell lines M1RW5 and c83-2C were grown in RPMI 1640 medium with supplements. Early passage exponentially growing cells were irradiated in suspension by a 18 MeV linear accelerator operating at 10 MeV and yielding a dose of 5 Gy/min to the cells. Replicates were irradiated with a single dose of X-rays at 0.5, 1.0, 2.0, 2.5, 3.0, 4.0, and 10.0 Gy. Cells were plated into a bilayer agar assay system within one hour after radiation. Cellular aggregate control plates were used to assess the single cell nature of the experimental plates. Colonies with more than 50 cells were counted by the Omnicon FAS-II image-analysis system. Classical radiation survival curves with initial shoulders were obtained for both melanoma cell lines. D_0 values that define the slope of the exponential line are 1.62 Gy and 1.3 Gy for M1RW5 and c83-2C respectively. Extrapolation numbers, n, that describe the extent of the shoulder are 1.3 for M1RW5 and 2.0 for c83-2C.

13 ACTIVATION/INACTIVATION OF CANCER CHEMOTHERAPEUTIC AGENTS BY RAT HEPATOCYTES CO-CULTURED WITH HUMAN TUMOR CELL LINES. M.C. Alley, G. Powis, P.L. Appel, K.L. Kooistra, and M.M. Lieber, Urology Research Lab and Developmental Oncology Research Lab, Mayo Clinic, Rochester, Minnesota

Various cell culture methodologies are employed in the study of mechanisms of anticancer drug action and in the search for new anticancer drugs. While colony formation assays provide sensitive indices of tumor cell proliferation and growth inhibition imposed by many chemotherapeutic agents, drugs which require metabolic activation lack activity in such assays. In the present study we have utilized freshly isolated rat hepatocytes for the activation of drugs which are metabolized by hepatic microsomal as well as extra-microsomal enzymes. Optimally, 10^5 hepatocytes in fluid media are placed over soft-agarose matrix containing 10^4 tumor-derived cells (e.g., A101D, A204, A549) within 35 mm culture dishes; drug and/or drug vehicle is added directly to the hepatocyte layer; and tumor cell colony formation is assessed following 7 to 10 days' incubation. Such a co-culture system appears to provide an efficient means to detect activation/inactivation of metabolizable agents: In the presence of hepatocytes, the IC_{50} for cyclophosphamide, indicine N-oxide, and procarbazine are markedly decreased, whereas the IC_{50} for AZQ, carmustine, and 5-fluorouracil are significantly increased. By contrast, the IC_{50} for actinomycin D, mitomycin C, and 6-mercaptopurine and other agents are unaffected by hepatocyte presence. Results suggest that hepatocyte/tumor cell co-cultures may be well-suited for assessing the relevancy of hepatic metabolism to the activity/inactivity of new anticancer drugs.

14 EFFECT OF HUMAN GROWTH HORMONE ON CLONING EFFICIENCY OF HEp-2 HUMAN TUMOR CELL LINE GROWN IN DOUBLE-LAYER AGAR. D. L. Elson, J. M. McMillin, S. J. Klusick, University of South Dakota School of Medicine, Sioux Falls, South Dakota

Human growth hormone, hGH, (Somatropin, National Institute of Arthritis, Diabetes, Digestive and Kidney Diseases) in various concentrations was incorporated into the plating media of double-layer agar cultures of HEp-2 (epidermoid carcinoma cells, Source: ATCC). The culture system was that described by Hamburger and Salmon.

$.5 \times 10^4$ HEp-2 cells were incorporated into 1 ml of the plating layer (which was CMRL-based with .3% agar) and layered over a 1 ml feeder layer (which was McCoy's based with .5% agar) in a 35x10mm petri dish. Tumor cell colony numbers in control plates and plates containing human growth hormone (.02, .002, .0002 IU/ml incorporated into the plating layer) were compared.

The average HEp-2 colony counts for 33 control plates was 422 ± 92 for a cloning efficiency of $8\% \pm 2\%$. The colony counts for .0002 IU/ml hGH plates were 286 ± 20 ($p < .005$), for .002 IU/ml hGH plates were 322 ± 35 ($p < .05$) and for .02 IU/ml, hGH plates were 307 ± 81 ($p < .005$).

We conclude that Human Growth Hormone in the concentrations tested does not improve the cloning efficiency of HEp-2 cells in the double-layer agar culture system. The apparent growth inhibition induced by hGH may warrant further study.

15 A STUDY OF CLONING EFFICIENCY AND CHEMOSENSITIVITY OF HEp-2 CELLS GROWN IN DOUBLE-LAYER AGAR CULTURE AFTER PRE-INCUBATION WITH SUB-LETHAL CONCENTRATIONS OF MTX (METHOTREXATE). D. L. Elson, S. J. Klusick, University of South Dakota School of Medicine, Sioux Falls, SD

Methotrexate in sub-lethal concentrations is known to synchronize cell populations at the G_1-S phase. We performed an experiment to test whether the pre-incubation of HEp-2 with MTX prior to plating in double-layer culture would influence the cloning efficiency or chemosensitivity of the cultures compared with controls.

$.5 \times 10^4$ HEp-2 (epidermoid carcinoma cells, Source: ATCC) were incubated with Methotrexate (.05, .25, .5 µg/ml) for various times (.5, 1, 2 hrs) prior to plating in the double-layer agar culture system described by Hamburger & Salmon. Colony numbers in control cultures and cultures pre-incubated with MTX were compared. In addition, the colony numbers of control cultures and MTX pre-incubated cultures following continuous exposure to 5FU (.1, 1, 10 μg/ml) and Adriamycin (.4, .04, .004 μg/ml), beginning 24 hrs after plating were compared. At .05μg/ml MTX and 1 hr pre-incubation, treated plates averaged 137 ± 30 colonies vs. 147 ± 30 ($p < .6$) colonies for control plates incubated in plating media only. At all other MTX concentrations and times of exposure, no stimulation of colony growth was seen. Colony growth inhibition from continuous exposure to various concentrations of 5FU and Adria beginning 24 hours after plating was similar in control cultures and MTX pre-incubated cultures.

We conclude that MTX pre-incubated HEp-2 cells, at the concentrations and times tested, have no better cloning efficiency in double-layer cultures than control cells and the chemosensitivity of both control and MTX pre-incubated cultures following continuous exposure to 5FU and Adria at various concentrations are equivalent.

16 Evaluation of Human Tumor Chemosensitivity in the Soft Agar Assays and its Clinical Correlation. Dominic Fan, Christine Schneider, Susan Fan, Helene Blank and Lee Roy Morgan. Dept. of Pharmacology, LSU Medical Center, New Orleans, LA

The soft agar tumor cloning assay developed by Hamburger and Salmon in 1976 has found its place in cancer research. The true clinical predicting ability of this assay has been tested in many laboratories and yet a general consensus is still to be reached. Many technical difficulties have held the assay at its developmental stage. In a ten month study we received 134 clinical tumor specimens of 26 classifications. We were able to plate out 122 of these tissues with a 76% success in colony growth. The cells were exposed to 3 log concentrations each of 4 commonly used anticancer agents. Retrospective correlations between soft agar chemosensitivity of tumor colonies and clinical drug responses were made possible in 31% of the patients. Evaluation of 45 in vitro and in vivo associations in this study indicated a combined sensitivity of 0.65 and a specificity of 0.68 for the soft agar assays. The results for the individual tumor types were as follows:

	Breast	Colorectal	Lung	Ovarian
Sensitivity	0.8	0.5	1.0	0.5
Specificity	0.5	0.6	1.0	1.0

(Supported in part by an ACS Grant #IN-150 and by a Cancer Research Grant from the Elsa U. Pardee Foundation of Midland, Michigan.)

17 SOFT AGAR HUMAN TUMOR DRUG SENSITIVITY TESTING: COMPARISON OF (^{3}H)-THYMIDINE INCORPORATION WITH COLONY FORMATION ASSESSED IN THE PRESENCE OF THE VIABLE STAIN, INT. C.A. Jones, C.B. Uhl, M.M. Lieber, Mayo Clinic, Rochester, MN

Initial reports (Friedman et al, 1982; Tanigawa et al, 1982) indicated a close correlation between (^{3}H)-dThd incorporation and colony counts for *in vitro* drug sensitivity testing of primary human tumors. The viable stain, INT, was recently (Alley et al, 1982) shown to improve detection of drug-induced decreases in colony formation in primary human tumor soft agar cultures when compared to non-stained cultures. In view of the above, *in vitro* drug sensitivity as determined by inhibition of (^{3}H)dThd incorporation or optical enumeration of INT-stained primary tumor colonies were compared. Soft agarose cultures from 112 primary human tumor specimens, 5 xenograft specimens, and 2 human tumor cell lines (SW-480, A1663) were exposed to drug and assessed by both assays. (^{3}H)dThd-uptake was determined after 72 hours (modification of Tanigawa et al, 1982) and colony counts were evaluated after 7-14 days. Active drugs produced $\geq$ 70% decreases in (^{3}H)-dThd uptake or colony number compared to non-drug treated controls. 63% and 33.5% of primary tumor specimens were evaluable by (^{3}H)dThd or CFA, respectively. 35 tumors were evaluable for drug sensitivity by both assays. Reasonable correlations between assays (r^2 = 0.67, 0.64) were obtained when drug D/R were performed with cell lines and xenografts, respectively. Poor correlations (r^2 = 0.21, 0.25) were obtained when drug tests were performed in primary tumor specimens with either low or high day 1 counts respectively. (^{3}H)dThd incorporation and enumeration of INT stained colonies do not correlate when measuring drug sensitivities of primary human tumor specimens.

18 IN VITRO SENSITIVITY OF CANINE TUMOR STEM CELLS TO ADRIAMYCIN D. Macy, B. Ensley, E. Gillette
Colorado State University, Fort Collins

Tissue samples from 30 canine tumors were prepared and assayed *in vitro* to 1, 5 and 10 mcg/ml adriamycin using soft agar culture. Tissue was obtained either from surgical or necropsy specimens and single cell suspensions were prepared. Cells (5×10^5) were incubated continuously in the presence of drug except for controls. Surviving colonies (30 cells or more per colony) were counted and percent survival calculated. Three different concentrations of drug were tested - 1, 5 and 10 mcg/ml. Some tumors did show a dose related response - decreasing survival as dose increased up to 10 mcg/ml. The results presented here are for the lowest drug concentration, 1 mcg/ml, which correlates more with the peak plasma levels reported for humans. Many tumors showed a moderate to marked resistance to adriamycin and some even exhibited enhanced growth in the presence of drug. Three tumors showed a marked sensitivity, i.e., survival at or below 30%.

19 USE OF THE L1210 CELL LINE AS A QUALITY CONTROL INDICATOR FOR THE CLONOGENIC ASSAY. J.J. Marx, Jr., T.K. Banerjee, S.K. Spencer, S. Trentlege. Marshfield Medical Foundation and Marshfield Clinic, Marshfield, WI

The human tumor stem cell assay (HTSCA) is used widely to test the responsiveness to a variety of therapeutic and experimental anticancer agents. The tumor heterogeneity, plating efficiency and relative cloning efficiency precludes any semblance of a representative quality control program. We have used the L1210 mouse leukemia cell line as an indicator cell to establish a quality control program in our laboratory as well as to investigate various aspects of the HTSCA. Cloning of this cell line was sufficiently reproducible to compare new reagents (i.e., new serum lots). The concentration of drugs that inhibited 50% of the clones was defined as the

ID_{50} and calculated for each agent. Using this drug dose the intra-assay variation was measured at 8.1% (N = 26) while the inter-assay variation was measured at 18.0% (N = 15). Furthermore, methods of exposure (i.e., continuous vs 1 hr) times were compared using this dose. There was little difference in response, except with those drugs thought to be cell-cycle phase specific. Reduced O_2 tension (8-10%) significantly enhanced cloning efficiency. Biotransformation of some drugs enhanced anticloning activity. These results are similar to estimates of activity in human tumor cultures. Thus the L1210 is a cell line suitable for routine quality control of the HTSCA as well as limited investigational studies.
(Supported by the Elsa U. Pardee Foundation, Marshfield Clinic and Marshfield Medical Foundation).

20 Predictability of Anticancer Drug Activity in Human Bladder Cancer Using Human Bladder Tumor Cell Lines. H.B. Niell, K.C. Webster and E.E. Smith. University of Tennessee Center for the Health Sciences, Memphis, TN

The tumor colony assay (TCA) developed by Hamburger and Salmon has been shown to be predictive of anticancer drug response in individual cancer patients. This TCA is potentially useful in screening anticancer drugs for activity in human bladder cancer. Due to the low growth rates of primary explants of bladder cancer, we are evaluating alternative sources of bladder tumor cells for use in drug screening. Numerous human bladder tumor cell lines (HBTCL) have been established in long term culture (Fogh, Natl. Cancer Inst. Monogr, 1978). Using a panel of HBTCL in the TCA, we have attempted to determine whether the patterns of drug sensitivity will correlate with the clinical activity of anticancer drugs in human bladder cancer. We are evaluating the use of a panel of 10 HBTCL (CUB-1, CUB-2, CUB-3, 647V, RT4, TCCSUP, T24, Scaber, J-82 and 639V) using 10 standard and investigational drugs. All cell lines produced colonies when plated at 100,000/plate with cloning efficiencies varying from 0.1-1%. To date, the following patterns of sensitivity have been noted:

	MTX	5FU	ADR	MITO	DDP	VLB	AMSA	VP-16	Bleo	Mitoxan
% in vitro	0	14%	57%	14%	14%	14%	57%	14%	42%	57%
% clinical	28%	35%	25%	25%	35%	15%	0	0	20%	un-known

In general, the anthracyclines were overpredicted for activity while methotrexate was unpredicted for activity. This panel of HBTCL resulted in drug sensivitity patterns that were not altogether consistent with known clinical activity in human bladder cancer.

21 SCREENING PLATINUM COMPLEXES AS POTENTIAL ANTI-TUMOR AGENTS WITH AN L1210 CLONOGENIC ASSAY.

Helen Ridgway, David P. Stewart, Nancy Michaelis, Michael J. Guthrie, and Robert J. Speer. Wadley Institutes of Molecular Medicine, Dallas, Texas

Murine leukemia L1210 is one of the best known models for selecting antitumor drugs that will have human utility. However, it is expensive and time-consuming. We have established our L1210 line in suspension culture and have employed these cells for a clonogenic assay to screen platinum coordination complexes for their antitumor activity. Preliminary experiments employing 1 hr incubation of the cells with drug concentrations varying from 0.1 - 50 μg/ml indicated that a clonogenic cell kill of ≥ 90% at 4 μg/ml correlated well with an increased life span ≥ 30%. Thirty-five platinum coordination complexes have been compared in vitro to the in vivo system. There were 18 true positive, 0 false positive, 15 true negative and 2 false negative results for an overall efficiency of 94%. The test was not limited to platinum(II) complexes but also correctly predicted results for 3 platinum(IV) complexes. Also it was able to distinguish between cisplatin and its inactive trans isomer. This procedure is simpler and more economical than animal testing. The compounds selected in this fashion can then be tested for toxicity and screened with a variety of fresh human explants in the human tumor stem cell assay as part of their preclinical evaluation.

22 ESTABLISHMENT AND CHARACTERISTICS OF A CYTOSINE ARABINOSIDE-RESISTANT SUBLINE OF HUMAN KB EPIDERMOID CARCINOMA : Takashi Tsuruo[1] and Saburo Sone[2] ([1]Cancer Chemotherapy Center, Japanese Foundation for Cancer Research, Tokyo 170 and [2]Department of Internal Medicine, The University of Tokushima, School of Medicine, Tokushima 770, Japan)

A subline of human KB epidermoid carcinoma resistant to cytosine arabinoside (KB/ara-C) was developed in vitro by forcing the KB parent cells (KB/p) to grow for at least 6 months in increasing dose of the drug (10^{-5} moles). The KB/ara-C cells had a 800-fold resistance when compared to the KB/p cells. Resistance to ara-C was stable for 3 months. The size and grwoth rate of KB/ara-C cells were almost same as parent cells. Drug levels in KB/ara-C cells treated in vitro with ara-C were twofold to threefold lower than were levels in similarly treated KB/p cells. KB/ara-C cells were also associated with about fivefold decrease in deoxycytidine kinase activity. In contrast, cytidine deaminase activity in the resistant cells was slightly higher than KB/p cells. KB/ara-C cells proved to be sensitivie to hydroxyurea, 6-thioguanine, adriamycin, 6-mercaptopurine and bleomycin. A marginal degree of cross-resistance was also observed to amethopterin, cisplatin and ACNU. Further studies on mechanism of resistance will be discussed.

This resistant human cell line system may provide a pathway for elucidating mechanism of drug resistance and for overcoming of such resistance in human tumor.

Supported by Grant-in-Aid for Cancer Research from the Ministry of Education, Science and Culture, Japan.

23 MULTIPLE DRUG SENSITIVITY TESTING WITH THE CLONOGENIC ASSAY. T.K. Banerjee, J.J. Marx, Jr. and S. Spencer. Marshfield Clinic and Marshfield Medical Foundation, Marshfield, WI

Tumor cell sensitivity to chemotherapeutic agents is tested in the clonogenic assay in order to select effective agents for treatment. Frequently patients are treated with multiple rather than single agents. We tested a variety of commonly used drug regimens for

synergistic activity using the L1210 cell line and a variety of human tumors. Using the ID_{50} (dose of drug which when cultured continuously with L1210 cells inhibits 50% of clones) drug concentration previously determined for the L1210 cell line, no increase (or decrease) in clone inhibition was seen when Adriamycin was tested in various combinations including cis-Platinum, Mitomycin C, Cytoxan and 6-thio TEPA. These results were interpreted as if no synergism (antagonism) was present since this is a homogeneous cell line. Human tumor cells are more heterogeneous and multiple drug sensitivities could be expected. Thus far in testing 11 different tumor samples (including 6 ovarian carcinomas) with 39 different combinations, no significant differences were seen when comparing the response to single agent versus multiple agents in vitro. Clones from 1/6 ovarian carcinomas showed a slight but consistent increase in sensitivity when Adriamycin and/or 5-FU was added to Mitomycin C. This difference was within the experimental variation of the technique.
(Supported by the Elsa U. Pardee Foundation, Marshfield Clinic and Marshfield Medical Foundation).

24 DESIGN AND IN VITRO TESTING OF BIOMODULATOR THERAPY FOR HUMAN MELANOMA. Marvin D. Bregman, Cindy Hahn, Diane J. Sander, and Frank L. Meyskens, Jr., Department of Internal Medicine, University of Arizona Health Sciences Center, Tucson, Arizona

Clonogenic assays have been used to screen biological modifiers in combination and measure the effect on suppression of the transformed phenotype (cloning efficiency) and retardation of growth (proliferative capacity). Two biomodulator classes were tested. Class I agents modulate genetic expression and results in general cellular changes. Class II agents act on or through a single enzyme (or pathway). Potential class I agents are glucocorticoids, interferon, retinoids, growth regulatory peptides and steroids. Potential class II agents are 5-fluorouracil, difluoromethylornithine (DFMO), methotrexate and 4-hydroxyanisole (4-HA).

Evidence for the concept of phenotypic heterogeneity to class I modulators will be presented. When class I agents were added in combination there was a large synergistic inhibition of clonogenic melanoma cells. Class II agents (such as DFMO and 4-HA) demonstrated significant inhibitory activity when tested as single agents. Combination of class I and II agents in which human pharmacological data already exists for each individual agents were also tested and therefore the combination may be available for clinical evaluation. Our in vitro results suggest dexamethasone, DFMO, interferon, and retinoic acid in combination may be an effective therapy for human melanoma.

25 In-vitro Phase II trial of 5FU and Folinic Acid (FA) in Non-Gastrointestinal Malignancy.

Budd GT and Neelon R, Cleveland Clinic; Cleveland, Ohio.

The cytotoxicity of 5FU for some human tumor cell lines is enhanced by simultaneous treatment with the reduced folate FA. Uncontrolled clinical trials of this combination in patients (pts) with metastatic colo-rectal cancer have produced response-rates that are greater than would be expected with 5FU alone. We report the results of an "in-vitro Phase II Trial" of 5FU and FA in non-gastrointestinal malignancies. Ten tumors from 9 pts were tested. Diagnoses were ovarian cancer (7), soft-tissue sarcoma (1), mesothelioma (1), and carcinoma of the renal pelvis (1). The clonogenic assay of Hamburger and Salmon was used, modified to provide continuous exposure to all test drugs throughout the assay period. Tumors were assayed against 5FU alone, equimolar concentrations of 5FU and FA, and FA alone at concentrations of 15 μM and 150 μM. FA alone did not affect clonogenicity, with colony counts relative to control averaging 1.09 and 1.06 for FA 15 μM and 150 μM, respectively. Defining as sensitive (S) a greater than 50% inhibition of colony formation, no tumor was S to 5FU 15 μM, 1 was S to 5FU 150 μM, 1 was S to 5FU/FA 15 μM, and 1 to 5FU/FA 150 μM. Thus, enhanced cytotoxicity by 5FU and FA as compared with 5FU alone was exhibited in only one ovarian tumor. These observations could result from a lack of potentiation of the cytotoxicity of 5FU by FA in the tumors tested or as an artifact of the in-vitro technique, such as might be produced by excess reduced folates or "rescue" by thymidine in the culture medium. Further studies to resolve these questions are under way.

26 CLONOGENIC CELL ASSAY IN SOFT AGAR OF HUMAN CERVICAL CARCINOMA. POTENTIAL APPLICATION TO RADIOTHERAPY. S. Casillo,* S. Lücke,* A. La Pera,* A. Vitturini,** and C. Biagini.** Istituto Tecnologie Biomediche CNR 00161 Roma (*) and Istituto Radiologia Universita Roma (**), Italy.

The *in vitro* human tumor stem cells assay system(Hamburger and Salmon,Science 197:461,1977)has provided a powerful new approach to the study of spontaneous human tumours.Studies have been conducted with the use of the *in vitro* assay of cellular clonogenic capacity to analyze the radiosensitivity of cervical Carcinoma cells treated *in vivo* with radiation.

Twenty patients with different histological types of Carcinoma of the Cervix were studied.After obtaining informed consent,biopsies for culture studies and histological examination were taken before and after the first course of radiotherapeutic treatment,ovoiding necrotic and infected areas.Although *in vitro* growth of colonies was observed in 15 cases(75%)and the number of colonies varied from 26 to 147 per $5x10^5$incubated cells,only for 11 patients *in vitro-in vivo* correlation was possible.The direct correlation of *in vitro* testing of tumor cell radiosensitivity with the actual patients response in these few cases is encouraging.However,further work is necessary to better understand the potentials and limitations of such clonogenic cell analisys tecniques in order to determine if they should play a role in patient radiotherapy. Supported by grants from project CNR"Control of Neoplastic Growth".

27 SCREENING OF PHASE I DRUGS IN THE HUMAN TUMOR CLONING SYSTEM (HTCS) TO PINPOINT AREAS OF EMPHASIS IN PHASE II STUDIES. B. Lathan, D.D. Von Hoff, T.J. Melink and D.L. Kisner. University of Texas Health Science Center, 7703 Floyd Curl Drive, San Antonio, Texas

One major application for the HTCS is the early testing of investigational anticancer drugs against various human tumor specimens. This testing could provide useful direction for Phase II clinical trials. The HTCS was utilized to screen for *in vitro* cytotoxic activity of six investigational drugs, i.e., the antifolate Trimetrexate (TMO), the bifunctional intercalating antibiotic Echinomycin (ECH), the antimetabolite 2-Fluoro-ara-AMP (2-FLAMP), the synthetic C-nucleoside Tiazofurin (TCAR), the vinca alkaloid Vinzolidine (VLZ), and the synthetic polyelectrolyte Carbetimer (CARB). All drugs were tested against a variety of tumor types. A specimen had to have at least 20 colonies on control plates to be evaluable. *In vitro* activity was defined as a $\leq$ 50 percent survival of tumor colony forming units. A summary of the overall *in vitro* response rates at approximately one-tenth the achievable peak plasma concentrations in man is shown below.

Drug	Concentration	# sensit/# evaluable	(%)
TMO	0.1 µg/ml	14/69	20
ECH	0.01 µg/ml	36/230	16
2-FLAMP	1.0 µg/ml	52/212	25
TCAR	1.0 µg/ml	32/190	17
VLZ	0.1 µg/ml	23/87	26
CARB	10.0 µg/ml	7/28	25

These results indicate that the tested compounds have different cytotoxic activity, with CARB, VLZ and 2-FLAMP having the highest *in vitro* activity. Sensitivities of specific tumor types have also been determined. It will be of interest to see how these *in vitro* results correlate with future Phase II clinical trials.

Supported by Contract #N01-CM-27542 and by Grants from Warner Lambert, Monsanto, and Eli Lilly.

28 ASSESSMENT OF SOME "NATURAL DRUGS" IN THE HUMAN TUMOR STEM CELL ASSAY. M. MATRAT, S.MARCHAL, B. WEBER, R. METZ. Centre A. Vautrin RN 74 54511 VANDOEUVRE LES NANCY FRANCE

About 20 % of our cancerous patients utilize "natural drugs" alone or generally during the same time they receive standard cytotoxic drugs. We were interested in testing these natural drugs in an in vitro system. 4 drugs were studied : Carzodelan, Conium, Viscum Album

(Iscador) and Tri-X from Dr Solomides. The content of these drugs is poorly known. Arsenic was detected in some of them. Standard anticancer drugs were tested for comparison. Two human cancer cell lines were grown in agar : MCF-7 from a breast cancer (H. Soule) and CAL-1 from a melanoma (J. Gioanni). Cells were plated in a two layer soft agar system (technique of Hamburger and Salmon with slight modifications) with continuous drug exposure. Cell sensitivity was defined as less than 30 % colony survival. A significant antitumor activity (on MCF-7) was observed with Carzodelan (stock solution diluted at 1/10), with Conium (cut-off concentration between 1/10 and 1/100), with Solomides Tri-X (1/100). Viscum Album was ineffective. We are checking out these results on CAL-1 melanoma. We were able to compare the relative efficiency of Carzodelan with Adriamycin. The concentration of the other drugs is unknown. The ratio of ID50% (inhibition of growth of TCFU to 50 %) of Adriamycin to Carzodelan was 2 10 . To get the same effect with Carzodelan as with Adriamycin it would be necessary to inject unfeasable very large amounts of Carzodelan. Patients believe that these drugs are really effective. They are wrong. Are others experiments warranted to prove it ? Viscum Album was forbidden by F.D.A. in cancer treatment patients because it was ineffective.

29 *IN VITRO* COMBINATION CHEMOTHERAPY FOR BRAIN TUMORS.

Mulne AF, Salgaller ML, Walson PD, Hayes JR, Ruymann FB, Columbus Children's Hospital, Columbus, Ohio.

A new chemotherapy approach to brain tumors, referred to as "8 in 1" therapy, consists of eight different drugs administered in one twelve hour time period. Although preliminary clinical results with the combination appear encouraging, an argument against multidrug chemotherapy has been the improved results of single agent therapy, such as CPDD. Using the Human Tumor Stem Cell Assay we have successfully cultured 59 brain tumors with 76% growth, 61% adequate for drug testing. In 16 of these tumors there were adequate colonies for *in vitro* testing of all eight drugs, singly and in combination.

In no instance was the effect of the combination completely additive. The % colony kill and drug ranking for each tumor varied greatly. No drug was consistently the best or worst. These data support the hypothesis of heterogeneity of malignant clones within brain tumors.

Although in all tumors tested at least one single drug showed colony counts below the combination, the post hoc multiple T test analysis showed no difference in results between the eight drug combination and the best drug. The combination was significantly better than either the control or least effective drug (critical value of t= 2.003, df= 60, p < .05).

Thus, until more information is available, the combination chemotherapy appears to have an equal chance of being successful in a given brain tumor population compared to the random choice of a single agent.

30 Antiproliferative Effect of Mismatched Double-Stranded RNA in the Human Tumor Stem Cell Assay (HTSCA). D.R. Strayer, J. Weisband, W.A. Carter and I. Brodsky, Cancer Institute, Hahnemann University, Philadelphia, PA

Double-stranded (ds) RNAs are inducers of the different molecular forms of human interferon, are activators of the interferon-associated intracellular mediators, as well as being direct inhibitors of tumor cell proliferation. Modification of certain structural features in dsRNA produce a mismatched analog of $rI_n \cdot rC_n$, $rI_n \cdot r(C_{12},U)_n$, which retains these properties while triggering much less toxicity. We have used the HTSCA to assist the early clinical development of $rI_n \cdot r(C_{12},U)_n$, Ampligen®, by identifying human tumors which were sensitive to its antiproliferative effects. Freshly sampled human solid tumors were cloned in soft agar in the presence of 250 μg/ml $rI_n \cdot r(C_{12},U)_n$. In vitro colony formation (>40 cells/colony) adequate to evaluate drug sensitivity (>30 colonies/5×10^5 cells) was obtained in 48 cases (66%). High in vitro drug sensitivity (S) was defined as <30% colony survival compared to control; >60% colony survival constituted resistance (R); 30-60% survival comprised an intermediately sensitive group (I). At 250 μg/ml 10 tumors were S, 18=I, and 20=R. The most sensitive tumor was renal cell carcinoma 8(47%)=S, 5(29%)=I, and 4(24%)=R. Other tumors which showed sensitivity (S or I) were carcinoid (5/5), glioblastoma (1/1), breast (4/6), melanoma (2/3), colorectal (1/4), and ovarian (1/5). These results suggest that different human solid malignancies vary in sensitivity to Ampligen® and the HTSCA may be useful in identification of human tumor types and individuals for clinical trials with Ampligen®, including patients resistant to interferon.

31 EFFECT OF EPIDERMAL GROWTH FACTOR (EGF) ON THE CLONING EFFICIENCY OF FRESH HUMAN TUMOR CELLS IN SOFT AGAR.

G. Umbach, V. Hug, G. Spitzer, B. Tomasovic, N. Merchant. Univ. of Texas System Cancer Center, M.D. Anderson Hospital, Houston, TX

The effect of EGF (25 ng/ml) on the capacity of fresh human tumor cells to form colonies in a bilayer soft agar system was evaluated in 41 specimens obtained from various types of malignancy. Colony growth (5 or more colonies measuring at least 100 um) occured in 46% of the assays without EGF and in 44% of assays with EGF. The addition of EGF increased the cloning efficiency by 50% or more in only 7% of (3/41) specimens, if changes in the range below 5 colonies are disregarded. These 3 specimens already showed growth of 5 or more colonies, so that we never observed the transformation of a non-growing specimen into a growing specimen through the addition of EGF. The number of samples evaluable for drug sensitivity studies (20 or more colonies) was identical (20%) in both groups. The mean number of colonies in the assays without EGF was 17.5 (SD $\pm$37.5) and in the assays with EGF 18.7 (SD $\pm$43.1). There was no statistically significant difference (p-value being 0.77) between those values. Our failure to show a major colony-stimulatory activity of EGF on fresh human cancer cells differs from results published by Hamburger et al and Pathak et al. This discrepancy could be due to differences in disaggregation methods (we routinely use a sequence of mechanical and high-dose overnight enzymatical disaggregation) that could influence the amount of extracellular matrix, cell to cell interaction, and the nature of the EGF receptor, all of which may be of importance in determining response to EGF. Supported by NIH, Max Kade Foundation.

32 CYTOTOXICITY OF ANTICANCER DRUGS ON GRANULOCYTE-MACROPHAGE COLONY-FORMING UNITS IN CULTURE (GM-CFUC) AND CHOICE OF IN VITRO DRUG CONCENTRATIONS FOR PREDICTIVE ASSAYS.

G. Umbach, V. Hug, G. Spitzer, B. Tomasovic, B. Drewinko. Univ. of Texas System Cancer Center, M.D. Anderson Hospital, Houston, TX

Adequate drug concentrations are important for the result of in vitro sensitivity or resistance. Bone marrow cells have been commonly used as a reference system to evaluate in-vitro drug-induced lethality in hematologic neoplasms. We evaluated the cytotoxic effects of 12 anticancer drugs on normal human GM-CFUC at continuous exposure in a bilayer soft agar system. The dose-survival pattern for 11 drugs followed a simple negative exponential curve that formed a straight line on a semilogarithmic plot. The Bleomycin curve, however consisted of an initially steep slope followed by a shallower slope. The D_0 (drug concentration required to kill 63% of GM-CFUC) for the various drugs in ug/ml are: Velban 0.002, Homoharringtonine 0.004, Adriamycin and Mitomycin 0.005, Elliptinium 0.026, Spirogermanium 0.135, Hydroxyperoxy-Cyclophospamide 0.167, Melphalan 0.178, Cis-Platinum 0.394, Fluoro-AMP 0.508, 5-Fluorouracil 0.529, Bleomycin 1.2 (by extrapolation). Bone marrow toxicity is one of the dose limiting factors of most drugs in vivo; since GM-CFUC are cultured under similar conditions as the tumor cells, we recommed that the above D_0 be used as references for selecting biologically meaningful in vitro drug concentrations. However, for non-myelosuppressive drugs, e.g. Bleomycin, this may not be entirely appropriate. This approach minimizes the risk of false in-vitro-sensitive results originating from testing at too high drug doses and of false in-vitro-resistant results from testing at too low doses. Supported by NIH, Max Kade Foundation.

33 HUMAN TUMOR CLONOGENIC CELL CULTURE (HTC^3) IN MONITORING TREATMENT RESULTS: NEW APPLICATIONS.

R.H.M.Verheijen*,W.J.K.Kirkels**,P.Kenemans*,
F.M.J.Debruyne**,G.P.Vooys*** & C.J.Herman****
*Dept. of Gynecology and Obstetrics & Dept. of Pathology, **Dept. of Urology,***Dept. of Pathology, St. Radboud Hospital Nijmegen;**** Lab. of the Delft Hospitals (SSDZ) Delft. THE NETHERLANDS.

Since the introduction of the HTC^3 the main application of this test has been in cytostatic drug testing: assessment of individual chemosensitivity and testing of new agents. Colony counting on a single time point is commonly used to measure the drug effect. Evaluation based on one day counts, however, does not take into account the growth pattern of the tumor. Automated colony counting allows accurate assessment of the temporal growth pattern. This provides means for: 1) adequate chemotherapy selection, 2) detection of viable tumor cells, and 3) evaluation of in vitro growth behavior.

TABLE I: in vitro growth and cytologic examination of abdominal washings at SLO for ovarian ca. (# of pts.)

	cytology +	cytology -
in vitro growth +	[3]	0
in vitro growth -	1	4

TABLE II: in vitro growth and histopathologic follow-up of patients with TCC of the bladder (# of pts.)

	follow-up +	follow-up -
in vitro growth +	10	[20]
in vitro growth -	1	15

Positive growth in vitro has been accurately determined using the temporal growth patterns. The presence of clonogenic cells can be demonstrated, indicating clonogenicity of the malignant cells seen at cytopathologic examination of peritoneal washings obtained at second look operations (SLO) in ovarian cancer (see box in table I). In patients without histopathologic evidence of recurrence of transitional cell carcinoma (TCC) of the bladder, positive growth may define a group at risk (see box in table II).
Conclusion: accurate assessment of in vitro tumor growth by using the temporal growth patterns, obtained at automated colony counting, allows detection of patients at with minimal viable disease and patients at risk.

34 THE CLONOGENICITY OF CELLS OBTAINED FROM BIOPSIES OF BENIGN PROSTATIC HYPERTROPHY (BPH).

F.R. Ahmann L. Woo, C. Guevara. Univ. of Arizona, Cancer Center

Because other workers have demonstrated that cells from benign parathyroid adenomas have clonogenic potential (Can Res 40:3694, 1980) and that BPH may have clonogenic potential (Cancer 50:1332, 1982), we have assayed single cell suspensions from transurethrally resected specimens of BPH for colony formation. Specimens were handled steriley and divided in the Pathology department so that adequate material was available for both research and pathologic study. Single cell suspensions were obtained via mechanical mincing, enzyme disaggregation with DNAse and collagenase, passage through sieves, and removal of debris via separation over a Ficoll-Hypaque gradient. 10^4 to 10^5 viable nucleated cells were plated in 35 mm Petri dishes in 1 ml of 0.3% agar layer over a 1 ml 0.5% agar underlayer containing an admixture of fetal bovine serum and growth factors. Colony formation (>20 cells) occurred in 6/8 specimens with colony counts per plate ranging from 20 to 120. The BPH colonies were characterized by the following stains: Wright-Giemsa, PAS (+), esterase (-), and acid phos (+). Ultrastructural studies with electron microscopy and cytogenetic analyses are ongoing. Preliminary evidence suggests that the clonogenic cells in BPH are epithelial in origin. BPH cells are clonogenic and this work offers the potential of developing an _in-vitro_ model to study at least one cellular population of this common disorder.

35 USE OF HUMAN TUMOR CLONOGENIC ASSAY (HTCA) IN DETECTING BONE MARROW METASTASIS IN NEUROBLASTOMA (NBL).

B Bostrom, NKC Ramsay, ME Nesbit, University of Minnesota, Minneapolis, MN

Eighteen bone marrow specimens were studied from 9 patients (pts), aged 5 days to 5 years with NBL (range 1 to 5 specimens per pt). The pts were staged as follows: stage IV (6 pts), stage I to III (3 pts). Two to 5 ml of heparinized bone marrow was centrifuged over ficoll-hypaque and plated following washing. One to 5 x 10^5 viable cells were plated in quadruplicate without dextran or 2-mercaptoethanol in a modification of the method of VonHoff et al. (Cancer Res 1980; 40:3591). Tumor colonies (>20 cells) were enumerated on day 7 to 14. A Wrights stained preparation of the tumor colonies in situ was examined to verify the histologic nature of the colonies. The results of the HTCA and simultaneous bone marrow aspirates and biopsies were compared as follows:

Stage IV	Specimens	Positive BM Aspirate &/or Biopsy	Pos HTCA	% Cloning Efficiency
diagnosis	2	2	2	.016,.183
relapse	2	2	1	.051
after therapy	11	9	5	.002,.002,.003, .004,.010
Stage I-III	3	0	0	
	18	13	8	

Five of 13 histologically positive bone marrow specimens had no growth in the HTCA. All had minimal residual tumor seen in aspirate smears. None of 5 histologically negative specimens had growth in the HTCA. This data suggests that HTCA is less sensitive then bone marrow aspirate and biopsy examinations in detecting small residual NBL in the bone marrow.

36 MARKED VARIATION OF CLONOGENIC GROWTH BETWEEN INCUBATORS AND THE LOSS OF WATER DURING INCUBATION

Julie A. Buckmeier, Stephen P. Thomson, Marvin D. Bregman, and Frank L. Meyskens, Jr., Department of Internal Medicine, University of Arizona Health Sciences Center, Tucson, Arizona

We noticed considerable variation between control plates of the same experiment placed in different incubators. Triplicate control plates of a murine

(Cloudman S91-CCL 53.1) and human melanoma cell line (81-46A) were placed in several different CO_2 incubators which were in routine use for clonogenic assays. Variation of clonogenic growth within a single incubator for the human and murine cell lines was low, with coefficients of variation 11 and 9.3%, respectively. Variation between incubators was significantly higher with C.V. of 70.3 and 25.9%, respectively (Fisher ratio $p > 0.025$). Several incubators gave consistantly higher growth, although their temperature, humidity, and CO_2 levels were similar to all other incubators. We also weighed plates prior to and following incubation and found a 5-40% weight loss in 12 days due to drying of the agar. Changes in agar appearance were not apparent until a 35-40% water loss occurred. Large water loss was associated with poor growth and was avoided by shutting the door heater off, filling the top tray, and the bottom of the incubator with water. The marked variation of clonogenic growth indicates the difficulty with comparing the absolute number of colonies from different incubators.

37 EFFECT OF FEEDING ON THE IN VITRO GROWTH OF HUMAN CLONOGENIC TUMOR CELLS

E.E.Holdener[1],P.Schnell[1],P.Spieler[2],M.Bessler[1]and H.J. Senn[1].Div of Oncol,Dept of Med[1] and Dept of Pathol[2], Kantonsspital,CH-9007 St.Gallen,Switzerland.

Successful cloning of a variety of human tumors is increasing,however,in vitro growth of the majority of human tumors is still disappointingly low and does not allow large scale drug testing.We compared the growth of clonogenic tumor cells in the Hamburger/Salmon system (Science 1977;197:461)with the growth in a one-way feeding system(modified from Park et al;Cancer Res 1977;37: 4595).Both double layer systems were set up in identical fashion.The cultures of the feeding system were fed on alternative days with 0.5ml feeding solution(FCS 10% in alpha medium with insulin 0.1U/ml,ascorbic acid 0.15mM and l-aspara. 50 µg). A total of 41 tumors showed growth in one or both culture systems.Significant increase ($p<0.05$;Wilcoxon,Mann and Whitney) of colony growth was found in 14.6% of tumor samples: lung 2,fallopian tube 1, pancreas 1,ovary 1,ACUP 1 and in ME-8/14 melanoma cell line. The number of colonies was significantly decreased

($p<0.05$) in the feeding system in 29.3%: ovary 3,stomach 1 melanoma 1,kidney 1,breast 2,ACUP 4 and in WiDr colon carcinoma cell line.No difference in colony growth was observed in 56.1% of the tumors. Our data suggest that feeding of the double layer system does not result in increased colony growth in the majority of tumor samples. Different composition of the feeding solution or feeding of individual growth factors,however, may result in increased tumor colony growth in vitro.
(Supported by SNF 3.873.0.79)

38 GROWTH KINETICS OF HUMAN CLONOGENIC TUMOR CELLS

E.E.Holdener[1], M.Bessler[1], P.Spieler[2], H.J.Senn[1]
Div.of Oncol.[1], Dept.of Med. and Pathol., KSSG, CH-9007 St.Gallen/Switzerland.

Differences in the kinetic of colony development may be crucial for the timing of drug exposure to obtain maximal cytostatic effect. Therefore we used the double layer soft agar system (Hamburger&Salmon 1977) to study the in vitro growth of human tumors over time. Qudruplets were set up for the different time points [8] during the 28 day culture period. Out of 151 human tumors, only 37 could be used for our study. Twentyeight exp. for 23 tumors ultimately grew $\geq$ 30 colonies of $\geq$72 µm in diameter per dish. The average plating eff. was 0.058% (0.006-0.27). In addition, 6 exp. with human tumor cell lines (WiDr colon, ME-8/14 melanoma) were performed.

Within the 28 day culture time we could observe 4 major growth patterns: 1) "up-slope" with max.colony number on day 28 (breast 1, kidney 2, ACUP 1); 2) "up-slope" with plateau beginning betw. day 8 and day 22 followed by a dedrease (ovary 1, stomach 1); 3) "up-slope" with max. betw day 8 and day 22 followed by a decrease (breast 2, ovary 6 melanoma 2, kidney 1, ACUP 3, ME-8/14 3); 4) "up-slope" with plateau and second growth phase(melanoma 1, lung 1, kidney 1, ACUP 1, WiDr 3). Four tumors with $>$100 clumps/ dish on day 0 showed an initial decrease (degenerating clumps) before definite growth. In 12/28 samples we could observe an initial lag phase up to 22 days. None of these growth patterns could be related to a specific tumor type. Our data suggest a high variability of in vitro growth pattern of human tumors. To know the individual growth kinetic may be of importance for the timing of drug exposure and be helpful in the analysis of drug sensitivity results.
(Supported by SNF 3.873.0.79).

39 SELECTIVITY OF THE HUMAN STEM CELL ASSAY (HTSCA) WHEN APPLIED IN THE ORAL AND FACIAL REGION.
C. Metelmann*, E. Wartenburg, H. Wolinsky, H.-R. Metelmann. *Institute for Medical Microbiology and the Dept. of Maxillofacial Surgery and Plastic Surgery of the Face, Klinikum Steglitz, Free University of Berlin, Berlin, FRG.

The HTSCA makes possible a largely selective cloning of malignant tumors in the oral and facial region by impairing the growth of other tissue. In none of the 30 different tissue samples from benign tumors, chronically proliferative inflammations and odontogenous germinal tissue was culturing successful (more than 5 clones per plate). In 4 cases, there were exactly 5 cell formations per plate, which according to definition, could not be regarded as successful culturing. Two odontogenous cysts, 1 ameloblastoma and 1 germinal saccule were invloved here.

However, 10 non-malignant tissue samples from the tonsil area (tonsillectomy material in chronic tonsillitis cases) showed growth of 5 to 12 typical clone-like cell formations per plate in 4 cases. Because of their immunological surface markers, these clone formations could be regarded as B-cell formations whose growth is attributed to mitogen stimulus of the serum fraction in the medium. Under routine microscopic examination, the cell formations can hardly be differentiated from malignant stem cell clones. The cloning results from tonsillar carcinomas thus become very unclear. The first experiments with mitogen-free media have so far only shown that the cloning capacity of both non-malignant and malignant cells is inhibited.

40 NORMAL, TRANSFORMED, AND MALIGNANT HUMAN GASTROINTESTINAL CELL CULTURES. Mary Pat Moyer, Ph.D. Department of Surgery, The University of Texas Health Science Center at San Antonio, Texas

Few appropriate normal cells are being used in comparative studies with cultured human tumor cells. This is particularly true of colorectal and gastric carcinomas, since no good techniques were available to culture normal gastrointestinal (GI) epithelial cells. Our recent success in culturing human colon, small intestine, and stomach cells (Moyer, Proc Soc Exp Biol Med 174:12-15, 1983; Moyer et al., In Vitro Models of Human Cancer, in press) has permitted comparing these cells to malignant cells and the initiation of studies on in vitro transformation by chemical and viral carcinogens. Normal cells treated with the oncogenic virus, simian virus 40 (SV40), or the chemical carcinogen azoxymethane displayed several phenotypic changes which suggested that they were "transformed" in vitro. Analyses of primary cultures and cell lines from 30 normal and 49 malignant GI tissues, under a variety of culture conditions, revealed that there was a general propensity for GI cells to grow as "islands" or clusters of cells, most frequently in suspension, but sometimes as monolayers. Use of enzymes for dissociation was often toxic and selected for fibroblasts, particularly when high (10% or greater) serum concentrations were used to supplement the medium. There was great variability in proportions of differentiated cell types in the normal or malignant cultured cell populations, but they produced mucous and other colon-specific antigens, responded to GI hormones, and displayed transepithelial transport. Collectively, these observations have provided the basis for development of procedures which yield a culture success rate of greater than 90% for both normal and malignant human GI cells.

41 PRELIMINARY EXPERIMENTS WITH GERM REDUCTION AND CLONING OF MICROBIOLOGICALLY CONTAMINATED CARCINOMAS OF THE ORAL CAVITY. E. Sanger, C. Metelmann*, E. Wartenberg, H.-R. Metelmann. Dept of Maxillofacial Surgery and Plastic Surgery of the Face, Klinikum Steglitz, *Institute for Medical Microbiology of the Free University of Berlin, Berlin, F.R.G.

Contamination of tissue samples with microorganisms is the greatest problem involved in the cloning of tumors of the oral cavity in the human tumor stem cell assay (HTSCA). 80% of all unsuccessful cloning attempts in our laboratory can be at least partially attributed to contamination.

In this situation, we performed our first tests with decontamination through implantation in experimental animals: (1) Implantation of contaminated nylon cotton pellets in the retroperitoneal space of CF1-mice leads to nearly complete germ reduction within 3 to 4 days. (2) In 10 tumors of the oral cavity with clinically evident contamination, germ reduction was achieved in 7 cases by short-term implantation and in 3 cases by additional application of conventional methods (lavage, antibiotics). (3) Cloning in the HTSCA was successful for 3 of the 7 tumors decontaminated by implantation. In 1 case in which germ reduction was achieved in the conventional manner and by implantation, cloning and cytostatic testing were successful; cis-platinum proved to be the most effective drug for both systems. This in vitro result prospectively correlated with the clinical behavior of the tumor.

42 INHIBITORY EFFECTS OF SODIUM AZIDE ON ADJACENT, UNTREATED WELLS IN A RAPID CHEMOSENSITIVITY ASSAY.
V.K. Sondak, S.U. Hildebrand-Zanki, and D.H. Kern. Surgical Oncology, John Wayne Clinic, Jonsson Comprehensive Cancer Center, UCLA School of Medicine, Los Angeles, CA 90024; and Surgical Service, Veterans Administration Medical Center, Sepulveda, CA

In soft-agar colony-forming assays of tumor chemosensitivity, it is often difficult to distinguish between true tumor-cell colonies and cell clumps, blood clots, or aggregates of nonmalignant cells. We used 4 mg/ml Sodium Azide (AZ) as a positive control to inhibit cell replication and allow quantitation of clumping artifacts. Only 18/35 (51%) colony-forming assays demonstrated > 50% inhibition (inh) by AZ, compared to 18/23 (78%) assays using a refrigerated control, suggesting that AZ induced counting artifacts. To eliminate culture artifacts, we adopted a rapid chemosensitivity assay that measured thymidine uptake rather than colony formation. AZ was effective in this system as a positive control. However, experiments with cell lines indicated that AZ-treated wells inhibited the growth of untreated control wells in the same 6-well plate. A retrospective review revealed that only 3/77 (4%) thymidine assays fulfilled criteria for growth in agar ($\geq$ 300 cpm ^{3}HTdR uptake and $\geq$ 80% inh of uptake by AZ) when control wells and AZ-treated wells were plated on the same 6-well plate, compared to 121/181 (68%) not adjacent to AZ. AZ clearly inhibited growth of tumor cells in adjacent wells. This effect was not observed with pharmacologic concentrations of anti-cancer drugs. Care must be taken in the selection of positive control substances in tumor chemosensitivity testing.

Supported by Grants CA 09010 and CA 12582, awarded by the National Cancer Institute, DHHS, and by VA Medical Research Services.

43 HUMAN TUMOR CLONOGENIC ASSAY (HTCA) IMPROVED BY NEW METHODS OF CELL ISOLATION, IDENTIFICATION AND CULTURE. Marcelo B.Sztein, Lawrence S.Lessin, and Oliver Alabaster.George Washington University,Washington D.C.

HTCA is limited by current methods of obtaining single cell suspensions, poor viability, an unknown proportion of plated tumor cells (TC), and low cloning efficiency (CE). Using solid and ascites specimens from breast, colon, and ovarian carcinomas, we compared mechanical and enzymatic disaggregation techniques. The highest cell yield and viability ($\geq$75%) was obtained by incubating pieces of soft solid tumors or ascites cells in Hank's balanced salt solution ($Ca^{++}$and Mg^{++}free HBSS) containing 0.05% trypsin, and 0.02% EDTA with continuous stirring for 30 min. at 37°C. For hard tumors, better results were obtained when trpsin-EDTA treatment was preceded by an incubation in HBSS containing 0.2% collagenase, 0.02% hyaluronidase, and 0.002% DNAase for 30 min. at 37°C. The absolute number of TC was determined by morphology, and by flow cytometric (FCM) analysis of DNA content. The fraction of TC was between 7-65% and these values were used to calculate the CE. The effect of heat inactivated fetal calf serum (FCS) - horse serum (HS), human AB serum, autologous serum, or ascites fluid, in various concentrations, were compared to the standard assay. 25-35% concentrations of all sera improved the CE by 1.5 - 4 fold. Autologous serum produced the highest CE in 8 of 11 cases. Under optimum conditions, CE was 0.01-1.4%. Different sera did not change drug sensitivity.

44 RECENT OBSERVATIONS ON ENHANCED CLONING OF MULTIPLE MYELOMA IN VITRO. Brian G. M. Durie, Judy A. Christiansen, Elizabeth E. Vela, Dept. of Internal Medicine and the Cancer Center, University of Arizona Health Sciences Center, Tucson, Arizona.

Several methods have been evaluated to increase the in vitro cloning efficiency of bone marrow myeloma cells. A critical determinant of improved growth was rigorous (2X) T-lymphocyte depletion of plated cells using neuraminidase treated sheep red blood cells. Addition of nonproliferating human fetal lung fibroblasts (MRC-5) in concentrations of 1000 or 5000/plate further enhanced in vitro myeloma cell proliferation.

	Myeloma Growth	Colonies
Unfractioned in agar	3/21(14%)	0/21(0%)
Nonadherent with BALB-CCM in agar	11/37(30%)	0/37(0%)
T-depleted in methylcellulose (MeC)	28/40(70%)	12/40(30%)
T-depleted plus MRC-5 (MeC)	22/25(88%)	15/25(60%)

Methylcellulose allowed greater cell-cell interaction of the remaining cell populations after T-lymphocyte depletion. Addition of MRC-5 cells increased myeloma growth by as much as 302% with a significant increase overall in myeloma colonies (60% versus 30%, $p < 0.05$) over T-depletion alone. The percentage of ^{3}(H) thymidine labeled myeloma cells in colonies also increased with MRC-5 cells (20% with MRC-5: 8% T-depletion alone). Cytogenetic analyses confirmed the presence of abnormal karyotypes. Addition of PHA-LCM enhanced both myeloma cell and normal marrow cell growth (CFU-C). Other factors which are currently being investigated include different sera and plasma additives to the media, transferrin and low oxygen conditions (6% O_2). Although further studies are necessary, rigorous T-lymphocyte depletion of patient samples plus addition of MRC-5 cells significantly improve myeloma colony growth over previous methods.

45 Radiation Sensitivity of Human Ovarian Carcinoma *in vitro*.

C. R. Johanson and R. N. Buick, Ontario Cancer Institute, Toronto, Canada.

Four cell lines of human ovarian carinoma (HEY, OW7, SKOV, and CaOV3) were studied in cell culture and their *in vitro* radiation survival curve parameters determined. Cell survival was assessed for colonies grown on plastic, semi-solid agar, or methylcellulose.

All four cell lines show similar survival curves on plastic, with Do's from 150 to 190 cGr and extrapolation numbers from 1.8 to 1.9. The Do's for HEY grown on agar and methylcellulose are higher at 225 cGr and 240 cGr respectively with n numbers of 1.5. Under hypoxic conditions Do's for HEY on both plastic and methylcellulose increase by a factor of 2.5. Colony size distributions for SKOV on plastic show a relationship between survival and colony size.

In conclusion, small but reproducible differences in radiation survival parameters occur depending on type of culture technology employed. No indication is seen of the previously reported high degree of radioresistance of human epithelial tumor cell lines grown in agar culture. The family of cell survival curves obtained following radiation dependent on colony size suggests a source of artifact.

Caution should be employed in the selection of culture technology and colony size cutoffs for assays of clonogenicity in vitro following radiation.

46 CORRELATION OF IN VITRO CHEMOSENSITIVITY DATA WITH IN VIVO GROWTH INHIBITION IN A HUMAN SQUAMOUS CELL CARCINOMA XENOGRAFT.

A. Krishan, C. Swinkin, and L. Welham. Comp. Cancer Center for the State of Florida and the Dept. of Oncology, University of Miami Medical School, Miami, Florida.

In soft agar assays, ID 50 values for cis-plt, ara-c, and HU were similar under 5 or 20% oxygen incubation. In contrast, ID 50 values for AdR, Bleo, and VLb were 3-30 fold lower under 5% oxygen incubation. Based on comparison of 1/10 achievable peak plasma levels and the ID 50 values in CFA, cis-plt and AdR were considered as active drugs for use in vivo. The ID 50 values for AdR (20% oxygen) and Bleo (5 & 20 % oxygen) were non-achievable in vivo. Several protocols and routes of drug administration were used in multiple tumor bearing mice. Serial measurements showed that AdR and cis-plt caused significant and prolonged tumor growth delays. VLb effects were short lived. Bleo effects were highly variable ranging from no effect to prolonged growth inhibition.

These observations confirm a positive correlation between CFA data and growth inhibition caused by AdR and cis-plt in this tumor model.
(Supported by NIH-CA 29360).

Index of First Authors

Contributors

Matti S. Aapro, M.D., Chef de Clinique, Division of Onco-Hematology, University Hospital of Geneva, Geneva, Switzerland

Reto Abele, M.D., Medecin-Adjoint, Division of Onco-Hematology, University Hospital of Geneva, Geneva, Switzerland

Frederick R. Ahmann, M.D., Assistant Professor of Hematology-Oncology, Department of Internal Medicine, Tucson Veterans Administration Medical Center; University of Arizona School of Medicine, Tucson, Arizona

Jaffer A. Ajani, M.D., Assistant Professor of Medicine, Department of Medical Oncology, University of Texas, M.D. Anderson Hospital and Tumor Institute, Houston, Texas

Pierre Alberto, M.D., Professor, Division of Onco-Hematology, Swiss Institute for Cancer Research, Lausanne, Switzerland

David S. Alberts, M.D., Professor of Hematology-Oncology, Departments of Internal Medicine and Pharmacology, University of Arizona College of Medicine, Tucson, Arizona

Michael C. Alley, Ph.D., Assistant Professor, Department of Pharmacology, Mayo Medical School and Mayo Clinic, Rochester, Minnesota

Muhyi Al-Sarraf, M.D., Division of Oncology, Wayne State University, Detroit, Michigan

Mammo Amare, M.D., Professor, Department of Medicine, University of Kansas Medical Center, Southwest Oncology Group, Kansas City, Kansas

Fraser Baker, Ph.D., Research Associate, Division of Medicine, University of Texas, M.D. Anderson Hospital and Tumor Institute at Houston, Houston, Texas

Michael E. Berens, Ph.D., Research Assistant, Division of Oncology, University Hospital, Zurich, Switzerland

Carl A. Bertelsen, M.D., Fellow, Surgical Oncology, Division of Surgical Oncology, John Wayne Clinic, Jonsson Comprehensive Cancer Center, University of California at Los Angeles School of Medicine, Los Angeles, California

Christian Bieglmayer, Ph.D., Biochemist, Second Department of Obstetrics and Gynecology, University of Vienna, Vienna, Austria

Jürgen Bier, M.D., D.D.S., Professor, Chief of Department of Maxillofacial Surgery, Institute of Medical Microbiology, Klinikum Steglitz, The Free University of Berlin, Berlin, F.R. Germany

Helene Blank, B.S., Research Assistant, Department of Pharmacology and Experimental Therapeutics, Louisiana State University Medical Center, New Orleans, Louisiana

George Blumenschein, M.D., Medical Breast Department, M.D. Anderson Hospital and Tumor Institute, Houston, Texas

Michael G. Brattain, Ph.D., Bristol-Baylor Laboratory, Department of Pharmacology, Baylor College of Medicine, Houston, Texas

Gerhardt Breitenecker, M.D., Pathologist, Second Department of Obstetrics and Gynecology, University of Vienna, Vienna, Austria

Isadore Brodsky, M.D., Professor, Department of Hematology/Oncology, Institute for Cancer and Blood Diseases, Barry Ashbee Leukemia Research Laboratories, Hahnemann University, Philadelphia, Pennsylvania

Nils Brünner, M.D., Research Assistant, University Institute of Pathological Anatomy, University of Copenhagen, Copenhagen, Denmark

Julie A. Buckmeier, B.S., Research Assistant, Department of Internal Medicine, University of Arizona, Health Sciences Center, Cancer Center, Tucson, Arizona

Ronald N. Buick, Ph.D., Associate Professor, Department of Medical Biophysics, University of Toronto and Ontario Cancer Institute, Toronto, Ontario, Canada

Francis J. Carey, M.D., Associate Chief of Staff for Research, St. Louis Veterans Administration Medical Center; Assistant Professor of Medicine, Department of Medicine, St. Louis University, St. Louis, Missouri

Gary Z. Carl, Ph.D., Professional Research Associate, Department of Pharmacology and Division of Oncology, College of Medicine, University of Saskatchewan, Saskatoon, Saskatchewan, Canada

Desmond N. Carney, M.D., Ph.D., Senior Investigator, National Cancer Institute, Navy Medical Oncology Branch, Division of Cancer Treatment, National Naval Medical Center, Bethesda, Maryland

William A. Carter, M.D., Professor, Department of Hematology/Oncology, Institute for Cancer and Blood Diseases, Barry Ashbee Leukemia Research Laboratories, Hahnemann University, Philadelphia, Pennsylvania

Barbara Chang, M.D., Chief, Hematology/Oncology, Department of Medical Service, Augusta Veterans Administration Medical Center, Augusta, Georgia

Darwin Chee, Ph.D., Director, Oncology Laboratories, Inc., Warwick, Rhode Island

Yung-Chi Cheng, Ph.D., Professor, Departments of Pharmacology and Medicine, University of North Carolina, Chapel Hill, North Carolina

Alberto C. Cillo, Ph.D., Charge de Recherche, Department of Cell Biology, Swiss Institute for Cancer Research, Lausanne, Switzerland

Gary M. Clark, Ph.D., Assistant Professor, Departments of Medicine and Oncology, University of Texas Health Science Center at San Antonio, San Antonio, Texas

Pamela S. Cohen, M.D., Fellow, Pediatric Hematology/Oncology, Departments of Pediatrics and Pathology, Stanford University School of Medicine, Stanford, California; Division of Hematology/Oncology, Children's Hospital at Stanford, Palo Alto, California

Laure Coulombel, M.D., Researcher, Terry Fox Laboratory, British Columbia Cancer Research Centre, Vancouver, British Columbia, Canada

John Crissman, M.D., Professor, Department of Pathology, Wayne State University, Detroit, Michigan

Peter Csaicsich, M.D., Gynecologist, Second Department of Obstetrics and Gynecology, University of Vienna, Vienna, Austria

Glen Cummings, Ph.D., Division of Oncology, Wayne State University, Detroit, Michigan

Bruce W. Dana, M.D., Assistant Professor of Medicine, Division of Hematology and Medical Oncology, Oregon Health Sciences University, Portland, Oregon

Judith C. Dean, R.N., Research Associate, Department of Hematology-Oncology, University of Arizona, Health Sciences Center, Cancer Center, Tucson, Arizona

Christian Dittrich, M.D., Attending Physician/Internist, Department of Chemotherapy, University of Vienna, Vienna, Austria

Robert T. Dorr, Ph.D., Research Assistant, Professor of Hematology-Oncology, University of Arizona, Health Sciences Center, Cancer Center, Tucson, Arizona

Benjamin Drewinko, M.D., Ph.D., Professor, Department of Laboratory Medicine, University of Texas, M.D. Anderson Hospital and Tumor Institute at Houston, Houston, Texas

Ian D. Dubé, Ph.D., Research Assistant, Terry Fox Laboratory, British Columbia Cancer Research Centre, Vancouver, British Columbia, Canada

Janice P. Dutcher, M.D., Associate Professor, Department of Medicine, Albert Einstein College of Medicine, Bronx, New York

Allen C. Eaves, M.D., Ph.D., Chief, Terry Fox Laboratory, British Columbia Cancer Research Centre, Vancouver, British Columbia, Canada

Connie J. Eaves, Ph.D., Research Associate, Terry Fox Laboratory, British Columbia Cancer Research Centre, Vancouver, British Columbia, Canada

Janine Einsphar, M.S., Research Associate, University of Arizona, Health Sciences Center, Cancer Center, Tucson, Arizona

Liv Endresen, M.S., Department of Clinical Pharmacology, the National Hospital, Oslo, Norway

Svend A. A. Engelholm, M.D., Professor, University Institute of Pathological Anatomy, University of Copenhagen, Copenhagen, Denmark

John Ensley, M.D., Division of Oncology, Wayne State University, Detroit, Michigan

Dominic Fan, Ph.D., Instructor, Department of Pharmacology and Experimental Therapeutics, Louisiana State University Medical Center, New Orleans, Louisiana

Susan Fan, B.S., Research Assistant, Department of Pharmacology and Experimental Therapeutics, Louisiana State University Medical Center, New Orleans, Louisiana

Robert Franco, M.D., Assistant Professor of Experimental Medicines, Division of Hematology/Oncology Department of Internal Medicine, University of Cincinnati, Cincinnati, Ohio

Ursina Frueh, Research Associate, Division of Oncology, University Hospital, Zürich, Switzerland

Elizabeth M. Gibby, Ph.D., Postdoctoral Fellow, Laboratory of Cellular Chemotherapy, Imperial Cancer Research Fund, London, England

Laura Gillen, B.S., Research Assistant, Department of Pharmacology and Experimental Therapeutics, Louisiana State University Medical Center, New Orleans, Louisiana

Armando E. Giuliano, Division of Surgical Oncology, John Wayne Clinic, Jonsson Comprehensive Cancer Center, University of California at Los Angeles School of Medicine, Los Angles, California

Bertil E. Glader, M.D., Ph.D., Associate Professor of Pediatrics, Department of Pediatrics and Pathology, Stanford University School of Medicine, Stanford, California, Division of Hematology/Oncology, Children's Hospital at Stanford, Palo Alto, California

Gary E. Goodman, M.D., Assistant Professor, Department of Medical Oncology, The Swedish Hospital Tumor Institute, Seattle, Washington

John A. Green, M.D., Visiting Fellow, Division of Cancer Treatment, National Cancer Institute, Bethesda, Maryland

Geoffrey L. Green, Ph.D., Ben May Laboratory for Cancer Research, University of Chicago, Chicago, Illinois

Charles D. Haas, M.D., Division of Oncology, Wayne State University, Detroit, Michigan

Anne W. Hamburger, Ph.D., Research Scientist, Cell Culture Department, American Type Culture Collection, Rockville, Maryland

Thomas C. Hamilton, Ph.D., Staff Fellow, Medicine Branch, Division of Cancer Treatment, National Cancer Institute, Bethesda, Maryland

L. Havelec, Ph.D., Associate Professor, Institute of Biostatistics, University of Vienna, Vienna, Austria

Margot Haynes, Medical Breast Department, M.D. Anderson Hospital and Tumor Institute, Houston, Texas

Robert A. Hickie, Ph.D., Professor of Pharmacology, Department of Pharmacology and Division of Oncology, College of Medicine, University of Saskatchewan, Saskatoon, Saskatchewan, Canada

Susanne U. Hildebrand-Zanki, M.S., Research Associate, Division of Surgical Oncology, John Wayne Clinic, Jonsson Comprehensive Cancer Center, University of California at Los Angeles School of Medicine, Los Angeles, California

Bridget T. Hill, Ph.D., Head, Laboratory of Cellular Chemotherapy, Imperial Cancer Research Fund, London, England

Victor E. Hofmann, M.D., Assistant Professor, Division of Oncology, University Hospital, Zürich, Switzerland

Heinrich Holzner, M.D., Institute of Pathology, University of Vienna, Vienna, Austria

Howard D. Homesley, M.D., Department of Obstetrics and Gynecology, Bowman Gray School of Medicine of Wake Forest University, Winston-Salem, North Carolina

Louise K. Hosking, Research Assistant, Laboratory of Cellular Chemotherapy, Imperial Cancer Research Fund, London, England

Yi Ji Hu, M.D., Postdoctoral Research Fellow, Department of Hematology, University of Texas, M.D. Anderson Hospital and Tumor Institute, Houston, Texas

Verena M. Hug, M.D., Assistant Professor of Medicine, Department of Medical Oncology, Medical Breast Department, University of Texas, M.D. Anderson Hospital and Tumor Institute, Houston, Texas

John Jacobs, M.D., Division of Otolaryngology, Wayne State University, Detroit, Michigan

Raimund Jakesz, M.D., Assistant Professor, First Department of Surgery, University of Vienna, Vienna, Austria

Herbert Janisch, M.D., Gynecologist, Chief, Second Department of Obstetrics and Gynecology, University of Vienna, Vienna, Austria

Michael E. Johns, M.D., Professor and Chairman, Department of Otolaryngology/Head and Neck Surgery, Johns Hopkins Hospital, Baltimore, Maryland; University of Virginia, Charlottesville, Virginia

Stephen E. Jones, M.D., Professor of Medicine, Department of Internal Medicine, Chief, Section of Hematology/Medical Oncology, Director, Multidisciplinary Oncology Clinics, University of Arizona, Health Sciences Center, Cancer Center, Tucson, Arizona

Dagmar K. Kalousek, M.D., Chief, Cytogenetics Laboratory, Terry Fox Laboratory, British Columbia Cancer Research Centre, Vancouver, British Columbia, Canada

Daniel E. Kenady, M.D., Assistant Professor of Surgery, Division of General Surgery, Department of Surgery, University of Kentucky, Lexington, Kentucky

David H. Kern, Ph.D., Chief, Human Tumor Cloning, Division of Surgical Oncology, John Wayne Clinic, Jonsson Comprehensive Cancer Center, University of California at Los Angeles School of Medicine, Los Angeles, California; Surgical Service, Veterans Administration Medical Center, Sepulveda, California

Julie A. Kish, M.D., Division of Oncology, Wayne State University, Detroit, Michigan

David J. Klaassen, M.D., Director, Cancer Clinic, Department of Pharmacology, Head, Division of Oncology, College of Medicine, University of Saskatchewan, Saskatoon, Saskatchewan, Canada

Fumio Kodama, M.D., Visiting Professor of Hematology-Oncology, University of Arizona, Health Sciences Center, Cancer Center, Tucson, Arizona

Roland Kolb, M.D., Professor, First Department of Surgery, University of Vienna, Vienna, Austria

Felix Krauer, M.D., Professor, Division of Onco-Hematology, University Hospital of Geneva, Geneva, Switzerland

Kristie L. Kreutzfeld, M.Sc., Lab Assistant, Department of Internal Medicine, Division of Hematology and Oncology, University of Arizona, Health Sciences Center, Cancer Center, Tucson, Arizona

Awtar Krishan, Ph.D., Professor of Oncology, Scientific Program Director, Department of Oncology, University of Miami Medical School, Miami, Florida

Bernd Lathan, M.D., Departments of Medicine and Oncology, University of Texas Health Science Center at San Antonio, San Antonio, Texas

Joyce Lebowitz, B.S., Research Assistant, Cancer Center, University of Arizona, Health Sciences Center, Tucson, Arizona

Martha B. Lee, M.S., Doctoral Candidate, Department of Biomathematics, Division of Surgical Oncology, John Wayne Clinic, Jonsson Comprehensive Cancer Center, University of California at Los Angeles School of Medicine, Los Angeles, California

Susan Leigh, R.N., Research Nurse, University of Arizona, Health Sciences Center, Cancer Center, Tucson, Arizona

Reinhart Lenzhofer, M.D., Internist, Department of Chemotherapy, University of Vienna, Vienna, Austria

Alan E. Levine, Ph.D., Assistant Professor, Bristol-Baylor Laboratory, Department of Pharmacology, Baylor College of Medicine, Houston, Texas

Michael M. Lieber, M.D., Department of Urology, Mayo Medical School and Mayo Clinic, Rochester, Minnesota

Michael P. Link, M.D., Assistant Professor of Pediatrics, Departments of Pediatrics and Pathology, Stanford University School of Medicine, Stanford, California; Division of Hematology/Oncology, Children's Hospital at Stanford, Palo Alto, California

Dan W. Luedke, M.D., Associate Professor of Internal Medicine, Department of Internal Medicine, St. Louis Veterans Administration Medical Center, St. Louis University Medical Center, St. Louis, Missouri

Thomas R. Maloney, M.D., Chief, Hematology/Oncology, Wilford Hall Medical Center, Lackland Air Force Base, Texas

Nannette R. Melnick, National Cancer Institute, Bethesda, Maryland

Paul Meltzer, M.D., Ph.D., Research Associate, Department of Pediatrics, University of Arizona, Health Sciences Center, Cancer Center, Tucson, Arizona

Nor Merchant, M.D., Resident Fellow, Department of Medical Oncology, University of Texas, M.D. Anderson Hospital and Tumor Institute, Houston, Texas

Claudia Metelmann, M.D., Microbiologist, Institute of Medical Microbiology, The Free University of Berlin, Berlin, F.R. Germany

Hans-Robert Metelmann, M.D., D.D.S., Surgeon, Department of Maxillofacial Surgery, Institute of Medical Microbiology, Klinikum Steglitz, The Free University of Berlin, Berlin, F.R. Germany

Frank L. Meyskens, Jr., M.D., Ph.D., Associate Professor of Internal Medicine, Department of Internal Medicine, Division of Hematology and Oncology, University of Arizona, Health Sciences Center, Cancer Center, Tucson, Arizona

Thomas P. Miller, Assistant Professor of Hematology-Oncology, Veterans Administration Hospital, University of Arizona, Health Sciences Center, Cancer Center, Tucson, Arizona

William T. Miller, VSE Corporation, Alexandria, Virginia

John D. Minna, M.D., Chief, National Cancer Institute, Navy Medical Oncology Branch, Division of Cancer Treatment, National Naval Medical Center, Bethesda, Maryland

Jazbieh Moezzi, M.D., Research Fellow, The Bob Hipple Laboratory for Cancer Research, Wright State University School of Medicine, Dayton, Ohio

Thomas E. Moon, Ph.D., Professor, Department of Internal Medicine, University of Arizona, Health Sciences Center, Cancer Center, Tucson, Arizona

Lee Roy Morgan, M.D., Ph.D., Chairman, Department of Pharmacology and Experimental Therapeutics, Louisiana State University Medical Center, New Orleans, Louisiana

Timothy M. Morgan, Ph.D., Section on Community Medicine (Biostatistics), Bowman Gray School of Medicine of Wake Forest University, Winston-Salem, North Carolina

Francis S. Morrison, M.D., Professor, Department of Medicine, University of Mississippi Medical Center, Southwest Oncology Group, Jackson, Mississippi

Donald L. Morton, M.D., Chief, Division of Surgical Oncology, John Wayne Clinic, Jonsson Comprehensive Cancer Center, University of California at Los Angeles School of Medicine, Los Angeles, California

Kurt Moser, M.D., Professor, Chief, Oncology Section, Department of Chemotherapy, University of Vienna, Vienna, Austria

Martin J. Murphy, Jr., Ph.D., Professor and Director, The Bob Hipple Laboratory for Cancer Research, Wright State University School of Medicine, Dayton, Ohio

Marion M. Nau, Chemist, National Cancer Institute, Navy Medical Oncology Branch, Division of Cancer Treatment, National Naval Medical Center, Bethesda, Maryland

Harvey B. Niell, M.D., Chief of Oncology, Department of Hematology/Oncology, Memphis Veterans Administration Medical Center, University of Tennessee, Memphis, Tennessee

Inge Nøhr, B.S., University Institute of Pathological Anatomy, University of Copenhagen, Copenhagen, Denmark

N. Odartchenko, M.D., Professor, of Cell Biology, Swiss Institute for Cancer Research, Lausanne, Switzerland

Sharon Olson, M.T., Research Assistant, University of Arizona, Health Sciences Center, Cancer Center, Tucson, Arizona

Robert F. Ozols, M.D., Ph.D., Senior Investigator, Attending Physician, Medicine Branch, Division of Cancer Treatment, National Cancer Institute, Bethesda, Maryland

Chan H. Park, M.D., Ph.D., Associate Professor, Department of Medicine, University of Kansas Medical Center, Southwest Oncology Group, Kansas City, Kansas

Linda Perrot, M.D., Research Associate, Veterans Administration Hospital, University of Arizona, Health Sciences Center, Cancer Center, Tucson, Arizona

Alexander Pihl, M.D., Ph.D., Norsk Hydro's Institute for Cancer Research, The Norwegian Radium Hospital, Oslo, Norway

Pei-Yu Pu, M.D., Guest Investigator, Department of Neurology, Memorial Sloan Kettering Cancer Center, New York, New York

Stephen R. Rodney, University of Arizona, Health Sciences Center, Cancer Center, Tucson, Arizona

Henrik Roed, M.D., Research Assistant, Department of Medicine, The Finsen Institute, Copenhagen, Denmark

Alfred M. Rogan, M.D., Visiting Fogarty International Fellow, Medicine Branch, Division of Cancer Treatment, National Cancer Institute, Bethesda, Maryland, Second Department of Obstetrics and Gynecology, University of Vienna, Vienna, Austria

Fred Rudolph, Ph.D., Department of Biochemistry, Rice University, Houston, Texas

Shail K. Sahu, Ph.D., Research Associate, Division of Medicine, University of Texas, M.D. Anderson Hospital and Tumor Institute at Houston, Houston, Texas

Sydney E. Salmon, M.D., Professor of Medicine, Section of Hematology and Oncology, Department of Internal Medicine, Director, Cancer Center, University of Arizona College of Medicine, Tucson, Arizona

Evelyne Satelhak, Research Associate, Department of Chemotherapy, University of Vienna, Vienna, Austria

Peter Schäefer, M.D., Medicin-Adjoint, Clinic of Gynecology, University Hospital of Geneva, Geneva, Switzerland

Karl Schieder, M.D., Gynecologist, Second Department of Obstetrics and Gynecology, University of Vienna, Vienna, Austria

Christine Schneider, B.S., Research Assistant, Department of Pharmacology and Experimental Therapeutics, Louisiana State University Medical Center, New Orleans, Louisiana

Ruth Serokman, M.Ph., Research Assistant, University of Arizona, Health Sciences Center, Cancer Center, Tucson, Arizona

Joan R. Shapiro, Ph.D., Assistant Professor, Department of Neurology, Memorial Sloan Kettering Cancer Center, New York, New York

William R. Shapiro, M.D., Head, Cotizas Laboratory, Attending Neurologist, Professor of Neurology, Department of Neurology, Memorial Sloan Kettering Cancer Center, New York, New York

Margaret Shatsky, Ph.D., Research Associate, Departments of Pediatrics and Pathology, Stanford University School of Medicine, Stanford, California, Division of Hematology/Oncology, Children's Hospital at Stanford, Palo Alto, California

Robert H. Shoemaker, Ph.D., Acting Head, Cell Culture Section, Division of Cancer Treatment, National Cancer Institute, Bethesda, Maryland

Richard M. Simon, National Cancer Institute, Bethesda, Maryland

Eva Singletary, M.D., Research Fellow, Department of Surgery, University of Texas, M.D. Anderson Hospital and Tumor Institute, Houston, Texas

Nancy J. Sipes, B.S., Research Assistant, Department of Internal Medicine, University of Arizona, Health Sciences Center, Cancer Center, Tucson, Arizona

Stephen D. Smith, M.D., Associate Professor of Pediatrics, Departments of Pediatrics and Pathology, Stanford University School of Medicine, Stanford, California, Division of Hematology/Oncology, Children's Hospital at Stanford, Palo Alto, California

Vernon K. Sondak, M.D., Fellow, Surgical Oncology, Division of Surgical Oncology, John Wayne Clinic, Jonsson Comprehensive Cancer Center, University of California at Los Angeles School of Medicine, Los Angeles, California

Mogens Spang-Thomsen, M.D., Professor, University Institute of Pathological Anatomy, University of Copenhagen, Copenhagen, Denmark

Gary Spitzer, M.D., Associate Professor of Medicine, Department of Hematology, University of Texas, M.D. Anderson Hospital and Tumor Institute, Houston, Texas

Karl-Hermann Spitzy, M.D., Professor, Head, Department of Chemotherapy, University of Vienna, Vienna, Austria

Robert Steininger, M.D., Surgeon, First Department of Surgery, University of Vienna, Vienna, Austria

David R. Strayer, M.D., Associate Professor, Department of Hematology/Oncology, Institute for Cancer and Blood Diseases, Barry Ashbee Leukemia Research Laboratories, Hahnemann University, Philadelphia, Pennsylvania

Earl A. Surwit, M.D., Associate Professor, Department of Obstetrics and Gynecology, University of Arizona College of Medicine, Tucson, Arizona

Nobuhiko Tanigawa, M.D., Associate Professor of Surgery, Second Department of Surgery, Fukui Medical College, Fukui, Japan

Howard Thames, Ph.D., Department of Biomathematics, University of Texas, M.D. Anderson Hospital and Tumor Institute at Houston, Houston, Texas

Stephen P. Thomson, B.S., Research Assistant, Department of Internal Medicine, Division of Hematology and Oncology, University of Arizona, Health Sciences Center, Cancer Center, Tucson, Arizona

Barbara Tomasovic, B.S., Research Scientist, Department of Hematology, University of Texas, M.D. Anderson Hospital and Tumor Institute, Houston, Texas

Tony Tong, M.S., Research Assistant, University of Arizona, Health Sciences Center, Cancer Center, Tucson, Arizona

Jeffrey M. Trent, Ph.D., Associate Professor of Medicine, Section of Hematology and Oncology, Department of Internal Medicine, University of Arizona, Director, Basic Research at the Cancer Center, College of Medicine, Tucson, Arizona

Kjell M. Tveit, M.D., Ph.D., Norsk Hydro's Institute for Cancer Research, The Norwegian Radium Hospital, Oslo, Norway

Peter R. Twentyman, Ph.D., Senior Scientist, Medical Research Council, Clinical Oncology and Radiotherapeutics Unit, Cambridge, England

Gunter E. Umbach, M.D., Fellow in Gynecology/Oncology, Department of Gynecology, University of Texas, M.D. Anderson Hospital and Tumor Institute, Houston, Texas

John M. Venditti, Ph.D., Director, Drug Evaluation Branch, National Cancer Institute, Bethesda, Maryland

Monika Vetterlein, Ph.D., Chief, Cell Culture Lab, Institute of Cancer Research, University of Vienna, Vienna, Austria

Lars Vindeløv, M.D., Research Assistant, Department of Medicine, The Finsen Institute, Copenhagen, Denmark

David T. Vistica, Ph.D., Senior Investigator, Division of Cancer Treatment, National Cancer Institute, Bethesda, Maryland

Daniel D. Von Hoff, M.D., Professor, Division of Oncology, Departments of Medicine and Oncology, University of Texas Health Science Center at San Antonio, The Cancer Therapy and Research Foundation of South Texas, San Antonio, Texas

Gerald A. Walls, M.Sc., Researcher, Medical Research Council, Clinical Oncology and Radiotherapeutics Unit, Cambridge, England

Bruce G. Ward, M.B.B.S., Clinical Research Fellow, Laboratory of Cellular Chemotherapy, Imperial Cancer Research Fund, London, England

Roger A. Warnke, M.D., Associate Professor of Pathology, Departments of Pediatrics and Pathology, Stanford University School of Medicine, Stanford, California; Division of Hematology/Oncology, Children's Hospital at Stanford, Palo Alto, California

Arthur Weaver, M.D., Department of Surgery, Wayne State University, Detroit, Michigan

Jan Weisband, B.S., Research Technician, Department of Hematology/Oncology, Institute for Cancer and Blood Diseases, Barry Ashbee Leukemia Research Laboratories, Hahnemann University, Philadelphia, Pennsylvania

Charles E. Welander, M.D., Assistant Professor, Department of Obstetrics and Gynecology, Bowman Gray School of Medicine of Wake Forest University, Winston-Salem, North Carolina

Richard D. H. Whelan, M.I. Biol., Research Associate, Laboratory of Cellular Chemotherapy, Imperial Cancer Research Fund, London, England

Peter H. Wiernik, M.D., Professor, Oncology Department, Albert Einstein College of Medicine, Bronx, New York

Mary K. Wolpert-DeFilippes, National Cancer Institute, Bethesda, Maryland

Jan Woodruff, B.S., Research Assistant, Division of Hematology and Medical Oncology, Oregon Health Sciences University, Portland, Oregon

Laurie A. Young, B.S., Research Assistant, Department of Internal Medicine, University of Arizona, Health Sciences Center, Cancer Center, Tucson, Arizona

Robert C. Young, M.D., Chief, Medicine Branch, Division of Cancer Treatment, National Cancer Institute, Bethesda, Maryland

Karim Zirvi, Ph.D., Assistant Professor of Medicine and Dentistry, Department of Surgery, New Jersey College of Medicine, Newark, New Jersey